AF389142

Advanced Machine Learning and Deep Learning Approaches for Remote Sensing

Advanced Machine Learning and Deep Learning Approaches for Remote Sensing

Editor

Gwanggil Jeon

MDPI • Basel • Beijing • Wuhan • Barcelona • Belgrade • Manchester • Tokyo • Cluj • Tianjin

Editor
Gwanggil Jeon
Department of Embedded
Systems Engineering,
Incheon National University,
Incheon, Republic of Korea

Editorial Office
MDPI
St. Alban-Anlage 66
4052 Basel, Switzerland

This is a reprint of articles from the Special Issue published online in the open access journal *Remote Sensing* (ISSN 2072-4292) (available at: https://www.mdpi.com/journal/remotesensing/special_issues/J938V8W2EM).

For citation purposes, cite each article independently as indicated on the article page online and as indicated below:

LastName, A.A.; LastName, B.B.; LastName, C.C. Article Title. *Journal Name* **Year**, *Volume Number*, Page Range.

ISBN 978-3-0365-7946-7 (Hbk)
ISBN 978-3-0365-7947-4 (PDF)

Contents

Preface to "Advanced Machine Learning and Deep Learning Approaches for Remote Sensing"

This book introduces advanced machine learning and deep learning techniques for remote sensing. A total of 17 research results are provided, and based on the research results introduced here, it is expected that development and research in the field of artificial intelligence-based remote sensing will become more active in the future.

Gwanggil Jeon

Editor

remote sensing

Editorial

Advanced Machine Learning and Deep Learning Approaches for Remote Sensing

Gwanggil Jeon [1,2]

[1] Department of Embedded Systems Engineering, Incheon National University, Incheon 22012, Republic of Korea; gjeon@inu.ac.kr
[2] Energy Excellence & Smart City Lab., Incheon National University, Incheon 22012, Republic of Korea

1. Introduction

Unlike field observation or field sensing, remote sensing is the process of obtaining information about an object or phenomenon without making physical contact. According to recent research results, technologies such as artificial intelligence-based deep learning show the potential to overcome the problems of image and video signal processing faced in remote sensing. These technologies generally require the help of high-speed image processing devices such as GPUs, and high computing performance is essential. Through the development of these devices, remote sensing technology, and aerial sensor technology, the scientific community can now monitor Earth with high-resolution images and secure huge quantities of earth observation data. These capacities stem from the fast, accurate and highly reliable technology based on artificial intelligence. The papers published in this Special Issue describe recent advances in big data processing and artificial intelligence-based technologies for remote sensing technologies. A total of 17 papers were published in this Special Issue.

2. Overview of Contributions

The most significant obstacle to optical remote sensing imaging is clouds. In the contribution by Ma et al., entitled "Cloud Removal from Satellite Images Using a Deep Learning Model with the Cloud-Matting Method", the authors introduce a technique for the removal of clouds from satellite images by paying attention to image overlap and using a method that considers ground surface reflection and cloud top reflection as a linear mixture of image elements [1]. To this end, a two-step convolutional neural network is used to extract cloud transparency information and then generate ground surface information for thin cloud regions. The authors test the proposed model on simulated and ALCD data sets. The model successfully recovers the surface information of the thin cloud region when thick and thin clouds coexist and does so without significantly damaging the information of the original image.

The use of semantic segmentation technology, being a core component of computer vision in remote sensing images, is currently widely applied. The majority of the remote sensing image semantic segmentation methods are based on CNN, but recently transformer-based technology is also widely applied. In the contribution by Li et al. "RCCT-ASPPNet: Dual-Encoder Remote Image Segmentation Based on Transformer and ASPP", the authors propose RCCT-ASPPNet, which includes a dual encoder structure of RCCT and ASPP [2]. The RCCT uses transformers to fuse global multiscale semantic information, and residual structures are used to connect inputs and outputs. ASPP, performed based on CNN, can extract contextual information about high-level semantics and spatial and channel information through the application of CBAM.

The SAR-ATR method uses unlabeled measured data and labeled simulated data to improve performance. This is due to the problem that there is not a significant quantity of labeled measurement data, and as such this method is currently widely used. In the

Citation: Jeon, G. Advanced Machine Learning and Deep Learning Approaches for Remote Sensing. *Remote Sens.* **2023**, 15, 2876. https://doi.org/10.3390/rs15112876

Received: 29 May 2023
Accepted: 30 May 2023
Published: 1 June 2023

contribution by Zhang et al., entitled "Azimuth-Aware Discriminative Representation Learning for Semi-Supervised Few-Shot SAR Vehicle Recognition", the authors propose a method for designing two AADR loss functions that suppress the intra-class variation of samples with large azimuth differences [3]. Through cosine similarity, they simultaneously magnify the difference between classes of samples with the same azimuthal angle in the feature embedding space. The unlabeled measurement data of the MSTAR dataset are assigned labels of a more similar category among the SARSIM and SAMPLE datasets.

Big data and parameter tuning are essential in the twin process of training and using convolutional neural networks. This process consumes an extensive temporal and computing resources. To improve this, this paper proposes a new lightweight model called FlexibleNet. The contribution by M. Awad, entitled "FlexibleNet: A New Lightweight Convolutional Neural Network Model for Estimating Carbon Sequestration Qualitatively Using Remote Sensing", proposed a scaling-based model of "width, depth and resolution" [4]. Unlike conventional methods that arbitrarily scale the "width, depth, and resolution" factors, FlexibleNet scales the network width, depth, and resolution uniformly using a fixed set of scaling factors. Experiments have shown that the FlexibleNet model exhibits higher robustness and lower parameter tuning requirements on smaller datasets compared to conventional models.

In the contribution by Ravishankar et al., published under the name of "Capacity Estimation of Solar Farms Using Deep Learning on High-Resolution Satellite Imagery", the authors propose a deep learning framework for detecting solar power plants via the application of semantic segmentation convolutional neural networks to satellite images [5]. They also propose a model that predicts the energy generation capacity of the detected solar power plant facility. According to their research results, the proposed deep learning model achieved high performance indicators by showing an accuracy of 96.87% and a Jaccard index value of 95.5%. In addition, the average error of the energy generation capacity prediction model was 4.5%. In this study, 23,000 images of 256×256 size were used.

In recent ocean studies, ocean wave parameters, such as SWH, are being actively predicted. Remote sensing has dramatically increased the available quantity of marine data, and artificial intelligence technologies have demonstrated the ability to process big data and derive meaningful insights from them. In the contribution by Atteia et al., entitled "Deep-Learning-Based Feature Extraction Approach for Significant Wave Height Prediction in SAR Mode Altimeter Data", the authors propose a deep learning-based hybrid approach for SWH prediction using satellite SAR data [6]. Several hybrid feature sets are created using the proposed approach and SWH is modeled using GPR and NNR. SAR mode altimeter data from Sentinel-3A missions, calibrated with field buoy data, were used to train and evaluate the SWH model.

There has been substantial progress in the segmentation of remote sensing images based on deep learning in recent years. However, existing remote sensing image segmentation techniques have two limitations: (1) object detection performance in various scales is poor in complex scene segmentation; (2) feature reconstruction for accurate segmentation is difficult. In order to improve this problem, the contribution by Ma et al., entitled "Deep-Separation Guided Progressive Reconstruction Network for Semantic Segmentation of Remote Sensing Images", proposed the use of a deep separation-induced progressive reconstruction network [7]. This study made two major contributions. First, the authors design a decoder composed of progressive reconstruction blocks that capture detailed features at various resolutions by utilizing multi-scale qualities obtained from different receptive fields. Second, they use deep features to detect objects of different scales by proposing a deep separation module that classifies various classes based on semantic features. On the basis of testing on two optical remote sensing image datasets, the proposed network shows the best performance among the comparison targets.

In the contribution by Yang et al., entitled "A Multi-Dimensional Deep-Learning-Based Evaporation Duct Height Prediction Model Derived from MAGIC Data", an EDH prediction network using MLP is proposed [8]. A multidimensional EDH prediction model is constructed using spatial and temporal "additional data" derived from meteorological measurements. The experimental results reveal the following. (1) Compared with the NPS model, the root mean square error of the weather-MLP-EDH model is 54%. (2) RMSE can be reduced through the contribution of spatial and temporal parameters. (3) The meteorological parameters can be appended to the multilayer-MLP-EDH model so that measurements can fit well at both large and small scales, and the error is improved by 77.51% compared to the NPS model. The proposed model can greatly improve the prediction accuracy of EDH.

Despite many advances in remote sensing imaging technology, remote sensing imaging struggles to meet application requirements due to its low resolution. In order to obtain high-resolution remote sensing images, the authors apply super-resolution techniques to restore and reconstruct remote sensing images. Super-resolution technology solves the quality degradation problem of remote sensing image acquisition systems and efficiently restores images. In the contribution by Wang et al., entitled "A Review of Image Super-Resolution Approaches Based on Deep Learning and Applications in Remote Sensing", a study on a super-resolution method in deep learning-based remote sensing images is conducted [9]. To this end, the research background of image super-resolution technology is explained, and details such as training and test data sets, image quality and model performance evaluation methods, and model design principles are explained.

The contribution by Wang et al., published un the title "Real-Time Vehicle Sound Detection System Based on Depthwise Separable Convolution Neural Network and Spectrogram Augmentation", proposes a lightweight model for intelligent sensor system and vehicle detection [10]. Vehicle detection is a binary problem that classifies vehicles or non-vehicles. Deep neural networks have shown high performance in many signal processing applications. However, the performance of deep neural networks depends on big data. Data abouts issues such as vehicle tracking are limited, making the application of data augmentation technology essential. The proposed algorithm applies mel spectrogram broadening before extracting MFCC features in order to improve the robustness of the system. As the results of the experiment, the final frame-level accuracy achieved was 94.64%, and 34% of the parameters were reduced after compression.

An image whose image quality is degraded due to atmospheric turbulence is additionally affected by noise. The added noise defeats basic signal processing techniques. Since conventional widely used optimization methods are performed under the assumption that there is no noise, noise removal and deblurring must be independently performed in advance in order to use these techniques. The contribution by Shu et al., entitled "Blind Restoration of Atmospheric Turbulence-Degraded Images Based on Curriculum Learning", proposes the use of an NSRN (noise suppression-based restoration network) for image degradation due to turbulence [11]. The noise suppression module is designed to learn low-order subspaces from turbulence-degraded images, the asymmetric U-NET module is used for blurry image deconvolution, and the fine deep back-projection (FDBP) module is used to reconstruct sharp images. It is used for multi-level functional fusion. They also propose an improved learning strategy to incrementally train the network with the purpose of achieving a good performance through a local-to-global, easy-to-difficult learning method. According to the experimental results, the method based on NSRN showed excellent performance with PSNR 30.1dB and SSIM 0.9.

Sea surface temperature (SST) joins the widely used physical parameters in oceanography and meteorology. In addition to direct measurement and remote sensing, models for SST data have been developed to obtain SST. Since the ocean is a comprehensive and complex dynamic system, the distribution and variability of SST are affected by a variety of factors. In the contribution by Guo et al., entitled "Prediction of Sea Surface Temperature by Combining Interdimensional and Self-Attention with Neural Networks," a multivariate long short-term memory (LSTM) model is proposed that uses wind speed and air pressure at sea level as inputs along with SST in order to overcome this problem and improve prediction accuracy [12]. In addition, for model optimization, a position encoding matrix and multi-dimensional input are studied. In addition, a self-attention strategy is adopted to smooth the data during the training process. According to the experimental results, the proposed model is superior to the LSTM alone model and the model with only SST as input.

In the contribution of Qu et al., submitted under the title "Mode Recognition of Orbital Angular Momentum Based on Attention Pyramid Convolutional Neural Network", the authors propose an OAM mode detection technique based on AP-CNN in order to solve the problem of lack of accuracy in existing OAM detection systems for vortex optical communication [13]. They introduce segmented image classification to exploit the low-level detailed features of the vortex beam superposition and the similar light intensity distribution of plane wave interferograms. ResNet18 is used as the backbone of AP-CNN, and a technique for the detection of subtle differences in light intensity in images is developed by adopting a dual path structure. According to the experimental results, AP-CNN improved accuracy by up to 7% and reduced false mode identification by 3% in the confusion matrix of superimposed vortex modes compared to ResNet18.

Improving the quality of low-light images is a key factor in the interpretation of the surface state of remote sensing images. In the contribution by Rasheed et al., entitled "An Empirical Study on Retinex Methods for Low-Light Image Enhancement", the authors aim to produce images with higher contrast, noise suppression, and better quality in their low-light versions [14]. Recently, an image enhancement method based on the Retinex theory has received a lot of attention. Therefore, the authors conduct a study to compare the Retinex-based low-light enhancement method with other state-of-the-art low-light enhancement methods and to determine the generalization ability and computational cost. They use experimental results to compare the robustness of Retinex-based methods with other low-light enhancement techniques using different test data sets. Various evaluation criteria are used to compare the results, and an average ranking system is proposed to rank quality enhancement methods.

Weather factors, such as bad weather, can occur when performing land classification through remote sensing, which is a major cause of poor sensing performance. This limitation can be reduced by several factors, such as low-quality aerial imagery and inefficient fusion of multimodal representations. Therefore, it is essential to build a reliable framework capable of robustly coding remote sensing images. In the contribution by Shi et al. on the multimodal convergence and attention mechanism, entitled "Towards Robust Semantic Segmentation of Land Covers in Foggy Conditions", the authors use HRNet techniques to extract basic features and then use the spectral and spatial representation learning module to extract spectral–spatial representations [15]. In addition, in order to bridge the gap between heterogeneous devices, the authors propose the use of a multimodal Representation fusion module.

Remote sensing images with high temporal and spatial resolution are important for monitoring land surface changes, vegetation changes, and natural disaster surveillance. However, it is difficult to directly obtain high-resolution remote sensing images, and thus the deployment of space–time convergence technology to obtain remote sensing images is receiving a lot of attention. In the contribution by Li et al., entitled "Enhanced Multi-Stream Remote Sensing Spatiotemporal Fusion Network Based on Transformer and Dilated Convolution", the authors propose a deep learning model with high accuracy and robustness to better extract spatiotemporal information from remote sensing images [16]. The proposed model is EMSNet, which extends the existing MSNet. Dilated convolution is used to extract temporal information and reduce parameters. The authors further adapt the improved transformer encoder to image fusion techniques and enhance it again to effectively extract spatiotemporal information. Experimental results show that EMSNet improved SSIM by 15.3% in the CIA dataset, ERGAS by 92.1% in the LGC dataset, and RMSE by 92.9% in the AHB dataset when compared to MSNet.

LFMC is an important indicator used to assess wildfire risk and fire spread rate. In the contribution by Xie et al. "Retrieval of Live Fuel Moisture Content Based on Multi-

Source Remote Sensing Data and Ensemble Deep Learning Model", the authors propose two ensemble models that combine deep learning models in order to further improve the inspection accuracy of LFMC [17]. One is a layered ensemble model based on LSTM, TCN and LSTM-TCN models, and the other is an Adaboost ensemble model based on an LSTM-TCN model. Measured LFMC data, MODIS, Landsat-8, and Sentinel-1 remote sensing data and auxiliary data, such as canopy height and land cover in wildfire-prone areas in the western United States, are selected as study subjects. As a result of the search, remote sensing data of different groups are compared. The experimental results suggest that the LFMC search accuracy is higher than that of single-source remote-sensing data since the use of multi-source data can incorporate the advantages of different types of remote-sensing data. The proposed ensemble model can better extract the non-linear relationship between LFMC and remote sensing data.

3. Conclusions

This Special Issue introduces 17 research findings on advanced machine learning and deep learning approaches for remote sensing. Based on the research results introduced here, it is expected that further development and research in the field of artificial intelligence-based remote sensing will yield results in the future.

Acknowledgments: I thank the authors who published their research results in this Special Issue and the reviewers who reviewed their papers. I also thank the editors for their hard work and perseverance in making this Special Issue a success.

Conflicts of Interest: The author declares no conflict of interest.

References

1. Ma, D.; Wu, R.; Xiao, D.; Sui, B. Cloud Removal from Satellite Images Using a Deep Learning Model with the Cloud-Matting Method. *Remote Sens.* **2023**, *15*, 904. [CrossRef]
2. Li, Y.; Cheng, Z.; Wang, C.; Zhao, J.; Huang, L. RCCT-ASPPNet: Dual-Encoder Remote Image Segmentation Based on Transformer and ASPP. *Remote Sens.* **2023**, *15*, 379. [CrossRef]
3. Zhang, L.; Leng, X.; Feng, S.; Ma, X.; Ji, K.; Kuang, G.; Liu, L. Azimuth-Aware Discriminative Representation Learning for Semi-Supervised Few-Shot SAR Vehicle Recognition. *Remote Sens.* **2023**, *15*, 331. [CrossRef]
4. Awad, M.M. FlexibleNet: A New Lightweight Convolutional Neural Network Model for Estimating Carbon Sequestration Qualitatively Using Remote Sensing. *Remote Sens.* **2023**, *15*, 272. [CrossRef]
5. Ravishankar, R.; AlMahmoud, E.; Habib, A.; de Weck, O.L. Capacity Estimation of Solar Farms Using Deep Learning on High-Resolution Satellite Imagery. *Remote Sens.* **2023**, *15*, 210. [CrossRef]
6. Atteia, G.; Collins, M.J.; Algarni, A.D.; Samee, N.A. Deep-Learning-Based Feature Extraction Approach for Significant Wave Height Prediction in SAR Mode Altimeter Data. *Remote Sens.* **2022**, *14*, 5569. [CrossRef]
7. Ma, J.; Zhou, W.; Qian, X.; Yu, L. Deep-Separation Guided Progressive Reconstruction Network for Semantic Segmentation of Remote Sensing Images. *Remote Sens.* **2022**, *14*, 5510. [CrossRef]
8. Yang, C.; Wang, J.; Shi, Y. A Multi-Dimensional Deep-Learning-Based Evaporation Duct Height Prediction Model Derived from MAGIC Data. *Remote Sens.* **2022**, *14*, 5484. [CrossRef]
9. Wang, X.; Yi, J.; Guo, J.; Song, Y.; Lyu, J.; Xu, J.; Yan, W.; Zhao, J.; Cai, Q.; Min, H. A Review of Image Super-Resolution Approaches Based on Deep Learning and Applications in Remote Sensing. *Remote Sens.* **2022**, *14*, 5423. [CrossRef]
10. Wang, C.; Song, Y.; Liu, H.; Liu, H.; Liu, J.; Li, B.; Yuan, X. Real-Time Vehicle Sound Detection System Based on Depthwise Separable Convolution Neural Network and Spectrogram Augmentation. *Remote Sens.* **2022**, *14*, 4848. [CrossRef]
11. Shu, J.; Xie, C.; Gao, Z. Blind Restoration of Atmospheric Turbulence-Degraded Images Based on Curriculum Learning. *Remote Sens.* **2022**, *14*, 4797. [CrossRef]
12. Guo, X.; He, J.; Wang, B.; Wu, J. Prediction of Sea Surface Temperature by Combining Interdimensional and Self-Attention with Neural Networks. *Remote Sens.* **2022**, *14*, 4737. [CrossRef]
13. Qu, T.; Zhao, Z.; Zhang, Y.; Wu, J.; Wu, Z. Mode Recognition of Orbital Angular Momentum Based on Attention Pyramid Convolutional Neural Network. *Remote Sens.* **2022**, *14*, 4618. [CrossRef]
14. Rasheed, M.T.; Guo, G.; Shi, D.; Khan, H.; Cheng, X. An Empirical Study on Retinex Methods for Low-Light Image Enhancement. *Remote Sens.* **2022**, *14*, 4608. [CrossRef]
15. Shi, W.; Qin, W.; Chen, A. Towards Robust Semantic Segmentation of Land Covers in Foggy Conditions. *Remote Sens.* **2022**, *14*, 4551. [CrossRef]

16. Li, W.; Cao, D.; Xiang, M. Enhanced Multi-Stream Remote Sensing Spatiotemporal Fusion Network Based on Transformer and Dilated Convolution. *Remote Sens.* **2022**, *14*, 4544. [CrossRef]
17. Xie, J.; Qi, T.; Hu, W.; Huang, H.; Chen, B.; Zhang, J. Retrieval of Live Fuel Moisture Content Based on Multi-Source Remote Sensing Data and Ensemble Deep Learning Model. *Remote Sens.* **2022**, *14*, 4378. [CrossRef]

Article

FlexibleNet: A New Lightweight Convolutional Neural Network Model for Estimating Carbon Sequestration Qualitatively Using Remote Sensing

Mohamad M. Awad

Remote Sesning Center, National Council for Scientific Research, Beirut 11072260, Lebanon; mawad@cnrs.edu.lb

Abstract: Many heavy and lightweight convolutional neural networks (CNNs) require large datasets and parameter tuning. Moreover, they consume time and computer resources. A new lightweight model called FlexibleNet was created to overcome these obstacles. The new lightweight model is a CNN scaling-based model (width, depth, and resolution). Unlike the conventional practice, which arbitrarily scales these factors, FlexibleNet uniformly scales the network width, depth, and resolution with a set of fixed scaling coefficients. The new model was tested by qualitatively estimating sequestered carbon in the aboveground forest biomass from Sentinel-2 images. We also created three different sizes of training datasets. The new training datasets consisted of six qualitative categories (no carbon, very low, low, medium, high, and very high). The results showed that FlexibleNet was better or comparable to the other lightweight or heavy CNN models concerning the number of parameters and time requirements. Moreover, FlexibleNet had the highest accuracy compared to these CNN models. Finally, the FlexibleNet model showed robustness and low parameter tuning requirements when a small dataset was provided for training compared to other models.

Keywords: peri-urban forests; lightweight convolutional neural network; FlexibleNet; carbon sequestration; remote sensing

Citation: Awad, M.M. FlexibleNet: A New Lightweight Convolutional Neural Network Model for Estimating Carbon Sequestration Qualitatively Using Remote Sensing. *Remote Sens.* **2023**, *15*, 272. https://doi.org/10.3390/rs15010272

Academic Editor: Dino Ienco

Received: 4 December 2022
Revised: 30 December 2022
Accepted: 31 December 2022
Published: 2 January 2023

1. Introduction

Since the advent of machine learning (ML) in the mid-twentieth century [1], it has played an important role in solving many complex problems such as image processing [2,3].

In the last decade, convolutional neural networks (CNNs), a sub-discipline of ML, have played an important role in advancing image processing such as segmentation, recognition, and classification sciences [4–7]. However, many networks suffered from huge computational resource and time requirements, such as ResNet50 [8], VGG16 [9], AlexNet [10], and GoogleNet [11]. Later, improvements to CNNs were introduced by reducing the number of layers and in turn reducing the number of parameters. The new generation of CNNs are called lightweight CNNs. The first lightweight model, SqueezeNet [12], showed classification accuracy close to AlexNet, and the number of parameters was only 1/510 compared to AlexNet. In addition to SqueezeNet, there are many lightweight models to mention, such as Xception [13], MobileNet [14], MobileNetV3 [15], ShuffleNet [16], and recently EfficientNet [17]. The last lightweight network has seven versions from B0 to B7.

However, some of these introduced lightweight CNN models still suffer from a growing amount of parameter tuning or inefficiency when there are insufficient samples [18]. Many researchers tried to improve some of these network models such as VGG16, ResNet50, and MobileNet by adding an auxiliary intermediate output structure named ElasticNet [19,20] that was directly connected to the network after each convolutional unit. Other researchers tried to improve the lightweight CNNs [21] by using MobileNet to extract deep and abstract image features. Each feature was then transformed into two features with two different convolutional layers. The transformed features were subjected to a Hadamard product operation to obtain an enhanced bilinear feature. Finally, an attempt was made to improve lightweight CNNs by introducing a model called DFCANet [22] for corn disease identification. The model consisted of dual feature fusion with coordinate attention (CA) and downsampling (DS) modules. The CA module suppressed the background noise and focused on the diseased area. In addition, the DS module was used for downsampling. The above models enhanced the existing CNN models or solved specific problems.

Carbon is one of many greenhouse gases that exist naturally in the Earth's system [23]. However, carbon dioxide emissions have increased abnormally because of using fossil fuels for energy and due to land use/cover (LULC) changes. The fast increase in the carbon dioxide concentration in the air is making a major contribution to possible climate change and in turn to natural disasters as well as environmental and economic losses in the future [24]. The world's total forest area is about 4 billion hectares, corresponding to about 31% of the total land area [25]. Forests that include one or mixed types of trees with different plants absorb air pollution and provide the oxygen we breathe through photosynthesis, which absorbs carbon dioxide and preserves it in the leaves and stems up to the roots. Planted forests and woodlots were found to have the highest CO_2 removal rates, ranging from 4.5 to 40.7 t CO_2 ha^{-1} year^{-1} during the first 20 years of growth [26,27].

Remote sensing data and methods are widely used to estimate carbon sequestration. Liu et al. [28] used airborne radar data to identify single-tree parameters such as the diameter at breast height (DBH) and tree height, and based on these measurements they estimated the AGB of single trees. Lizuka and Tateishi [29] used Landsat 8 and lso/Palsar to estimate forest tree volumes and tree ages. They used the extracted information to estimate carbon sequestration, and the verification was based on the collected field samples. Castro-Magnani et al. [30] used MODIS gross primary productivity (GPP) and net primary productivity (NPP) [31] to estimate carbon sequestration in the AGB. Later, they calculated the socio-economic benefit of sequestering carbon. Published research [32] has used airborne light detection and ranging (LiDAR) to acquire the vertical structure parameters of coniferous forests to construct two prediction models of aboveground carbon density (ACD). One is a plot-averaged height-based power model, and the other is a plot-averaged daisy-chain model. The correlation coefficients were significantly higher than that of the traditional percentile model. A paper published by Kanniah et al. [33] utilized different vegetation indices (Vis) and very high resolution WorldView-2 images to estimate carbon sequestration in an urban area. One of the Vis correlated strongly with the collected field data. However, the forest consisted of single tree species, which made the authors' research work simple. Uniyal et al. [34] estimated carbon sequestration using Landsat 8 and support vector machine (SVM) [35], random forest [36], k-nearest neighbor (kNN) [37], and the eXtreme gradient boosting (XGBoost) [38]. The authors used a huge number of variables extracted from Landsat image as inputs and field-collected data as training samples, and based on the R squared (coefficient of determination) they concluded that machine-learning-algorithm regressions are better than a linear regression. Zhang et al. [39] compared a convolutional neural network (CNN) to SVM and RF for estimating carbon sequestration in forests' AGB from Sentinel-2, Sentinel-1, and lso/Palsar. The authors used more than 67 variables to train the algorithms. The results showed that the CNN was better than RF and SVM at estimating carbon sequestered above the surface.

A literature review showed different attempts to estimate carbon sequestration using LIDAR data, which is limited by the technology's availability and cost and the size of the covered area. Some researchers used only one type of remote sensing optical data to extract vegetation indices (Vis) to compare some machine learning algorithms in estimating carbon sequestration. Other researchers used only optical images to calculate Vis and to estimate carbon sequestration in urban areas. Researchers deployed both optical and radar data without using machine learning to estimate carbon sequestration. One successful study combined multiple types of radar and optical data to compare machine learning algorithms, including a CNN, in estimating carbon sequestration in forests' AGB. However, this led to the need to calculate a large number of variables, and it demanded huge computation resources. It is also known that a CNN alone is more effective in detecting patterns than estimating specific information [40,41]. Moreover, all the above research shared one objective, which was quantitatively estimating carbon sequestration by AGB.

The objectives and the contributions of this research are the following: (1) creating a new lightweight CNN model (FlexibleNet); (2) testing the new model (FlexibleNet) for qualitatively estimating carbon sequestration in peri-urban forests' AGB; and (3) creating new datasets that combine multispectral satellite images and multicriteria themes with different sizes. These datasets and python programs are available on GitHub.

Many issues make the new model better than other lightweight CNN models. First, the new model's flexibility arises from its ability to adapt to changes in tuning many parameters, such as the image dimension, dataset size, and layer depth and width. Second, the model uses only three extracted features from Sentinel-2 as inputs compared to the multi-input for other CNN models. Third, the new lightweight model can qualitatively measure carbon sequestration in peri-urban forests. Fourth, it is more efficient in dealing with small datasets.

After the introduction section, the second section describes the data, the third section contains the methods, the fourth section presents the experimental results, and the final section provides our conclusions.

2. Data

2.1. Area of Study and Field Survey

The border of the study area is specified by a red square in Figure 1. It is located in the El-Bared river basin in the northeast of Lebanon. The selection was based on many criteria that included the diversity of the forest types, forest densities, the existence of urban economic activities, the pressure exerted by the residents on the forest cover (cutting and burning), the ease of accessibility to the area (specific spots), and the existence of local authority support for fieldwork.

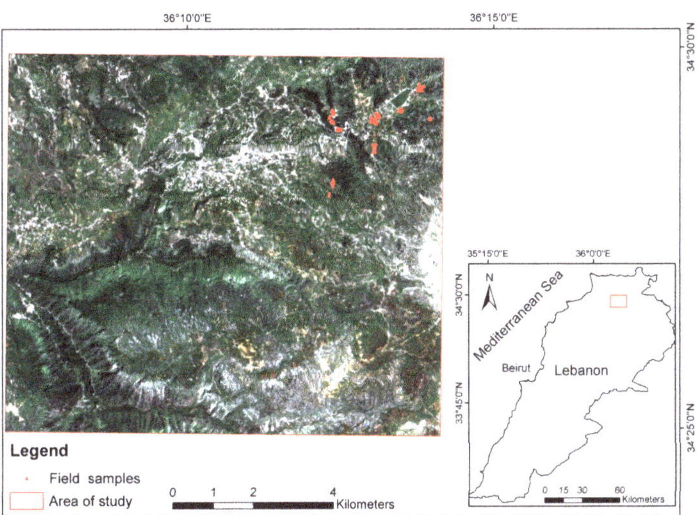

Figure 1. Study area.

The area of study occupies about 106.5 km^2 of different land cover types such as fruit trees, urban (including touristic facilities), forests, grasslands, etc. The highest elevation in the area of study is 1500 m, and the landform is flat to moderately steep (a slope less than 30%).

One can notice in Figure 1 that the field samples that were collected in the northeastern part of the study area. The selection of the field sampling area was based on having different forest types such as pine *"Pinus brutia"*, cedar *"Cedrus Libani"*, fir *"Abies cilicica"*, juniper *"Juniperus excelsa"*, and oak *"Quercus Cerris"* (Figure 2a–e).

(a)

(b)

(c)

(d)

(e)

Figure 2. Forest types: (**a**) *Quercus Cerris*, (**b**) *Abies cilicica*, (**c**) *Cedrus Libani*, (**d**) *Juniperus excelsa*, (**e**) *Pinus brutia*.

The sample collection was a random process, and it depended on the ease of accessibility to the investigated area. Table 1 shows the species type, the number of collected samples, the average height, and the average diameter at breast height (DBH). The cedars' cover was very small compared to other forest covers, and the authorities prohibited access to these trees because they were located in a reservation and they are national symbol.

Table 1. Information about the collected field samples.

Type	Number of Samples	Average DBH (cm)	Average Height (Meters)
Quercus Cerris	17	119	15
Pinus brutia	19	125	12
Abies cilicica	46	237	17
Juniperus excelsa	32	225	8

2.2. Data Type and Source

In this research, we deployed Sentinel-2 data, which is considered to be important and free optical remote sensing satellite data. Sentinel-2A and Sentinel-2B were launched in June 2015 and March 2017, respectively [42]. Sentinel-2 is an optical remote sensing satellite. It has a spatial resolution that varies between 10 m and 60 m depending on the wavelength. Sentinel-2A has a temporal resolution of 10 days, which can become 5 days with the combination of Sentinel-2B and another optical satellite with the same specifications as Sentinel-2A. The clipped image has a size of 1115 × 955 pixels and consists of bands 3, 4, and 8, which correspond to green, red, and near infrared. These bands were selected for two reasons: they have the highest spatial resolution, and they are representative of the crops' photosynthesis process. To extract the required area, we used Google Earth Engine's (GEE) Sentinel-2 dataset and computation facilities. One Sentinel-2 image was selected in May 2020 for two reasons: to reduce the cloud cover effect (less than 5% of the image size) and to obtain the maximum vegetation cover (deciduous and coniferous trees, grasslands, and agricultural lands).

Moreover, a vector layer representing the global canopy height for the year 2020 at a 10 m resolution [43] was used in the canopy density model (Figure 3).

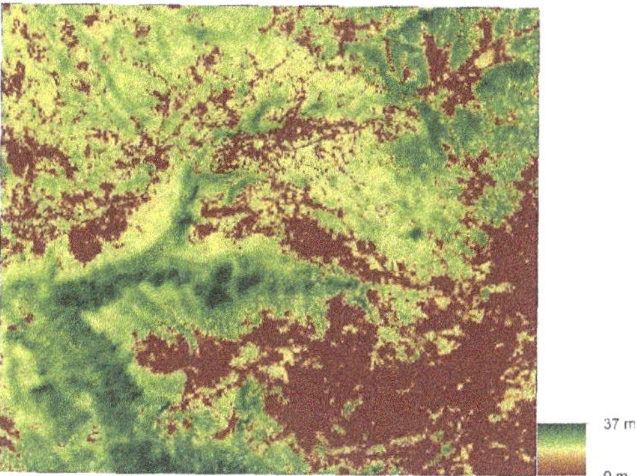

Figure 3. Canopy height map.

3. Methods

The following flowchart (Figure 4) shows the different tasks that were implemented in this research to qualitatively estimate carbon sequestration in ABG forests using the new lightweight CNN model (FlexibleNet) and the training and Sentinel-2 image datasets.

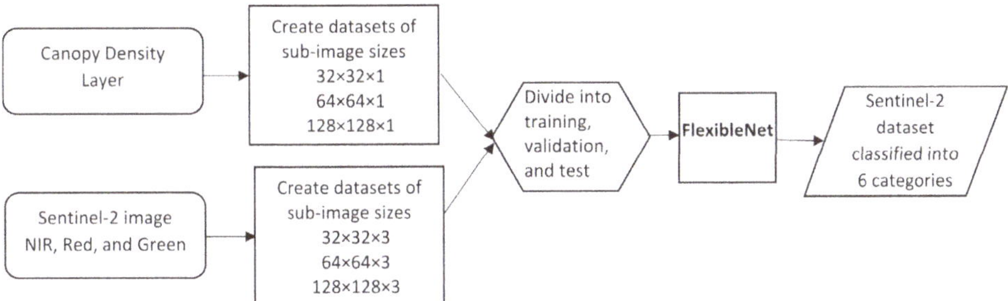

Figure 4. The general process for qualitatively assessing forests' AGB carbon sequestration capacities.

3.1. Canopy Density Model (CDM)

An adapted model created by Abdollahnejad et al. [44] incorporated different indices from Sentinel-2 images and the thermal band of Landsat to create a canopy density model. The adapted model combined different resolutions, which lowered the credibility and efficiency of the final product. Moreover, the model neglected the canopy heights, which can successfully differentiate between forests and other vegetation types.

Both the Sentinel-2 image (level 2) and the canopy height layer were obtained using the Google Earth Engine (GEE) platform. Scripts were written in the Java language to retrieve the needed data. Normally, the acquired Sentinel-2 image is level 2, which is an image that is corrected geometrically and atmospherically. Three indices were created from the Sentinel-2 image using the following equations:

$$AVI = [(NIR + 1) \times (1 - Red) \times (NIR - Red)]^{1/3} \tag{1}$$

$$BI = \frac{(NIR + Green) - Red}{(NIR + Green) + Red} \tag{2}$$

$$SI = \sqrt{(1 - Green) \times (1 - Red)} \tag{3}$$

where *AVI* is the advanced vegetation index, *BI* is the bare soil index, and *SI* is the canopy shadow index. Moreover, *NIR*, *Red*, and *Green* represent the three different spectrums and the bands B2, B3, and B8 in the Sentinel-2 image. *AVI* was modified to provide values between −1 and 1. The modification included replacing 256 with 1 and normalizing the bands. *BI* ranged between 0 and 1, where 0 meant complete bare soil or no vegetated area and 1 meant completely covered by vegetation. Finally, *SI* was modified by replacing 256 with 1, and the bands were normalized. *SI* values ranged between 0 and 1, where the maximum value indicated the highest canopy shadow.

These themes, including the canopy heights, were classified into six categories using natural break classification (Jenks) [45]. The classes were based on natural groupings inherent in the data. Normally, the classification process identifies breakpoints by picking the class breaks that best group similar values and maximize the differences between classes. Finally, a spatial analysis that included mathematical operations was deployed to obtain the canopy density theme. The above processes were combined according to the following flowchart (Figure 5).

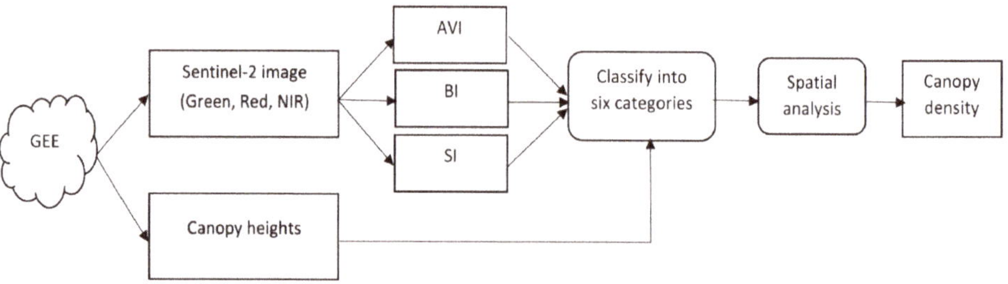

Figure 5. Canopy density model.

Further investigation in the future to improve the canopy density layer may include higher-spatial-resolution satellite images and time series of NDVI to separate deciduous forest trees from evergreens, which could further enhance research work.

3.2. The New Lightweight Convolutional Neural Network Model (FlexibleNet)

CNNs are collections of neurons that are ordered in inter-related layers, with convolutional, pooling, and fully connected layers [46]. CNNs require less preprocessing, and they are the most effective learning algorithms for realizing image structures. Moreover, it was proven that CNNs excel in image classification, recognition, and retrieval [47].

Normally, a simple CNN model consists of one or many of the following layers: 1—convolutional layer, 2—pooling layer, 3—activation layer, and a fully connected layer.

In this research, we created a new lightweight CNN model (FlexibleNet) to reduce the resource and training dataset requirements (Figure 6). The performance of the new model was tested in the qualitative classification of carbon sequestration. Our new model is a CNN scaling-based model (width, depth, and resolution). The depth corresponds to the number of layers in a network. The width is associated with the number of neurons in a layer or, more pertinently, the number of filters in a convolutional layer. The resolution is simply the height and width of the input image. Unlike the conventional practice, which arbitrarily scales these factors, FlexibleNet uniformly scales the network width, depth, and resolution with a set of fixed scaling coefficients.

We combined different strategies to improve the FlexibleNet performance. These strategies were spatial exploitation and varying the depth. Spatial exploitation includes parameters such as the number of processing units (neurons), filter size, and activation function. We assumed that varying the CNN's depth can better approximate the target function with a number and can improve feature representations and network performance.

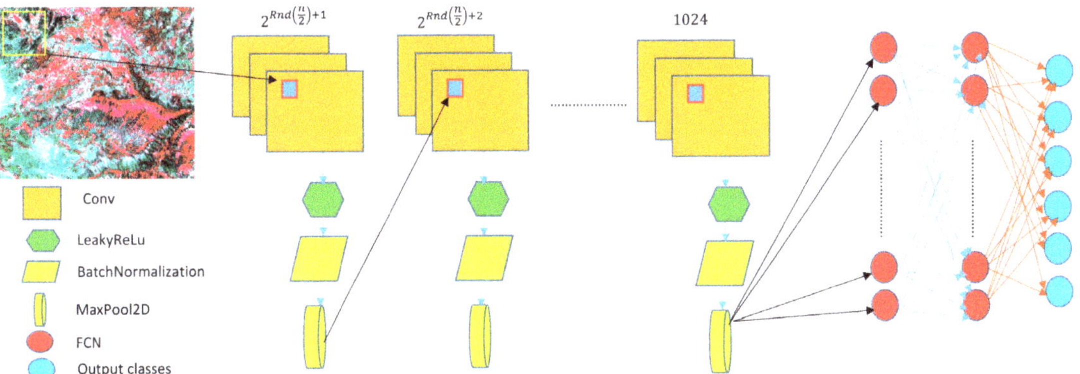

Figure 6. The inner structure of the FlexibleNet model.

The spatial exploitation included changes to the filter size and the activation function. Moreover, the depth of the FlexibleNet network or the number of convolutional layers varied when the dimensions of the features changed. The variations in the width and depth were based on the variation in the resolution of the image. The following equations depict the changes to filter size:

$$Im = \sum_{i=1}^{m} d_i \times w_i \times c_i \;\; \rightarrow d = w = 2^n \;\; \rightarrow f = Rnd\left(\frac{n}{2}\right) \times Rnd\left(\frac{n}{2}\right) \qquad (4)$$

where Im is the original image and m is the number of sub-images of size $d \times w \times c$, where d is the number of rows, w is the number of columns, and c is the number of channels. Moreover, n is the exponent, f is the filter size, and $Rnd()$ is the round function ($d, w, c,$ and $n \in \mathbb{Z}^+$). If the image (Im) has an uneven size, zeros are padded to the columns and/or rows to make them even.

The number of filters for each convolution layer can be set up based on the following rules:

$$f_m = \begin{cases} Initial\;\; 2^{Rnd\left(\frac{n}{2}\right)+1} \;\; where\; n \geq 5 \\ m = Rnd\left(\frac{n}{2}\right) + 2; \quad 2^m \;\; \rightarrow m = m + 1 \\ Final \quad 1024 \end{cases} \qquad (5)$$

where f_m represents the filter sizes. These rules work as follows: Suppose I have a sub-image of size 32×32. Then, $n = 5$. This means that the initial filter is $f_0 = 16$. Next, the filter size is obtained by calculating m $f_m = 32, 64, 128, 256,$ and 512, where $m = 6, 7, 8,$ and 9 and the final filter size is 1024 (maximum) with $m = 10$.

Then, the leaky rectified linear activation function (LReLU) is used [48], which is a modification of the ReLU activation function. It has the same form as the ReLU, but it will leak some positive values to 0 if they are close enough to zero (Equation (6)). It is a variant of the ReLU activation function. Normally, ReLU is half-rectified (from the bottom). ReLU(p) is zero when p is less than zero, and ReLU(p) is equal to p when p is above or equal to zero.

$$LReLU(p) = \max(0.01 \times p, p) \qquad (6)$$

The number of layers or the depth of the network (Lay_{depth}) can be computed as indicated in Equation (7). It is noticeable that the depth reached unity when the dimensions of the image were >18. The creation of Equation (7) was based on the assumptions that a sub-image cannot be less than 16×16 and that the maximum sub-image size is the image itself. Adapting to the increase in the sub-image size requires decreasing the network depth by one level (the number of convolution layers) each time the sub-image increases. The depth starts from $Rnd\left(\frac{n}{2}\right) + 7$ convolution layers to one layer, where the size of the image is the image itself, assuming it may reach infinity as a size.

$$Lay_{depth} = \begin{cases} Rnd\left(\frac{n}{2}\right) + 7 & 4 \leq n \\ Rnd\left(\frac{n}{2}\right) + 5 & 5 \leq n \leq 6 \\ Rnd\left(\frac{n}{2}\right) + 3 & 7 \leq n \leq 8 \\ Rnd\left(\frac{n}{2}\right) + 1 & 9 \leq n \leq 10 \\ Rnd\left(\frac{n}{2}\right) - 1 & 11 \leq n \leq 12 \\ Rnd\left(\frac{n}{2}\right) - 3 & 13 \leq n \leq 14 \\ Rnd\left(\frac{n}{2}\right) - 5 & 15 \leq n \leq 16 \\ Rnd\left(\frac{n}{2}\right) - 7 & 17 \leq n \leq 18 \\ 1 & n > 18 \end{cases} \tag{7}$$

In addition to the Lay_{depth} size, there is a fixed number of three dense layers (DL). According to [49], the dense layer is an often-used layer that contains a deeply connected neural network layer. DL is a hidden layer associated with one node in the next layer.

Figure 7a–c show the FlexibleNet structure for three different scales based on the above rules, where $n = 32, 256,$ and 512. One can notice that as the scale increases, the depth decreases. This strategy can help reduce the computation requirements (processing power and memory size).

3.3. Estimating Carbon Sequestration for the Collected AGB Samples

The measured trees were used to compute the volume of the AGB using Equations (8) and (9). Where Vm^3 is the volume of wood in cubic meters, Hm is the height of the tree, DBH is the diameter at breast height, and Bm^2 is the base area in square meters. Lee et al. [50] suggested Table 2 to help in the calculation process of carbon sequestration in the ABG. The carbon content usually uses a value of 0.5, which means that wood is about 50% carbon. We used the model created by Lizuka and Tateishi [29] to estimate carbon sequestration per hectare (CS_{ha}) (Equation (10)). Fc = 44/12 converts the carbon value to the carbon dioxide sequestration value, where 12 and 44 represent the molecular masses of carbon and carbon dioxide, respectively.

$$Vm^3 = Bm^2 \times Hm \tag{8}$$

$$Bm^2 = \pi \times \left(\frac{DBH}{2}\right)^2 \tag{9}$$

$$CS_{ha} = Vm^3 \times Be \times Bd \times Cc \times Fc \tag{10}$$

Table 2. Coefficients for calculating carbon sequestration by forest type.

Type of Forest	Bulk Density (Bd) (Tons/m^3)	Biomass Expansion (Be)	Carbon Content (Cc)
Coniferous	0.47	1.651	0.5
Deciduous	0.80	1.720	0.5
Mixed	0.635	1.685	0.5

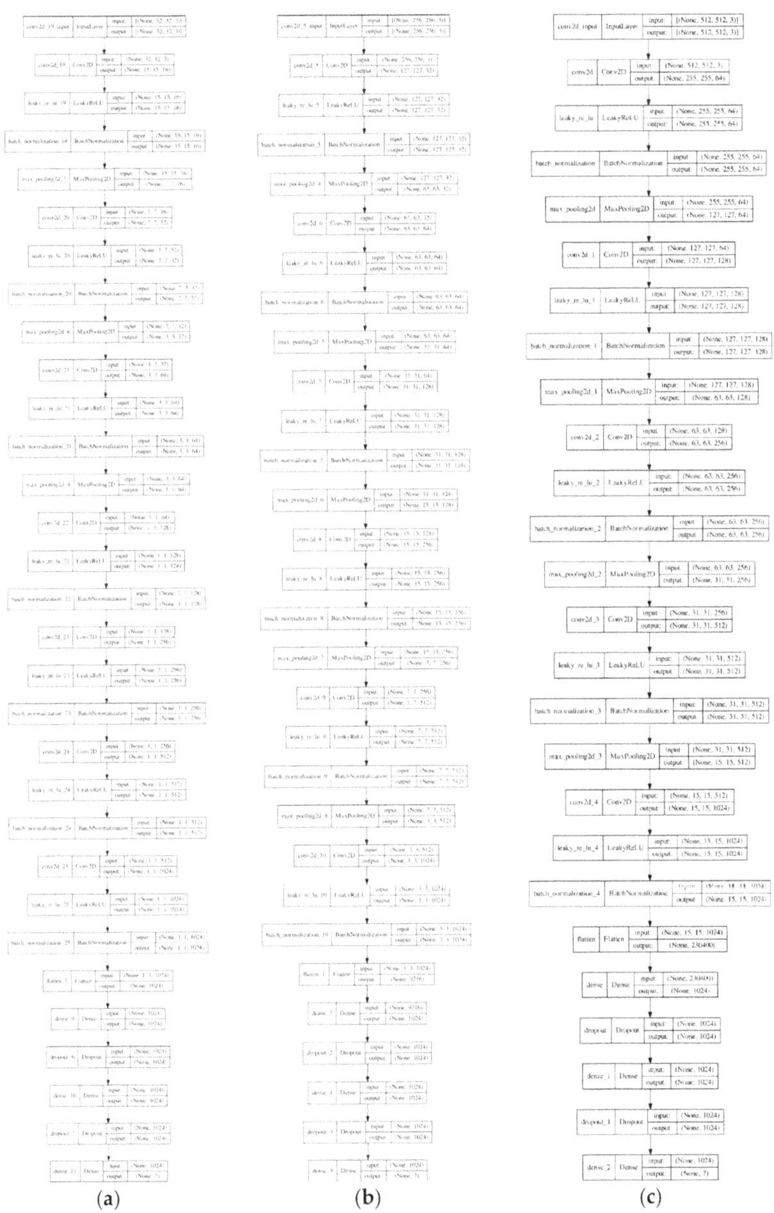

(a)   (b)   (c)

Figure 7. FlexibleNet with different scales and depths: (**a**) 32, 7; (**b**) 256, 6; (**c**) 512, 5.

4. Results

For this section, we created different datasets of Sentinel-2 sub-images to prove the efficiency of the new lightweight CNN model (FlexibleNet) in qualitatively estimating carbon dioxide sequestration. The collected samples of trees' characteristics, as shown in Table 1, were used to calculate CS_{ha} using Equations (8)–(10). Then, these values were converted to qualitative values using Sturges' rule [51]. Since the samples represent trees' characteristics, the "no carbon" class was removed. Figure 8 represents the distribution of the samples according to five classes (very low, low, moderate, high, and very high). These created qualitative samples were used to verify the credibility of the canopy density dataset using a confusion matrix [52] before using it in the training of the new model (Table 3). The accuracy computed from the matrix using estimated versus measured values was 92.1%.

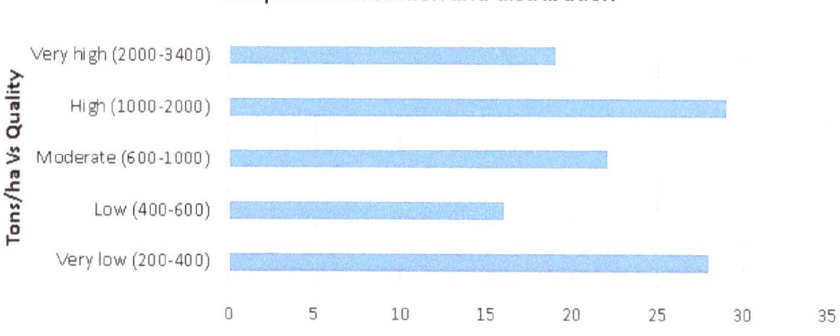

Figure 8. Trees samples classification.

Table 3. Confusion matrix.

Measured/Estimated	Very Low	Low	Moderate	High	Very High
Very low	23	2	1	2	0
Low	1	15	0	0	0
Moderate	0	1	21	0	0
High	1	1	0	27	0
Very high	0	0	0	0	19

We created different datasets that consisted of tiled sub-images with three different sizes of 32 × 32 (1050 images), 64 × 64 (270 images), and 128 × 128 (72 images) and three bands representing different spectrums (green, red, and near infrared). The other datasets consisted of the same size and number of tiles but only represented canopy densities with six classes (no carbon, very low, low, moderate, high, and very high). A script was written in the Python language to classify the Sentinel-2 sub-images into six classes based on the computed canopy density statistics (Algorithm 1). The script takes every computed sum (arr) for each canopy density sub-image and compares it to the created criteria (criteria) based on Sturges' rule.

The sums of the pixel values of all canopy density sub-images were calculated. Next, these sums' maximum, minimum, and average were computed. Then, they were used with Struges' rule to classify the Sentinel-2 sub-images into six classes. After that, the datasets were split into 80% training and 20% validation samples. Figure 9 shows examples of the original Sentinel 2 sub-images (false color) and their counterpart canopy density classes. The colors in the canopy density images signify that very low is dark brown, low is light brown, moderate is light green, high is green, and very high is dark green.

These samples were used as part of the training and validation datasets to check the efficiency of the FlexibleNet model.

Algorithm 1 A python script to classify Sentinel-2 sub-images.

```
if( (arr[i] ≤ 0) and file_exists):
# it checks if the sum is less or equal to zero and if the image exists in the folder
before copying it to the no carbon folder
shutil.copy(filename,dest1)
# copy image to no carbon folder (dest1)
if((arr[i] >0 and arr[i] ≤ criteria *2) and file_exists):
shutil.copy(filename,dest2)
# copy to folder very low
if((arr[i] > criteria *2 and arr[i] ≤ criteria *3) and file_exists):
shutil.copy(filename,dest3)
# copy to folder low
if((arr[i] > criteria *3 and arr[i] ≤ criteria *4) and file_exists):
shutil.copy(filename,dest4)
# copy to folder moderate
if((arr[i] > criteria *4 and arr[i] ≤ criteria *5) and file_exists):
shutil.copy(filename,dest5)
# copy to folder high
if((arr[i] > criteria *5 and arr[i] ≤ maxval) and file_exists):
shutil.copy(filename,dest6)
# copy to folder very high
```

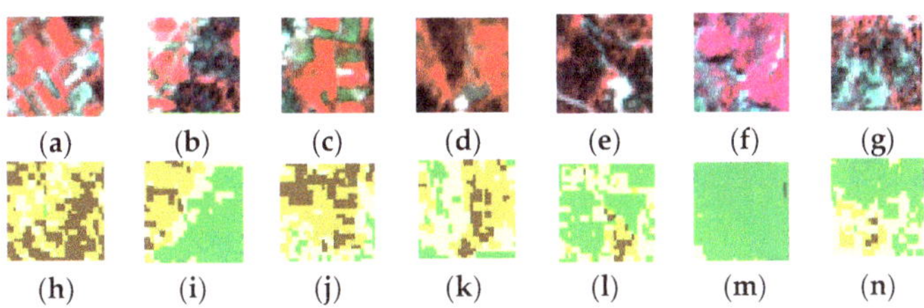

Figure 9. Different sub-images showing (**a–g**) the original Sentinel-2 images and (**h–n**) canopy density (very low, low, medium, high, and very high).

The FlexibleNet was compared with four popular and well-known convolutional neural networks: the large model ResNet50 [8], the lightweight models Xception [13] and MobileNetV3-Large [15], and the EfficientNet [17]. These models were selected based on their popularity, efficiency, and availability.

All the models, including FlexibleNet, were run using "Jupyter Notebook" on Amazon SageMaker cloud computing facilities that had 16 GB of memory capacity and two Intel Xeon Scalable processors with 3.3 GHz speed. Moreover, these models were run for a maximum of 100 epochs, and each epoch had several steps (number of steps per epoch = (total number of training samples)/batch size). We deployed a stochastic gradient descent (SGD) optimizer in FlexibleNet with an initial learning rate of 0.001. SGD is an iterative method for optimizing an objective function with suitable smoothness properties. SGD replaces the actual gradient (calculated from the entire dataset) with an estimate thereof (calculated from a randomly selected subset of the data). Especially in the high-dimensional optimization problem, this reduces the very high computational burden, achieving faster iterations in return for a lower convergence rate [53]. The learning rate of 0.001 was selected based on previous research conducted by Asif et al. [54].

In the first experiment, the datasets of 32 × 32 were used to compare these models. The outcomes of these models are shown in Table 4, and the behaviors of these models during the run process are shown in Figure 10a–j.

Table 4. Summary of the outcomes of the experiments using an image resolution of 32 × 32.

Model Name	Number of Parameters (Millions)	Time Requirement (Minutes)	Accuracy %	Lowest Loss Value
FlexibleNet	5.52	13.3	98.81	0.042
ResNet50	26.38	77	96.41	0.1074
EfficientNetB5	31.30	28.4	52	1.1
MobileNetV3-Large	6.23	13.3	68.69	0.7122
Xception	21.58	62	66.96	0.83

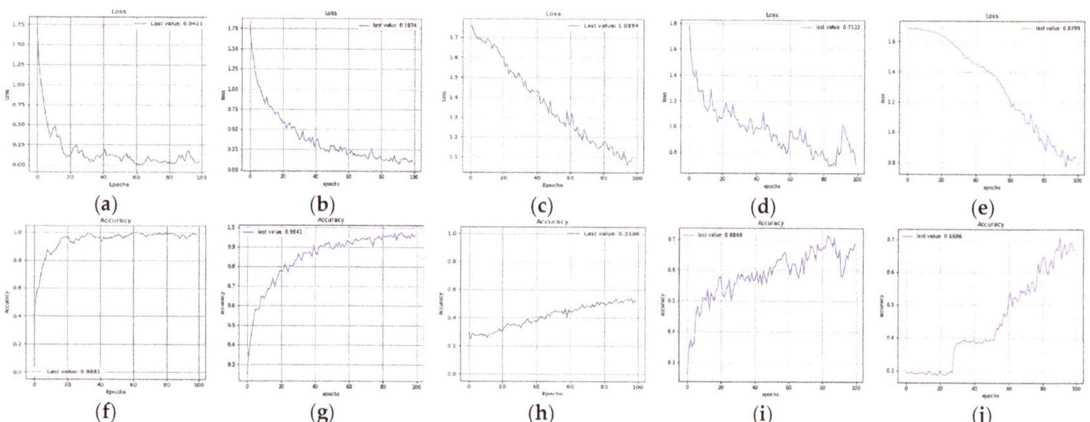

Figure 10. Loss and accuracy of sub-images of size 32 × 32 processed by (**a,f**) FlexibleNet, (**b,g**) ResNet50, (**c,h**) EfficientNetB5, (**d,i**) MobileNetV3-Large, and (**e,j**) Xception.

The new model had total parameters equal to 5.52 million. The total time was 13.3 min, with 8 s for each iteration. The accuracy of the final trained model was about 98.81%, and the final loss was 0.042.

The ResNet50 model was run for 100 iterations (epochs), with a total number of parameters equal to 26.38 million. ResNet50 took 77 min, with an accuracy of about 96.41%, and the final loss was 0.1074. The accuracy of ResNet50 was lower than that of FlexibleNet. This proved the reliability and efficiency of the new model.

EfficientNet was also tested using the same datasets. The number of iterations was 100, and the number of parameters was 31.3 million. It took the model 88.4 min to complete the iterations (epochs). The lowest loss was 1.1, and the highest accuracy was 52%. This proved that FlexibleNet is more efficient and accurate than the lightweight EfficientNet.

The lightweight network models MobileNetV3-Large and MobileNetV3-Small are normally targeted for high- and low-resource use cases. These models are then adapted and applied to object detection and semantic segmentation. MobileNetV3-Small is more suitable for mobile phone operating systems. MobileNetV3-Large is 3.2% more accurate in ImageNet classification while reducing latency by 15% compared to MobileNetV2 [55]. The implemented MobileNetV3-Large had 6.23 million total parameters, and it was run for 100 iterations. It took the model 13.3 min to complete the iterations (epochs). The lowest loss was 0.7122, and the highest accuracy was 68.69%. This proved that FlexibleNet was more efficient and accurate than the lightweight MobileNetV3-Large.

The second experiment was conducted in the same area of study, but the datasets had an image resolution of 64 × 64 pixels. Figure 11 shows different 64 × 64 sub-images alongside corresponding canopy sub-images. We placed constraints on running the FlexibleNet model and the other tested models to avoid falling into the overfitting problem because of the lack of a large dataset of images [56].

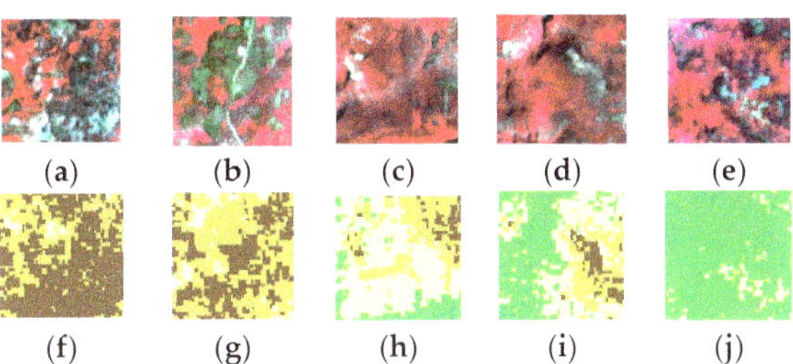

Figure 11. Sentinel-2 sub-images (64 × 64): (**a–e**) original and (**f–j**) canopy density (very low (dark brown) to very high (dark green)).

During the fitting process of these models, the loss function was tested. The fitting process was terminated when several iterations completed and the minimum loss value did not change. Table 5 shows the outcomes of testing the different models on different image resolutions. First, it is noticeable that the number of parameters increased for FlexibleNet, ResNet50, and EfficientNetB5. Nevertheless, the time requirement decreased for all models except EfficientNetB5. Finally, FlexibleNet was the only model with the highest accuracy and the lowest loss function value, as shown in Figure 12a–j.

Table 5. Summary of the outcomes of the experiments using an image resolution of 64 × 64.

Model Name	Number of Parameters (Millions)	Time Requirement (Minutes)	Accuracy %	Lowest Loss Value	Total Iterations
FlexibleNet	8.4	5	98.25	0.0457	60
ResNet50	32.6	13	96.74	0.0877	51
EfficientNetB5	32.9	40	93.06	0.1936	100
MobileNetV3-Large	6.23	4	90.22	0.3422	74
Xception	21.58	22	31.52	1.718	100

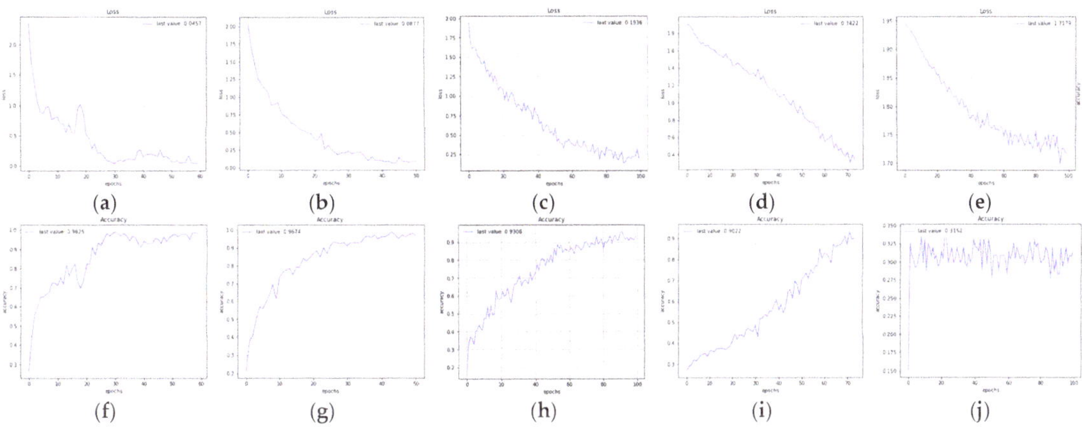

Figure 12. Loss and accuracy of 64 × 64 sub-images processed by (**a,f**) FlexibleNet, (**b,g**) ResNet50, (**c,h**) EfficientNetB5, (**d,i**) MobileNetV3-Large, and (**e,j**) Xception.

The final experiment was conducted in the same area of study, but the image resolution was 128 × 128 pixels, which resulted in smaller datasets. Figure 13 shows different 128 × 128 sub-images alongside corresponding canopy density sub-images.

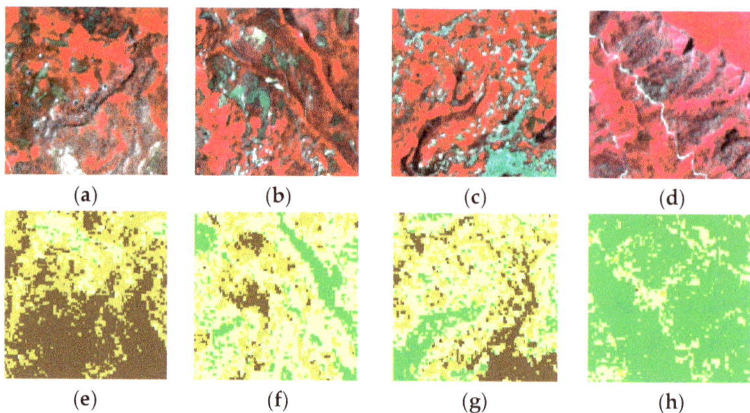

Figure 13. Sentinel-2 sub-images (128 × 128): (**a–d**) original and (**e–h**) canopy density.

We placed constraints on running the FlexibleNet model and the other tested models to avoid falling into the overfitting problem because of a lack of a large dataset of images [56]. The loss function was tested, and the fitting process was terminated when the minimum loss value did not change after a specific number of iterations.

Table 6 lists the results of running different models. Again, FlexibleNet and MobileNetV3-Large showed stable numbers of parameters, even when the dimensions of the image increased from 64 × 64 to 128 × 128. However, FlexibleNet was the fastest, and it had the highest accuracy and lowest loss value compared to the other models. In this experiment, FlexibleNet showed robustness in dealing with very small datasets (72 images), whereas the others failed to deal with the problem. Many adjustments were made (such as duplicating the dataset) to overcome the limited size of the dataset and make the other models run smoothly. The performances of these models (accuracy and loss) are shown in Figure 14a–j.

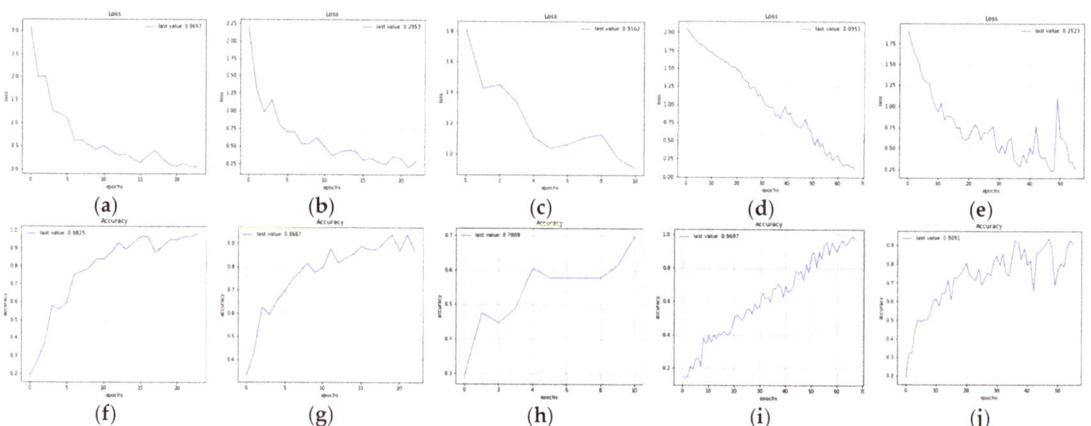

Figure 14. Loss and accuracy of 128 × 128 sub-images processed by (**a,f**) FlexibleNet, (**b,g**) ResNet50, (**c,h**) EfficientNetB5, (**d,i**) MobileNetV3-Large, and (**e,j**) Xception.

Table 6. Summary of the outcomes of the experiments using an image resolution of 128 × 128.

Model Name	Number of Parameters (Millions)	Time Requirement (Minutes)	Accuracy %	Lowest Loss Value	Total Iterations
FlexibleNet	8.4	0.8	98.25	0.0657	24
ResNet50	57.8	7.7	86.87	0.2953	23
EfficientNet	62.7	6	70.09	0.9102	11
MobileNetV3-Large	6.23	8.1	96.97	0.0951	68
Xception	55.1	20.53	90.91	0.2523	56

5. Conclusions

There were many advantages of deploying the new lightweight convolutional neural network model, FlexibleNet. First, we obtained the highest accuracy in qualitatively classifying Sentinel-2 images into different carbon sequestration classes. Second, the FlexibleNet model had the lowest loss values compared to the other models. Third, except for MobileNetV3-Large, the new model used the lowest number of parameters and required the lowest time. In the first experiment, the FlexibleNet model was the best one because it had the lowest number of parameters compared to the other models, including MobileNetV3-Large. In the second and third experiments, the MobileNetV3-Large model was slightly better than the FlexibleNet model, but both were stable concerning the number of parameters when the problem size changed. One disadvantage of the FlexibleNet model was its inability to overcome the MobileNetV3 model in reducing the number of parameters in all experiments. The FlexibleNet model is the first version of a series that will include enhancements to many existing features in the new model, including reducing the parameter requirements. It is also expected to be used to conduct more experiments on other complex problems, such as using tropical forest datasets.

Funding: This research received no external funding.

Data Availability Statement: All data and programs were placed on the website https://github.com/users/ma850419/FlexibleNet (accessed on 30 December 2022).

Conflicts of Interest: The authors declare no conflict of interest.

References

1. Fradkov, A. Early History of Machine Learning. *IFAC-Pap. OnLine* **2020**, *53*, 1385–1390. [CrossRef]
2. Wang, P.; Wang, L.; Leung, H.; Zhang, G. Super-Resolution Mapping Based on Spatial–Spectral Correlation for Spectral Imagery. *IEEE Trans. Geosci. Remote Sens.* **2021**, *59*, 2256–2268. [CrossRef]
3. Awad, M. Cooperative evolutionary classification algorithm for hyperspectral images. *J. Appl. Remote Sens.* **2020**, *14*, 016509. [CrossRef]
4. Masi, G.; Cozzolino, D.; Verdoliva, L.; Scarpa, G. Pansharpening by Convolutional Neural Networks. *Remote Sens.* **2016**, *8*, 594. [CrossRef]
5. Girshick, R.; Donahue, J.; Darrell, T.; Malik, J. Rich Feature Hierarchies for Accurate Object Detection and Semantic Segmentation. In Proceedings of the IEEE Conference on Computer Vision and Pattern Recognition, Columbus, OH, USA, 23–28 June 2014; pp. 580–587. [CrossRef]
6. Awad, M.M.; Lauteri, M. Self-Organizing Deep Learning (SO-UNet)—A Novel Framework to Classify Urban and Peri-Urban Forests. *Sustainability* **2021**, *13*, 5548. [CrossRef]
7. Sylvain, J.; Drolet, G.; Brown, N. Mapping dead forest cover using a deep convolutional neural network and digital aerial photography. *ISPRS J. Photogramm. Remote Sens.* **2019**, *156*, 14–26. [CrossRef]
8. Sarwinda, D.; Paradisa, R.; Bustamam, A.; Anggia, P. Deep Learning in Image Classification using Residual Network (ResNet) Variants for Detection of Colorectal Cancer. *Procedia Comput. Sci.* **2021**, *179*, 423–431. [CrossRef]
9. Tao, J.; Gu, Y.; Sun, J.; Bie, Y.; Wang, H. Research on VGG16 convolutional neural network feature classification algorithm based on Transfer Learning. In Proceedings of the 2nd China International SAR Symposium (CISS), Shanghai, China, 3–5 November 2021; pp. 1–3. [CrossRef]
10. Singh, I.; Goyal, G.; Chandel, A. AlexNet architecture based convolutional neural network for toxic comments classification. *J. King Saud Univ. Comput. Inf. Sci.* **2022**, *34*, 7547–7558. [CrossRef]

11. Szegedy, C.; Liu, W.; Jia, Y.; Sermanet, P.; Reed, S.; Anguelov, D.; Erhan, D.; Vanhoucke, V.; Rabinovich, A. Going deeper with convolutions. In Proceedings of the 2015 IEEE Conference on Computer Vision and Pattern Recognition (CVPR), Boston, MA, USA, 7–12 June 2015; pp. 1–9. [CrossRef]

12. Iandola, F.N.; Han, S.; Moskewicz, M.W.; Ashraf, K.; Dally, W.J.; Keutzer, K. SqueezeNet: AlexNet-level accuracy with 50x fewer parameters and <0.5 MB model size. *arXiv* **2016**, arXiv:1602.07360. [CrossRef]

13. Chollet, F. Xception: Deep Learning with Depth wise Separable Convolutions. In Proceedings of the IEEE Conference on Computer Vision and Pattern Recognition (CVPR), Honolulu, HI, USA, 21—26 July 2017; pp. 1800–1807. [CrossRef]

14. Yuan, H.; Cheng, J.; Wu, Y.; Zeng, Z. Low-res MobileNet: An efficient lightweight network for low-resolution image classification in resource-constrained scenarios. *Multimed. Tools Appl.* **2022**, *81*, 38513–38530. [CrossRef]

15. Howard, A.; Sandler, M.; Chu, G.; Chen, L.C.; Chen, B.; Tan, M.; Wang, W.; Zhu, Y.; Pang, R.; Vasudevan, V. Searching for MobileNetV3. In Proceedings of the IEEE International Conference on Computer Vision, Seoul, Republic of Korea, 27–28 October 2019; pp. 1314–1324. [CrossRef]

16. Ma, N.; Zhang, X.; Zheng, H.T.; Sun, J. Shufflenet v2: Practical guidelines for efficient CNN architecture design. In Proceedings of the European conference on computer vision (ECCV), Munich, Germany, 8–14 September 2018; pp. 116–131. [CrossRef]

17. Tan, M.; Le, Q.V. (2019) EfficientNet: Rethinking Model Scaling for Convolutional Neural Networks. In Proceedings of the 36th International Conference on Machine Learning, Long Beach, CA, USA, 10–19 June 2019; pp. 6105–6114.

18. Sarker, I.H. Deep Learning: A Comprehensive Overview on Techniques, Taxonomy, Applications and Research Directions. *SN Comput. Sci.* **2021**, *2*, 420. [CrossRef] [PubMed]

19. Zhou, Y.; Bai, Y.; Bhattacharyya, S.; Huttunen, H. Elastic Neural Networks for Classification. In Proceedings of the 2 IEEE International Conference on Artificial Intelligence Circuits and Systems (AICAS), Taiwan, China, 18–20 March 2019; pp. 251–255. [CrossRef]

20. Bai, Y.; Bhattacharyya, S.; Happonen, A.; Huttunen, H. Elastic Neural Networks: A Scalable Framework for Embedded Computer Vision. In Proceedings of the 26th European Signal Processing Conference (EUSIPCO), Rome, Italy, 3–7 September 2018; pp. 1472–1476. [CrossRef]

21. Yu, D.; Xu, Q.; Guo, H.; Zhao, C.; Lin, Y.; Li, D. An Efficient and Lightweight Convolutional Neural Network for Remote Sensing Image Scene Classification. *Sensors* **2020**, *20*, 1999. [CrossRef] [PubMed]

22. Chen, Y.; Chen, X.; Lin, J.; Pan, R.; Cao, T.; Cai, J.; Yu, D.; Cernava, T.; Zhang, X. DFCANet: A Novel Lightweight Convolutional Neural Network Model for Corn Disease Identification. *Agriculture* **2022**, *12*, 2047. [CrossRef]

23. Kawamiya, M.; Hajima, T.; Tachiiri, K.; Watanabe, S.; Yokohata, T. Two decades of Earth system modeling with an emphasis on Model for Interdisciplinary Research on Climate (MIROC). *Prog. Earth Planet. Sci.* **2020**, *7*, 64. [CrossRef]

24. Deng, L.; Zhu, G.Y.; Tang, Z.S.; Shangguan, Z.P. Global patterns of the effects of land-use changes on soil carbon stocks. *Glob. Ecol. Conserv.* **2016**, *5*, 127–138. [CrossRef]

25. Food And Agriculture Organization of the United Nations (FAO). *Global Forest Resources Assessment 2015—How Are the World's Forests Changing?* 2nd ed.; FAO: Rome, Italy, 2016; p. 54.

26. Bernal, B.; Murray, L.T.; Pearson, T.R.H. Global carbon dioxide removal rates from forest landscape restoration activities. *Carbon Balance Manag.* **2018**, *13*, 22. [CrossRef]

27. Kim, H.; Kim, Y.H.; Kim, R.; Park, H. Reviews of forest carbon dynamics models that use empirical yield curves: CBM-CFS3, CO2FIX, CASMOFOR, EFISCEN. *For. Sci. Technol.* **2015**, *11*, 212–222. [CrossRef]

28. Liu, F.; Tan, C.; Zhang, G.; Liu, J.X. Single-wood parameters and biomass airborne LiDAR estimation of Larix olgensis. *Trans. Chin. Soc. Agric.* **2013**, *44*, 219–224.

29. Lizuka, K.; Tateishi, R. Estimation of CO_2 Sequestration by the Forests in Japan by Discriminating Precise Tree Age Category using Remote Sensing Techniques. *Remote Sens.* **2015**, *7*, 15082–15113. [CrossRef]

30. Castro-Magnani, M.; Sanchez-Azofeifa, A.; Metternicht, G.; Laakso, K. Integration of remote-sensing based metrics and econometric models to assess the socio-economic contributions of carbon sequestration in unmanaged tropical dry forests. *Environ. Sustain. Indic.* **2021**, *9*, 100100. [CrossRef]

31. Costanza, R.; de Groot, R.; Braat, L.; Kubiszewski, I.; Fioramonti, L.; Sutton, P.; Farber, S.; Grasso, M. Twenty years of ecosystem services: How far have we come and how far do we still need to go? *Ecosyst. Serv.* **2017**, *28*, 1–16. [CrossRef]

32. Hao, H.; Li, W.; Zhao, X.; Chang, Q.; Zhao, P. Estimating the Aboveground Carbon Density of Coniferous Forests by Combining Airborne LiDAR and Allometry Models at Plot Level. *Front. Plant Sci.* **2019**, *10*, 917. [CrossRef] [PubMed]

33. Kanniah, K.; Muhamad, N.; Kang, C. Remote sensing assessment of carbon storage by urban forest, IOP Conference Series: Earth and Environmental Science. In Proceedings of the 8th International Symposium of the Digital Earth (ISDE8), Kuching, Malaysia, 26–29 August 2013; Volume 18.

34. Uniyal, S.; Purohit, S.; Chaurasia, K.; Rao, S.; Amminedu, E. Quantification of carbon sequestration by urban forest using Landsat 8 OLI and machine learning algorithms in Jodhpur, India. *Sci. Direct Urban For. Urban Green.* **2022**, *67*, 127445. [CrossRef]

35. Foody, G.; Mathur, A. A relative evaluation of multiclass image classification by support vector machines. *IEEE Trans. Geosci. Remote Sens.* **2004**, *42*, 1335–1343. [CrossRef]

36. Gwal, S.; Singh, S.; Gupta, S.; Anand, S. Understanding forest biomass and net primary productivity in Himalayan ecosystem using geospatial approach. *Model. Earth Syst. Environ.* **2020**, *6*, 10. [CrossRef]

37. Kimes, D.; Nelson, R.; Manry, M.; Fung, A. Review article: Attributes of neural networks for extracting continuous vegetation variables from optical and radar measurements. *Int. J. Remote Sens.* **1998**, *19*, 2639–2663. [CrossRef]
38. Sagi, O.; Rokach, L. Approximating XGBoost with an interpretable decision tree. *Inf. Sci.* **2021**, *572*, 522–542. [CrossRef]
39. Zhang, F.; Tian, X.; Zhang, H.; Jiang, M. Estimation of Aboveground Carbon Density of Forests Using Deep Learning and Multisource Remote Sensing. *Remote Sens.* **2022**, *14*, 3022. [CrossRef]
40. Liu, W.; Wang, Z.; Liu, X.; Zeng, N.; Liu, Y.; Alsaadi, F.E. A survey of deep neural network architectures and their applications. *Neurocomputing* **2017**, *234*, 11–26. [CrossRef]
41. LeCun, Y.; Bengio, Y.; Hinton, G. Deep learning. *Nature* **2015**, *521*, 436–444. [CrossRef]
42. Li, J.; Roy, D.A. global analysis of Sentinel-2A, Sentinel-2B and Landsat-8 data revisit intervals and implications for terrestrial monitoring. *Remote Sens.* **2017**, *9*, 902. [CrossRef]
43. Lang, N.; Jetz, W.; Schindler, K.; Wegner, A. High-resolution canopy height model of the Earth. *arXiv* **2022**, arXiv:2204.08322.
44. Abdollahnejad, A.; Panagiotidis, D.; Surový, P. Forest canopy density assessment using different approaches—Review. *J. For. Sci.* **2017**, *63*, 107–116.
45. Chen, J.; Yang, S.; Li, H.; Zhang, B.; Lv, J. Research on Geographical Environment Unit Division Based on The Method of Natural Breaks (Jenks), The International Archives of the Photogrammetry. *Remote Sens. Spat. Inf. Sci.* **2013**, *3*, 47–50, 2013 ISPRS/IGU/ICA Joint Workshop on Borderlands Modelling and Understanding for Global Sustainability 2013, Beijing, China.
46. Mamoshina, P.; Vieira, A.; Putin, E.; Zhavoronkov, A. Applications of Deep Learning in Biomedicine. *Mol. Pharm.* **2016**, *13*, 1445–1454. [CrossRef]
47. Khan, A.; Sohail, A.; Zahoora, U.; Qureshi, A.S. A Survey of the Recent Architectures of Deep Convolutional Neural Networks. *Artif. Intell. Rev.* **2020**, *53*, 5455–5516. [CrossRef]
48. Xu, B.; Wang, N.; Chen, T.; Li, M. Empirical Evaluation of Rectified Activations in Convolutional Network. *arXiv* **2015**, arXiv:1505.00853.
49. Josephine, V.L.; Nirmala, A.P.; Allur, V. Impact of Hidden Dense Layers in Convolutional Neural Network to enhance Performance of Classification Model, IOP Conference Series: Materials Science and Engineering. In Proceedings of the 4th International Conference on Emerging Technologies in Computer Engineering: Data Science and Blockchain Technology (ICETCE 2021), Jaipur, India, 3–4 February 2021; Volume 1131.
50. Lee, D.; Park, C.; Tomlin, D. Effects of land-use-change scenarios on terrestrial carbon stocks in South Korea. *Landsc. Ecol. Eng.* **2015**, *11*, 47–59. [CrossRef]
51. Scott, D. Sturges' rule. *WIREs Comput. Stat.* **2009**, *1*, 303–306. [CrossRef]
52. Belavkin, R.; Pardalos, P.; Principe, J. Value of Information in the Binary Case and Confusion Matrix. *Phys. Sci. Forum* **2022**, *5*, 5008. [CrossRef]
53. Bottou, L.; Bousquet, O. The Tradeoffs of Large Scale Learning. In *Optimization for Machine Learning*; Sra, S., Nowozin, S., Stephen, J.W., Eds.; MIT Press: Cambridge, UK, 2012; pp. 351–368. ISBN 978-0-262-01646-9.
54. Asif, A.; Waris, A.; Gilani, S.; Jamil, M.; Ashraf, H.; Shafique, M.; Niazi, I.K. Performance Evaluation of Convolutional Neural Network for Hand Gesture Recognition Using EMG. *Sensors* **2020**, *20*, 1642. [CrossRef] [PubMed]
55. Sandler, M.; Howard, A.; Zhu, M.; Zhmoginov, A.; Chen, L.C. MobileNetV2: Inverted Residuals and Linear Bottlenecks. In Proceedings of the IEEE/CVF Conference on Computer Vision and Pattern Recognition, Salt Lake City, UT, USA, 18–23 June 2018; pp. 4510–4520. [CrossRef]
56. Goodfellow, I.; Bengio, Y.; Courville, A. *Deep Learning (Adaptive Computation and Machine Learning Series)*; The MIT Press: Cambridge, MA, USA, 2016; p. 800.

Article

Cloud Removal from Satellite Images Using a Deep Learning Model with the Cloud-Matting Method

Deying Ma [1,2], Renzhe Wu [1,*], Dongsheng Xiao [2] and Baikai Sui [1]

[1] Faculty of Geosciences and Environmental Engineering, Southwest Jiaotong University, Chengdu 611756, China
[2] School of Civil Engineering and Geomatics, Southwest Petroleum University, Chengdu 610500, China
* Correspondence: mrwurenzhe@my.swjtu.edu.cn

Abstract: Clouds seriously limit the application of optical remote sensing images. In this paper, we remove clouds from satellite images using a novel method that considers ground surface reflections and cloud top reflections as a linear mixture of image elements from the perspective of image superposition. We use a two-step convolutional neural network to extract the transparency information of clouds and then recover the ground surface information of thin cloud regions. Given the poor balance of the generated samples, this paper also improves the binary Tversky loss function and applies it on multi-classification tasks. The model was validated on the simulated dataset and ALCD dataset, respectively. The results show that this model outperformed other control group experiments in cloud detection and removal. The model better locates the clouds in images with cloud matting, which is built based on cloud detection. In addition, the model successfully recovers the surface information of the thin cloud region when thick and thin clouds coexist, and it does not damage the original image's information.

Keywords: improved Tversky loss; two-step convolution model; cloud detection; cloud matting; cloud removal

1. Introduction

In recent years, optical satellite remote sensing has become the primary survey and monitoring means for disaster relief, geology, environment, and engineering construction, which has introduced great convenience to the development of human science. However, clouds are an unavoidable dynamic feature in optical remote sensing images. Global cloud coverage in mid-latitude regions is about 35% [1], and global surface cloud coverage ranges from 58% [2] to 66% [3]. High-quality images are not available almost all year round, especially in areas with high water vapor content changes [4]. Clouds reduce the reliability of remote sensing images and increase the difficulty of data processing [5].

Cloud detection is the first step in image de-clouding and restoration, which has received much attention from researchers. There are many methods concerning the detection of clouds and cloud shadows [6–13]. These methods can be divided into temporal and non-temporal solutions in terms of the number of images or non-deep learning solutions and deep learning [11–14] in terms of detection schemes. Foga et al. [15] summarized thirteen commonly used cloud detection methods and five cloud shadow detection methods. They found that the accuracy of each cloud removal method has its advantages and disadvantages within different scenarios. Deep learning-based methods mainly segment clouds in remote sensing images non-linearly with their solid-fitting ability. In the early years, scholars used fully connected neural networks [16,17] for cloud detection. In recent years, they primarily use convolutional neural networks [18,19] that are more suitable for image processing. Mahajan et al. [20] investigated the main cloud detection methods from 2004 to 2018, and they found that neural networks can largely compensate for the limitations of existing algorithms. The cloud detection scheme treats the detection process as a pixel

Citation: Ma, D.; Wu, R.; Xiao, D.; Sui, B. Cloud Removal from Satellite Images Using a Deep Learning Model with the Cloud-Matting Method. *Remote Sens.* **2023**, *15*, 904. https://doi.org/10.3390/rs15040904

Academic Editor: Gwanggil Jeon

Received: 18 November 2022
Revised: 30 January 2023
Accepted: 4 February 2023
Published: 6 February 2023

classification, and it obtains a high-quality mask file but ignores the ground surface's information under the cloud. In order to solve the problem in which the inaccurate mask file produces unsatisfactory results during cloud removal, Lin et al. [21] used the RTCR method and the augmented Lagrange multiplier. However, in most cases, the signals received by remote sensing imaging sensors are a superposition of the surface reflection signal and the cloud reflection signal [22,23]. Simple classification methods only locate and identify clouds in images but cannot estimate cloud amounts and recover surface information. Li et al. [24] suggested a hybrid cloud detection algorithm by utilizing various algorithms to their full potential. Clouds in images are usually mixed with surface information, and different transparency leads to different superposition patterns. Therefore, it is better to detect clouds by using a hybrid image element decomposition method.

Although there is a strong interconnection between cloud detection and cloud removal, studies have always been conducted separately [22,23]. Many scholars use deep learning techniques for single-image cloud removal. The widely used dark channel method has excellent mathematical derivations [25]. However, its applicability may be limited due to the imaging difference between satellite images and other images. Moreover, there are some errors in transmittance estimations and the dark channel prior, so the images are prone to dimming or are even distorted after cloud removal. The k-nearest neighbor (KNN) matting [26] method falls under nonlocal matting. It assumes that the transparency of a pixel can be described by weighting the transparency values of nonlocal pixels with a similar appearance, such as matching the color and texture. The goal is to allow the transparency value to propagate in nonlocal pixels. This includes laborious computations due to the comparison with the nonlocal images. KNN matting improves nonlocal matting by only considering the first K neighbors in the high-dimensional feature space. It reduces the amount of computation by only considering similarities between the color and the position in their feature space. The drawback of this method is that it requires a priori trimap as input and usually leaks pixels. Defining a general feature space with few parameters is difficult. Closed-form matting [27] assumes that the reflectivity of the foreground and background is the same in the local range of the sliding window and solves the transmittance formula using the color-line model and the ridge regression optimization algorithm. However, the clouds are easily overcorrected, and the solution requires an accurate trilateral as an a priori input, which significantly limits the application of closed extinction methods. The conditional generation countermeasure network (CGAN) [28] can reconstruct damaged information well when entities are still visible. However, the number of objects in remote-sensing images greatly increased. Therefore, the generative countermeasure network exhibits noticeable distortions in the thick cloud area. Isola et al. [29] proposed an image-to-image translation method (Pix2pix) based on CGAN to achieve image-to-image generation, providing a new method for image de-clouding restoration. Ramjyothi et al. [30] used GAN to repair the ground cover information under clouds in remote sensing images. Pan et al. [22] and Emami et al. [31] introduced spatial attention to GAN to control the redundancy of the model. Wen et al. [32] used a residual channel attention network for cloud removal. Via the solid-fitting ability of deep learning, the models can effectively learn the difference between the features of clouded and cloudless images, and then they can directly restore the absolute brightness value of the surface using image reconstruction. Cloud removal based on generative adversarial networks for reconstructing surface information is one of the trending research topics in recent years. However, the biggest drawback of deep learning is that "it cannot admit that it does not know when thin and thick clouds coexist". The output of the models meets high metrics. However, there is a big difference between created images and real images. The commonly used cloud removal solutions for satellite images, especially for Sentinel-2, include Sen2cor [33], Fmask [34], and S2cloudless [35]. Qiu et al. introduced Global Surface Water Occurrence (GSWO) data based on Fmask3.3 and the global digital elevation model (DEM), and then proposed the use of Fmask4 to improve the accuracy by 7.2% compared to the Sen2cor algorithm specified by the European Space Agency (ESA) in version 2.5.5. Housman et al.

proposed the cloud detection method S2cloudless by selecting ten bands of Sentinel-2 based on the XGBoost and LightGBM tree learning algorithms for model inference, which is the primary tool for Sentinel-hub cloud product production.

Other than the two-step "detection-removal" methods, Ji et al. [36] proposed a BC smooth low-rank plus group sparse model to detect and remove clouds at the same time.

Cloud removal methods for a single image rarely consider cloud transparency information. Surface information is often recovered approximately by interpolation or by mapping convolutional layers based on relevant samples. In most cases, the information about areas under thick clouds is completely lost. Cloud removal operations for such regions using interpolation or mapping methods introduce significant errors and sometimes result in useless images.

From the preceding description, this paper carries out experiments from a new cloud detection paradigm by simulating the mixing relationship between surface information and clouds, establishing a linear model based on the image superposition model. We propose an integrated method for cloud detection, transparency estimation, and cloud removal. This method can distinguish the foreground and background of mixed image elements based on single-band images in order to achieve cloud removal in a single satellite image. Considering that, the transparency of clouds varies in different bands of remote sensing images, the reflected signal of clouds in the RGB channel is the same, and the blue band is more sensitive to thin clouds. To promote the application of the model to multiple bands in order to enhance the applicability and generalization ability of the model, this paper uses the Sentinel-2 blue band for cloud-matting attempts. This idea mainly came from applying deep learning in image-matting methods, which assume that the image's foreground and background are mixed by transparency information. The classic linear superposition formula is shown in Equation (1) [37]. Image I can be decomposed into a linear combination of foregrounds, F, and backgrounds, B:

$$I = \alpha F + (1 - \alpha)B, \ \alpha \in [0, 1] \tag{1}$$

where α is the cloud's opacity ($\alpha = [0, 1]$). The convolutional neural network can acquire deeper feature information about the target [19], so the alpha matte of the foreground image estimated using the convolutional neural network can better remove the background information and extract the foreground information out of the image [38–40]. Given the poor balance of the generated samples, this paper also improves the binary Tversky loss function for multi-classification tasks. Through the improved Tversky loss function, we can automatically balance the weight of multi-class samples in the complex and changeable generated samples and focus the model's attention on a specific class or multi-class samples. In this manner, we can improve the prediction results of hard segmentation, effectively distinguish thin and thick cloud regions, and recover cloud and shadow regions based on cloud transparency information.

2. Methodology

2.1. Remote Sensing Imaging Process

The cloud removal model proposed in this paper is a deep-learning-based assumption. Therefore, we simplify atmospheric transport operations by not considering the scattering of particles in the air as well as aerosols. As shown in Figure 1, cloud occlusion between the satellite and the ground surface results in a superposition between the reflected energy from the ground surface and the reflected energy from the cloud's top in the final reflected energy obtained.

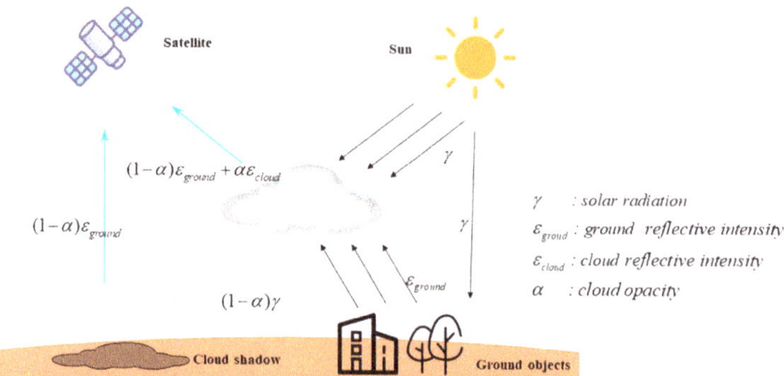

Figure 1. A schematic diagram of the remote sensing imaging process. The reflected energy received by sensors is a linear superposition of the reflected energy from the cloud's top and the reflected energy from the surface for given cloud transparency.

Different solar incidence angles form shadows that weaken or completely cover surface information. The pixel composition of the reflected signal intensity received by the remote sensing imaging system is as follows:

$$\begin{cases} \varepsilon = (1-\alpha)\varepsilon_{ground} + \alpha\varepsilon_{cloud} & ,(a) \\ \quad\quad \varepsilon = (1-\alpha)\varepsilon_{ground} & ,(b) \end{cases} \tag{2}$$

where (a) represents the received reflective intensity of area with clouds and (b) represents the received reflective intensity of area with cloud shadows. α is the cloud opacity ($\alpha = [0,1]$), ε_{ground} is the reflection intensity at the surface, and ε_{cloud} is the reflection intensity at the top of the cloud. Note that we assume constant solar irradiation with respect to clouds in the remote sensing image and a fixed cloud brightness (random sampling 4000–6000). Cloud brightness and transparency can be better balanced using sample generation based on Equations (1) and (2).

2.2. Model and Algorithm

The automatic generation of cloud-trimap is the first part of our proposed model, followed by the generation of a cloud-matting mask and cloud removal and, finally, the refinement and optimization of the cloud-matting mask and cloud removal's result. Our model contains two convolutional networks, as shown in Figure 2. The first convolutional network (green) is the T-Net (Trimap generate network) and the other network (blue) is the M-Net (Matting network). The T-Net is a semantic segmentation model that detects clouds on satellite images. This model generates a cloud-trimap, which can classify the image into opaque clouds, transparent clouds (uncertain regions), and non-clouds. The M-Net is an end-to-end pixel estimation model that uses a multi-output method to fuse the model's feature extraction results to estimate cloud transparency and residuals between the recovered and original images. Both T-Net and M-Net encoders adopt the Atrous spatial pyramid pooling (ASPP) structure at the bottom layer to represent more scaled information of image features with fewer parameters. The entire model can significantly improve prediction accuracies by model fusion and residual calculation. B, F, and U in Figure 2 represent the background, foreground, and conflicted regions, respectively. The output's results are not activated using the Softmax function because the loss in the T-Net training process contains cross-entropy errors. We can obtain B_S and U_S by using the same method and, obviously, $F_S + B_S + U_S = 1$ where 1 denotes the pixel value of each image element in the feature map.

$$F_S = \frac{\exp(F)}{\exp(F) + \exp(B) + \exp(U)} \tag{3}$$

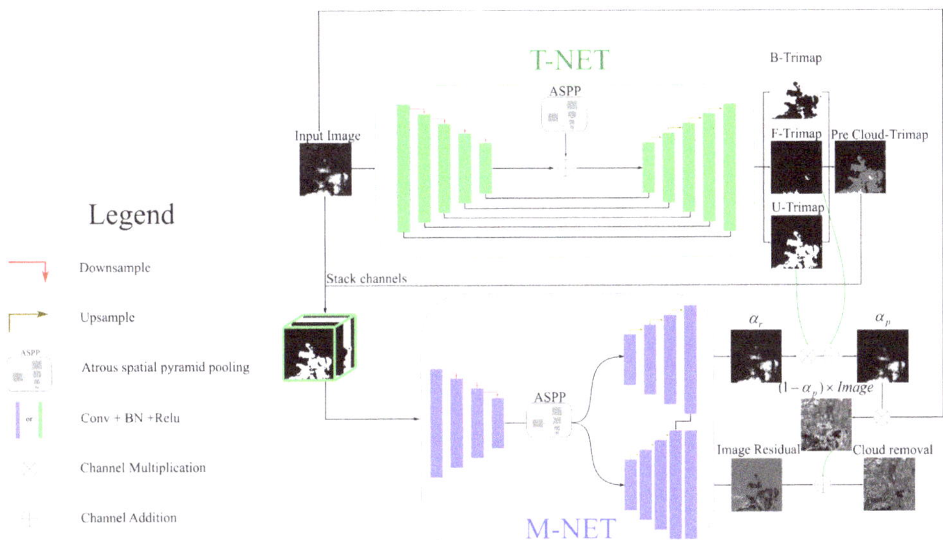

Figure 2. Two-step cloud-matting model.

The output result of picking M-Net contains two parts. The first part, α_r, mainly predicts the transparency information of clouds in the image. When the pixel is located in an uncertain region, this part is very likely to have transparent clouds. Otherwise, α_r can be filtered.

$$\alpha_p = F_S + U_S \alpha_r \tag{4}$$

α_p is the refinement of α_r; when U_S shifts towards 1, then F_S shifts towards 0. At the same time, α_p shifts towards α_r. When F_S shifts towards 1, then U_S shifts towards 0 and α_p shifts towards F_S ($\alpha_p \to 1$). This simple filtering method can improve the prediction result's confidence level. It also effectively shields background information interference and directs the model's attention to the region where the image's elements are mixed. The residuals between the predicted image and the cloud-free image are output by the M-Net, allowing the recovery of images using cloud transparency and the preservation of the original image's features.

Taking the derivative of both sides of the Equation (1), we have the following.

$$\frac{\partial B}{\partial \alpha} = \frac{I - F + (1 - \alpha)\partial I / \partial \alpha}{(1 - \alpha)^2} \tag{5}$$

From Equation (5), we can observe that when $(1 - \alpha)$ shifts towards zero, even a slight perturbation will result in a colossal mistake. The bottom map's recovery is prone to distortion. Permitting M-Net to directly recover the original surface's information—which is obscured by clouds—is unreliable without considering cloud transparency information. The prediction results need to be masked for regions with poor reliability (the mask's threshold in Figure 2 is $\alpha \geq 0.9$). It is worth noting that the M-Net model's input is the channel's superimposed feature map of both the T-Net's input and output, and Softmax is used to activate the T-Net's output and project the feature value to [0,1].

Complex problems can be simplified by employing the two-step method. Compared to the commonly used one-step method, the two-step method can fix a portion of the parameters while training another portion, resulting in a smoother model optimization process and faster training convergence. Interpretability improves over time, resulting in more accurate predictions.

For both T-Net and M-Net, we adopt the classic end-to-end (Sequence to Sequence, Seq2Seq) structure, extract features via the encoder, and then fuse features via the decoder. Due to a large number of parameters in the T-Net, a residual connection is adopted. Since fewer parameters exist in the M-Net, the encoder and decoder channel stacking are adopted to minimize information loss between feature maps. The encoding and decoding process will still reduce the image's sharpness when restored, so we will output the residuals to recover the image to the maximum extent.

2.3. Loss Function

The model is evaluated using a combined loss function.

Pre-training T-Net: Our T-Net primarily uses the cross-entropy error as the error function, according to Chen et al. [41]. The cross-entropy function is calculated using the following formula.

$$L_{cross} = -\sum_{i=1}^{n} x_i \log \hat{x}_i \tag{6}$$

In Equation (6), $[x_i, \hat{x}_i]$ denotes the pixel value of the predicted cloud-trimap and the real cloud-trimap, respectively. On the one hand, using only L_{cross} to generate cloud-trimap is unsatisfactory because T-Net's input categories are unbalanced, and the output's results are biased towards the background and conflicted regions, ignoring foreground information. On the other hand, due to the sample generation scheme used in the paper, it is challenging to add weights directly, making T-Net convergence difficult. The Tversky function was created to solve the problem of unbalanced medical image classification between focal and non-focal regions in machine learning by balancing the proportion of false positives and false negatives in training [42], resulting in a higher callback rate and a better balance between accuracy and sensitivity for the function. Therefore, we improve the binary classification Tversky function to solve the problem of unbalanced T-Net samples. In the binary classification problem, Tversky loss incorporates the benefits of focal loss [43] and Dice loss [44,45], and it is applied to the image segmentation study with the following formula transformation.

$$L_{Tversky} = 1 - \frac{\sum_{i=1}^{n} P_{x_i} P_{\hat{x}_i} + S}{\sum_{i=1}^{n} \left[(2P_{x_i} + 1)(1 - \alpha_1 P_{\hat{x}_i}) + \alpha_1 P_{x_i} \right] + S} \tag{7}$$

In the neural network's training process, P_{x_i} is the foreground probability of labeled pixels, $P_{\hat{x}_i}$ is the foreground probability of predicted pixels, and α_1 is the weight of control parameters to balance the samples. We usually set $0 < S < 10^{-6}$ to ensure that the equation holds, and $L_{Tversky}$ is the corresponding loss function.

The Tversky weight balance function is designed for binary classification problems and cannot directly apply to multiclassification problems. It is difficult to express the model error with a fixed weight because the first step of our model generates a cloud-trimap of images associated with multiple classifications. In addition, the trimap of each set of images is uncertain. In this paper, we improve the Tversky loss function by assuming that one or more classes of weights have a negative balance of significance. We build the automatic balance loss function with the classification corresponding to the unique thermal encoding channel.

$$TP_k = \sum_{k=1}^{m_0} \sum_{i=1}^{n} P_{x_i}^k \times P_{\hat{x}_i}^k$$

$$FP_k = \sum_{i=1}^{n} \left(\sum_{j=1}^{m_1} P_{x_i}^j \times \sum_{k=1}^{m_0} P_{\hat{x}_i}^k \right) \tag{8}$$

$$FN_k = \sum_{i=1}^{n} \left(\sum_{k=1}^{m_0} P_{\hat{x}_i}^k \times \sum_{j=1}^{m_1} P_{x_i}^j \right)$$

$$
\begin{cases}
L_{Tversky} = 1 - \sum\limits_{k=1}^{m_0} \dfrac{TP_k}{(TP_k + \beta FP_k + (1-\beta)FN_k + S)m_0} & \sum\limits_{k=1}^{n} TP_k > 0 \\[2ex]
L_{Tversky} = \sum\limits_{k=1}^{m_0} \dfrac{FP_k + FN_k}{(M \times N)m_0} & \sum\limits_{k=1}^{n} TP_k = 0
\end{cases}
\tag{9}
$$

In Equations (8) and (9), m_0 is the image channel of interest after one-hot encoding; m_1 is the remaining channels included in the one-hot encoding; n is the number of pixels of the image; $[P_x, P_{\hat{x}}]$ corresponds to the predicted classification and labeled classification, respectively; k and j represent the kth channel and jth channel of the image, respectively; β is the weight balance parameter; $[TP_k, FP_k, FN_k]$ denotes the true positive rate, false positive rate, and false negative rate of the attention channel, respectively; $[M, N]$ is the training sample size of the image; $L_{Tversky}$ is the loss value; S is the factor that prevents the denominator from proceeding to zero.

We effectively extend the dichotomous classification method to multi-categorization scenario applications by improving the Tversky loss function. The method does not require obtaining the sample's share in advance. It can automatically balance the sample's weights based on the samples' distribution characteristics, which can still effectively adjust the model's attention in the case of there being significant differences in the number of multi-categorization samples, ensuring that the model's optimization process does not favor the more dominant category.

It is worth noting that when TP_k is 0, the loss function $L_{Tversky}$ degrades significantly. To compensate for the loss in model training, we concentrate on optimizing the loss function's balance to be applied to any multiclassification model. To compensate for the model's training degeneracy, we focus on optimizing the loss function's balance to increase its applicability. $L_{Tversky}$ directs the gradient optimizer towards the channel of interest for iterative optimization using the improved Tversky loss function. As the number of training increases, the number of false positives increases. The $L_{Tversky}$ gradient direction shifts to reduce both false positives and false negatives for iterative optimizations. The T-Net loss function is calculated by adding L_{cross} and $L_{Tversky}$.

$$
L_{T-Net} = 0.5(L_{cross} + L_{Tversky})
\tag{10}
$$

Freeze T-Net and training M-Net: we fixed the weight of the T-Net network to train the M-Net after several rounds of iterative T-Net output results converged. The final output of the model contains two parts: cloud transparency estimation, α_p, and the recovered image, I_{pre}. We express the accuracy of α_p as $L_{||\alpha||^2}$ and the reconstruction error as L_c. The multi-scale expression of the distribution of $L_{ms-ssim}$ and the image element error in terms of I_{pre} are included. The α_p error function can be expressed as follows.

$$
L_{||\alpha||^2} = \sqrt{\sum_{i=1}^{n}(x_i - \hat{x}_i)^2}
\tag{11}
$$

$$
L_c = \sqrt{\sum_{i=1}^{n}(c_i - \hat{c}_i)^2}
\tag{12}
$$

In Equations (11) and (12), $[x_i, \hat{x}_i]$ represents the pixel values of predicted α_p and actual α, respectively, and $[c_i, \hat{c}_i]$ represents the pixel values of the synthetic cloud removal image and the actual cloud-free remote sensing image pixel values, respectively. The synthetic cloud removal image is generated from the actual background image and α_p, according to Equation (1).

We introduce MS-SSIM as the I_{pre} error function; MS-SSIM is an image quality evaluation method that merges image details at different resolutions. It can evaluate two images based on their brightness, contrast, and structural similarity. The MS-SSIM loss function is calculated as shown in Equation (13).

$$L_{ms-ssim} = 1 - \prod_{m=1}^{M} \left(\frac{2\mu_p\mu_g + c_1}{\mu_p^2 + \mu_g^2 + c_1} \right)^{\beta_m} \left(\frac{2\sigma_{pg} + c_2}{\sigma_p^2 + \sigma_g^2 + c_2} \right)^{\gamma_m} \tag{13}$$

M represents the scale factor, $[\mu_p, \mu_g]$ denotes the mean value between the predicted feature map and the actual image, $[\sigma_p, \sigma_g]$ denotes the standard deviation between the predicted image and the actual image, σ_{pg} denotes the covariance between the predicted image and the actual image, $[\beta_m, \gamma_m]$ denotes the importance between the two multiplicative terms, and $[c_1, c_2]$ is a constant term used to prevent the divisor from being 0. It is worth noting that the cloud occupation is usually tiny in remote sensing images. Therefore, the loss value obtained by calculating the global error function is small and cannot guide the optimization correctly.

We record the cloud-trimap output by the T-Net as the weight of the loss function, ω, to solve the problem that M-Net's error cannot be optimized to calculate the feature mat's local error (reduce the background error weight). Therefore, the $L_{ms-ssim}$ error function is $[\mu_p, \mu_g, \sigma_p, \sigma_g, \sigma_{pg}] = \omega[\mu_p, \mu_g, \sigma_p, \sigma_g, \sigma_{pg}]$. As shown in Equation (14), the M-Net loss function combines the error functions of $L_{||\alpha||^2}$, L_c, and $L_{ms-ssim}$. w denotes the significant coefficient, which ensures that the image is similar to the actual image and promotes the image's element value to be more similar, and it will decrease as the number of iterations increases.

$$L_{M-Net} = w(L_{||\alpha||^2} + L_c) + (1-w)L_{ms-ssim} \tag{14}$$

3. Experiments

3.1. Datasets

Existing cloud datasets are primarily designed for cloud detection, and they are accompanied by a mask for distinguishing clouds from other regions, which cannot be used for cloud-matting operations. As a result, we need simulated remote sensing cloud images as the model's data driver. Therefore, in this paper, we refer to traditional matting sample generation cases such as the alphamatting.com dataset [46], portrait image matting dataset [47], classical remote sensing image cloud detection dataset, L7Irish [48], and L8SPARCS [49]. Cloud matting samples were obtained from the blue band of the Sentinel-2 satellite, and the samples were pooled into one image as the actual label of cloud transparency; the cloud-free Sentinel-2 blue band image was used as the base image according to Equation (1) to build the training and validation dataset required for the study.

We used Equation (2) to assume that the absolute brightness of clouds is consistent within a specific range, and the cloud's transparency primarily determines the variation of cloud light and darkness; thus, at first, we used the Sentinel-2 images from the sea to produce a normalized alpha layer based on the color range. We created a cloud-trimap based on the transparency threshold and added an offset (50–150 pixels) to simulate cloud shadows on this foundation. Secondly, we selected multi-scene Sentinel-2 images with few clouds in different areas and at different times. Then, we used a slice index to rank and build a cloudy area mask one by one in order to obtain a cloud-free remote sensing image base map. Thirdly, the base image was randomly cropped to the specified size, and then training and validation samples were generated using the cloud transparency image, shadow image, and random cloud brightness. Finally, we generated a total of 50,000 samples, of which 20% were used as the validation set, 5% were used as the prediction set, and 75% were used as the training set. Figure 3 depicts the dataset construction scheme and the result.

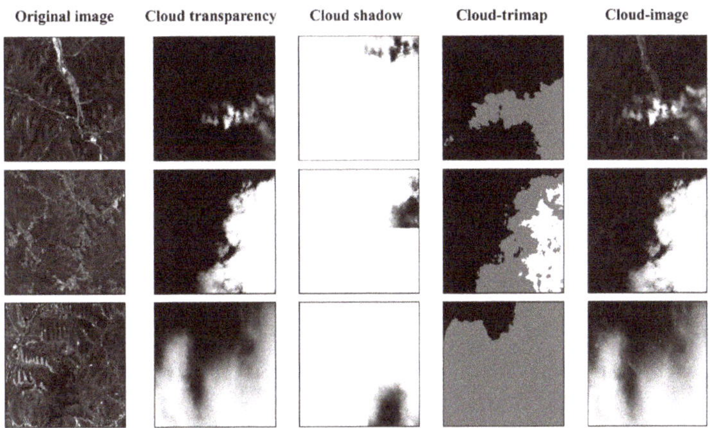

Figure 3. Cloud-matting dataset generation. Columns 1–3: The cloud-free image base map with Sentinel-2 Band2. The cloud transparency information notation is α. The cloud shadow notation is $f(1-\alpha)$, which is randomly generated according to cloud transparency, and f represents the offset calculation. The fourth column is the trimap image of the cloud, and we set $\alpha > 0.9$ relative to the cloud-trimap foreground. The fifth column is the composite image with clouds.

Cloud-trimap is obtained using a 3×3 sliding window image expansion calculation method based on cloud transparency. This aims to increase the tolerance of cloud detection by incorporating all information on image elements that may be clouded into cloud-trimap, and then they are further discriminated by the M-Net.

3.2. Evaluation Metrics

Our evaluation task involves cloud detection and cloud removal. The confusion matrix statistics of precision, recall, and accuracy were used for the former. The specific calculation is shown in Figure 4. For the latter, we used two methods to verify the results. 1. RMSE is used to verify the accuracy of the alpha calculation directly, and 2. SSIM is used to calculate the difference between the structural features of the predicted image and the real image. 3. The peak signal-to-noise ratio (PSNR) is also used. 4. The root mean square error (RMSE) is used to directly count the pixel difference between the predicted and actual images.

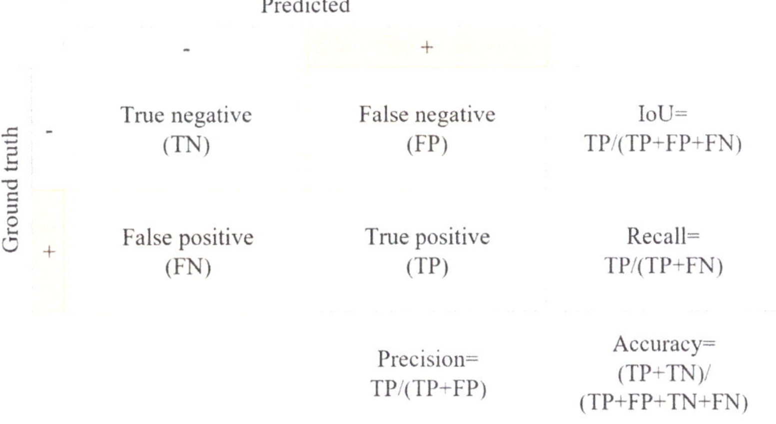

Figure 4. Confusion matrix applied to the evaluation index.

3.3. Implementation Details Evaluation Metrics

We compared and validated the cloud-matting method against the generated dataset and the Sentinel-2 classification dataset ALCD established by Baetens et al. [50]. For better verification, we also used three cloud detection methods and four de-clouding algorithms to demonstrate its effectiveness.

The cloud detection methods used for comparisons include S2cloudless, which is based on XGBoost and LightGBM tree gradient-boosting machine learning algorithms used by Sentinel hub, the ESA's (European Space Agency) atmospheric correction tool Sen2cor2.09 [51], and the USGS's (United States Geological Survey) remote sensing image classification tool FMASK4.0. The other four cloud removal methods are as follows: dark channel based on prior features, SpA-GAN based on attention mechanism, KNN-Image-matting based on non-local similarity, and closed-form-matting based on image local smoothness and color line model assumption.

We first validate the cloud detection performance on the ALCD dataset. As shown in Figure 5, S2cloudless ($p = 0.5$), Sen2cor, FMASK4.0, and T-Net can effectively locate clouds in the Sentinel-2 images. FMASK4.0 and T-Net detection results are more consistent with the actual distribution of thin clouds. Sen2cor and S2cloudless tend to miss some thin cloud features. Although S2cloudless can extract thin clouds better, as the threshold decreases, it will lead to many misclassifications.

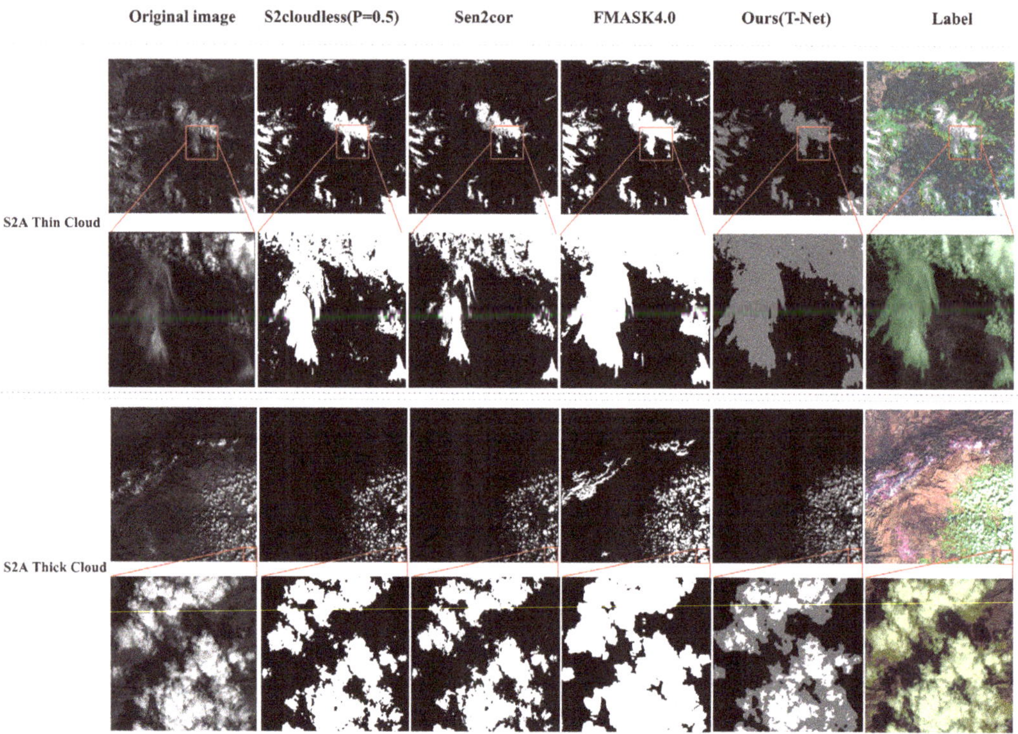

Figure 5. Cloud detection comparison experiments. The first to sixth columns are Band-2 image information, S2cloudless cloud detection results with a probability greater than 0.5, Sen2cor-2.09 cloud detection results, FMASK4.0 cloud detection results, T-Net trisection prediction results, and ALCD Tags.

S2cloudless (p = 0.5) and Sen2cor extracted more refined results and higher differentiation between clouds and snow in thick cloud regions, whereas FMASK4.0 has a high number of misclassifications due to the lower differentiation between clouds and snow. The T-Net's results are moderately granular compared to S2cloudless (p = 0.5) and Sen2cor. The T-net model can effectively distinguish thick clouds from thin clouds in trimap because the expansion factor is used in the training process. The T-Net distinguishes clouds and snow better because we build the corresponding bottom map information to enhance the difference between clouds and snow. The misclassification can be effectively reduced in areas where clouds and snow are separated. We calculated five groups of indicators based on thick and thin clouds to compare the cloud detection accuracy of the four models further, and the results are shown in Table 1.

Table 1. Comparison of the accuracy of four cloud detection methods.

Methods	Sen2cor	S2cloudless	Fmask4.0	Ours-TNet	Label
Precision (thin cloud)	0.6837	0.7712	0.7762	0.7981	
Recall (thin cloud)	0.9632	0.9400	0.9271	0.9445	
Accuracy (thin cloud)	0.9458	0.9560	0.9550	0.9596	
IoU (thin cloud)	0.6663	0.7351	0.7315	0.7551	
Cloud content (thin cloud)	9.4740	12.975	17.271	16.315	15.815
Precision (thick cloud)	0.6658	0.7172	0.7699	0.8019	
Recall (thick cloud)	0.8835	0.8757	0.8643	0.8665	
Accuracy (thick cloud)	0.9409	0.9448	0.9477	0.9596	
IoU (thick cloud)	0.6122	0.6509	0.6868	0.7254	
Cloud content (thick cloud)	4.4960	5.7000	13.400	10.810	12.190

Even if two repeat-pass images are used, obtaining the same surface reflection information is hard. We use a simulated dataset to assess the robustness and accuracy of the cloud removal algorithm. As described in the Introduction section, closed form-matting is similar to our scheme proposed in this paper. Therefore, we emphatically describe the difference from the other three models, such as dark channel, SpA-GAN, and KNN image matting. The cloud removal results are shown in Figure 6. Since most de-clouding models are built using RGB color images, this paper creates a set of corresponding RGB cloud images. The image data types are converted using an alpha superposition operation, resulting in differences in image color parameters in human vision. However, the actual image element's reflection signals are unaffected.

Dark-channel, closed-form matting, and cloud matting can filter out thin clouds well for image recovery when there are thick or cirrus clouds in the image. In Figure 6, we can see that dark-channel, SpA-GAN, closed-form matting, and cloud matting show a better cloud removal effect when only thin clouds appear in the image. We rank the overall cloud removal effect as our cloud-matting method > SpA-GAN > Dark channel > Closed form mating > KNN image mating. However, it is worth noting the following.

1. When using the dark-channel for remote sensing image de-clouding operations, the estimated projection size is often inversely proportional to the overall brightness of the remote sensing image, resulting in a weakening of feature brightness and a reduction in the overall brightness of the image.

2. Although the SpA-GAN used in this paper performed migration learning on the generated dataset, the results are unsatisfactory. The model's inference results are close to fitting adjacent image elements. This method is better for de-clouding restoration in thin cloud regions, but in thick cloud regions the model tends to generate image elements with similar characteristics to the entire image, resulting in significant distortions.

3. Both dark-channel and SpA-GAN process the entire image, so regardless of the presence of clouds, both models modify the pixel values of the original image, resulting in pixel distortions in the de-clouded image and making them unsuitable for quantitative, qualitative remote sensing and other studies.

4. KNN image matting and closed-form matting perform de-clouding by estimating the transparency of clouds. However, these methods require a substantial amount of prior manual inputs, such as accurate trimap and maximum reflected brightness of cloud tops. The accuracy of the two models drops significantly or even fails when only thin clouds are in the image. KNN image matting and closed-form matting are limited for cloud removal over remote sensing images.

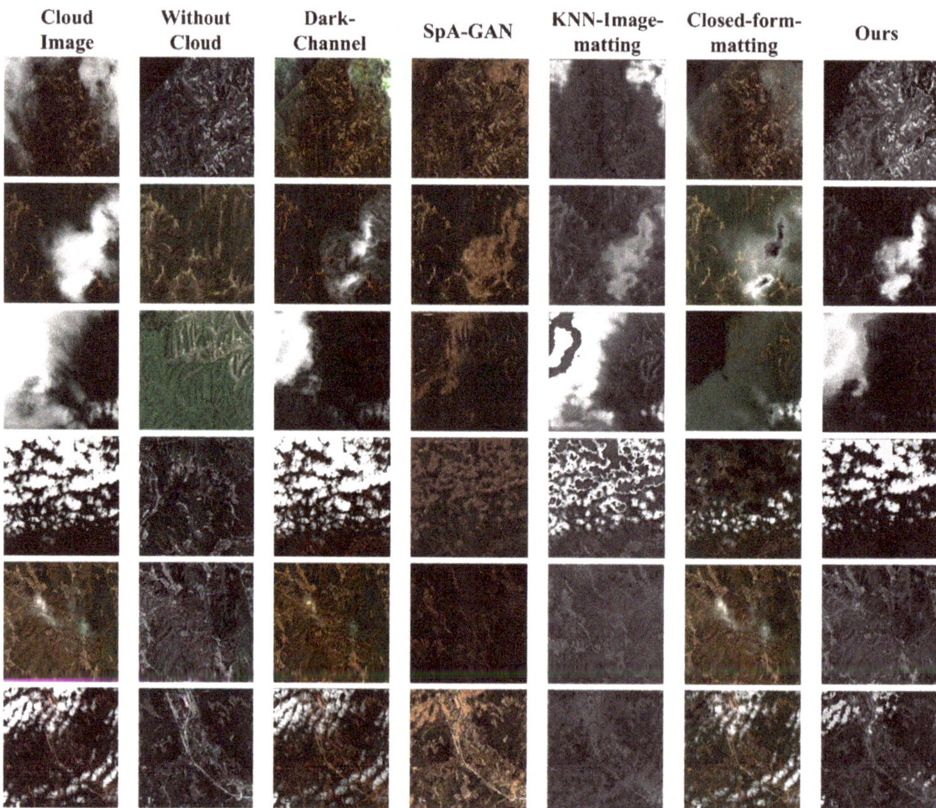

Figure 6. Comparison of cloud removal results of five models. The first to seventh columns are remote sensing images with clouds, remote sensing images without clouds, dark-channel, SpA-GAN, KNN image matting, closed-form matting, and our proposed cloud-matting method, respectively.

We observed that the cloud transparency estimation image, i.e., opacity image α, can be obtained using the de-clouded image as the background (Figure 7). Because image α only has brightness variations, and it is no longer disturbed by the image's background, it can more intuitively reflect the effect of model de-clouding processes. The better the effect of model de-clouding, the closer the brightness variation of the opacity image relative to the label it represents. The degree of damage to the original image during model de-clouding processes is represented by the purity of the opacity image.

According to Equation (2), $\alpha = (\varepsilon - \varepsilon_{ground})/(\varepsilon_{cloud} - \varepsilon_{ground})$. Theoretically, the calculated α is greater than 0. The brightness of estimated α from the dark Channel is closest to $\hat{\alpha}$, but the background of the α layer is disorderly. Most features are on this layer, resulting in the serious distortion of thick cloud areas. The α obtained by SpA-GAN is less stable, with significant variations in lightness, darkness, and purity, leading to image distortion as well.

KNN image matting is more accurate for α locations. The background of the obtained α is purer than SpA-GAN. The estimation of the transparency probability value has a large offset, making it difficult to recover the image accurately. Closed-form matting estimates the brightness of α, and the purity of α is quite close to the label. However, there will be an underestimation, leading to poor cloud removal effects.

Although there are various methods for single remote sensing image de-clouding, cloud matting can better maintain the original image element information and is less likely to cause image distortion, as shown in the above comparison. We used 640 sets of sliced image pairs to evaluate the restored images in terms of RMSE, SSIM, and PSNR to compare the effect of the five models further. Table 2 shows the results of the evaluation. The SpA-GAN and cloud-matting methods produce the most accurate de-clouding results and cloud transparency. The SpA-GAN metric results are very similar to cloud matting, especially in the mean and minimum values of image de-clouding recovery, which are significantly higher than other methods. However, this is the metric trap of SpA-GAN, which employs a Nash equilibrium-trained model. SpA-GAN uses the model obtained by Nash equilibrium training. Rather than removing the cloud by using the model, the better explanation is that SpA-GAN creates a pixel to minimize the loss function via the generator. Therefore, the model's accuracy is often high, but the results can be better. As shown in Figures 6 and 7, the result of SpA-GAN in the fourth row of the image is PSNR(Image) = 20.669, while the result of our cloud-matting is PSNR(Image) = 2.820. The image element information in the thick cloud region is completely covered. The thicker the cloud, the lower the reliability of the cloud's removal result. It is impossible to remove thick clouds by using only one image. The results of SpA-GAN have significant errors, but the overall brightness and structure of the image are very similar to the original one, which lead to large errors. In contrast, cloud-matting results have a higher confidence level. It performs thin cloud removal well in the presence of both thick and thin clouds without damaging the original image.

Table 2. Comparison of five cloud removal methods. The optimal value, average value, and worst value of the cloud removal result are represented by green, blue, and red, respectively.

Metrics	Dark-Channel	SpA-GAN	KNN Image Matting	Closed-Form Matting	Ours
	0.0233	0.0121	0.0073	0.0065	0.0025
RMSE (Image)	0.1234	0.1098	0.8620	0.1429	0.2121
	0.3396	0.3788	7.1633	1.1419	3.2967
	0.8198	0.9959	0.9922	0.9942	0.9992
SSIM (Image)	0.4115	0.8321	0.6153	0.7418	0.8120
	0.1542	0.2570	0.0276	0.1404	0.1040
	32.6296	44.1723	42.6939	43.6871	51.8999
PSNR (Image)	19.3394	26.7704	11.1344	20.0632	23.8369
	9.3797	8.4318	-17.1023	-1.1526	-10.3616
	0.0059	0.0071	0.0129	0.0159	0.0067
RMSE (Alpha)	0.0803	0.0314	0.2382	0.1141	0.0263
	0.2993	0.0793	0.8259	0.6057	0.0791
	0.9928	0.9941	0.9893	0.9953	0.9967
SSIM (Alpha)	0.8171	0.8616	0.7537	0.8588	0.9810
	0.4960	0.6412	0.0000	0.4268	0.9350
	44.5602	43.1151	37.7872	35.9338	43.3984
PSNR (Alpha)	23.8009	30.5172	17.0365	21.1993	32.7192
	10.4768	23.1798	1.6613	4.3540	22.0270

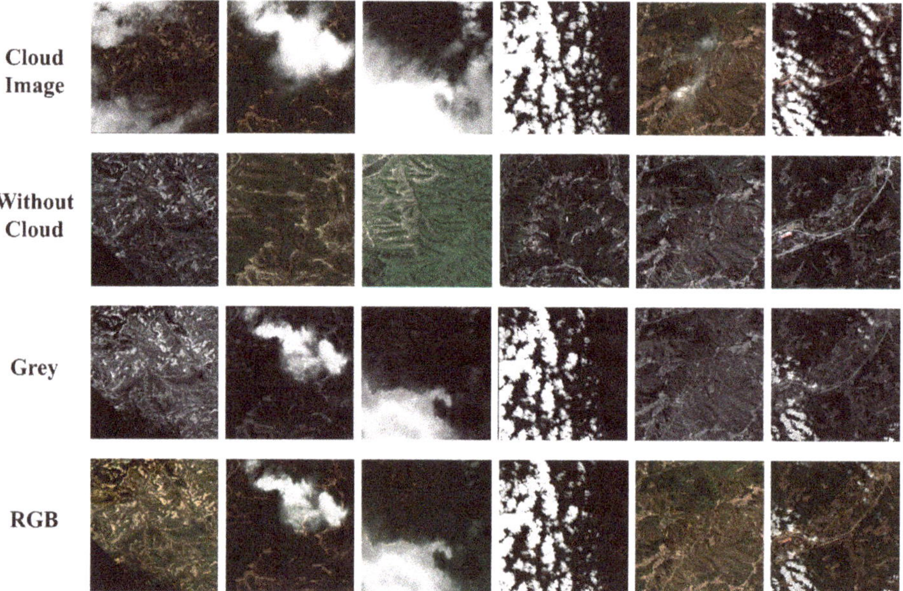

Figure 7. Comparison of estimated cloud transparency images.

4. Discussion

Tversky is an efficient and excellent balance loss function for two-class samples. This paper expanded it to multi-class applications. It is suitable for the dataset that we generated and can be effectively applied to other types of negative balance sample research without manual work. Setting the weights can automatically balance the weights of the samples.

Dark channel transmittance estimation has the drawback of not adhering to the imaging mechanism of remote sensing images, causing the image to be enhanced or weakened depending on the brightness of the pixels. The approximate pixels will still be output, resulting in a sharp drop in the model's reliability.

Generally speaking, GAN added a discriminator based on CNN, which makes the generated image close to the domain of the target image through the Nash equilibrium principle. Therefore, GAN has an additional constraint than CNN. The discriminator calculates the distance between the generated and target image domains, which leads to the fact that the results obtained by GAN on this basis are more in line with human vision. The disadvantage of the methods of CGANs is that the generated data points conform to image distribution characteristics. Furthermore, commonly used single-image cloud removal methods will damage the original remote sensing image's reflection information, resulting in inconsistent brightness changes in the input and output images.

It is not reasonable to directly apply the SpA-GAN to pin the clouded images to cloudless images. SpA-GAN is an image translation network with an attention mechanism that can be well applied to image restoration tasks. However, if SpA-GAN removes clouds from a single remote sensing image, the obtained cloud removal results must be over-corrected. Since the model learns the mapping from cloudy to cloudless images, it must generate pixels similar to the target domain (cloudless image) in the cloud coverage area. However, clouds in remote sensing images usually cover multiple entities rather than a part of them, which is troublesome for image restoration tasks. Therefore, SpA-GAN and other generated countermeasures networks will output a pixel deception discriminator subject to the target domain. It is challenging to locate these overcorrected pixels, resulting in errors in the cloudless image. Since the characteristics of the cloud are similar to a noise, the discriminator considers the output image true as long as it detects that the generated image conforms to the reasonable noise distribution. However, the generator can easily acquire noise signal and deceive the discriminator. The loss function value provided by the discriminator then is almost meaningless. SpA-GAN degenerates into a CNN network that only relies on the generator and image similarity loss.

In contrast, our cloud-matting model is of great significance for cloud removal. As long as the cloud can be accurately segmented from remote sensing images, cloud removal can be completed without damaging the image surface information. There are many mature methods in the field of cloud detection. The model structure adopted by our method is simple. The model includes only an essential multi-scale image segmentation analysis. Therefore, the accuracy has much room to improve in the future. In the following study, we will perform the following: 1. The model will be improved and trained with the more reliable and advanced backbone. 2. The difficulty of model training and migration will be reduced by combining two-step and one-step methods. 3. The image base maps of heterogeneous areas will be collected to improve the cloud removal results of the model.

5. Conclusions

In this paper, based on the principle of image superposition, we studied the cloud removal of remote sensing images from a new perspective and discussed the principles, advantages, and disadvantages of various single-image cloud removal methods. A set of simulated cloud map generation schemes have been established and is open source. The following conclusions can be drawn from the research findings.

1. The traditional cloud removal models for a single image can only restore the surface information covered by thin clouds. The model's reliability is significantly reduced when thick and thin clouds coexist.

2. Our cloud-matting scheme only takes the reflection intensity at the top of the cloud into consideration, which is more in line with the imaging mechanism of remote sensing images.

3. Our cloud-matting scheme uses cloud detection to restore surface information based on cloud opacity. It is easily mathematically interpretable, and it does not affect the original cloud-free areas.

4. The experiment results show that our cloud-matting method outperforms other methods. It is worth noting that the GAN image element's reconstruction ability is powerful in the cloud removal index, but it can easily appear "fabricated" when thick and thin clouds coexist.

5. Using deep learning combined with cloud matting to remove clouds from a single remote sensing image can effectively establish a cloud mask and show good anti-interference performances when thick clouds and thin clouds coexist without damaging the surface information of the original image. Cloud removal with a combination model is a valuable research direction, and we will continue to work in this direction.

Author Contributions: Conceptualization, D.M. and R.W.; methodology, D.M. and R.W.; software, D.M. and R.W.; validation, D.M., R.W. and B.S.; formal analysis, D.M. and R.W.; investigation, D.M. and R.W.; resources, R.W.; data curation, D.M. and R.W.; writing—original draft preparation, D.M. and R.W.; writing—review and editing, D.M. and R.W.; visualization, D.M. and R.W.; supervision, R.W.; project administration, D.M. and D.X.; funding acquisition, D.M. and D.X. All authors have read and agreed to the published version of the manuscript.

Funding: This research was jointly funded by the National Natural Science Foundation of China 491 (Grant No. 51774250) and the Sichuan Science and Technology Program (Grant No. 2022NSFSC1113, 23QYCX0053).

Institutional Review Board Statement: Not applicable.

Informed Consent Statement: Not applicable.

Data Availability Statement: Not applicable.

Conflicts of Interest: The authors declare no conflict of interest.

References

1. Ju, J.; Roy, D.P. The Availability of Cloud-Free Landsat ETM+ Data over the Conterminous United States and Globally. *Remote Sens. Environ.* **2008**, *112*, 1196–1211. [CrossRef]
2. Rossow, W.B.; Schiffer, R.A. Advances in Understanding Clouds from ISCCP. *Bull. Am. Meteorol. Soc.* **1999**, *80*, 2261–2287. [CrossRef]
3. Zhang, Y.; Rossow, W.B.; Lacis, A.A.; Oinas, V.; Mishchenko, M.I. Calculation of Radiative Fluxes from the Surface to Top of Atmosphere Based on ISCCP and Other Global Data Sets: Refinements of the Radiative Transfer Model and the Input Data. *J. Geophys. Res. Atmos.* **2004**, *D19*, 109. [CrossRef]
4. Wu, R.; Liu, G.; Zhang, R.; Wang, X.; Li, Y.; Zhang, B.; Cai, J.; Xiang, W. A Deep Learning Method for Mapping Glacial Lakes from the Combined Use of Synthetic-Aperture Radar and Optical Satellite Images. *Remote Sens.* **2020**, *12*, 4020. [CrossRef]
5. Stubenrauch, C.J.C.J.; Rossow, W.B.W.B.; Kinne, S.; Ackerman, S.; Cesana, G.; Chepfer, H.; di Girolamo, L.; Getzewich, B.; Guignard, A.; Heidinger, A.; et al. Assessment of Global Cloud Datasets from Satellites: Project and Database Initiated by the GEWEX Radiation Panel. *Bull. Am. Meteorol. Soc.* **2013**, *94*, 1031–1049. [CrossRef]
6. Lin, B.; Rossow, W.B. Precipitation Water Path and Rainfall Rate Estimates over Oceans Using Special Sensor Microwave Imager and International Satellite Cloud Climatology Project Data. *J. Geophys. Res. Atmos.* **1997**, *102*, 9359–9374. [CrossRef]
7. Lubin, D.; Harper, D.A. Cloud Radiative Properties over the South Pole from AVHRR Infrared Data. *J. Clim.* **1996**, *9*, 3405–3418. [CrossRef]
8. Hahn, C.J.; Warren, S.G.; London, J. The Effect of Moonlight on Observation of Cloud Cover at Night, and Application to Cloud Climatology. *J. Clim.* **1995**, *8*, 1429–1446. [CrossRef]
9. Hagolle, O.; Huc, M.; Pascual, D.V.; Dedieu, G. A Multi-Temporal Method for Cloud Detection, Applied to FORMOSAT-2, VENµS, LANDSAT and SENTINEL-2 Images. *Remote Sens. Environ.* **2010**, *114*, 1747–1755. [CrossRef]
10. Guosheng, L.; Curry, J.A.; Sheu, R.S. Classification of Clouds over the Western Equatorial Pacific Ocean Using Combined Infrared and Microwave Satellite Data. *J. Geophys. Res.* **1995**, *100*, 13811–13826. [CrossRef]
11. Ackerman, S.A.; Holz, R.E.; Frey, R.; Eloranta, E.W.; Maddux, B.C.; McGill, M. Cloud Detection with MODIS. Part II: Validation. *J. Atmos. Ocean. Technol.* **2008**, *25*, 1073–1086. [CrossRef]
12. Zhu, Z.; Woodcock, C.E. Object-Based Cloud and Cloud Shadow Detection in Landsat Imagery. *Remote Sens. Environ.* **2012**, *118*, 83–94. [CrossRef]
13. Scaramuzza, P.L.; Bouchard, M.A.; Dwyer, J.L. Development of the Landsat Data Continuity Mission Cloud-Cover Assessment Algorithms. *IEEE Trans. Geosci. Remote Sens.* **2011**, *50*, 1140–1154. [CrossRef]
14. Zou, Z.; Li, W.; Shi, T.; Shi, Z.; Ye, J. Generative Adversarial Training for Weakly Supervised Cloud Matting. *Proceedings of the IEEE Int. Conf. Comput. Vis.* **2019**, *2019*, 201–210. [CrossRef]
15. Foga, S.; Scaramuzza, P.L.; Guo, S.; Zhu, Z.; Dilley, R.D.; Beckmann, T.; Schmidt, G.L.; Dwyer, J.L.; Joseph Hughes, M.; Laue, B. Cloud Detection Algorithm Comparison and Validation for Operational Landsat Data Products. *Remote Sens. Environ.* **2017**, *194*, 379–390. [CrossRef]
16. Shi, M.; Xie, F.; Zi, Y.; Yin, J. Cloud Detection of Remote Sensing Images by Deep Learning. In Proceedings of the 2016 IEEE International Geoscience and Remote Sensing Symposium (IGARSS), Beijing, China, 10–15 July 2016; pp. 701–704.
17. le Goff, M.; Tourneret, J.-Y.; Wendt, H.; Ortner, M.; Spigai, M. Deep Learning for Cloud Detection. In Proceedings of the 8th International Conference of Pattern Recognition Systems (ICPRS 2017), Madrid, Spain, 11–13 July 2017; IET: Stevenage, UK, 2017; pp. 1–6.
18. He, Q.; Sun, X.; Yan, Z.; Fu, K. DABNet: Deformable Contextual and Boundary-Weighted Network for Cloud Detection in Remote Sensing Images. *IEEE Trans. Geosci. Remote Sens.* **2021**, 1–16. [CrossRef]
19. Jeppesen, J.H.; Jacobsen, R.H.; Inceoglu, F.; Toftegaard, T.S. A Cloud Detection Algorithm for Satellite Imagery Based on Deep Learning. *Remote Sens. Environ.* **2019**, *229*, 247–259. [CrossRef]
20. Mahajan, S.; Fataniya, B. Cloud Detection Methodologies: Variants and Development—A Review. *Complex Intell. Syst.* **2019**, *6*, 251–261. [CrossRef]
21. Lin, J.; Huang, T.Z.; Zhao, X.L.; Chen, Y.; Zhang, Q.; Yuan, Q. Robust thick cloud removal for multitemporal remote sensing images using coupled tensor factorization. *IEEE Trans. Geosci. Remote Sens.* **2022**, *60*, 1–16. [CrossRef]
22. Pan, X.; Xie, F.; Jiang, Z.; Yin, J. Haze Removal for a Single Remote Sensing Image Based on Deformed Haze Imaging Model. *IEEE Signal Process. Lett.* **2015**, *22*, 1806–1810. [CrossRef]
23. Mitchell, O.R.; Delp, E.J.; Chen, P.L. Filtering to Remove Cloud Cover in Satellite Imagery. *IEEE Trans. Geosci. Electron.* **1977**, *15*, 137–141. [CrossRef]
24. Li, F.F.; Zuo, H.M.; Jia, Y.H.; Wang, Q.; Qiu, J. Hybrid Cloud Detection Algorithm Based on Intelligent Scene Recognition. *J. Atmos. Ocean. Technol.* **2022**, *39*, 837–847. [CrossRef]
25. He, K.M.; SUN, J.T.X.O. Single Image Haze Removal Using Dark Channel Prior. *IEEE Trans. Pattern Anal. Mach. Intell.* **2011**, *33*, 2341.
26. Chen, Q.; Li, D.; Tang, C.-K. KNN Matting. *IEEE Trans. Pattern Anal. Mach. Intell.* **2013**, *35*, 2175–2188. [CrossRef]
27. Levin, A.; Lischinski, D.; Weiss, Y. A closed-form solution to natural image matting. *IEEE Trans. Pattern Anal. Mach. Intell.* **2007**, *30*, 228–242. [CrossRef]
28. Mirza, M.; Osindero, S. Conditional Generative Adversarial Nets. *arXiv* **2014**, arXiv preprint. arXiv:1411.1784.
29. Isola, P.; Zhu, J.-Y.; Zhou, T.; Efros, A.A. Image-to-Image Translation with Conditional Adversarial Networks. In Proceedings of the IEEE Conference on Computer Vision and Pattern Recognition, Hawai, HI, USA, 20–26 June 2017; pp. 5967–5976.

30. Ramjyothi, A.; Goswami, S. Cloud and Fog Removal from Satellite Images Using Generative Adversarial Networks (Gans). 2021. Available online: https://hal.science/hal-03462652 (accessed on 17 November 2022).
31. Emami, H.; Aliabadi, M.M.; Dong, M.; Chinnam, R.B. Spa-gan: Spatial attention gan for image-to-image translation. *IEEE Trans. Multimed.* **2020**, *23*, 391–401. [CrossRef]
32. Wen, X.; Pan, Z.; Hu, Y.; Liu, J. An effective network integrating residual learning and channel attention mechanism for thin cloud removal. *IEEE Geosci. Remote Sens. Lett.* **2022**, *19*, 1–5. [CrossRef]
33. Qiu, S.; Zhu, Z.; He, B. Fmask 4.0: Improved Cloud and Cloud Shadow Detection in Landsats 4–8 and Sentinel-2 Imagery. *Remote Sens. Environ.* **2019**, *231*, 111205. [CrossRef]
34. Frantz, D.; Haß, E.; Uhl, A.; Stoffels, J.; Hill, J. Improvement of the Fmask Algorithm for Sentinel-2 Images: Separating Clouds from Bright Surfaces Based on Parallax Effects. *Remote Sens. Environ.* **2018**, *215*, 471–481. [CrossRef]
35. Housman, I.W.; Chastain, R.A.; Finco, M.V. An Evaluation of Forest Health Insect and Disease Survey Data and Satellite-Based Remote Sensing Forest Change Detection Methods: Case Studies in the United States. *Remote Sens.* **2018**, *10*, 1184. [CrossRef]
36. Ji, T.Y.; Chu, D.; Zhao, X.L.; Hong, D. A unified framework of cloud detection and removal based on low-rank and group sparse regularizations for multitemporal multispectral images. *IEEE Trans. Geosci. Remote Sens.* **2022**, *60*, 1–15. [CrossRef]
37. Fattal, R. Single image dehazing. *ACM Trans. Graph. (TOG)* **2008**, *27*, 1–9. [CrossRef]
38. Sun, Y.; Tang, C.-K.; Tai, Y.-W. Semantic Image Matting. In Proceedings of the IEEE/CVF Conference on Computer Vision and Pattern Recognition, Nashville, TN, USA, 20–25 June 2021; pp. 11120–11129.
39. Chen, Q.; Ge, T.; Xu, Y.; Zhang, Z.; Yang, X.; Gai, K. Semantic Human Matting. In Proceedings of the 2018 ACM Multimedia Conference, Seoul, Republic of Korea, 22–26 October 2018; pp. 618–626. [CrossRef]
40. Xu, N.; Price, B.; Cohen, S.; Huang, T. Deep Image Matting. In Proceedings of the IEEE Conference on Computer Vision and Pattern Recognition, Honolulu, HI, USA, 21–26 July 2017; pp. 2970–2979.
41. Chen, H.; Han, X.; Fan, X.; Lou, X.; Liu, H.; Huang, J.; Yao, J. Rectified cross-entropy and upper transition loss for weakly supervised whole slide image classifier. In *International Conference on Medical Image Computing and Computer-Assisted Intervention*; Springer: Cham, Switzerland, 2019; pp. 351–359.
42. Salehi, S.S.M.; Erdogmus, D.; Gholipour, A. Tversky Loss Function for Image Segmentation Using 3D Fully Convolutional Deep Networks. In Proceedings of the Lecture Notes in Computer Science (Including Subseries Lecture Notes in Artificial Intelligence and Lecture Notes in Bioinformatics), Quebec City, QC, Canada, 10 September 2017; Springer International Publishing: Berlin/Heidelberg, Germany, 2017; Volume 10541, pp. 379–387.
43. Lin, T.-Y.; Goyal, P.; Girshick, R.; He, K.; Dollar, P. Focal Loss for Dense Object Detection. In Proceedings of the IEEE International Conference on Computer Vision, Venice, Italy, 22–29 October 2017; Volume 2017, pp. 2999–3007.
44. Li, X.; Sun, X.; Meng, Y.; Liang, J.; Wu, F.; Li, J. Dice Loss for Data-Imbalanced NLP Tasks. *arXiv* **2020**, arXiv:1911.02855.
45. Wang, L.; Wang, C.; Sun, Z.; Chen, S. An Improved Dice Loss for Pneumothorax Segmentation by Mining the Information of Negative Areas. *IEEE Access* **2020**, *8*, 167939–167949. [CrossRef]
46. Rhemann, C.; Rother, C.; Wang, J.; Gelautz, M.; Kohli, P.; Rott, P. A Perceptually Motivated Online Benchmark for Image Matting. In Proceedings of the 2009 IEEE Conference on Computer Vision and Pattern Recognition, CVPR, Miami, FL, USA, 20–25 June 2009; pp. 1826–1833.
47. Shen, X.; Tao, X.; Gao, H.; Zhou, C.; Jia, J. Deep Automatic Portrait Matting. In Proceedings of the Computer Vision–ECCV 2016: 14th European Conference, Amsterdam, The Netherlands, 11–14 October 2016; Springer International Publishing: Berlin/Heidelberg, Germany, 2016; Volume 9905, ISBN 9783319464473.
48. Irish, R.R.; Barker, J.L.; Goward, S.N.; Arvidson, T. Characterization of the Landsat-7 ETM+ Automated Cloud-Cover Assessment (ACCA) Algorithm. *Photogramm. Eng. Remote Sens.* **2006**, *72*, 1179–1188. [CrossRef]
49. Hughes, M.J.; Hayes, D.J. Automated Detection of Cloud and Cloud Shadow in Single-Date Landsat Imagery Using Neural Networks and Spatial Post-Processing. *Remote Sens.* **2014**, *6*, 4907–4926. [CrossRef]
50. Baetens, L.; Desjardins, C.; Hagolle, O. Validation of Copernicus Sentinel-2 Cloud Masks Obtained from MAJA, Sen2Cor, and FMask Processors Using Reference Cloud Masks Generated with a Supervised Active Learning Procedure. *Remote Sens.* **2019**, *11*, 433. [CrossRef]
51. Louis, J.; Debaecker, V.; Pflug, B.; Main-Knorn, M.; Bieniarz, J.; Mueller-Wilm, U.; Cadau, E.; Gascon, F. Sentinel-2 SEN2COR: L2A Processor for Users. In Proceedings of the Living Planet Symposium 2016, Spacebooks Online, Prague, Czech Republic, 9–13 May 2016; Volume SP-740, pp. 1–8, ISBN 978-92-9221-305-3.

remote sensing

MDPI

Article

Azimuth-Aware Discriminative Representation Learning for Semi-Supervised Few-Shot SAR Vehicle Recognition

Linbin Zhang [1], Xiangguang Leng [1], Sijia Feng [1], Xiaojie Ma [1], Kefeng Ji [1], Gangyao Kuang [1] and Li Liu [2,*]

[1] The State Key Laboratory of Complex Electromagnetic Environment Effects on Electronics and Information System, National University of Defense Technology (NUDT), Changsha 410073, China
[2] The College of System Engineering, National University of Defense Technology (NUDT), Changsha 410073, China
* Correspondence: liuli_nudt@nudt.edu.cn

Abstract: Among the current methods of synthetic aperture radar (SAR) automatic target recognition (ATR), unlabeled measured data and labeled simulated data are widely used to elevate the performance of SAR ATR. In view of this, the setting of semi-supervised few-shot SAR vehicle recognition is proposed to use these two forms of data to cope with the problem that few labeled measured data are available, which is a pioneering work in this field. In allusion to the sensitivity of poses of SAR vehicles, especially in the situation of only a few labeled data, we design two azimuth-aware discriminative representation (AADR) losses that suppress intra-class variations of samples with huge azimuth-angle differences, while simultaneously enlarging inter-class differences of samples with the same azimuth angle in the feature-embedding space via cosine similarity. Unlabeled measured data from the MSTAR dataset are labeled with pseudo-labels from categories among the SARSIM dataset and SAMPLE dataset, and these two forms of data are taken into consideration in the proposed loss. The few labeled samples in experimental settings are randomly selected in the training set. The phase data and amplitude data of SAR targets are all taken into consideration in this article. The proposed method achieves 71.05%, 86.09%, and 66.63% under 4-way 1-shot in EOC1 (Extended Operating Condition), EOC2/C, and EOC2/V, respectively, which overcomes other few-shot learning (FSL) and semi-supervised few-shot learning (SSFSL) methods in classification accuracy.

Keywords: semi-supervised learning; few-shot learning; SAR target recognition; discriminative representation learning

Citation: Zhang, L.; Leng, X.; Feng, S.; Ma, X.; Ji, K.; Kuang, G.; Liu, L. Azimuth-Aware Discriminative Representation Learning for Semi-Supervised Few-Shot SAR Vehicle Recognition. *Remote Sens.* **2023**, *15*, 331. https://doi.org/10.3390/rs15020331

Academic Editor: Gwanggil Jeon

Received: 14 November 2022
Revised: 28 December 2022
Accepted: 28 December 2022
Published: 5 January 2023

1. Introduction

As a longstanding and challenging problem in Synthetic Aperture Radar (SAR) imagery interpretation, SAR Automatic Target Recognition (SAR ATR) has been an active research field for several decades. SAR ATR plays a fundamental role in various civil applications including prospecting and surveillance, and military applications such as border security [1]. (Armored) vehicle recognition [2–4] in SAR ATR aims at giving machines the capability of automatically identifying the classes of interested armored vehicles (such as tank, artillery and truck), which is the focus of this work. Recently, high-resolution SAR images are increasingly easier to produce than before, offering great potential for studying fine-grained, detailed SAR vehicle recognition. Despite decades of effort by researchers, including the recent successful preliminary attempts presented by deep learning [5–8], as far as we know, the problem of SAR vehicle recognition remains an underexploited research field with the following significant challenges [9].

- **The lack of large, realistic, labeled datasets.** Existing SAR vehicle datasets, i.e., the Moving and Stationary Target Acquisition and Recognition (MSTAR) dataset [10], are too small and relatively unrealistic, and cannot represent the complex characteristics of SAR vehicles [1] including imaging geometry, background clutter, occlusions, and speckle noise and true data distributions, but are very easy for many

---

machine-learning methods to achieve high performance with abundant training samples. Certainly, such SAR vehicle datasets are hard to create due to the non-cooperative application scenario and the high cost of expert annotations. Therefore, label-efficient learning methods deserve attention in such a context. In other words, in the recognition missions of SAR vehicle targets, labeled SAR images of armored vehicles are usually difficult to obtain and interpret in practice, which leads to an insufficient sample situation in this field [11].

- **Large intra-class variations and small inter-class variations.** Variations in imaging geometry, such as the imaging angle including azimuth angle and depression angle, imaging distance and background clutter, lead to remarkable effects on the vehicle appearance in SAR images (examples shown in Figure 1), causing large intra-class variation. The aforementioned variations in imaging conditions can also cause vehicles of different classes to manifest highly similar appearances (examples shown in Figure 1), leading to small inter-class variations. Thus, SAR vehicle recognition demands robust yet highly discriminative representations that are difficult to learn, especially from a few labeled samples.

- **The more difficult recognition missions among extended standard operation (EOCs).** In MSTAR standard operation condition (SOC), the training samples and testing samples are only different in the depression angles, which are $17°$ and $15°$. When it comes to EOCs, different from the SOC, the variations in the depression angles and the configuration or versions of targets lead to obvious imaging behaviors among SAR targets. Thus, the recognition missions among EOCs are much more difficult than SOC in the MSTAR dataset. This phenomenon also exists in the few-shot recognition missions.

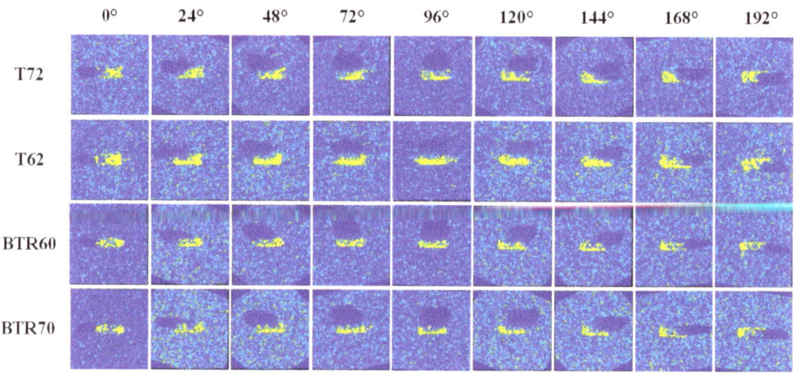

Figure 1. The samples of four categories among the MSTAR SOC under the azimuth-angle normalization. According to its azimuth angle, the SAR image from each category is selected every 24 degrees. To ensure the continuity of the samples based on azimuth angles, the image of an adjacent azimuth angle is chosen if there is a vacancy of the particular degree value.

Recently, in response to the aforementioned challenges, FSL [12] has been introduced to recognition missions of SAR ATR, aiming to elevate the recognition rate through a few labeled data. The lack of training data suppresses the performance of those CNN-based SAR target classification methods, which achieve a high recognition accuracy when the labeled data are sufficient [2]. To handle this challenge, simulated SAR images generated from auto-CAD models and the mechanism of electromagnetic scattering are introduced into the SAR ATR to elevate the recognition accuracy [13–15]. Although some common information can be transferred from labeled simulated data, there still exists huge differences between simulated data and measured data. The surroundings of the imaging target, the disturbance of the imaging platform, and even the material of vehicles make it hard to simulate the samples in the real environment. Because of this, some scholars are willing to leverage unlabeled measured data, instead of simulated data, in their algorithms, which launches the settings of semi-supervised SAR ATR [16–19].

Building upon our previous study in [11], this paper presents the first study of SSFSL in the field of SAR ATR, aiming to improve the model by making use of labeled simulated data and unlabeled measured data. Besides leveraging these data, information on azimuth angle is regarded as a kind of significant knowledge in digging discriminative representation in this paper.

When there are enough labeled SAR training samples, the feature-embedding space based on the azimuth angle of a category is approximately complete from $0°$ to $359°$. Hence, under this situation, the influence of a lack of several azimuth angles on recognition rate is limited. Nevertheless, if there are only an extremely small number of labeled samples, their azimuth angle will dominate the SAR vehicle recognition results. Figure 1 shows four selected categories of SAR images from the MSTAR SOC within the azimuth-angle normalization [11]. It is obvious that the SAR vehicle images with huge differences in azimuth angles from the same category own quite different backscattering behaviors, which can be considered to be high intra-class diversity. When the difference of the azimuth angles of samples is over 50 degrees, the backscattering behaviors, including the shadow area and target area of the target, are dissimilar in accordance with the samples in the same row of Figure 1. In the meantime, the SAR vehicle images with the same or adjacent azimuth angles from different categories share similar backscattering behaviors, which is the inter-class similarity. The samples in the same column in Figure 1 are homologous in the appearance of the target area and shadow area, especially when vehicle types are approximate; for instance, the group of BTR60 and BTR70, and T62 and T72. These two properties among SAR images cause confusion in representation learning and mistakes in classification results.

To solve this problem, an azimuth-aware discriminative representation (AADR) learning method is proposed, and this algorithm can grasp the distinguishable information through azimuth angles among both labeled simulated data and unlabeled measured data. The motivation of the method is to design a specific loss to let the model study not only the category information but also the azimuth-angle information. For suppressing the intra-class diversity, the pairs of SAR samples from the same category within huge azimuth-angle differences are selected, and their absolute value of cosine similarity of representations will be adjusted from zero-near value to one-near value. Simultaneously, to enlarge inter-class differences, samples from different categories with the same azimuth angle are selected and their feature vectors will be pulled from approximately overlap to near orthogonality in the metric manner of cosine similarity. Following this idea, the azimuth-aware regular loss (AADR-r) and its variant azimuth-aware triplet loss (AADR-t) are proposed, and the details will be introduced in Section 3. Furthermore, the cross-entropy loss from the labeled simulated datasets and the KL divergence of pseudo-labels from the unlabeled measured dataset (MSTAR) are also considered in the proposed loss. After experiencing the modification through the proposed loss, the algorithm is used to learn the discriminative representation from the few-shot samples and be tested among the query set.

Based on the baseline in SSFSL, there is no overlap between categories among the source domain and the target domain. The number of simulated data in the source domain is abundant, whereas there are an extremely small number of measured samples with labels and enough unlabeled measured data in the target domain. According to the settings of SSFSL, samples from the support and query sets are distinguished by different depression angles, and the unlabeled data are only chosen from the samples in the support set.

Extensive contrast experiments and ablation experiments were carried out to show the performance of our method. In general, the three contributions of the paper are summarized below:

- Due to the lack of large and realistic labeled datasets among SAR vehicle targets, for the first time, we propose the settings of semi-supervised few-shot SAR vehicle recognition, which takes both unlabeled measured data and labeled simulated data into consideration. In particular, simulated datasets act as the source domain in FSL, while the measured dataset MSTAR serves as the target domain. Additionally, the unlabeled data in MSTAR dataset are available in the process of model training. This configuration is really close to the active task in few-shot SAR vehicle recognition that labeled simulated data, and unlabeled measured data can be obtained easily.

- An azimuth-aware discriminative representation loss is proposed to learn the similarity of representations of intra-class samples with large azimuth-angle differences among the labeled simulated datasets. The representation pairs are considered to be feature vector pairs, which are pulled close to each other in the direction of the vector. Meanwhile, the inter-class differences of samples with the same azimuth angle are also expanded by the proposed loss in the feature-embedding space. The well-designed cosine similarity works as the distance to make representation pairs in the inter-class be orthogonal to each other.

- tlo information and phase data knowledge are adopted in the stage of SAR vehicle data pre-processing. Moreover, the variants of azimuth-aware discriminative representation loss achieve 47.7% (10-way 1-shot SOC), 71.05% (4-way 1-shot EOC1), 86.09% (4-way 1-shot EOC2/C), and 66.63% (4-way 1-shot EOC2/V), individually. Plenty of contrast experiments with other FSL methods and SSFSL methods prove that our proposed method is effective, especially in three EOC datasets.

There are five sections in this paper. In Section 2, the semi-supervised learning and its applications in SAR ATR, FSL and its applications in SAR ATR, and SAR target recognition based on azimuth angle are introduced in the related work. The settings of SSFSL among SAR target classification is presented in Section 3.1. Then, in Section 3.2, the whole framework of the proposed AADR-r is shown. After that, AADR-t is described in Section 3.3. Then, in Section 4, experimental results among SOC and three EOCs are demonstrated in diagrams and tables. Sufficient contrast experiments, ablation experiments, and implementation details are introduced and analyzed in Section 5. Finally, this paper is concluded, and future work is designed in Section 6.

2. Related Work

2.1. Semi-Supervised Learning and Its Applications in SAR Target Recognition

(1) Semi-supervised learning: Semi-supervised learning uses both labeled and unlabeled data to perform certain learning tasks. In contrast to supervised learning, it permits the harnessing of large amounts of unlabeled data available in many cases [20]. Generally, there are three representative approaches for semi-supervised learning—generative models [21,22], conditional entropy minimization [23], and pseudo-labeling [24]. Among the methods of generative models, various auto-encoders [25,26] were proposed by adding consistency regularization losses computed on unlabeled data. However, all unlabeled examples were encouraged to make confident predictions on some classes in the approaches of conditional entropy minimization [27]. The means of pseudo-labeling [28], which was adopted in this article, imputes approximate classes on unlabeled data by making predictions from a model trained only on labeled data.

(2). Semi-supervised SAR target recognition: According to the classification of methods in semi-supervised learning, the methods of semi-supervised SAR target recognition can also be divided into three parts. A symmetric auto-encoder was used to extract node features and the adjacency matrix is initialized using a new similarity measurement method [16]. The methods with generative adversarial networks were also popular in solving the semi-supervised SAR target recognition [29]. In [30], the pseudo-labeling and the consistency regularization loss were both adopted, and these unlabeled samples with pseudo-labels were mixed with the labeled samples and trained together in the designed loss to improve recognition performance. Multi-block mixed (MBM) in [31], which could effectively use the unlabeled samples, was used to interpolate a small part of the training image to generate new samples. In addition, semi-supervised SAR target recognition under limited data was also studied in [18,19]. Kullback–Leibler (KL) divergence was introduced to minimize the distribution divergence between the training and test data feature representations in [18]. The dataset attention module (DAM) was proposed to add the unlabeled data into the training set to enlarge the limited label training set [32].

2.2. Few-Shot Learning and Its Applications in SAR Target Recognition

(1). Few-shot learning: Currently, few-shot learning is proposed to learn a classifier from the base dataset and adapt with extremely limited supervised information of each class. The methods to solve the few-shot learning problems are generally divided into metric-based and optimization-based. The metric-based methods tend to classify the samples by judging the distance between the query-set image and the support-set image, such as matching networks [33], prototypical networks [34], deep nearest-neighbor neural network (DN4) [35]. Optimization-based algorithms designed novel optimization functions [36], better initialization of training models [37] and mission-adapted loss [38] to improve the rapid adaptability to new tasks, which could be regard as common solutions in few-shot learning methods.

(2). Semi-supervised few-shot learning: When there are only a few labeled examples among novel classes, it is intuitive to use extra unlabeled data to improve the learning [39]. This leads to the setting of semi-supervised few-shot learning. Prototypical networks were improved by Ren et al. [40] to produce prototypes for the unlabeled data. Liu et al. [41] constructed a graph between labeled and unlabeled data and used label propagation to obtain the labels of unlabeled data. By adding the confident prediction of unlabeled to the labeled training set in each round of optimization, Li et al. [42] applied self-training in semi-supervised few-shot learning. In [43], a simple and effective solution was proposed to tackle the extreme domain gap by self-training a source domain representation on unlabeled data from the target domain.

(3). Few-shot SAR target recognition: Few-shot SAR target recognition [44–53] has had more and more emphasis placed on it in recent years. An AG-MsPN [9] was proposed to consider both complex-value information of SAR data and the prior attribute information of the targets. The connection-free attention module and Bayesian-CNN were proposed to transfer common features from the electro-optical domain to the SAR domain for SAR image classification in the extreme few-shot case [54]. The Siamese neural network [55] was also ameliorated to cope with the problems of few-shot SAR target recognition [51]. The MSAR [45] with a meta-learner and a base-learner could learn a good initialization as well as a proper update strategy. The inductive inference and the transductive inference were adopted in the hybrid inference network (HIN) [49] to distinguish the samples in the embedding space. These methods divided the MSTAR dataset into query set and support set and the performance is not reflected on the whole MSTAR dataset. DKTS-N was proposed to take SAR domain knowledge into consideration and evaluated among the whole categories in the MSTAR dataset, but the performance of DKTS-N among MSTAR EOCs was not pleasant according to [11].

2.3. SAR Target Recognition Based on Azimuth Angle

The information on azimuth angle, which is a kind of important domain knowledge in SAR images, has been applied in the algorithms for a long time. Usually, a series of SAR images with regular azimuth angles are input into the network, which is named multiview or multi-aspect [56–60]. In [56], every input multiview SAR image was first examined by sparse representation-based classification to evaluate its validity for multiview recognition. Then, the selected views were jointly recognized with joint sparse representation. Multiview similar-angle target images were used to generate a joint low-rank and sparse multiview denoising dictionary [57]. MSRC-JSDC learned a supervised sparse model from training samples by using sample label information, rather than directly employing a predefined one [61]. A residual network (ResNet) and bidirectional long short-term memory (BiLSTM) network was proposed to learn the azimuth-angle information among SAR images [58]. However, to exploit the spatial and temporal features contained in the SAR image sequence simultaneously, this article proposed a sequence SAR target classification method based on the spatial-temporal ensemble convolutional network (STEC-Net) [59]. The authors in [60] adopted a parallel network topology with multiple inputs and the features of input SAR images from different azimuth angles would be learned layer by layer. Although these above-mentioned methods made full use of the azimuth angles, a certain number of SAR images with different azimuth angles were required, which was impossible in extremely few-shot SAR target recognition. In this article, the discriminative representation information among different samples is refined from specially designed loss during model training.

3. Proposed Method

To cope with the challenge of semi-supervised few-shot SAR target recognition, the AADR framework is proposed within three stages in Figure 2. In this section, the settings of SSFSL are illustrated first. Then, the whole framework of AADR-r is introduced. Finally, the variant loss AADR-t will be described in detail.

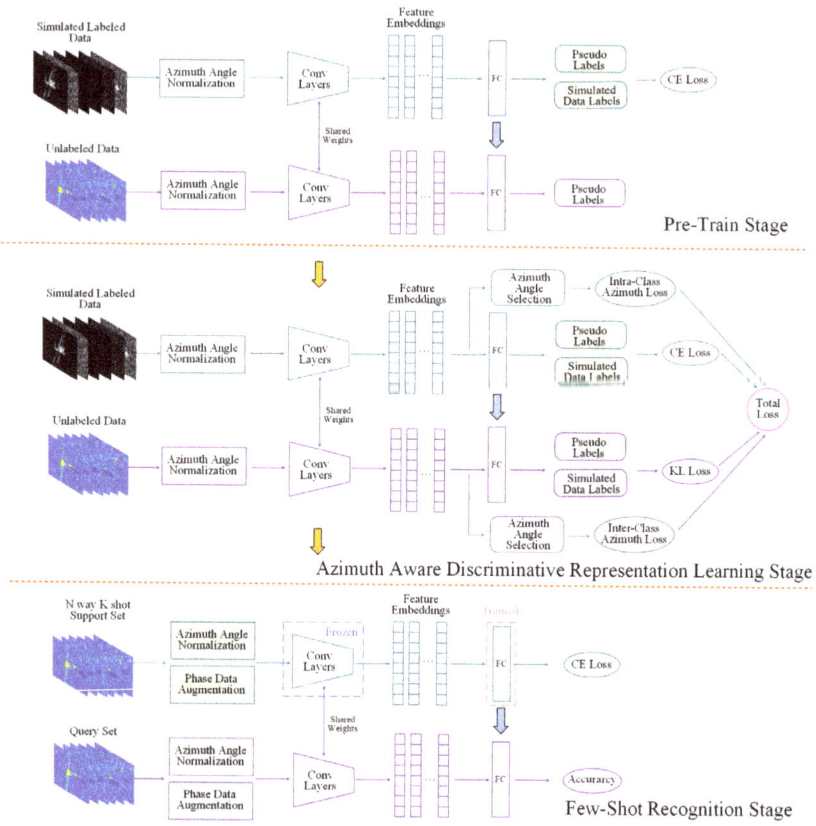

Figure 2. The whole framework of the azimuth-aware discriminative representation framework with regular loss.

3.1. Problem Setting

Initially, the definition of terminology used in semi-supervised few-shot SAR target recognition can be written as follows: a huge labeled simulated dataset

$D_{sim} = \left\{ (x_i, l_i, \alpha_i) | i = 1, 2, \ldots, p, l_i \in C_{sim} \right\}$. x_i is the image in the labeled simulated dataset D_{sim}. l_i is the label of x_i and the α_i is the azimuth angle of the x_i. C_{sim} is the categories among the simulated dataset. The measured MSTAR datasets are divided into D_{mea}^{train} and D_{mea}^{test}, according to the popular baseline. The train dataset and test dataset can be formulated by

$$D_{mea}^{train} = \left\{ (x_i^{train}, l_i^{train}, \alpha_i^{train}) | i = 1, 2, \ldots, r, l_i^{train} \in C_{mea}^{train} \right\} \tag{1}$$

$$D_{mea}^{test} = \left\{ (x_j^{test}, l_j^{test}, \alpha_j^{test}) | j = 1, 2, \ldots, s, l_j^{test} \in C_{mea}^{test} \right\} \tag{2}$$

$x_i^{train}, l_i^{train}, \alpha_i^{train}$ are the image, label and azimuth angle of the image among the training measured dataset, respectively, while $x_j^{test}, l_j^{test}, \alpha_j^{test}$ are the image, label and azimuth angle of the image among the testing measured dataset. C_{mea}^{train} and C_{mea}^{test} are the categories in the training and testing dataset and satisfy the relationship of $C_{mea}^{train} \supseteq C_{mea}^{test}$. Actually, in the experiment settings of MSTAR SOC and EOC1, the relationship between categories among training and testing sets is $C_{mea}^{train} = C_{mea}^{test}$, while in the experimental settings of EOC2/C and EOC2/V, there are more categories in the training set than categories in the testing set, so the relationship changes to $C_{mea}^{train} \supset C_{mea}^{test}$. The unlabeled measured dataset $D_{unlabel}$ is the same dataset as the D_{mea}^{train} without the label information, $D_{unlabel} = \left\{ (x_j, \alpha_j) | j = 1, 2, \ldots, r, \right\}$.

N-way K-shot indicates that there are N categories and each category contains K labeled samples. In most times, K is set to 1 or 5 in the experiments of MiniImageNet [36] and Ominiglot [62]. At the pre-train stage, all the labeled data are sampled from the D_{sim}, and the unlabeled data $D_{unlabel}$ are the same samples as the D_{mea}^{train} but without the labels. In addition, D_{sim} and $D_{unlabel}$ are involved in the azimuth-aware discriminative representation learning stage. It should be noted that there is not any FSL setting at either the first or the second stages. At the few-shot recognition stage, N-way K-shot samples are randomly selected from the D_{mea}^{train}, which act as the support set. All the data in D_{mea}^{test} compose the query set.

3.2. The Whole Framework with AADR-r

The whole framework of the azimuth-aware discriminative representation method is illustrated in Figure 2. At the pre-training stage, after the processing of azimuth-angle normalization, the simulated labeled data are fed into the deep neural network [6] within cross-entropy loss.

$$\min_{\theta} \sum_{(x_i, l_i, \alpha_i) \in D_{sim}} L_{ce}(f_\theta(x_i), l_i) \tag{3}$$

In Formula (3), a trained model ϕ with parameters θ_0 and classifier are achieved. Due to the plenty of labeled data in the labeled simulated dataset, the recognition accuracy among D_{sim} is perfect, which is shown in the first row in the pre-training stage in Figure 2. ϕ embeds the input image x into $\mathbb{R}^d$. The input dim of the classifier is d, and the output dim is the number of classes among labeled simulated dataset C_{sim}. Then, the trained model ϕ with parameters θ_0 and the classifier is adopted to classify the unlabeled data in D_{mea}^{train} with the pseudo-labels C_{sim}. After that, the pseudo-label of each unlabeled data are achieved, which is shown in Formula (4) and described in the second row in Figure 2. Every pseudo-label C_{sim} of the unlabeled image is fixed thorough the whole azimuth-aware discriminative representation learning stage. Formula (5) describes the process of feature extraction and classification, which appears many times in the training stage with the changing parameters θ.

$$l_j = f_{\theta_0}(x_j), l_j \epsilon C_{sim}, \forall x_j \epsilon D_{unlabel} \tag{4}$$

$$f_\theta(x_j) = classifier(\phi(x_j)) \tag{5}$$

The azimuth-aware discriminative representation learning module works at the second stage. Both simulated labeled data and unlabeled measured data experience azimuth-angle normalization before being fed into the network. Among the simulated labeled data, the information on labels and azimuth angles is accessible, so it is feasible to calculate the difference in azimuth angles between the samples from the same category. In the article, the gap is set to 50 degrees. The selected pair of samples among simulated labeled datasets obeys the rule according to Formula (6). The number of pairs in P_{sim} is $N_{sim} = Card(P_{sim})$, which is involved in Formula (7). The *cos* distance restricts the directions among representations of selected pairs. Before optimization, the *cos* distance of samples from the same category with huge differences in azimuth angle is relatively close to zero. However, through the designed azimuth-aware discriminative representation learning regular loss in simulated labeled dataset modules, the above-mentioned situation can be alleviated and the result of *cos* distance is guided to one. The cross-entropy loss of recognition among simulated labeled datasets is also taken into consideration to restrict the adaptation of θ in the whole optimization.

$$P_{sim} = \left\{ \begin{matrix} (x_{i_1}, l_{i_1}, \alpha_{i_1}), (x_{i_2}, l_{i_2}, \alpha_{i_2}) | 1 \leq i_1 < i_2 \leq p, \\ 50 < |\alpha_{i_1} - \alpha_{i_2}| < 310, l_{i_1} = l_{i_2} \epsilon C_{sim} \end{matrix} \right\} \tag{6}$$

$$L_{sim} = \sum_{(x_i, l_i, \alpha_i) \epsilon D_{sim}} L_{ce}(f_\theta(x_i), l_i) + \frac{\lambda}{N_{sim}} \sum_{1 \leq i_1 < i_2 \leq p} \frac{1}{|\cos(\phi(x_{i_1}), \phi(x_{i_2}))| + 1} \tag{7}$$

When it comes to the unlabeled measured dataset $D_{unlabel}$, although the samples are selected from the D_{mea}^{train}, only the pseudo-labels from the pre-training stage can be achieved, which is l_j in Formula (8). The azimuth angles of samples are known. It is worth noting that, among the MSTAR dataset, the azimuth angles of any two samples in one category are different. Therefore, in $D_{unlabel}$, if there are two samples with the same azimuth angle, then these two samples must come from two different categories. Actually, before optimization, the *cos* distance of these two samples may be closer to 1 than 0, because of the similar backscattering behaviors. The purpose of the designed loss in Formula (9) is to make the model distinguish data from various categories within the same azimuth angle, through which the proposed loss enlarges the inter-class differences of samples and lets the feature vectors of these sample pairs be orthogonal to each other as far as possible. The selected samples pair $P_{unlabel}$ is described in Formula (8) and the number of pairs in $P_{unlabel}$ is $N_{unlabel} = Card(P_{unlabel})$. λ is the hyper-parameter in the loss. The KL loss introduces noise during training by encouraging the model to learn the representations that emphasize the groupings induced by the pseudo-labels among the unlabeled measured samples. The total loss of the second stage is shown in Formula (10). In ablation experiments, the effects of different parts of loss will be discussed and the results are shown in the corresponding tables.

$$P_{unlabel} = \left\{ \begin{matrix} (x_{j_1}, l_{j_1}, \alpha_{j_1}), (x_{j_2}, l_{j_2}, \alpha_{j_2}) | \\ 1 \leq j_1 < j_2 \leq r, \alpha_{j_1} = \alpha_{j_2} \end{matrix} \right\} \tag{8}$$

$$L_{mea} = \sum_{(x_j, \alpha_j) \epsilon D_{unlabel}} L_{KL}(f_\theta(x_j), l_j) + \frac{\lambda}{N_{unlabel}} \sum_{1 \leq j_1 < j_2 \leq r} |\cos(\phi(x_{j_1}), \phi(x_{j_2}))| \tag{9}$$

$$L_{total} = L_{sim} + L_{mea} \tag{10}$$

In terms of the few-shot recognition stage, it is composed of a training and testing process, as shown in Figure 2. N way K shot labeled samples are randomly selected from the D_{mea}^{train}, acting as the support set, and all the samples in D_{mea}^{test} comprise the query set. The parameters θ are reserved from the second stage and frozen in this stage. Because the output categories in the fully connected classifier are different between the second stage and the third stage, the parameter of the classifier needs adapting through the few labeled data. After that, the feature extractor and the fully connected classifier are tested through the query set. The operation in the third stage is repeated 600 times and the average recognition rate and variance are recorded.

3.3. The Variants of AADR-t

In fact, the motivation of our designed AADR-r is similar to the triplet loss but without the anchor samples. To compare with the standard triplet loss, the AADR-t is designed to minimize the distance between an anchor sample and a positive sample with the same category, and maximize the distance between the anchor sample and a negative sample of a different category [63]. The differences between AADR-r and AADR-t are the selection rules of sample pairs and the loss function, which all exist in the second stage. It is worth noticing that, in this article, the selection of anchor sample $(x_i^a, l_i^a, \alpha_i^a)$, its hard negative sample $(x_i^n, l_i^n, \alpha_i^n)$ and its hard positive sample $(x_i^p, l_i^p, \alpha_i^p)$ take the azimuth angle into consideration. The hard negative sample shares the same azimuth angle as the anchor samples, but they are from different categories. The hard positive sample is the same class as the anchor sample but with a huge difference in azimuth angle. The details of selection rules are shown in Formulas (11) and (12). The categories among unlabeled samples are from pseudo-labels in the simulated dataset. The numbers of the triplet group in P_{sim}^t and $P_{unlabel}^t$ can be expressed through $N_{sim}^t = Card(P_{sim}^t)$ and $N_{unlabel}^t = Card(P_{unlabel}^t)$. Unlike the max operation and settings of margin in the raw triplet loss, the proposed AADR-t expands the cosine distance among the anchor sample and hard negative sample, and pulls in the cosine distance between the anchor sample and hard positive sample in both simulated dataset and unlabeled dataset, as shown in Formulas (13) and (14). The total triplet loss is composed of the loss in the simulated dataset and the unlabeled dataset, which is similar to Formula (10).

$$
P_{sim}^t = \left\{ \begin{array}{l} (x_i^a, l_i^a, \alpha_i^a), (x_i^p, l_i^p, \alpha_i^p), (x_i^n, l_i^n, \alpha_i^n) | \\ 1 \le i \le p, 50 < \left| \alpha_i^a - \alpha_i^p \right| < 310, \\ \alpha_i^n = \alpha_i^a, l_i^a = l_i^p \ne l_i^n, l_i^a, l_i^p, l_i^n \epsilon C_{sim} \end{array} \right\}
\tag{11}
$$

$$
P_{unlabel}^t = \left\{ \begin{array}{l} (x_j^a, l_j^a, \alpha_j^a), (x_j^p, l_j^p, \alpha_j^p), (x_j^n, l_j^n, \alpha_j^n) | \\ 1 \le j \le r, 50 < \left| \alpha_j^a - \alpha_j^p \right| < 310, \\ \alpha_j^n = \alpha_j^a, l_j^a = l_j^p \ne l_j^n, l_j^a, l_j^p, l_j^n \epsilon C_{sim} \end{array} \right\}
\tag{12}
$$

$$
L_{sim}^t = \sum_{(x_i^a, l_i^a, \alpha_i^a) \epsilon D_{sim}} L_{ce}(f_\theta(x_i^a), l_i^a) +
$$
$$
\frac{\lambda}{N_{sim}^t} \sum_{1 \le i \le p} |\cos(\phi(x_i^a), \phi(x_i^n)| - \left| \cos(\phi(x_i^a), \phi(x_i^p) \right|
\tag{13}
$$

$$
L_{mea}^t = \sum_{(x_j^a, l_j^a, \alpha_j^a) \epsilon D_{unlabel}} L_{KL}(f_\theta(x_j^a), l_j^a) +
$$
$$
\frac{\lambda}{N_{unlabel}^t} \sum_{1 \le j \le r} \left| \cos(\phi(x_j^a), \phi(x_j^n) \right| - \left| \cos(\phi(x_j^a), \phi(x_j^p) \right|
\tag{14}
$$

$$
L_{total}^t = L_{sim}^t + L_{mea}^t
\tag{15}
$$

4. Experiments

To test the validity of AADR for semi-supervised few-shot SAR vehicle classification, extensive experiments were performed under the experimental settings that the public simulated SARSIM dataset and the simulated part of Synthetic and Measured Paired Labeled Experiment (SAMPLE) dataset were combined as the D_{sim}. The public MSTAR dataset was recognized as the D_{mea}. Actually, the few-shot labeled data are sampled from D_{mea}^{train}, and the data in the query set D_{mea}^{test} are from different depression angles or different types. The unlabeled data are all from D_{mea}^{train}. Take MSTAR SOC (Standard Operating Condition) as an example: the few-shot labeled data comprise the support set, while the unlabeled measured data are selected from the set of $17°$ depression angle in MSTAR SOC, while the samples within $15°$ depression angle in MSTAR SOC compose the query set.

Contrast experiments with traditional classifiers, other advanced FSL approaches, and semi-supervised learning approaches were conducted. Additionally, ablation experiments with different dimensions of features, various base datasets, and errors in azimuth-angle estimation are also involved in our work.

Without the phase data, the data in the SAMPLE dataset and the SARSIM dataset only experience the gray-image adjustment and azimuth-angle normalization. However, the samples in the MSTAR database experience the phase data augmentation as in [11].

The feature extractor network in Figure 2 contains four fully convolutional blocks [64], which own a 3×3 convolution layer with 64, 128, 256 and 512 filters, relatively, 2×2 max-pooling layer, a batch-normalization layer, and a RELU (0.5) nonlinearity layer. We use the SGD with momentum optimizer with momentum 0.9 and weight decay 1×10^{-4}. All experiments were run on a PC with an Intel single-core i9 CPU, four Nvidia GTX-2080 Ti GPUs (12 GB VRAM each), and 128 GB RAM. The PC operating system was Ubuntu 20.04. All experiments were conducted using the Python language on the PyTorch deep-learning framework and CUDA 10.2 toolkit.

4.1. Datasets

(1). SARSIM: The public SARSIM dataset [15] contains seven kinds of vehicles (humvee 9657 and 3663, bulldozer 13,013 and 8020, tank 65,047 and 86,347, bus 30,726 and 55,473, motorbike 3972 and 3651_Suzuki, Toyota car and Peugeot 607, and truck 2107 and 2096). Every image is simulated in the identical situation to MSTAR and for $5°$ azimuth-angle interval at the following depression angles ($15°$, $17°$, $25°$, $30°$, $35°$, $40°$, and $45°$), so there are 72 samples in each category under a certain depression angle.

(2). SAMPLE: The public SAMPLE dataset [65,66] is released by Air Force Research Laboratory with both measured and simulated data in 10 sorts of armored vehicle (tracked cargo carrier: M548; military truck: M35; wheeled armored transport vehicle: BTR70; self-propelled artillery: ZSU-23-4; tanks: T-72, M1, and M60; tracked infantry fighting vehicle: BMP2 and M2; self-propelled howitzer: 2S1). The azimuth angles of the samples, which are 128x128 pixel, in the SAMPLE dataset are from $10°$ to $80°$ and their depression angles are from $14°$ to $17°$. For every measured target, a corresponding synthetic image is created with the same sensor and target configurations, but with totally different background clutter. In order to make the categories in the few-shot recognition stage and pre-training stage different, in most experiment settings in this article, only the synthetic images in the SAMPLE dataset are leveraged and combined with the SARSIM dataset to expand the richness of categories in the base dataset.

(3). MSTAR: In recent years, the MSTAR SOC dataset [10], including ten kinds of military vehicles during the Soviet era (military truck: ZIL-131; tanks: T-72 and T-62; bulldozer: D7; wheeled armored transport vehicle: BTR60 and BTR70; self-propelled howitzer: 2S1; tracked infantry fighting vehicle: BMP2; self-propelled artillery: ZSU-23-4; armored reconnaissance vehicle: BRDM2), was remarkable for verifying the algorithm performance among SAR vehicle classification missions. Imaged under the airborne X-band radar, the samples in this dataset were HH polarization mode within the resolution of 0.3×0.3 m. Targets, whose depression angles were $17°$, were for the support set and consisted of the unlabeled measured data, and $15°$ were for testing, whose numbers among each category were shown in Table 1. The EOC1 (large depression variation) contained four kinds of target (ZSU-23-4, T-72, BRDM-2 and 2S1). The depression angle of the training and testing set were $17°$ and $30°$, relatively. The targets in the EOC2/C (configuration variation) were various in parts of the vehicle, including explosive reactive armor (ERA) and an auxiliary gasoline tank. The EOC2/V (version variation) corresponded to the target version variation and shared the identical support set to the EOC2/C, but with a different query set, which is displayed in Table 2.

Table 1. Categories among MSTAR SOC.

SOC	17° Support Set	15° Support Set
2S1	299	274
BMP2	233	196
BRDM2	298	274
BTR60	256	195
BTR70	233	196
D7	299	274
T62	299	274
T72	232	196
ZIL131	299	274
ZSU-23-4	299	274

Table 2. Categories among MSTAR EOCs.

Target	Support Set	EOC1	Query Set
2S1	299	2S1-b01	288
BRDM2	298	BRDM2-E71	287
T72	691	T72-A64	288
ZSU-23-4	299	ZSU-23-4-d08	288

Target	Support Set	EOC2/C	Query Set
BMP2-9563	233	T72-S7	419
BRDM2-E71	298	T72-A32	572
BTR70-c71	233	T72-A62	573
T72-SN132	232	T72-A63	573
		T72-A64	573

Target	Support Set	EOC2/V	Query Set
BMP2-9563	233	T72-SN812	426
		T72-A04	573
BRDM2-E71	298	T72-A05	573
		T72-A07	573
BTR70-c71	233	T72-A10	567
		BMP2-9566	428
T72-SN132	232	BMP2-C21	429

4.2. Experimental Results

4.2.1. Experiments in SOC

Comparative experiments including classical classifiers (CC), FSL methods and SSFSL methods are shown in Table 3, under the FSL setting among 10-way K-shot ($K = 1, 2, 5, 10$). The average recognition rate and variance of 600 random experiments for each setting are displayed in Table 3. CC algorithms include LR (logistic regression) [67], DT (decision tree) [68], SVM (support vector machine) [69], GBC (gradient-boosting classifier) [70] and RF (random forest) [71]. These methods share the same feature extractors as the AADR with individual classifiers. The average recognition rate of algorithms in SSFSL is higher than in FSL and CC in Table 3. Although the recognition rates of classical classifiers are unsatisfactory in few-shot conditions, some of them achieve a higher result than SSFSL in the settings of 10-way 10-shot. Our proposed AADR-r and AADR-t obtain a relatively better recognition rate in few-shot settings ($K \leq 5$), which are only a little lower than DKTS-N. The DKTS-N outstrips all the other methods in the settings of both few-shot and limited data in SOC for the following reason. The advantage of DKTS-N is learning the global and local features. The samples in training and testing sets in MSTAR SOC are similar because of the approximate depression angle 17° and 15°. Hence, the global and local features between the two sets are close and easy to be matched through Earth's mover distance and nearest-neighbor classifiers in DKTS-N. However, highly different configurations and versions of armored vehicle lead to huge discrepancies in local features, which influence the scattering characteristics among SAR images. Therefore, the performance in the EOCs of DKTS-N decreases, which is the restriction of this metric-learning-based algorithm. The proposed AADR, an optimization-based method, overcomes the difficulties and shows an overwhelming performance in EOCs.

Table 3. Few-shot classification accuracy of SOC among CC, FSL and SSFSL algorithms.

	Algorithm	SOC (10-Way)			
		1-Shot	2-Shot	5-Shot	10-Shot
CC	SVM [69]	38.75 ± 0.45	50.32 ± 0.41	67.49 ± 0.34	77.99 ± 0.27
	LR [67]	41.96 ± 0.35	52.82 ± 0.37	69.06 ± 0.32	79.52 ± 0.24
	DT [68]	18.54 ± 0.47	26.02 ± 0.44	40.70 ± 0.43	50.31 ± 0.38
	GBC [70]	34.64 ± 0.41	36.49 ± 0.40	38.72 ± 0.39	47.56 ± 0.36
	RF [71]	18.64 ± 0.50	24.79 ± 0.43	39.96 ± 0.38	51.86 ± 0.35
FSL	DeepEMD [72]	36.19 ± 0.46	43.49 ± 0.44	53.14 ± 0.40	59.64 ± 0.39
	DeepEMD grid [73]	35.89 ± 0.43	41.15 ± 0.41	52.24 ± 0.37	56.04 ± 0.31
	DeepEMD sample [73]	35.47 ± 0.44	42.39 ± 0.42	50.34 ± 0.39	52.36 ± 0.28
	DN4 [35]	33.25 ± 0.49	44.15 ± 0.45	53.48 ± 0.41	64.88 ± 0.34
	Prototypical Network [34]	40.94 ± 0.47	54.54 ± 0.44	69.42 ± 0.39	78.01 ± 0.29
	Relation Network [74]	34.23 ± 0.47	41.89 ± 0.42	54.32 ± 0.37	64.45 ± 0.32
	DKTS-N [11]	49.26 ± 0.48	58.51 ± 0.42	72.32 ± 0.32	84.59 ± 0.24
SSFSL	ICI [75]	49.18 ± 0.54	54.31 ± 0.46	57.82 ± 0.35	63.92 ± 0.22
	EP [76]	44.74 ± 0.64	47.82 ± 0.57	53.20 ± 0.46	57.16 ± 0.30
	PPSML [77]	36.56 ± 0.48	46.19 ± 0.34	59.56 ± 0.23	73.36 ± 0.16
	STARTUP [43]	36.19 ± 0.33	49.81 ± 0.32	65.27 ± 0.26	74.47 ± 0.20
	STARTUP (no SS) [43]	37.96 ± 0.37	51.61 ± 0.39	67.17 ± 0.30	75.47 ± 0.19
	ConvT [78]	42.57 ± 0.79	54.37 ± 0.62	75.16 ± 0.21	88.63 ± 0.22
	our AADR-r $\lambda = 0.5$	46.84 ± 0.43	57.00 ± 0.37	69.12 ± 0.27	78.19 ± 0.20
	our AADR-t $\lambda = 0.7$	47.70 ± 0.45	58.37 ± 0.38	69.91 ± 0.29	78.77 ± 0.19

4.2.2. Experiments in EOCs

Due to the huge differences among SAR vehicle images, the FSL missions are harder in EOCs than in SOC. However, most of the SSFSL methods are better than FSL methods in the results of both SOC setting and EOCs settings, which means the usage of unlabeled data is beneficial for the FSL among SAR vehicles. In addition, the awareness of the azimuth angle also helps the model to grasp the important domain knowledge among SAR vehicles and overcome the intra-class diversity and inter-class similarity in few-shot conditions. From Table 4, it is obvious that our proposed AADR-r and AADR-t do a good job in EOCs, and the recognition results are much higher than other FSL methods and SSFSL methods. Instead of comparing the metric distances between the features, the model optimization through designed loss performs well in a large difference in depression angle, vehicle version, and configuration.

A similar process with different losses causes different results such that the accuracy of AADR-r exceeds AADR-t in most times. In fact, the categories of the anchor sample, hard negative sample, and hard positive sample among unlabeled data are generated by the trained model in the pre-training stage, according to Figure 2. Thus, the pseudo-labels participate in the loss and influence the result. For instance, the anchor sample and its hard negative sample are from the different categories, which are the pseudo-labels among the simulated data. However, if these two samples are from the same actual category in D_{mea}^{train}, this will lead to the wrong training in the second stage. Therefore, the results of AADR-t contain more uncertainties than AADR-r.

Table 4. Few-shot classification accuracy of EOCs.

	EOC1 (4-Way)			
	Algorithm	**1-Shot**	**2-Shot**	**5-Shot**	**10-Shot**
FSL	DeepEMD [72]	56.81 ± 0.99	62.8 ± 0.78	65.16 ± 0.61	67.58 ± 0.49
	DeepEMD grid [73]	55.95 ± 0.43	57.46 ± 0.41	63.81 ± 0.37	65.72 ± 0.31
	DeepEMD sample [73]	49.65 ± 0.44	54.00 ± 0.42	58.19 ± 0.39	60.34 ± 0.28
	DN4 [35]	46.59 ± 0.83	51.41 ± 0.69	58.11 ± 0.49	62.15 ± 0.43
	Prototypical Network [34]	53.59 ± 0.93	56.57 ± 0.53	61.94 ± 0.48	65.13 ± 0.43
	Relation Network [74]	43.21 ± 1.02	46.93 ± 0.81	54.97 ± 0.56	38.62 ± 0.49
	DKTS-N [11]	61.91 ± 0.91	63.94 ± 0.73	67.43 ± 0.48	71.09 ± 0.41
SSFSL	ICI [75]	57.90 ± 1.03	61.02 ± 0.84	64.31 ± 0.61	65.49 ± 0.45
	EP [76]	51.46 ± 0.85	55.81 ± 0.72	57.62 ± 0.59	58.20 ± 0.49
	PPSML [77]	65.01 ± 0.96	74.32 ± 0.79	79.56 ± 0.61	84.23 ± 0.46
	STARTUP [43]	52.83 ± 0.60	60.20 ± 0.52	69.23 ± 0.40	74.07 ± 0.26
	STARTUP (no SS) [43]	63.33 ± 0.67	70.99 ± 0.58	76.34 ± 0.35	77.77 ± 0.25
	ConvT [78]	59.57 ± 0.76	64.06 ± 0.88	68.17 ± 0.38	74.80 ± 0.20
	our AADR-r $\lambda = 0.5$	71.05 ± 0.74	76.00 ± 0.57	82.52 ± 0.38	85.83 ± 0.33
	our AADR-t $\lambda = 0.7$	70.02 ± 0.69	75.43 ± 0.59	81.13 ± 0.42	83.61 ± 0.33

	EOC2/C (4-Way)			
	Algorithm	**1-Shot**	**2-Shot**	**5-Shot**	**10-Shot**
FSL	DeepEMD [72]	38.39 ± 0.86	45.65 ± 0.75	54.53 ± 0.60	62.13 ± 0.50
	DN4 [35]	46.13 ± 0.69	51.21 ± 0.62	58.14 ± 0.54	63.08 ± 0.51
	Prototypical Network [34]	43.59 ± 0.84	51.17 ± 0.78	59.15 ± 0.70	64.15 ± 0.61
	Relation Network [74]	42.13 ± 0.90	48.24 ± 0.82	53.12 ± 0.71	36.28 ± 0.59
	DKTS-N [11]	47.26 ± 0.79	53.61 ± 0.70	62.23 ± 0.56	68.41 ± 0.51
SSFSL	ICI [75]	69.85 ± 1.73	73.62 ± 1.44	80.26 ± 1.10	85.32 ± 0.93
	EP [76]	81.74 ± 1.35	86.36 ± 1.02	89.68 ± 0.81	93.77 ± 0.65
	PPSML [77]	46.67 ± 1.66	50.83 ± 1.31	60.85 ± 1.09	71.32 ± 0.85
	STARTUP [43]	67.22 ± 1.47	79.54 ± 1.41	89.95 ± 0.86	95.95 ± 0.50
	STARTUP (no SS) [43]	69.42 ± 1.29	80.33 ± 1.16	91.46 ± 0.94	96.38 ± 0.46
	ConvT [78]	44.32 ± 0.65	51.93 ± 0.82	64.12 ± 0.34	89.74 ± 0.18
	our AADR-r $\lambda = 0.5$	83.78 ± 1.19	90.41 ± 0.71	95.69 ± 0.34	97.02 ± 0.17
	our AADR-t $\lambda = 0.7$	82.52 ± 1.06	87.38 ± 0.77	90.65 ± 0.56	92.22 ± 0.38

	EOC2/V (4-Way)			
	Algorithm	**1-Shot**	**2-Shot**	**5-Shot**	**10-Shot**
FSL	DeepEMD [72]	40.92 ± 0.76	49.12 ± 0.65	58.43 ± 0.51	67.64 ± 0.42
	DN4 [35]	47.00 ± 0.72	52.21 ± 0.61	58.87 ± 0.55	63.93 ± 0.52
	Prototypical Network [34]	45.13 ± 0.72	52.86 ± 0.65	62.07 ± 0.52	67.71 ± 0.40
	Relation Network [74]	40.24 ± 0.91	46.32 ± 0.82	54.22 ± 0.68	35.13 ± 0.52
	DKTS-N [11]	48.91 ± 0.70	55.14 ± 0.58	65.63 ± 0.49	70.18 ± 0.42
SSFSL	ICI [75]	50.75 ± 1.38	56.44 ± 1.12	68.19 ± 1.01	84.00 ± 0.83
	EP [76]	51.33 ± 1.22	55.48 ± 1.07	61.62 ± 0.89	64.16 ± 0.65
	PPSML [77]	46.74 ± 1.14	51.66 ± 0.96	61.09 ± 0.87	71.43 ± 0.61
	STARTUP [43]	50.94 ± 1.06	61.04 ± 1.01	68.40 ± 0.74	75.07 ± 0.42
	STARTUP (no SS) [43]	53.63 ± 0.98	63.14 ± 0.91	71.89 ± 0.67	80.18 ± 0.47
	ConvT [78]	42.27 ± 0.89	58.27 ± 0.68	68.05 ± 0.52	83.55 ± 0.25
	our AADR-r $\lambda = 0.5$	66.63 ± 1.21	73.99 ± 0.96	81.41 ± 0.45	84.64 ± 0.32
	our AADR-t $\lambda = 0.7$	62.77 ± 1.24	69.42 ± 0.92	77.77 ± 0.55	83.10 ± 0.41

5. Discussion

In this section, the value of hyper-parameter λ in the total loss is discussed, which determines the proportion of the azimuth-aware discriminative representation learning loss. Then, the influence of different compositions of categories among base datasets is analyzed in this subsection. Moreover, the dimension configuration of the feature extractor is also discussed. The azimuth angles are accurate in the processing of normalization, but in this subsection, the angle errors are taken into account in both MSTAR SOC and EOCs.

5.1. The Influence of Loss Modules

In Tables 5 and 6, the influence of different loss modules and their parameters on the recognition rate is shown, and other experiment settings are the same. Both 10-way recognition in SOC and 4-way recognition in EOCs are conducted. Azimuth-aware in the table is the proposed module in this article, which uses azimuth angle to suppress intra-class diversity of samples with huge azimuth-angle differences and enlarged inter-class differences of samples. The $\times$ in the table means that the related loss is not in the total loss. The SimCLR module is proposed in [23] and widely leveraged in semi-supervised learning. It encourages augmentations such as cropping, adding noise, and flipping. KL and CE indicate the Kullback–Leibler divergence and the cross-entropy loss, respectively. "-r" and "-t" represent the "AADR-r" and "AADR-t" and the hyper-parameter λ ranges from 0.1 to 0.7.

Table 5. Influence of different loss module among SOC (10-way).

	Azimuth-Aware	KL	CE	SimCLR	1-Shot	2-Shot	5-Shot	10-Shot
	$\times$	✓	✓	✓	36.19 ± 0.33	49.81 ± 0.32	65.27 ± 0.26	74.47 ± 0.20
	$\times$	✓	✓	$\times$	37.96 ± 0.37	51.61 ± 0.39	67.17 ± 0.30	75.47 ± 0.19
	-r ($\lambda = 0.7$)	✓	✓	$\times$	46.23 ± 0.47	55.26 ± 0.40	68.84 ± 0.26	77.32 ± 0.19
	-r ($\lambda = 0.5$)	✓	✓	$\times$	46.84 ± 0.43	57.00 ± 0.37	69.12 ± 0.27	78.19 ± 0.20
	-r ($\lambda = 0.3$)	✓	✓	$\times$	43.92 ± 0.42	53.65 ± 0.38	68.88 ± 0.29	77.86 ± 0.19
SOC	-r ($\lambda = 0.1$)	✓	✓	$\times$	41.68 ± 0.41	53.79 ± 0.36	67.06 ± 0.27	76.13 ± 0.20
	-r ($\lambda = 0.7$)	$\times$	✓	$\times$	38.94 ± 0.41	48.17 ± 0.37	59.95 ± 0.27	68.32 ± 0.22
	-t ($\lambda = 0.7$)	✓	✓	$\times$	47.70 ± 0.45	58.37 ± 0.38	69.91 ± 0.29	78.77 ± 0.19
	-t ($\lambda = 0.5$)	✓	✓	$\times$	41.94 ± 0.43	51.13 ± 0.39	63.27 ± 0.28	71.21 ± 0.21
	-t ($\lambda = 0.3$)	✓	✓	$\times$	43.76 ± 0.43	53.56 ± 0.39	65.34 ± 0.29	74.77 ± 0.22
	-t ($\lambda = 0.1$)	✓	✓	$\times$	41.17 ± 0.37	55.17 ± 0.34	68.05 ± 0.27	76.19 ± 0.21

The loss in the first row in each experiment setting is the result of STARTUP [43] and the loss in the second row is the result of STARTUP (no SS) [43]. Although the SimCLR module is beneficial to the classification rate in optical image datasets, it is obvious that the total loss without SimCLR (no SS) shows a better performance. Actually, the targets are in the center of the images and with the behaviors of backscatterings, which is different from the optical images. The operations in the SimCLR module, such as cropping, adding noise, and flipping, are not suitable for the SAR vehicle images. For instance, the crop operation may cut the key part of the SAR vehicles and the added noise is not reasonable according to the SAR imaging mechanism. The AADR-r is more stable than its variants AADR-t because the anchor samples among AADR-t, involving the pseudo-labels, which are the classification results of the unlabeled data, participate in the triplet loss. Every unlabeled sample actually owns its real label. If different unlabeled samples, which are from the same real category, are classified into different pseudo-labels, the results of AADR-t will be poorly influenced. The λ indicates the proportion of the azimuth-aware module in the total loss, and a fixed λ cannot be competent to all experimental settings. Comparatively, the result of $\lambda = 0.5$ in AADR-r is better. When it comes to the contribution of KL divergence, which is an important part of the semi-supervised learning with pseudo-labeling, it is easy to see that the absence of KL in the AADR-r with $\lambda = 0.7$ decreases a lot, compared to the raw contrast version.

Table 6. Influence of different loss module among EOCs (4-way).

	Azimuth-Aware	KL	CE	SimCLR	1-Shot	2-Shot	5-Shot	10-Shot
	$\times$	$\checkmark$	$\checkmark$	$\checkmark$	52.83 ± 0.60	60.20 ± 0.52	69.23 ± 0.40	74.07 ± 0.26
	$\times$	$\checkmark$	$\checkmark$	$\times$	63.33 ± 0.67	70.99 ± 0.58	76.34 ± 0.35	77.77 ± 0.25
	-r ($\lambda = 0.7$)	$\checkmark$	$\checkmark$	$\times$	66.98 ± 0.61	72.17 ± 0.46	76.60 ± 0.31	81.80 ± 0.29
	-r ($\lambda = 0.5$)	$\checkmark$	$\checkmark$	$\times$	71.05 ± 0.74	76.00 ± 0.57	82.52 ± 0.38	85.83 ± 0.33
	-r ($\lambda = 0.3$)	$\checkmark$	$\checkmark$	$\times$	70.62 ± 0.68	75.91 ± 0.56	80.11 ± 0.31	83.70 ± 0.28
EOC1	-r ($\lambda = 0.1$)	$\checkmark$	$\checkmark$	$\times$	68.31 ± 0.68	74.50 ± 0.51	81.74 ± 0.30	84.87 ± 0.21
	-r ($\lambda = 0.7$)	$\times$	$\checkmark$	$\times$	60.36 ± 0.41	65.87 ± 0.74	70.32 ± 0.52	71.47 ± 0.38
	-t ($\lambda = 0.7$)	$\checkmark$	$\checkmark$	$\times$	70.02 ± 0.69	75.43 ± 0.59	81.13 ± 0.42	83.61 ± 0.33
	-t ($\lambda = 0.5$)	$\checkmark$	$\checkmark$	$\times$	61.34 ± 0.94	66.07 ± 0.80	70.35 ± 0.55	72.08 ± 0.42
	-t ($\lambda = 0.3$)	$\checkmark$	$\checkmark$	$\times$	65.34 ± 0.93	70.90 ± 0.79	76.34 ± 0.56	78.86 ± 0.43
	-t ($\lambda = 0.1$)	$\checkmark$	$\checkmark$	$\times$	65.13 ± 0.57	69.24 ± 0.52	77.72 ± 0.41	82.09 ± 0.32
	$\times$	$\checkmark$	$\checkmark$	$\checkmark$	67.22 ± 1.47	79.54 ± 1.41	89.95 ± 0.86	95.95 ± 0.50
	$\times$	$\checkmark$	$\checkmark$	$\times$	69.42 ± 1.29	80.33 ± 1.16	91.46 ± 0.94	96.38 ± 0.46
	-r ($\lambda = 0.7$)	$\checkmark$	$\checkmark$	$\times$	76.32 ± 1.46	83.64 ± 1.11	91.15 ± 0.63	94.81 ± 0.43
	-r ($\lambda = 0.5$)	$\checkmark$	$\checkmark$	$\times$	83.78 ± 1.19	90.41 ± 0.71	95.69 ± 0.34	97.02 ± 0.17
	-r ($\lambda = 0.3$)	$\checkmark$	$\checkmark$	$\times$	81.60 ± 1.49	89.13 ± 1.02	95.41 ± 0.49	97.02 ± 0.30
EOC2/C	-r ($\lambda = 0.1$)	$\checkmark$	$\checkmark$	$\times$	86.09 ± 1.13	92.75 ± 0.75	97.00 ± 0.48	99.00 ± 0.24
	-r ($\lambda = 0.7$)	$\times$	$\checkmark$	$\times$	70.05 ± 2.34	80.86 ± 1.65	89.95 ± 1.12	90.27 ± 1.06
	-t ($\lambda = 0.7$)	$\checkmark$	$\checkmark$	$\times$	82.52 ± 1.06	87.38 ± 0.77	90.65 ± 0.56	92.22 ± 0.38
	-t ($\lambda = 0.5$)	$\checkmark$	$\checkmark$	$\times$	82.45 ± 2.19	90.04 ± 1.41	94.16 ± 0.81	95.63 ± 0.56
	-t ($\lambda = 0.3$)	$\checkmark$	$\checkmark$	$\times$	82.50 ± 2.20	90.13 ± 1.42	94.40 ± 0.80	95.65 ± 0.58
	-t ($\lambda = 0.1$)	$\checkmark$	$\checkmark$	$\times$	68.83 ± 1.14	80.28 ± 1.04	90.16 ± 0.74	95.25 ± 0.45
	$\times$	$\checkmark$	$\checkmark$	$\checkmark$	50.94 ± 1.06	61.04 ± 1.01	68.40 ± 0.74	75.07 ± 0.42
	$\times$	$\checkmark$	$\checkmark$	$\times$	53.63 ± 0.98	63.14 ± 0.91	71.89 ± 0.67	80.18 ± 0.47
	-r ($\lambda = 0.7$)	$\checkmark$	$\checkmark$	$\times$	56.15 ± 1.20	61.50 ± 0.99	74.66 ± 0.66	83.92 ± 0.39
	-r ($\lambda = 0.5$)	$\checkmark$	$\checkmark$	$\times$	66.63 ± 1.21	73.99 ± 0.96	81.41 ± 0.45	84.64 ± 0.32
	-r ($\lambda = 0.3$)	$\checkmark$	$\checkmark$	$\times$	61.94 ± 1.32	69.22 ± 1.14	79.06 ± 0.68	84.99 ± 0.40
EOC2/V	-r ($\lambda = 0.1$)	$\checkmark$	$\checkmark$	$\times$	60.93 ± 1.04	66.36 ± 0.90	74.40 ± 0.62	81.46 ± 0.40
	-r ($\lambda = 0.7$)	$\times$	$\checkmark$	$\times$	53.21 ± 1.5	60.87 ± 1.26	69.53 ± 0.89	74.28 ± 0.71
	-t ($\lambda = 0.7$)	$\checkmark$	$\checkmark$	$\times$	62.77 ± 1.24	69.42 ± 0.92	77.77 ± 0.55	83.10 ± 0.41
	-t ($\lambda = 0.5$)	$\checkmark$	$\checkmark$	$\times$	57.76 ± 1.65	64.61 ± 1.28	73.11 ± 0.89	78.47 ± 0.58
	-t ($\lambda = 0.3$)	$\checkmark$	$\checkmark$	$\times$	58.04 ± 1.64	65.20 ± 1.26	74.23 ± 0.88	79.75 ± 0.57
	-t ($\lambda - 0.1$)	$\checkmark$	$\checkmark$	$\times$	54.62 ± 1.11	62.98 ± 0.96	74.34 ± 0.68	82.24 ± 0.43

5.2. Estimation Errors of Azimuth Angle

Figure 3 illustrates the accuracy of SOC and EOCs with various azimuth-angle estimation errors from 1-shot to 30-shot in 10-way and 4-way. The five-set of experiments shares the same configurations and parameters but with random estimation errors within a given range. The given range $\pm\alpha°$ indicates that the estimation errors of azimuth-angle range from $-\alpha°$ to $\alpha°$. $\pm0°$ shows that the estimation error of azimuth angle is approximate to zero and achieves the highest recognition rate in the figure, which is regarded as the baseline. From the figure, when the estimation azimuth-angle errors are less than $\pm5°$, there are almost 1% decreases in comparison to the baseline in SOC, EOC1, and EOC2/V. This demonstrates that our AADR is impressive under low estimated errors of azimuth angle. If the random errors ascend to $\pm 20°$, the recognition rates will witness a nearly 10% drop in SOC and EOC1. However, in the result of EOC2/C and EOC2/V, within the changes among weapon configurations and versions, the impact of estimated azimuth error on the accuracy is relatively tiny. According to these results, the estimated errors of azimuth angle have a more marked influence on the large variation of depression angle between the source tasks and the target tasks, than the changes in version or weapon configuration.

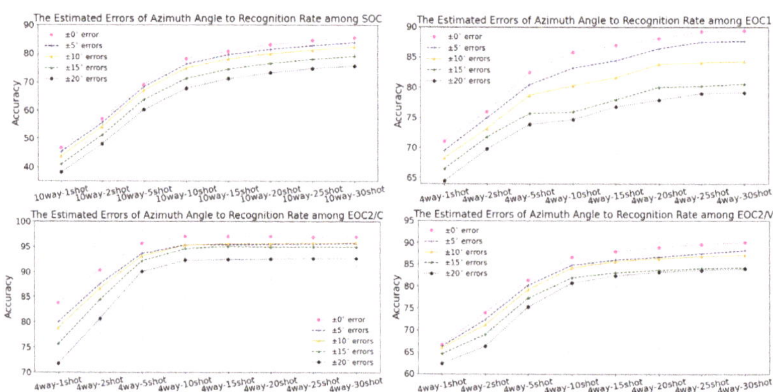

Figure 3. The line charts of estimated azimuth-angle errors to the accuracy among SOC and EOCs.

6. Conclusions

To sum up, we put forward the AADR to deal with the task of few-shot SAR target classification, especially in the situations of a huge difference between support sets and query sets. The use of unlabeled measured data and labeled simulated data are one of the key means to elevate the recognition rate in a fresh semi-supervised manner. Additionally, azimuth-aware discriminative representation learning is also an available way to cope with the intra-class diversity and inter-class similarity among vehicle samples. In general, a large number of experiments showed that AADR was more impressive than other FSL algorithms.

There are still some flaws in the proposed methods. First, due to the optimization-based design, the classifier of the fully connected layer in AADR is not pleasant when the number of labeled data is over 10. According to Figure 3, as the number increases, the elevation of performance is limited. Hence, how to use more labeled data is significant to making the AADR powerful in situations of both few-shot and limited data. Second, the hyper-parameter λ in the loss, which indicates the proportion of azimuth-aware module, is fixed in the current algorithm. From the results, it is hard to determine a certain value of λ that can fit four experiments. Thus, a self-adaptation λ in the loss, related to the training epochs and learning rate, can guide the gradient descent in a better way.

Author Contributions: Conceptualization, L.Z., X.L. and L.L.; methodology, L.Z.; software, L.Z.; validation, L.Z., S.F. and X.M.; resources, K.J.; data curation, L.Z.; writing—original draft preparation, L.Z.; writing—review and editing, L.L.; visualization, L.Z.; supervision, L.L.; project administration, K.J. and G.K.; funding acquisition, X.L., K.J., G.K. and L.L. All authors have read and agreed to the published version of the manuscript.

Funding: This research was funded by National Key Research and Development Program of China grant number 2021YFB3100800 and the National Natural Science Foundation of China under Grant 61872379, 62001480 and Hunan Provincial Natural Science Foundation of China under Grant 2018JJ3613, 2021JJ40684.

Data Availability Statement: MSTAR dataset used in this work can be downloaded at https://www.sdms.afrl.af.mil/index.php?collection=registration (accessed on 28 December 2022).

Conflicts of Interest: The authors declare no conflict of interest.

Abbreviations

The following abbreviations are used in this manuscript:

SAR	Synthetic Aperture Radar
ATR	Automatic Target Recognition
AADR	Azimuth-Aware Discriminative Representation
EOC	Extended Operating Condition
SOC	Standard Operating Condition
FSL	Few-shot Learning
SSFSL	Semi-supervised Few-shot Learning
MSTAR	Moving and Stationary Target Acquisition and Recognition
KL	Kullback–Leibler
MBM	Multi-block mixed
DAM	Dataset attention module
DN4	Deep Nearest-Neighbor Neural Network
HIN	Hybrid Inference Network
ResNet	Residual Network
BiLSTM	Bidirectional Long Short-term Memory
STEC-Net	Spatial-temporal Ensemble Convolutional Network
SAMPLE	Synthetic and Measured Paired Labeled Experiment

References

1. Kechagias-Stamatis, O.; Aouf, N. Automatic target recognition on synthetic aperture radar imagery: A survey. *IEEE Aerosp. Electron. Syst. Mag.* **2021**, *36*, 56–81. [CrossRef]
2. Chen, S.; Wang, H.; Xu, F.; Jin, Y. Target classification using the deep convolutional networks for SAR images. *IEEE Trans. Geosci. Remote Sens.* **2016**, *54*, 4806–4817. [CrossRef]
3. Tian, Z.; Zhan, R.; Hu, J.; Zhang, J. SAR ATR based on convolutional neural network. *J. Radars* **2016**, *5*, 320–325.
4. Guo, W.; Zhang, Z.; Yu, W.; Sun, X. Perspective on explainable sar target recognition. *J. Radars* **2020**, *9*, 462–476.
5. Liu, L.; Ouyang, W.; Wang, X.; Fieguth, P.; Chen, J.; Liu, X.; Pietikainen, M. Deep learning for generic object detection: A survey. *Int. J. Comput. Vis.* **2019**, *128*, 261–318. [CrossRef]
6. Schmidhuber, J. Deep learning in neural networks: An overview. *Neural Netw.* **2015**, *61*, 85–117.
7. Krizhevsky, A.; Sutskever, I.; Hinton, G.E. Imagenet classification with deep convolutional neural networks. *Commun. ACM* **2017**, *60*, 84–90. [CrossRef]
8. Simonyan, K.; Zisserman, A. Very deep convolutional networks for large-scale image recognition. *arXiv* **2014**, arXiv:1409.1556.
9. Wang, S.; Wang, Y.; Liu, H.; Sun, Y. Attribute-guided multi-scale prototypical network for few-shot SAR target classification. *IEEE J. Sel. Top. Appl. Earth Obs. Remote Sens.* **2021**, *14*, 12224–12245. [CrossRef]
10. Keydel, E.R.; Lee, S.W.; Moore, J.T. MSTAR extended operating conditions: A tutorial. In *Algorithms for Synthetic Aperture Radar Imagery III*. Proc. SPIE. 1996. Available online: https://www.sdms.afrl.af.mil/index.php?collection=registration (accessed on 28 December 2022).
11. Zhang, L.; Leng, X.; Feng, S.; Ma, X.; Ji, K.; Kuang, G.; Liu, L. Domain knowledge powered two-stream deep network for few-shot SAR vehicle recognition. *IEEE Trans. Geosci. Remote Sens.* **2022**, *60*, 1–15. [CrossRef]
12. Lake, B.M.; Salakhutdinov, R.; Tenenbaum, J.B. Human-level concept learning through probabilistic program induction. *Science* **2015**, *350*, 1332–1338. [CrossRef] [PubMed]
13. He, Q.; Zhao, L.; Ji, K.; Kuang, G. SAR target recognition based on task-driven domain adaptation using simulated data. *IEEE Geosci. Remote Sens. Lett.* **2021**, *19*, 1–5. [CrossRef]
14. Zhang, C.; Wang, Y.; Liu, H.; Sun, Y.; Hu, L. SAR target recognition using only simulated data for training by hierarchically combining CNN and image similarity. *IEEE Geosci. Remote Sens. Lett.* **2022**, *19*, 1–5. [CrossRef]
15. Malmgren-Hansen, D.; Kusk, A.; Dall, J.; Nielsen, A.A.; Engholm, R.; Skriver, H. Improving SAR automatic target recognition models with transfer learning from simulated data. *IEEE Geosci. Remote Sens. Lett.* **2017**, *14*, 1484–1488. [CrossRef]
16. Wen, L.; Huang, X.; Qin, S.; Ding, J. Semi-supervised SAR target recognition with graph attention network. In Proceedings of the EUSAR 2021; 13th European Conference on Synthetic Aperture Radar, Online, 29 March–1 April 2021; VDE: Frankfurt, Germany, 2021; pp. 1–5.

17. Liu, X.; Huang, Y.; Wang, C.; Pei, J.; Huo, W.; Zhang, Y.; Yang, J. Semi-supervised SAR ATR via conditional generative adversarial network with multi-discriminator. In Proceedings of the 2021 IEEE International Geoscience and Remote Sensing Symposium IGARSS, Brussels, Belgium, 11–16 July 2021; pp. 2361–2364.

18. Wang, N.; Wang, Y.; Liu, H.; Zuo, Q. Target discrimination method for SAR images via convolutional neural network with semi-supervised learning and minimum feature divergence constraint. *Remote Sens. Lett.* **2020**, *11*, 1167–1174. [CrossRef]

19. Zhang, Y.; Guo, X.; Ren, H.; Li, L. Multi-view classification with semi-supervised learning for SAR target recognition. *Signal Process.* **2021**, *183*, 108030. [CrossRef]

20. Engelen, J.E.V.; Hoos, H.H. A survey on semi-supervised learning. *Mach. Learn.* **2020**, *109*, 373–440. [CrossRef]

21. Odena, A. Semi-supervised learning with generative adversarial networks. *arXiv* **2016**, arXiv:1606.01583.

22. Salimans, T.; Goodfellow, I.; Zaremba, W.; Cheung, V.; Radford, A.; Chen, X. Improved techniques for training gans. *arXiv* **2016**, arXiv:1606.03498

23. Chen, T.; Kornblith, S.; Norouzi, M.; Hinton, G. A simple framework for contrastive learning of visual representations. In Proceedings of the International Conference on Machine Learning, PMLR, Sun, CA, USA, 12–18 July 2020; pp. 1597–1607.

24. Zhai, X.; Oliver, A.; Kolesnikov, A.; Beyer, L. S4l: Self-supervised semi-supervised learning. In Proceedings of the IEEE/CVF International Conference on Computer Vision, Seoul, Republic of Korea, 27 October–2 November 2019; pp. 1476–1485.

25. Rasmus, A.; Berglund, M.; Honkala, M.; Valpola, H.; Raiko, T. Semi-supervised learning with ladder networks. *Adv. Neural Inf. Process. Syst.* **2015**, *28*, 3532–3540.

26. Kingma, D.P.; Mohamed, S.; Rezende, D.J.; Welling, M. Semi-supervised learning with deep generative models. *Adv. Neural Inf. Process. Syst.* **2014**, *4*, 3581–3589.

27. Grandvalet, Y.; Bengio, Y. Semi-supervised learning by entropy minimization. *Adv. Neural Inf. Process. Syst.* **2005**, *17*, 529–536.

28. Lee, D.-H. Pseudo-label: The simple and efficient semi-supervised learning method for deep neural networks. In Proceedings of the ICML 2013 Workshop: Challenges in Representation Learning (WREPL), Atlanta, GA, USA, 16–21 June 2013.

29. Gao, F.; Ma, F.; Wang, J.; Sun, J.; Yang, E.; Zhou, H. Semi-supervised generative adversarial nets with multiple generators for SAR image recognition. *Sensors* **2018**, *18*, 2706. [CrossRef] [PubMed]

30. Tian, Y.; Zhang, L.; Sun, J.; Yin, G.; Dong, Y. Consistency regularization teacher–student semi-supervised learning method for target recognition in SAR images. *Vis. Comput.* **2021**, *38*, 4179–4192. [CrossRef]

31. Tian, Y.; Sun, J.; Qi, P.; Yin, G.; Zhang, L. Multi-block mixed sample semi-supervised learning for sar target recognition. *Remote Sens.* **2021**, *13*, 361. [CrossRef]

32. Gao, F.; Shi, W.; Wang, J.; Hussain, A.; Zhou, H. A semi-supervised synthetic aperture radar (SAR) image recognition algorithm based on an attention mechanism and bias-variance decomposition. *IEEE Access* **2019**, *7*, 108617–108632. [CrossRef]

33. Vinyals, O.; Blundell, C.; Lillicrap, T.P.; Kavukcuoglu, K.; Wierstra, D. Matching networks for one shot learning. *Neural Inf. Process. Syst.* **2016**, *29*, 3637–3645.

34. Snell, J.; Swersky, K.; Zemel, R.S. Prototypical networks for few-shot learning. *Neural Inf. Process. Syst.* **2017**, *30*, 4077–4087.

35. Li, W.; Wang, L.; Xu, J.; Huo, J.; Luo, J. Revisiting local descriptor based image-to-class measure for few-shot learning. In Proceedings of the IEEE/CVF Conference on Computer Vision and Pattern Recognition, Long Beach, CA, USA, 15 June–20 June 2019; pp. 7260–7268.

36. Ravi, S.; Larochelle, H. Optimization as a model for few-shot learning. In Proceedings of the International Conference on Learning Representations, Toulon, France, 24–26 April 2017.

37. Finn, C.; Abbeel, P.; Levine, S. Model-agnostic meta-learning for fast adaptation of deep networks. In Proceedings of the International Conference on Machine Learning, Sydney, NSW, Australia, 6–11 August 2017; pp. 1126–1135.

38. Shi, D.; Orouskhani, M.; Orouskhani, Y. A conditional triplet loss for few-shot learning and its application to image co-segmentation. *Neural Netw.* **2021**, *137*, 54–62. [CrossRef]

39. Yu, Z.; Chen, L.; Cheng, Z.; Luo, J. Transmatch: A transfer-learning scheme for semi-supervised few-shot learning. In Proceedings of the IEEE/CVF Conference on Computer Vision and Pattern Recognition, Seattle, WA, USA, 14–19 June 2020; pp. 12856–12864.

40. Ren, M.; Triantafillou, E.; Ravi, S.; Snell, J.; Swersky, K.; Tenenbaum, J.B.; Larochelle, H.; Zemel, R.S. Meta-learning for semi-supervised few-shot classification. *arXiv* **2018**, arXiv:1803.00676.

41. Liu, Y.; Lee, J.; Park, M.; Kim, S.; Yang, E.; Hwang, S.J.; Yang, Y. Learning to propagate labels: Transductive propagation network for few-shot learning. *arXiv* **2018**, arXiv:1805.10002.

42. Li, X.; Sun, Q.; Liu, Y.; Zhou, Q.; Zheng, S.; Chua, T.-S.; Schiele, B. Learning to self-train for semi-supervised few-shot classification. *Adv. Neural Inf. Process. Syst.* **2019**, *32*, 10276–10286.

43. Phoo, C.P.; Hariharan, B. Self-training for few-shot transfer across extreme task differences. *arXiv* **2020**, arXiv:2010.07734.

44. Wang, L.; Bai, X.; Zhou, F. Few-shot SAR ATR based on conv-BiLSTM prototypical networks. In Proceedings of the 2019 6th Asia-Pacific Conference on Synthetic Aperture Radar (APSAR), Xiamen, China, 26–29 November 2019.

45. Fu, K.; Zhang, T.; Zhang, Y.; Wang, Z.; Sun, X. Few-shot SAR target classification via metalearning. *IEEE Trans. Geosci. Remote Sens.* **2021**, *60*, 1–14. [CrossRef]

46. Che, J.; Wang, L.; Bai, X.; Liu, C.; Zhou, F. Spatial-Temporal Hybrid Feature Extraction Network for Few-shot Automatic Modulation Classification. *IEEE Trans. Veh. Technol.* **2022**, *71*, 13387–13392. [CrossRef]

47. Wang, Y.; Gui, G.; Lin, Y.; Wu, H.; Yu, C.; Adachi, F. Few-Shot Specific Emitter Identification via Deep Metric Ensemble Learning. *IEEE Internet Things J.* **2022**, *9*, 24980–24994. [CrossRef]

48. Yang, R.; Xu, X.; Li, X.; Wang, L.; Pu, F. Learning relation by graph neural network for SAR image few-shot learning. In Proceedings of the IGARSS 2020—2020 IEEE International Geoscience and Remote Sensing Symposium, Waikoloa, HI, USA, 26 September–2 October 2020; pp. 1743–1746.

49. Wang, L.; Bai, X.; Gong, C.; Zhou, F. Hybrid inference network for few-shot SAR automatic target recognition. *IEEE Trans. Geosci. Remote Sens.* **2021**, *59*, 9257–9269. [CrossRef]

50. Rostami, M.; Kolouri, S.; Eaton, E.; Kim, K. SAR image classification using few-shot cross-domain transfer learning. In Proceedings of the Conference on Computer Vision and Pattern Recognition Workshops, Long Beach, CA, USA, 16–20 June 2019.

51. Tang, J.; Zhang, F.; Zhou, Y.; Yin, Q.; Hu, W. A fast inference networks for SAR target few-shot learning based on improved siamese networks. In Proceedings of the IGARSS 2019—2019 IEEE International Geoscience and Remote Sensing Symposium, Yokohama, Japan, 28 July–2 August 2019.

52. Lu, D.; Cao, L.; Liu, H. Few-shot learning neural network for sar target recognition. In Proceedings of the 2019 6th Asia-Pacific Conference on Synthetic Aperture Radar (APSAR), Xiamen, China, 26–29 November 2019.

53. Zhai, Y.; Deng, W.; Lan, T.; Sun, B.; Ying, Z.; Gan, J.; Mai, C.; Li, J.; Labati, R.D.; Piuri, V. MFFA-SARNET: Deep transferred multi-level feature fusion attention network with dual optimized loss for small-sample SAR ATR. *Remote Sens.* **2020**, *12*, 1385. [CrossRef]

54. Tai, Y.; Tan, Y.; Xiong, S.; Sun, Z.; Tian, J. Few-shot transfer learning for SAR image classification without extra SAR samples. *IEEE J. Sel. Top. Appl. Earth Obs. Remote Sens.* **2022**, *15*, 2240–2253. [CrossRef]

55. Koch, G.; Zemel, R.; Salakhutdinov, R. Siamese neural networks for one-shot image recognition. In Proceedings of the ICML Deep Learning Workshop, Lille, France, 6–11 July 2015; Volume 2.

56. Ding, B.; Wen, G. Exploiting multi-view SAR images for robust target recognition. *Remote Sens.* **2017**, *9*, 1150. [CrossRef]

57. Huang, Y.; Liao, G.; Zhang, Z.; Xiang, Y.; Li, J.; Nehorai, A. Sar automatic target recognition using joint low-rank and sparse multiview denoising. *IEEE Geosci. Remote. Sens. Lett.* **2018**, *15*, 1570–1574. [CrossRef]

58. Zhang, F.; Fu, Z.; Zhou, Y.; Hu, W.; Hong, W. Multi-aspect SAR target recognition based on space-fixed and space-varying scattering feature joint learning. *Remote Sens. Lett.* **2019**, *10*, 998–1007. [CrossRef]

59. Xue, R.; Bai, X.; Zhou, F. Spatial–temporal ensemble convolution for sequence SAR target classification. *IEEE Trans. Geosci. Remote Sens.* **2021**, *59*, 1250–1262. [CrossRef]

60. Pei, J.; Huang, Y.; Huo, W.; Zhang, Y.; Yang, J.; Yeo, T.-S. Sar automatic target recognition based on multiview deep learning framework. *IEEE Trans. Geosci. Remote Sens.* **2018**, *56*, 2196–2210. [CrossRef]

61. Ren, H.; Yu, X.; Zou, L.; Zhou, Y.; Wang, X. Joint supervised dictionary and classifier learning for multi-view sar image classification. *IEEE Access* **2019**, *7*, 165127–165142. [CrossRef]

62. Lake, B.M.; Salakhutdinov, R.; Gross, J.; Tenenbaum, J.B. One shot learning of simple visual concepts. *Cogn. Sci.* **2011**, *33*, 2568–2573.

63. Schroff, F.; Kalenichenko, D.; Philbin, J. Facenet: A unified embedding for face recognition and clustering. In Proceedings of the IEEE Conference on Computer Vision and Pattern Recognition, Boston, MA, USA, 7–12 June 2015; pp. 815–823.

64. Long, J.; Shelhamer, E.; Darrell, T. Fully convolutional networks for semantic segmentation. *IEEE Trans. Pattern Anal. Mach. Intell.* **2015**, *39*, 640–651.

65. Scarnati, T.; Lewis, B. A deep learning approach to the synthetic and measured paired and labeled experiment (sample) challenge problem. In *Algorithms for Synthetic Aperture Radar Imagery XXVI*; International Society for Optics and Photonics: Bellingham, WA, USA, 2019; Volume 10987, p. 109870G.

66. Lewis, B.; Scarnati, T.; Sudkamp, E.; Nehrbass, J.; Rosencrantz, S.; Zelnio, E. A SAR dataset for ATR development: The synthetic and measured paired labeled experiment (sample). In *Algorithms for Synthetic Aperture Radar Imagery XXVI*; International Society for Optics and Photonics: Bellingham, WA, USA, 2019; Volume 10987, p. 109870H.

67. Allison, P.D. *Logistic Regression Using the Sas System: Theory and Application*; SAS Publishing: Cary, NC, USA, 1999.

68. Safavian, S.R.; Landgrebe, D. A survey of decision tree classifier methodology. *IEEE Trans. Syst. Man, Cybern.* **1991**, *21*, 660–674. [CrossRef]

69. Saunders, C.; Stitson, M.O.; Weston, J.; Holloway, R.; Bottou, L.; Scholkopf, B.; Smola, A. Support vector machine. *Comput. EnCE* **2002**, *1*, 1–28.

70. Burez, J.; Poel, D. Handling class imbalance in customer churn prediction. *Expert Syst. Appl.* **2008**, *36*, 4626–4636. [CrossRef]

71. Liaw, A.; Wiener, M. Classification and regression by randomforest. *R News* **2002**, *2*, 18–22.

72. Zhang, C.; Cai, Y.; Lin, G.; Shen, C.S. Deepemd: Few-shot image classification with differentiable earth mover's distance and structured classifiers. In Proceedings of the Conference on Computer Vision and Pattern Recognition, Seattle, WA, USA, 14–19 June 2020.

73. Zhang, C.; Cai, Y.; Lin, G.; Shen, C. Deepemd: Differentiable earth mover's distance for few-shot learning. *arXiv* **2020**, arXiv:2003.06777v3.

74. Sung, F.; Yang, Y.; Zhang, L.; Xiang, T.; Torr, P.H.; Hospedales, T.M. Learning to compare: Relation network for few-shot learning. In Proceedings of the IEEE Conference on Computer Vision and Pattern Recognition, Salt Lake City, UT, USA, 18–22 June 2018; pp. 1199–1208.

75. Wang, Y.; Zhang, L.; Yao, Y.; Fu, Y. How to trust unlabeled data instance credibility inference for few-shot learning. *IEEE Trans. Pattern Anal. Mach. Intell.* **2021**, *44*, 6240–6253. [CrossRef]

76. Rodríguez, P.; Laradji, I.; Drouin, A.; Lacoste, A. Embedding propagation: Smoother manifold for few-shot classification. In Proceedings of the European Conference on Computer Vision, Glasgow, UK, 23–28 August 2020; Springer: Berlin/Heidelberg, Germany, 2020; pp. 121–138.
77. Zhu, P.; Gu, M.; Li, W.; Zhang, C.; Hu, Q. Progressive point to set metric learning for semi-supervised few-shot classification. In Proceedings of the 2020 IEEE International Conference on Image Processing (ICIP), Abu Dhabi, United Arab Emirates, 25–28 October 2020; pp. 196–200.
78. Wang, C.; Huang, Y.; Liu, X.; Pei, J.; Zhang, Y.; Yang, J. Global in local: A convolutional transformer for SAR ATR FSL. *IEEE Geosci. Remote. Sens. Lett.* **2022**, *19*, 1–5. [CrossRef]

remote sensing

MDPI

Article

RCCT-ASPPNet: Dual-Encoder Remote Image Segmentation Based on Transformer and ASPP

Yazhou Li [1,2], Zhiyou Cheng [1,2], Chuanjian Wang [1,2,*], Jinling Zhao [1,2] and Linsheng Huang [1,2]

[1]   National Engineering Research Center for Analysis and Application of Agro-Ecological Big Data, Anhui University, Hefei 230601, China
[2]   School of Internet, Anhui University, Hefei 230039, China
*   Correspondence: wcj_si@ahu.edu.cn

Abstract: Remote image semantic segmentation technology is one of the core research elements in the field of computer vision and has a wide range of applications in production life. Most remote image semantic segmentation methods are based on CNN. Recently, Transformer provided a view of long-distance dependencies in images. In this paper, we propose RCCT-ASPPNet, which includes the dual-encoder structure of Residual Multiscale Channel Cross-Fusion with Transformer (RCCT) and Atrous Spatial Pyramid Pooling (ASPP). RCCT uses Transformer to cross fuse global multiscale semantic information; the residual structure is then used to connect the inputs and outputs. ASPP based on CNN extracts contextual information of high-level semantics from different perspectives and uses Convolutional Block Attention Module (CBAM) to extract spatial and channel information, which will further improve the model segmentation ability. The experimental results show that the mIoU of our method is 94.14% and 61.30% on the datasets Farmland and AeroScapes, respectively, and that the mPA is 97.12% and 84.36%, respectively, both outperforming DeepLabV3+ and UCTransNet.

Keywords: remote image; deep learning; semantic segmentation; CNN; multiscale feature fusion; Transformer

Citation: Li, Y.; Cheng, Z.; Wang, C.; Zhao, J.; Huang, L. RCCT-ASPPNet: Dual-Encoder Remote Image Segmentation Based on Transformer and ASPP. *Remote Sens.* **2023**, *15*, 379. https://doi.org/10.3390/rs15020379

Academic Editors: Gwanggil Jeon and Richard Gloaguen

Received: 13 October 2022
Revised: 31 December 2022
Accepted: 3 January 2023
Published: 7 January 2023

1. Introduction

With the continuous development of artificial intelligence technology, computer vision has attracted much attention as one of the important research areas. Unlike the other fields of computer vision, such as image classification, object detection, and instance segmentation, current mainstream deep learning-based image semantic segmentation research aims to densely predict each pixel of an image using algorithms in which each pixel is labeled with its own category, thus achieving the goal of assigning semantic information to each identical pixel in the image [1]. The result of deep semantic segmentation gives computers a more detailed and accurate understanding of images and has a wide range of application needs in the fields of autonomous driving [2,3], face segmentation [4–7], and medical imaging [8–11].

Due to rapid advances in aerospace and sensor technology, it is easy and fast to obtain high-resolution satellite imagery and aerial imagery. Remote image semantic segmentation is one of the core contents of computer vision research [12–14]. With the active development of deep learning, remote image deep learning semantic segmentation networks have been continuously proposed, such as the FCN [15], UNet [16], SegNet [17], DeepLab [18–21], PSPNet [22], SETR [23], UCtransNet [24] models, etc. Some of these are based on convolutional neural network (CNN) and some are based on Transformer, which are explained below in terms of these two aspects.

1.1. Remote Image Segmentation Method Based on CNN

The CNN-based semantic segmentation method is one of the mainstream methods and mainly utilizes the encoder–decoder structure. The encoder typically uses convolutional neural networks and downsampling to reduce the resolution and to extract image feature maps, while the decoder aims to transfer the low-resolution image and feature into image segmentation maps to achieve a pixel-level prediction, often using deconvolution [25] for upsampling, and the last layer of the network structure is mostly softmax classifiers to classify each pixel.

FCN replaces the fully connected layer at the end of the CNN with a convolutional layer and then upsamples it to obtain an image of the same size as the input. UNet also has an encoder–decoder structure (same as FCN), in which feature extraction is carried out in the first half and upsampling is carried out in the second half, and the skip connection layer in UNet merges low-level location information with deep-level semantic information. Similar to UNet, SegNet uses an encoder–decoder structure, but the encoder and decoder use different technologies. In addition, the encoder part of SegNet uses the first 13 layers of the VGG16 [26] convolutional network, each encoder layer corresponds to a decoder layer, and the output of the final decoder is fed into a softmax classifier to generate class probabilities for each pixel independently. DeepLabv1 [18] is based on two innovations: dilated convolution [27] and fully connected conditional random field. DeepLabv2 differs by proposing Atrous Spatial Pyramid Pooling (ASPP) [19], and DeepLabv3 [20] is based on further optimization of ASPP by adding convolution, BN operation, etc. DeepLabv3+ [21] is based on the structure of DeepLabV3 by adding an upsampling decoder module to optimize the accuracy of edges. These methods are widely used in remote image segmentation tasks and have obtained effective performance. However, the traditional CNN-based encoder–decoder network will lose some spatial resolution after a series of downsampling in the encoder stage, which affects the performance of semantic segmentation algorithms.

1.2. Remote Image Segmentation Method Based on Transformer

Transformer [28] was originally used in the field of natural language processing. Transformer is essentially an encoder–decoder structure. Transformer is based on the attention mechanism, which can solve the long-distance dependence problem. The attention mechanism has a better memory and can remember longer distance information. The most important thing is that attention supports parallelized computation, which is very suitable for remote images semantic segmentation. The transformer model is completely based on the attention mechanism, and it completely discards the structure of CNN.

Recently, some scholars used Transformer in semantic segmentation. Zheng et al. [23] proposed the SETR model for semantic segmentation, in which CNN is not used and so the resolution of the image is not degraded. Transformer cuts the image into multiple small pieces and encodes the ordering to achieve sequence-to-sequence encoding using attention mechanisms. Cao et al. [29] combined UNet with Transformer to extract multi-scale features. Although Transformer has achieved good performance in some semantic segmentation tasks, it has some limitations, such as larger model parameters and less segmentation capability than CNN. UCTransNet incorporates different feature layers of CNN into Transformer, which provides a new idea for multi-scale feature fusion. However, its CNN layer structure is simple, feature fusion is relatively single, the encoding and decoding methods are complex, and the ability to express various application scenarios of remote imaging is not sufficient.

1.3. Remote Image Segmentation Method Based on CNN and Transformer

To solve the problems caused by a single encoder, we concatenate CNN-based and Transformer-based network structures in order to compensate for the shortcomings of a single structure in remote image segmentation. First, we propose the Residual Multi-scale Channel Cross-Fusion with Transformer (RCCT) module as one of our encoding structures based on the multi-scale feature cross fusion approach of UCTransNet. Unlike UCTransNet, RCCT takes the first three feature layers of ResNet50 [30] as input and performs a cross-fusion of features, which capture the relationship between different feature layers in a Transformer way in order to obtain multi-scale semantic information. The output of RCCT is then concatenated into a whole feature layer and finally takes residual concatenation with the three input feature layers. Second, To enhance the segmentation capability of the model on a remote image, the fourth feature layers of ResNet50 are input into the ASPP module, followed by connecting a Convolutional Block Attention Module (CBAM) [31] as the second encoding structure. The dual–encoder structure, which is called RCCT-ASPPNet, can effectively represent the global contextual information in the image and increase the receptive field, with a comprehensive performance higher than that of a single encoder.

1.4. Contributions

This paper addresses some challenges in the field of semantic segmentation of remote image by proposing a dual-encoder RCCT-ASPPNet. The main contributions of this paper can be illustrated in the following points.

First, an efficient remote image segmentation method based on CNN and Transformer is proposed. We design a Transformer-based RCCT structure. The first three feature layers of resnet50 are used as the input of RCCT, and the dependencies between each feature layer are learned in a Transformer cross-fusion manner. Then, we use the residual structure to link the fused input feature layer with the fused output feature layer.

Second, we not only extract features by transformer but also utilize the CNN-based ASPP module to obtain larger receptive field information, while adding channel attention and spatial attention after ASPP to learn deeper semantics. With the dual-encoder structures, we alleviate the problems of small targets, multiple scales, and diverse and complex categories in remote image.

Finally, we tested the method proposed on two datasets. The AeroScapes is a public dataset, which has a variety of perspectives, complex scenes, and more categories. The Farmland is a self-made dataset, which is top-down view, and the data have small objects. The experimental results show the effectiveness of our method. Our method has a further improvement in semantic segmentation of remote image with an mIoU of 94.14% and 61.30% on the datasets Farmland and AeroScapes [32] respectively.

1.5. Article Structure

In the Introduction section, an overview of deep learning, semantic segmentation techniques is provided, and the semantic segmentation method based on CNN and Transformer is introduced. The Methods and Data section explains the theoretical approach behind the model proposed in this paper and some parameter settings, in addition to the dataset used in this paper. The Results section mainly describes the ablation experiment and provides a comparison of different models. The Discussion section analyzes the experimental results, and their advantages and disadvantages. The Conclusions section summarizes the remote image semantic segmentation method proposed in this paper; in addition, the shortcomings of the method of this paper and the next step are explained.

2. Methods and Data

2.1. RCCT-ASPPNet Model Overview

Figure 1 shows a general description of the proposed approach. We use a two-layer encoder structure, including RCCT and ASPP-CBAM. The RCCT encoding module uses Transformer as the backbone network, and Transformer is used to cross fuse each feature layer to learn the feature relationship between the layers in an end-to-end way. The input and output are connected with residuals in a feature fusion way. The ASPP-CBAM encoding module combines ASPP with channel and spatial attention mechanisms, extracts feature maps of different receptive fields for concatenation, and then uses CBAM as an attention layer to learn the importance of channel and spatial importance.

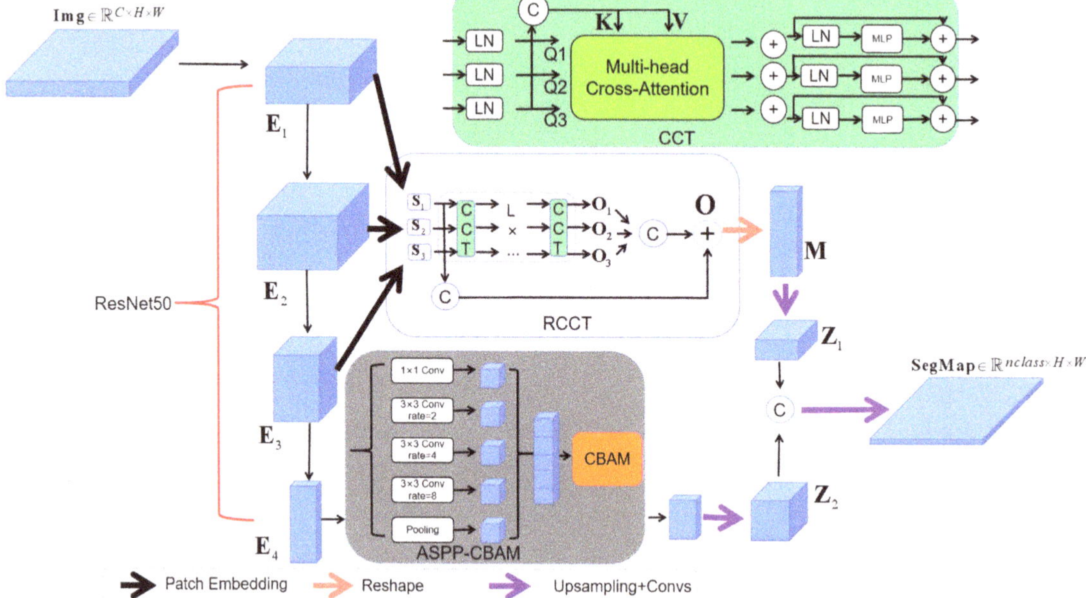

Figure 1. RCCT-ASPPNet model structure.

2.2. Residual Multi-Scale Channel Cross-Fusion Transformer (RCCT)

To address multi-scale feature fusion, we propose a RCCT module that takes advantage of long-dependency modeling in Transformer to fuse features from multi-scale encoders. The RCCT module has four steps: multi-scale feature embedding, multi-headed channel-based cross-attention, multi-layer perceptron (MLP), and residual feature fusion.

2.2.1. Multi-Scale Feature Embedding

Given as an input, the multi-scale feature embedding is the first three feature layers of ResNet50 $\mathbf{E}_i \in \mathbb{R}^{\frac{H \times W}{2^{2 \times (i+1)}} \times C_i}$, $(i = 1, 2, 3)$, where C_i is the number of channel dimensions, and $C_1 = 256$, $C_2 = 512$, and $C_3 = 1024$. The standard Transformer accepts a sequence of token embeddings as input. To process 2D features, we reshape the feature $\mathbf{E}_i$ as a flattened block sequence $\mathbf{S}_i \in \mathbb{R}^{N \times C_i}$, $(i = 1, 2, 3)$, where $(\frac{H}{2^{i+1}}, \frac{W}{2^{i+1}})$ is the resolution of the original feature; (p_i, p_i) is the resolution of each feature block; $p_1 = 16$, $p_2 = 8$, $p_3 = 4$; and N (Equation (1)) is the number of feature blocks generated, that is, the effective input sequence length of the Transformer. In this process, we keep the channel size N constant. Position embedding $\mathbf{S}_{pos} \in \mathbb{R}^{N \times C_i}$ is also added to the feature block to retain the spatial location information between the input feature blocks (Equation (2)).

$$N = \frac{\frac{H \times W}{2^{i+1} \times 2^{i+1}}}{p_i^2},$$

(1)

$$\mathbf{S}_i = \mathbf{S}_i + \mathbf{S}_{pos}.$$

(2)

Then, we fuse the three embedded layers as the key and value (Equation (3)).

$$\mathbf{S}_\Sigma = Concat(\mathbf{S}_1, \mathbf{S}_2, \mathbf{S}_3).$$

(3)

2.2.2. Residual Channel Cross-Fusion Transformer

From Figure 1, we know that $\mathbf{S}_i$ is input into the multi-head channel cross attention module, followed by an MLP with a residual structure; the CCT module has been used L times; and we obtained the three outputs of the CCT module $\mathbf{O}_i$. Finally, the fused input feature layer is connected to the fused output feature layer in a residual manner, with $\mathbf{O}$ as the final output feature layer of RCCT. In this way, we learn the dependencies between the different input feature layers in a Transformer cross-fusion way.

As shown in Figure 2, the multi-head cross attention module contains four inputs, that is, three embedded layers as the query matrix and one integrated embedded layer $\mathbf{S}_\Sigma$ as the key $\mathbf{K}$ and the value $\mathbf{V}$.

$$\mathbf{Q}_i = \delta(\mathbf{S}_i)W_{\mathbf{Q}_i}, \mathbf{K} = \mathbf{S}\sum W_{\mathbf{K}}, \mathbf{V} = \mathbf{S}\sum W_{\mathbf{V}}, \tag{4}$$

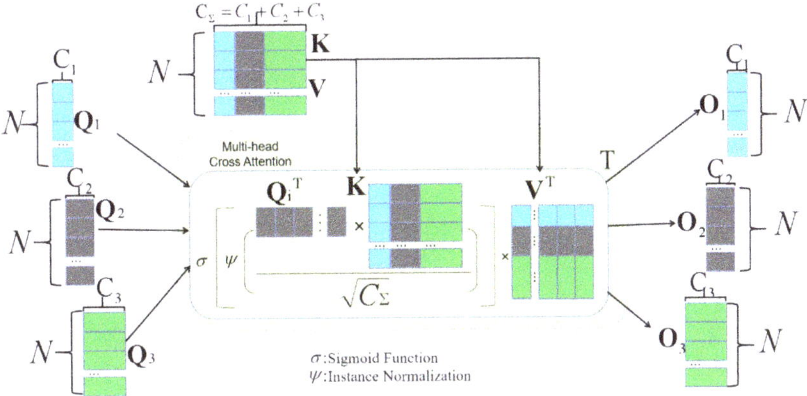

Figure 2. Multi-head cross attention.

$W_{\mathbf{Q}_i} \in \mathbb{R}^{C_i \times C_i}$, $W_{\mathbf{K}} \in \mathbb{R}^{C_\Sigma \times C_\Sigma}$, and $W_{\mathbf{V}} \in \mathbb{R}^{C_\Sigma \times C_\Sigma}$ are different input weights and $\delta(\cdot)$ represents layer normalization. Meanwhile, $\mathbf{Q}_i \in \mathbb{R}^{N \times C_i}$, $\mathbf{K} \in \mathbb{R}^{N \times C_\Sigma}$, and $\mathbf{V} \in \mathbb{R}^{N \times C_\Sigma}$. The formula for cross attention is described as follows:

$$Attention(\mathbf{Q}_i, \mathbf{K}, \mathbf{V}) = \left\{ \sigma \left[\psi \left(\frac{\mathbf{Q}_i^{\mathrm{T}}\mathbf{K}}{\sqrt{C_\Sigma}} \right) \right] \mathbf{V}^{\mathrm{T}} \right\}^{\mathrm{T}}, \tag{5}$$

$\psi(\cdot)$ and $\sigma(\cdot)$ represent the instance normalization and softmax function, respectively. We adopt instance normalization, which can normalize each instance matrix of multi-head attention so that the gradient can spread smoothly. $Attention(\mathbf{Q}_i, \mathbf{K}, \mathbf{V}) \in \mathbb{R}^{N \times C_i}$ is the same size as the input $\mathbf{Q}_i$. Because we have H_n heads' attention, the output result after multi-head cross attention is calculated as follows:

$$MHAttention_i = \left[\begin{array}{c} Attention\left(\mathbf{Q}_i^1, \mathbf{W}, \mathbf{K}\right) + Attention\left(\mathbf{Q}_i^2, \mathbf{W}, \mathbf{K}\right) \\ +,\ldots,+ Attention\left(\mathbf{Q}_i^{H_n}, \mathbf{W}, \mathbf{K}\right) \end{array} \right] / H_n, \tag{6}$$

where H_n is the number of heads. Then, combining MLP and residual operation, the output is as follows:

$$\mathbf{O}_i = \delta(MHAttention_i) + MLP(\mathbf{Q}_i + MHAttention_i), \tag{7}$$

The operation in Equation (7) is repeated L times to establish an L-layer Transformer, in which the output of the L-layer is $\mathbf{O}_i \in \mathbb{R}^{C_i \times N}$. In this paper, H_n and L are both set to 4. The output of the last layer is a multi-scale residual fusion, and the formula is as follows:

$$\mathbf{O} = Concat(\mathbf{S}_1, \mathbf{S}_2, \mathbf{S}_3) + Concat(\mathbf{O}_1, \mathbf{O}_2, \mathbf{O}_3), \tag{8}$$

$\mathbf{O} \in \mathbb{R}^{C_\Sigma \times N}$ is the final output of the RCCT module.

Matrix $\mathbf{O}$ obtains the feature layer $\mathbf{M} \in \mathbb{R}^{C_\Sigma \times \sqrt{N} \times \sqrt{N}}$ through the reshape operation, and we obtain $\mathbf{Z}_1$ through upsampling and the convolutional operation.

2.3. CBAM Module

In the output feature map of the ASPP module, CBAM, as shown in Figure 3, infers the attention map in turn along two independent dimensions (channel and spatial) and then multiplies the attention map and the input feature map for adaptive feature optimization. Now, the channel and spatial attention modules are distinguished.

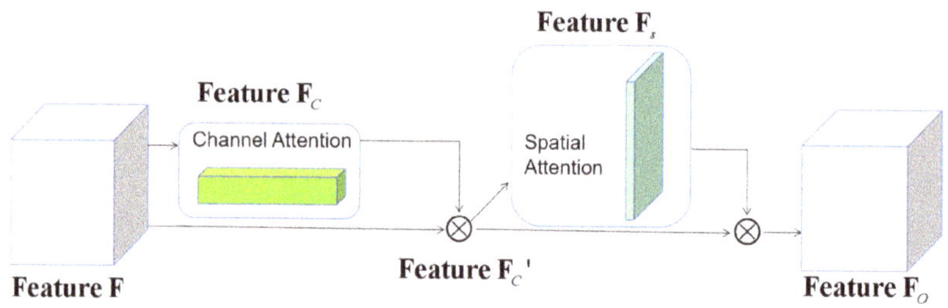

Figure 3. CBAM sketch map.

Figure 4 shows the structure of the channel attention module CA. It models the importance of each feature channel and then enhances or suppresses different channels. The output $\mathbf{F}$ of ASPP is used as the input of channel attention. $\mathbf{F}$ goes through the maximum pooling layer and the average pooling layer. The two outputs are connected to the same MLP, and their parameters are shared. The two outputs of MLP are added, and a sigmoid function is finally connected to obtain $\mathbf{F}_C$.

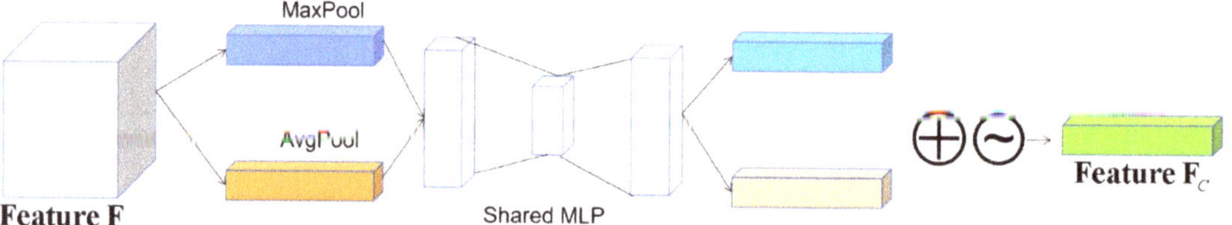

Figure 4. Channel attention module.

Figure 5 depicts the structure of the SA module. Not all regions in the image contribute equally to the task. Only the regions related to the task need to be concerned. The SA model aims to find the most important part of the network for processing. $\mathbf{F}_C{}'$ is the input of SA, $\mathbf{F}_C{}'$ also goes through the maximum pooling layer and the average pooling layer, and their outputs are concatenated on the channel dimension. Then, a convolution layer and a sigmoid function are connected to obtain feature $\mathbf{F}_S$.

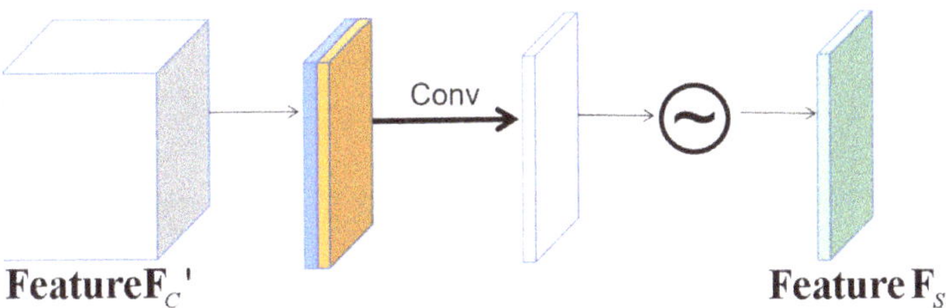

Figure 5. Spatial attention module.

2.4. Dual Encoders of ASPP-CBAM and RCCT

ASPP samples the given input in parallel with a convolution of different dilation rates, which is equivalent to capturing the context of the image at multiple scales. The last feature layer $\mathbf{E}_4 \in \mathbb{R}^{\frac{H \times W}{2^{2 \times (4+1)}} \times C_4}$ of ResNet50 is used as the input of ASPP. $\mathbf{E}_4$ went through four different convolution operations with different dilation rates to extract the features under different receptive fields. ASPP includes a 1×1 convolution layer and three 3×3 dilated convolution, dilated rate = {2, 4, 8}, and pooling layers, as shown in Figure 1. The five feature layers are merged as the output of ASPP and enter the CBAM attention layer. Then, after the upsampling and convolution operations, we obtain the ASPP-CBAM output layer $\mathbf{Z}_2$. Finally, the output layer is merged with the output layer of RCCT $\mathbf{Z}_1$ by the following equation:

$$\mathbf{Z} = Concat(\mathbf{Z}_1, \mathbf{Z}_2), \tag{9}$$

2.5. Data

2.5.1. Self-Made Dataset

The data come from the agricultural remote sensing image (Farmland) taken by UAV, as shown in Figure 6, which is divided into six categories: grassland, construction land, cultivated land, forest land, garden land, and other lands. The image is from an overhead perspective, and the difference between target sizes is large. The drone is a DJI M300 RTK with a flight altitude of 5 km, and a DJI P1 camera with an image size of 8192×5460 pixels and a sensor size of 35.9 mm $\times$ 24 mm, with 45 million effective pixels and an image element size of 4.4μm. In addition, the data were collected in November 2021.

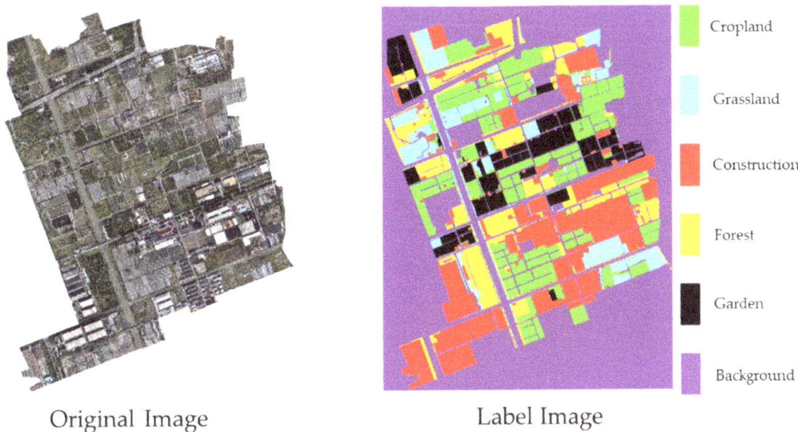

Figure 6. Farmland including raw and labeled images.

To enhance the generalization ability of the model, this study adopts the method of data enhancement. The original remote images and labeled images were first generated by cropping multiple images of size 512×512 pixels, expanding the data set using random rotation, adding noise, flipping, and using other data enhancement methods, as shown in Figure 7. A total of 5000 images were generated, which were then divided into the training and validation sets according to a 4:1 ratio.

Figure 7. Data enhancement methods.

2.5.2. AeroScapes Dataset

The AeroScapes semantic segmentation dataset includes images captured from 5 m to 50 m height using commercial UAVs. This dataset provides 3269 720 pixel $\times$ 1280 pixel resolution images and real land surface labels for 12 classes. Figure 8 shows the information of 11 categories and background categories. The dataset has a variety of perspectives, target scales vary greatly, and there are many small targets.

Figure 8. AeroScapes dataset.

3. Results

3.1. Experimental Environment and Parameter Setting

We use Python as the deep learning framework, JetBrains PyCharm 2021 as the development platform, and Python 3.8 as the development language. All models are trained and tested on computers configured with Intel Core (TM) i7-10700K CPUs and NVIDIA GeForce GTX 3090 Ti graphics cards.

The model uses a poly [33] strategy to reduce the learning rate. The formula is as follows:

$$lr = initial\_lr \times (1 - \frac{epo}{maxepo})^{power},\qquad(10)$$

The initial learning rate *initial_lr* is set to 0.001, and *power* is set to 0.9. The maximum number of iterations *maxepo* is 300, and *epo* represents the current number of iterations. In this study, the batch size is set to 8, and the Stochastic gradient descent SGD [34] optimizer is used to optimize the poly algorithm and network parameters. The model's backbone network Resnet50 uses the trained weight of the dataset ImageNet [35] as the initial weight.

3.2. Evaluation Index and Loss Function

To quantify the effect of the evaluation model, this study uses the most common evaluation index in the field of semantic segmentation: mean Intersection over Union (*mIoU*).

$$mIoU = \frac{1}{c+1}\sum_{i=0}^{c}\frac{p_{ii}}{\sum_{j=0}^{c}p_{ij} + \sum_{j=0}^{c}p_{ij} - p_{ii}},\tag{11}$$

where p_{ii} represents the number of pixels predicted by category i to category i; $c+1$ is the total number of categories; and p_{ij} is the number of pixels predicted by category i to category j.

We also use mean Pixel Accuracy (*mPA*) as another evaluation index.

$$mPA = \frac{1}{c+1}\sum_{i=0}^{c}\frac{p_{ii}}{\sum_{j=0}^{c}p_{ij}},\tag{12}$$

The loss function of Lovasz Softmax [36] is usually used to evaluate semantic segmentation using the Jaccard index [37], also known as the IoU index. In Equation (13), $y*$ is the true label, and y is the predicted value. The Jaccard index of category c is defined as follows:

$$J_c(y*,y) = \frac{|\{y* = c\} \cap \{y = c\}|}{|\{y* = c\} \cup \{y = c\}|},\tag{13}$$

The corresponding loss function is as follows:

$$L_{J_c}(y*,y) = 1 - J_c(y*,y),\tag{14}$$

Figure 9 shows the trend of the training set and validation set of the proposed method on the Farmland dataset with the number of iterations. The curve decreases more rapidly in the first 50 iterations of the loss value and stabilizes after the number of iterations exceeds 250.

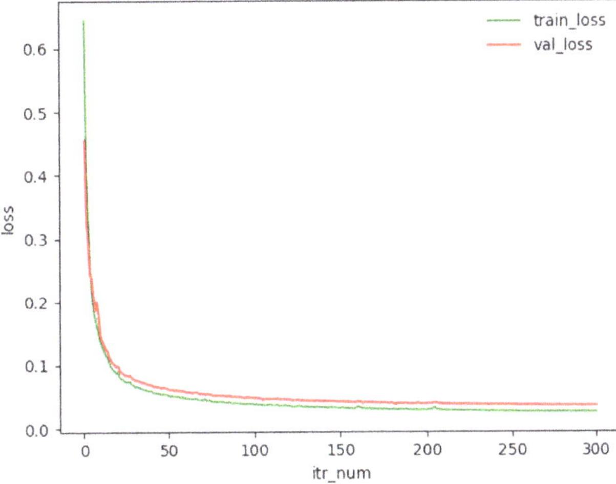

Figure 9. Lovasz Softmax loss function change curve.

3.3. Ablation Experiment of RCCT Module with Different Feature Combinations

To verify the influence of input E_i of different scales on the model, four groups of RCCT module inputs with different combinations are designed for experimental comparison. The four groups of experiments were combined with the ASPP module.

The following conclusions can be drawn through the comparison of four groups of experiments in Table 1:

Table 1. Comparison of different feature combination inputs of the RCCT module.

Input Feature	Farmland		AeroScapes	
	mIoU (%)	mPA (%)	mIoU (%)	mPA (%)
E_1, E_2, E_3	93.97	96.83	60.86	84.20
E_1, E_2	93.46	96.75	60.47	83.61
E_1, E_3	**94.11**	**97.10**	**61.19**	**84.58**
E_2, E_3	94.00	96.97	61.05	84.55

Not all feature inputs can increase the performance of the model, and some features can reduce the accuracy of the model. For example, the mIoU and mPA of the combination E_1, E_2, and E_3 are lower than that of combination E_1 and E_3 and the combination E_2 and E_3, indicating that, when feature input E_1 and E_2 are used together, they will have negative effects. Considering that the combination effect of E_1 and E_3 is better than that of E_2 and E_3, it indicates that E_1 and E_3 are the optimal combination, that is, the effect of E_1 is greater than that of E_2. Therefore, E_2 needs to be removed from the final RCCT module feature input, thus improving the model performance and reducing the model parameters.

A high-level feature is necessary. From the experimental results, combination E_1 and E_2 has the worst effect. As long as E_3 appears in the combination, its mIoU will be 0.5% or higher than that without E_3. In the combination with E_3, mIoU has little difference. Therefore, high-level feature E_3 plays a crucial role in the model.

3.4. Ablation Experiment of Different Attention Combinations in the ASPP Module

The combination of the ASPP module with SA and CA can effectively increase the mIoU and mPA of the model. We have verified the effect of different attentions on the module (Table 2). The input combination of the RCCT module in the experiment is E_1, E_2, and E_3.

Table 2. ASPP module: different attention combinations.

Attention Combination	Farmland		AeroScapes	
	mIoU (%)	mPA (%)	mIoU (%)	mPA (%)
ASPP	93.97	96.83	60.86	84.20
ASPP + CA	94.12	97.06	61.22	84.22
ASPP+ SA	94.02	96.87	61.13	84.25
ASPP + CBAM	**94.14**	**97.12**	**61.30**	**84.36**

From the experiment results in Table 2, the CA and SA modules are more effective than the ASPP without the attention module. The mIoU and mPA values are higher when both attention modules act on ASPP simultaneously than when one attention module is used alone. Therefore, CBAM can effectively improve the performance of ASPP by paying attention to the feature information under different visual fields in space and channel.

3.5. Ablation Experiment of Dual Encoders

To compare the effects of the two encoders on the model, experiments using RCCT module alone and ASPP module alone are designed. Table 3 shows the results. The RCCT used an E_1, E_2, and E_3 input combination in the experiment, and the ASPP did not use the CBAM module.

Table 3. Comparison of two encoder modules.

Module	Farmland		AeroScapes	
	mIoU (%)	mPA (%)	mIoU (%)	Mpa (%)
RCCT	91.52	93.68	59.72	82.18
ASPP	92.72	94.81	60.35	80.65
RCCT + ASPP	**93.97**	**96.83**	**60.86**	**84.20**

Table 3 shows the experimental results of the two independent encoders. The accuracy of the dual encoding structure of RCCT and ASPP is greater than that of either encoding structure, illustrating that both Transformer-based multiscale feature fusion and ASPP with different dilation rates are important components of semantic segmentation of remote images. From another aspect, the accuracy of the Transformer-based RCCT module alone is lower than that of the ASPP module in Farmland, reflecting that ASPP is more capable than RCCT at overhead angle semantic segmentation tasks. Moreover, the difference in mIoU between RCCT and ASPP on AeroScapes is small, but for mPA, RCCT is better than ASPP, so RCCT performs better on multi-angle and more categories datasets.

3.6. Comparative Experiment of Different Network Models

We compared our method with some mainstream remote image semantic segmentation methods, including FCN-8s, UNet, DeepLabV3+, SETR, and UCTransNet. Table 4 shows the experimental results of various models in our dataset and AeroScapes dataset.

Table 4. Comparison of different networks.

Network Model	Farmland		AeroScapes		Model Parameters (M)
	mIoU (%)	mPA (%)	mIoU (%)	mPA (%)	
FCN-8s	92.21	96.43	40.23	78.69	80
UNet	89.38	95.06	42.38	50.41	124
DeepLabV3+	92.80	96.28	59.63	67.07	170
SETR	49.53	64.82	30.63	37.38	348
UCTransNet	92.82	93.27	52.33	81.67	363
RCCT-ASPPNet	**94.14**	**97.12**	**61.30**	**84.36**	411

Table 4 shows that the mIoU of DeepLabV3+ is 92.80% and 59.63% in Farmland and AeroScapes datasets, respectively, which is the best performance among the CNN-based network models. However, the mPA of FCN-8s is 78.69% in AeroScapes datasets, and it is better than DeepLabV3+, but the mIoU of FCN-8s is only 40.23%. Moreover, UCTransNet performs best among Transformer-based models. Our RCCT-ASPPNet network model outperforms the other models with mIoU of 94.14% and 61.30% and mPA of 97.12% and 84.36% for the Farmland and AeroScapes datasets, respectively, which is a dual-encoder structure based on CNN and Transformer.

4. Discussion

4.1. Visual Analysis

Figure 10 shows the effect of Farmland data prediction. The performance of CNN-based FCN-8s, UNet, and DeepLabV3+ is evidently better than that of Transformer-based SETR and UCTransNet. RCCT-ASPPNet combines CNN and Transformer, and its performance is relatively good in Farmland. Figure 11 shows that the prediction results of FCN-8s, UNet, DeepLabV3+, and SETR are relatively poor in AeroScapes. UCTransNet and RCCT-ASPPNet have relatively good prediction effects in (l). Therefore, the improved Transformer model is better than the traditional model. Although RCCT-ASPPNet showed some misdividing in (k), overall, the segmentation effect of RCCT-ASPPNet in different views and small targets was better than UCTransNet. By comparing the three lines of pictures in (g), (h), and (i), our model handles the best details in terms of the prediction results under people's tilt angle of view and top angle.

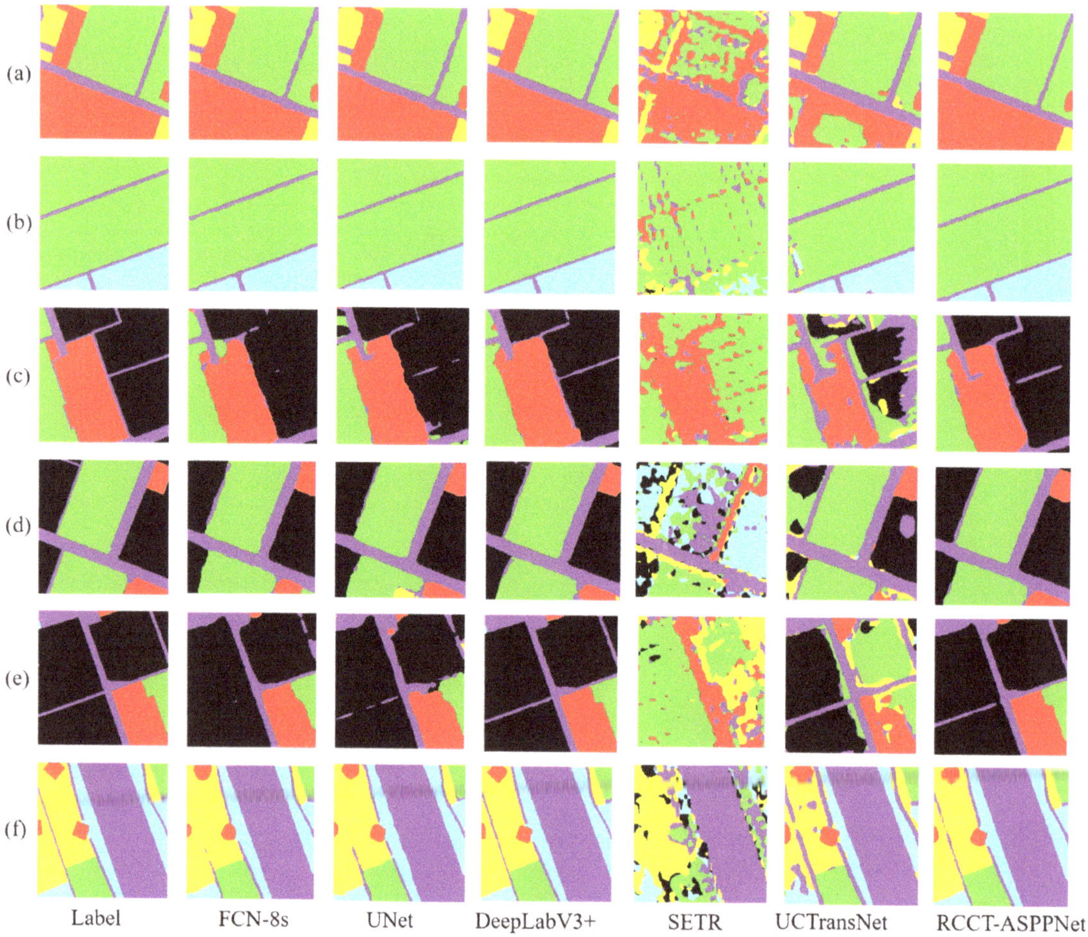

Figure 10. Farmland data prediction results, from left to right, are real labels, FCN, UNet, DeeoLabV3+, SETR, UCTransNet, and RCCT-ASPPNet, from top to bottom, (**a–f**) are the selected 6 sets of Farmland test images.

Combining the performance of the above two datasets, the CNN-based model is more effective at processing the top-view images in the Farmland dataset. However, for the multiple views and more categories in the AeroScapes dataset, the CNN model does not perform well. Moreover, UCTransNet performs the opposite. The combination of Transformer and ASPP can compensate for the shortcomings of each.

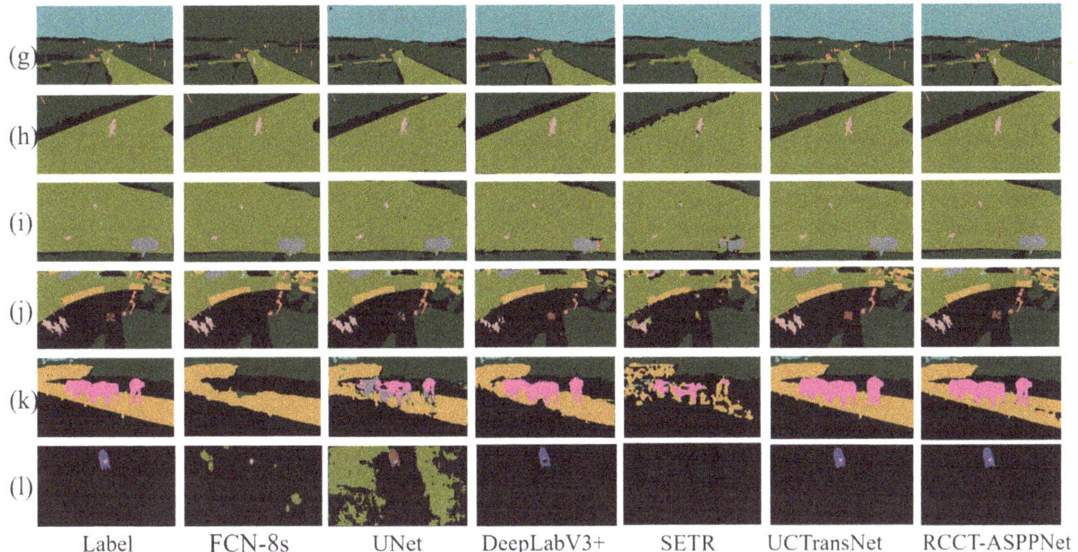

Figure 11. AeroScapes data prediction results, from left to right, are real labels, FCN, UNet, Deeo-LabV3+, SETR, UCTransNet, and RCCT-ASPPNet, from top to bottom, (**g**–**l**) are the selected 6 sets of AeroScapes test images.

4.2. Analysis of Experimental Results

4.2.1. Analysis between Network Models

The performance of the six models on Farmland and AeroScapes is compared, as presented in Table 4. Among all the models, SETR has the worst results on both datasets because the simple Transformer model does not combine multiscale features and loses many low-level semantic features. In addition, the SETR model is more homogeneous and difficult to segment remote images in complex situations. The FCN-8s network model uses multiscale feature fusion, which can effectively learn multiple features while up-sampling using deconvolution, preserving the spatial information of the original input image. However, FCN-8s performs poorly in AeroScapes images with multiple categories and views, with only 40.23% mIoU. The reason is that FCN-8s has difficulty extracting higher-level features and the multiscale fusion approach is simply concatenated, which does not sufficiently learn the relationships among the feature layers. The ASPP module of DeepLabV3+ convolution with higher-level features at different dilate rates extracts the feature information under different fields of view and fuses the lower-level feature layers. The mIoU of DeepLabV3+ on Farmland and AeroScapes is 92.80% and 59.63%, respectively, which achieved good results but at the same time ignored the cross information between feature layers; the mPA of DeepLabV3+ on AeroScapes is only 67.07%. UCTransNet used Transformer to cross-fuse all input feature layers, which solved the problem of insufficient feature fusion, and it has a mPA metric of 81.67% on dataset AeroScapes, which exceeds most network models; this shows that UCTransNet has better performance in handling multi-category and multi-angle datasets. However, the different fields of view of feature layers are equally important; therefore, UCTransNet does not perform as well as the CNN-based model on Farmland. RCCT-ASPPNet cross-fuses some input features and then uses the residual method to connect the front and back feature layers as an encoder. Meanwhile, RCCT-ASPPNet uses the ASPP module to process the high-level features and CBAM to learn the channel and spatial information, which is the second encoder. RCCT-ASPPNet considers the feature fusion method and the field of view information of the feature layer to achieve optimal results on both Farmland and AeroScapes.

In this paper, a dual encoder model was proposed. The first encoder is used to cross-fuse the first three feature layers of ResNet50 using Transformer to learn multi-scale information; it can learn the dependencies of a feature layer with other feature layers. In addition, we added a residual module before and after fusion to prevent gradient disappearance; the second encoder module uses ASPP to process the highest feature layer of ResNet50 to obtain a larger receptive field and uses CBAM to learn its channel attention and spatial attention. In this paper, we used the common evaluation metrics of semantic segmentation, mIoU and mPA, to measure the accuracy of the model. RCCT-ASPPNet outperforms other semantic segmentation models on both Farmland and Aeroscapes in Table 4; in addition, we can see from Figures 10 and 11 that the algorithm in this paper has better segmentation performance in handling small targets and multi-scale objects and using one object in multiple views.

4.2.2. Analysis of Ablation Experiments

By introducing two encoders, the RCCT module and the ASPP module, the experimental design in Table 3 shows that the effect of double coding is more accurate than that of single coding. In addition, this study was designed for the influence of different input feature combinations of RCCT modules on the experimental results. Table 1 shows that the feature combination of $\mathbf{E}_1$ and $\mathbf{E}_3$ is optimal, and the mIoU and mPA on Farmland reach 94.11% and 97.10%, while, 61.19% and 84.58%, respectively, on AeroScapes, so not all features are effective combinations. From another aspect, the experimental results show that the $\mathbf{E}_3$ feature is essential. The experimental design in Table 2 can obtain the impact of different attention mechanisms on the ASPP module. When the attention mechanism is not applicable, mIoU and mPA are 93.97% and 96.83% of the lowest value on Farmland, the same for AeroScapes. When CA or SA is used alone, mIoU and mPA show slight increases, whereas when CBAM is used, the index reaches the highest value. Therefore, CBAM has a certain effect on the ASPP module.

5. Conclusions

In this work, we proposed an effective RCCT-ASPPNet network model for the semantic segmentation of remote image. We used a dual-encoder structure, including a residual multiscale channel cross-fusion Transformer to address multiscale feature fusion and ASPP to address information extraction at different scales on a single feature layer. Extensive experiments evaluated that the proposed model can effectively alleviate the problems of remote images with large-scale variations, small target objects, and diverse viewpoints. RCCT-ASPPNet outperforms the CNN-based DeepLabV3+ and Transformer-based UC-TransNet. Compared with other state-of-the-art remote image semantic segmentation methods, RCCT-ASPPNet's accuracy has a first-class performance.

Although our experimental results have achieved good results, the effects on other data sets are unclear, so we will study the performance of this algorithm on each data set later. From another aspect, Table 4 shows that our model parameters are larger than those of other algorithms, which is very unfriendly for the real-time segmentation. Therefore, future work should balance the accuracy and efficiency of the model.

Author Contributions: Conceptualization, Y.L., Z.C. and C.W.; methodology, Y.L., Z.C. and C.W.; formal analysis, C.W. and J.Z.; data curation, C.W.; writing—original draft preparation, Y.L.; writing—review and editing, Z.C., J.Z. and L.H.; visualization, L.H.; supervision, Z.C. and C.W.; funding acquisition, C.W. and L.H. All authors have read and agreed to the published version of the manuscript.

Funding: This research was funded by Natural Science Foundation of China, grant number 31971789, and Excellent Scientific Research and Innovation Team (2022AH010005), and National Key Research and Development Project, grant number 2017YFB050420.

Data Availability Statement: https://github.com/ishann/aeroscapes (accessed on 26 November 2022); https://pan.baidu.com/s/1wK4qCwqfMOTec2bl7JrMmA?pwd=abcd (accessed on 26 November 2022).

Acknowledgments: We thank all editors and reviewers for their valuable comments and suggestions, which improved this manuscript.

Conflicts of Interest: The authors declare no conflict of interest.

References

1. Garcia-Garcia, A.; Orts-Escolano, S.; Oprea, S.; Villena-Martinez, V.; Garcia-Rodriguez, J. A Review on Deep Learning Techniques Applied to Semantic Segmentation. *arXiv* **2017**, arXiv:1704.06857.
2. Liu, H.; Ye, Q.; Wang, H.; Chen, L.; Yang, J. A Precise and Robust Segmentation-Based Lidar Localization System for Automated Urban Driving. *Remote. Sens.* **2019**, *11*, 1348. [CrossRef]
3. Lai, C.; Yang, Q.; Guo, Y.; Bai, F.; Sun, H. Semantic Segmentation of Panoramic Images for Real-Time Parking Slot Detection. *Remote Sens.* **2022**, *14*, 3874. [CrossRef]
4. Mekyska, J.; Espinosa-Duro, V.; Faundez-Zanuy, M. Face segmentation: A comparison between visible and thermal images. In Proceedings of the 44th Annual 2010 IEEE International Carnahan Conference on Security Technology, San Jose, CA, USA, 5–8 October 2010; pp. 185–189. [CrossRef]
5. Khan, K.; Khan, R.U.; Ahmad, K.; Ali, F.; Kwak, K.-S. Face Segmentation: A Journey from Classical to Deep Learning Paradigm, Approaches, Trends, and Directions. *IEEE Access* **2020**, *8*, 58683–58699. [CrossRef]
6. Masi, I.; Mathai, J.; AbdAlmageed, W. Towards Learning Structure via Consensus for Face Segmentation and Parsing. In Proceedings of the IEEE/CVF Conference on Computer Vision and Pattern Recognition, Seattle, WA, USA, 13–19 June 2020; pp. 5507–5517. [CrossRef]
7. Wang, Y.; Dong, M.; Shen, J.; Wu, Y.; Cheng, S.; Pantic, M. Dynamic Face Video Segmentation via Reinforcement Learning. In Proceedings of the IEEE/CVF Conference on Computer Vision and Pattern Recognition, Seattle, WA, USA, 13–19 June 2020; pp. 6959–6969.
8. Abdelrahman, A.; Viriri, S. Kidney Tumor Semantic Segmentation Using Deep Learning: A Survey of State-of-the-Art. *J. Imaging* **2022**, *8*, 55. [CrossRef]
9. Arbabshirani, M.R.; Dallal, A.H.; Agarwal, C.; Patel, A.; Moore, G. Accurate Segmentation of Lung Fields on Chest Radiographs Using Deep Convolutional Networks. In Proceedings of the Medical Imaging: Image Processing, Orlando, FL, USA, 11–16 February 2017; pp. 37–42.
10. Dai, P.; Dong, L.; Zhang, R.; Zhu, H.; Wu, J.; Yuan, K. Soft-CP: A Credible and Effective Data Augmentation for Semantic Segmentation of Medical Lesions. *arXiv* **2022**. [CrossRef]
11. Wang, J.; Valaee, S. From Whole to Parts: Medical Imaging Semantic Segmentation with Very Imbalanced Data. In Proceedings of the 2019 IEEE Global Communications Conference (GLOBECOM), Waikoloa, HI, USA, 9–13 December 2019; pp. 1–6. [CrossRef]
12. Neupane, B.; Horanont, T.; Aryal, J. Deep Learning-Based Semantic Segmentation of Urban Features in Satellite Images: A Review and Meta-Analysis. *Remote. Sens.* **2021**, *13*, 808. [CrossRef]
13. Peng, B.; Zhang, W.; Hu, Y.; Chu, Q.; Li, Q. LRFFNet: Large Receptive Field Feature Fusion Network for Semantic Segmentation of SAR Images in Building Areas. *Remote. Sens.* **2022**, *14*, 6291. [CrossRef]
14. Li, Y.; Si, Y.; Tong, Z.; He, L.; Zhang, J.; Luo, S.; Gong, Y. MQANet: Multi-Task Quadruple Attention Network of Multi-Object Semantic Segmentation from Remote Sensing Images. *Remote. Sens.* **2022**, *14*, 6256. [CrossRef]
15. Shelhamer, E.; Long, J.; Darrell, T. Fully convolutional networks for semantic segmentation. *IEEE Trans. Pattern Anal. Mach. Intell.* **2017**, *39*, 640–651. [CrossRef]
16. Ronneberger, O.; Fischer, P.; Brox, T. U-Net: Convolutional Networks for Biomedical Image Segmentation. In *Medical Image Computing and Computer-Assisted Intervention—MICCAI 2015*; Lecture Notes in Computer Science; Navab, N., Hornegger, J., Wells, W.M., Frangi, A.F., Eds.; Springer: Cham, Switzerland, 2015; Volume 9351, pp. 234–241.
17. Badrinarayanan, V.; Kendall, A.; Cipolla, R. Segnet: A deep convolutional encoder-decoder architecture for image segmentation. *IEEE Trans. Pattern Anal. Mach. Intell.* **2017**, *39*, 2481–2495. [CrossRef]
18. Chen, L.-C.; Papandreou, G.; Kokkinos, I.; Murphy, K.; Yuille, A.L. Semantic Image Segmentation with Deep Convolutional Nets and Fully Connected CRFs. *arXiv* **2016**, arXiv:1412.7062.
19. Chen, L.-C.; Papandreou, G.; Kokkinos, I.; Murphy, K.; Yuille, A.L. DeepLab: Semantic Image Segmentation with Deep Convolutional Nets, Atrous Convolution, and Fully Connected CRFs. *IEEE Trans. Pattern Anal. Mach. Intell.* **2018**, *40*, 834–848. [CrossRef]
20. Chen, L.-C.; Papandreou, G.; Schroff, F.; Adam, H. Rethinking Atrous Convolution for Semantic Image Segmentation. *arXiv* **2017**, arXiv:1706.05587.
21. Chen, L.-C.; Zhu, Y.; Papandreou, G.; Schroff, F.; Adam, H. Encoder-Decoder with Atrous Separable Convolution for Semantic Image Segmentation. In *Computer Vision–ECCV 2018*; Lecture Notes in Computer Science; Ferrari, V., Hebert, M., Sminchisescu, C., Weiss, Y., Eds.; Springer: Cham, Switzerland, 2018; Volume 11211, pp. 833–851.
22. Zhao, H.; Shi, J.; Qi, X.; Wang, X.; Jia, J. Pyramid Scene Parsing Network. In Proceedings of the 2017 IEEE Conference on Computer Vision and Pattern Recognition (CVPR), Honolulu, HI, USA, 21–26 July 2017; pp. 6230–6239.

23. Zheng, S.; Lu, J.; Zhao, H.; Zhu, X.; Luo, Z.; Wang, Y.; Fu, Y.; Feng, J.; Xiang, T.; Torr, P.H.; et al. Rethinking Semantic Segmentation from a Sequence-to-Sequence Perspective with Transformers. In Proceedings of the IEEE/CVF Conference on Computer Vision and Pattern Recognition, Nashville, TN, USA, 20–25 June 2021; pp. 6877–6886. [CrossRef]
24. Wang, H.; Cao, P.; Wang, J.; Zaiane, O.R. UCTransNet: Rethinking the Skip Connections in U-Net from a Channel-Wise Perspective with Transformer. *Proc. Conf. AAAI Artif. Intell.* **2022**, *36*, 2441–2449. [CrossRef]
25. Dumoulin, V.; Visin, F. A Guide to Convolution Arithmetic for Deep Learning. *arXiv* **2018**, arXiv:1603.07285.
26. Simonyan, K.; Zisserman, A. Very Deep Convolutional Networks for Large-Scale Image Recognition. *arXiv* **2015**, arXiv:1409.1556.
27. Yu, F.; Koltun, V. Multi-Scale Context Aggregation by Dilated Convolutions. *arXiv* **2016**, arXiv:1511.07122, 615.
28. Vaswani, A.; Shazeer, N.; Parmar, N.; Uszkoreit, J.; Jones, L.; Gomez, A.N.; Kaiser, L.; Polosukhin, I. Attention Is All You Need. *Adv. Neural Inf. Process. Syst. arXiv* **2017**. [CrossRef]
29. Cao, H.; Wang, Y.; Chen, J.; Jiang, D.; Zhang, X.; Tian, Q.; Wang, M. Swin-Unet: Unet-like Pure Transformer for Medical Image Segmentation. *arXiv* **2021**, arXiv:2105.05537.
30. He, K.; Zhang, X.; Ren, S.; Sun, J. Deep Residual Learning for Image Recognition. In Proceedings of the 2016 IEEE Conference on Computer Vision and Pattern Recognition (CVPR), Las Vegas, NV, USA, 26 June–1 July 2016; pp. 770–778.
31. Woo, S.; Park, J.; Lee, J.Y.; Kweon, I.S. CBAM: Convolutional block attention module. In Proceedings of the European Conference on Computer Vision, Munich, Germany, 8–14 September 2018. [CrossRef]
32. Nigam, I.; Huang, C.; Ramanan, D. Ensemble Knowledge Transfer for Semantic Segmentation. In Proceedings of the 2018 IEEE Winter Conference on Applications of Computer Vision (WACV), Lake Tahoe, NV, USA, 12–15 March 2018; pp. 1499–1508.
33. Liu, W.; Rabinovich, A.; Berg, A.C. ParseNet: Looking Wider to See Better. *arXiv* **2015**, arXiv:1506.04579.
34. Robbins, H.; Monro, S. A Stochastic Approximation Method. *Ann. Math. Stat.* **1951**, *22*, 400–407. [CrossRef]
35. Deng, J.; Dong, W.; Socher, R.; Li, L.-J.; Li, K.; Li, F.-F. ImageNet: A Large-Scale Hierarchical Image Database. In Proceedings of the 2009 IEEE Conference on Computer Vision and Pattern Recognition, Miami, FL, USA, 20–25 June 2009; pp. 248–255.
36. Berman, M.; Triki, A.R.; Blaschko, M.B. The Lovász-Softmax Loss: A Tractable Surrogate for the Optimization of the Inter-section-over-Union Measure in Neural Networks. In Proceedings of the IEEE Conference on Computer Vision and Pattern Recognition, Salt Lake City, UT, USA, 18–22 June 2018; pp. 4413–4421.
37. Jaccard, P. The Distribution of The Flora in The Alpine Zone. *New Phytol.* **1912**, *11*, 37–50. [CrossRef]

Article

Capacity Estimation of Solar Farms Using Deep Learning on High-Resolution Satellite Imagery

Rashmi Ravishankar [1,*,†], Elaf AlMahmoud [2,†], Abdulelah Habib [2] and Olivier L. de Weck [1]

[1] Massachusetts Institute of Technology, Cambridge, MA 02139, USA
[2] King Abdulaziz City for Science and Technology (KACST), Riyadh 12354, Saudi Arabia
*   Correspondence: rashmir@mit.edu; Tel.: +1-617-2299577
†   These authors contributed equally to this work.

Abstract: Global solar photovoltaic capacity has consistently doubled every 18 months over the last two decades, going from 0.3 GW in 2000 to 643 GW in 2019, and is forecast to reach 4240 GW by 2040. However, these numbers are uncertain, and virtually all reporting on deployments lacks a unified source of either information or validation. In this paper, we propose, optimize, and validate a deep learning framework to detect and map solar farms using a state-of-the-art semantic segmentation convolutional neural network applied to satellite imagery. As a final step in the pipeline, we propose a model to estimate the energy generation capacity of the detected solar energy facilities. Objectively, the deep learning model achieved highly competitive performance indicators, including a mean accuracy of 96.87%, and a Jaccard Index (intersection over union of classified pixels) score of 95.5%. Subjectively, it was found to detect spaces between panels producing a segmentation output at a sub-farm level that was better than human labeling. Finally, the detected areas and predicted generation capacities were validated against publicly available data to within an average error of 4.5% Deep learning applied specifically for the detection and mapping of solar farms is an active area of research, and this deep learning capacity evaluation pipeline is one of the first of its kind. We also share an original dataset of overhead solar farm satellite imagery comprising 23,000 images (256 × 256 pixels each), and the corresponding labels upon which the machine learning model was trained.

Keywords: convolutional neural network; deep learning; computer vision; solar farm; solar panel; capacity estimation; photovoltaics; remote sensing; optical remote sensing

Citation: Ravishankar, R.; AlMahmoud, E.; Habib, A.; de Weck, O.L. Capacity Estimation of Solar Farms Using Deep Learning on High-Resolution Satellite Imagery. *Remote Sens.* **2023**, *15*, 210. https://doi.org/10.3390/rs15010210

Academic Editors: Gwanggil Jeon and Silvia Liberata Ullo

Received: 11 October 2022
Revised: 28 November 2022
Accepted: 15 December 2022
Published: 30 December 2022

1. Introduction

1.1. Motivation

The sharp increase in photovoltaic panel adoption has resulted in photovoltaic installations becoming a key contribution to renewable energy production, first through residential deployment, and subsequently through commercial solar farms. The reasons for this significant rise include the global push for renewables (the UN Sustainable Development Goals being a recognizable example [1]), coupled with the steadily decreasing cost of each unit of electricity produced (the global average cost of renewable energy has dropped by 89% for solar equipment since 2009 [2]). Figure 1 shows the official numbers and targets of various countries over the last decade. There has been a clear exponential trend over the last two decades, going from 0.3 GW in 2000 to 3.5 GW in 2009 to 63.5 GW in 2019, and a forecast to reach 4240 GW by 2040 [2]). Currently, the global solar capacity doubles every 18 months [2]. It is estimated that at least $400 million is being invested annually into commercial solar energy generation [3]. The International Solar Alliance has 180 member countries as of 2022, and has committed one trillion dollars as an investment target [4].

The generation behavior of renewables such as solar and wind reflects the uncertainty and complexity of the natural world. The inherent decentralized nature of the deployment has resulted in a dearth of traceable data to better understand the demographic, geographic, and regional trends. Satellite imagery provides an opportunity to track this inherently decentralized deployment at scale and with granularity, objectivity, and in potential real-time, which could be an instrumental tool that informs both policymakers and industries of the state of PV deployment by region. Enhancing the diffusion of PV solar energy generation is aligned with the UN Sustainable Development Goals (SDGs), specifically goal 7—to "Ensure access to affordable, reliable, sustainable and modern energy for all". Detailed asset-level data, including the spatial arrangement of installations, are particularly required to address the challenges of generation and planning faced by electricity system operators and electricity market operators and participants.

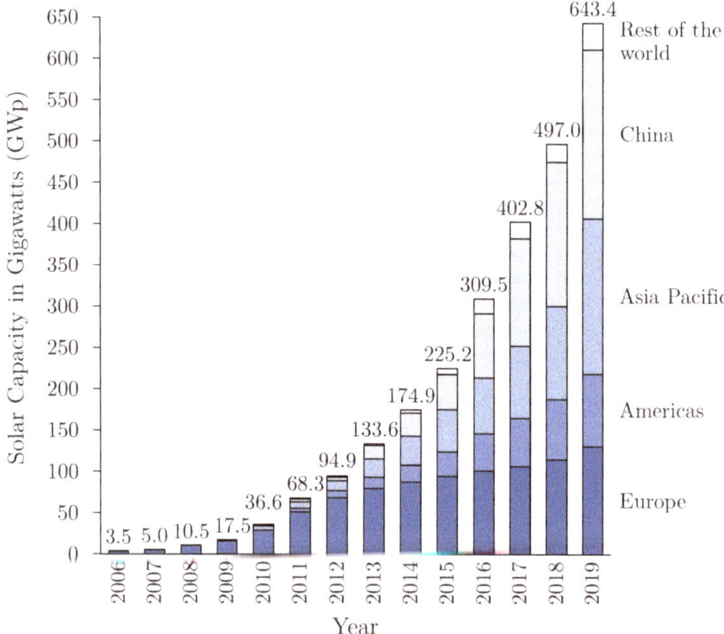

Figure 1. Growth of global photovoltaic capacity has been exponential over the last two decades, from 0.3 GW in 2000 to 63.5 GW in 2019, roughly doubling every 18 months [2,3].

Existing databases of solar generating capacity are insufficient to address databasing needs because they are either aggregated (for example, those of the IEA2, IRENA3, or BP1), limited in geographical scope (for example, Google OpenPV, DeepSolar [5], or SolarNet [6]), or are not geospatially localized (for example, S&P Global World Electric Power Plant Database [2]), and/or are not publicly available for the research and policy community (for example, IHS's Electric Plants).

This work aims to scientifically develop and test a globally generalizable approach for the detection and capacity evaluation of medium- and large-scale photovoltaic solar farms with state-of-the-art accuracy. This can be considered as a segmentation or pixel-level classification problem showing great potential for applying deep learning techniques to analyze remote sensing tasks. According to SolarNet [6], solar farm detection is more challenging than rooftop solar panel detection, because of the confusing backgrounds in which they are found. We use remote sensing and deep learning to detect solar farms—both their existence and precise boundaries—to estimate the energy generation capacities of individual facilities in an accurate manner, using publicly available satellite data and limited computational expense. These values are used to triangulate self-reported information, validate capacity figures, and even to identify real-world inefficiencies.

1.2. Previous Work

Identifying, understanding, and mapping renewables deployment is a topic that has gained interest in recent years. A variety of methods have been proposed to detect first residential and subsequently commercial photovoltaics from remote sensing images. Admittedly, rooftop detection is the more interesting case, given that they are more dispersed and not reported, but as commercial solar deployment becomes more widespread, the latter problem has developed into one of both intellectual and practical interest.

Stanford Deepsolar [5] kick-started interest in this field by proposing a deep learning framework to map residential rooftop solar panels for the US. DeepSolar utilized transfer learning to train a CNN classifier on imagery from Google Static Maps, and detected over 1.47 million PV installations in urban areas throughout the US with a precision of 93.1% and a recall of 88.5%. However, commercial solar deployment was not addressed by DeepSolar. Prior to that, rule-based efforts at detecting PV installations have not been able to achieve very high levels of precision and recall [7].

More recently, SolarNet [6] proposed an expectation maximization attention network to recognize solar farms on satellite imagery in China. In their paper, the authors compare the two most popular networks, UNet and EMANet, and combine the strengths of both to come up with their own SolarNet, which is a combination of the two. SolarNet was limited by geography and did not evaluate the capacity, or report semantic segmentation evaluation metrics such as the Jaccard Index. The detections by SolarNet and by Kruitwagen et al. [3] were at the bounding box/convex hull level for each solar farm. This is useful to achieve an upper bound on true solar capacity, but tends to overestimate the true solar capacity of an installation. Prior efforts also did not make the underlying data sets fully public.

1.3. Problem Statement

In this research, we seek to answer the following questions:

1. How do we best use deep learning to extract detected polygon areas containing solar farms from satellite imagery?
2. Apart from verifying the existence and geographic location of a solar farm, can we estimate the number of individual panels?
3. What is the best way to use this information to predict how much solar energy is generated annually?

We show how to extract this information from satellite imagery and to validate both the detected areas and generation capacities against publicly available data, including the electricity generation data reported by solar farm management.

1.4. Contributions

In this paper, we propose, optimize, and validate a deep learning based framework to detect and map solar farms across different geographies using a state-of-the-art semantic segmentation convolutional neural network-based pipeline. Semantic segmentation enables the precise localization of solar panel areas from satellite imagery for a more accurate estimate of the deployment area. As a final step in the pipeline, we develop a multi-step capacity evaluation model to estimate the number of panels and the energy generation capacity of the detected solar energy facilities.

The final question of the problem statement addresses the real world consequential information that can be extracted from the output polygons of the model. We develop a capacity evaluation model that starts where the deep learning problem ends, and demonstrate on some sample solar farms, verifying against real-world reported data. Deep learning applied specifically for the detection and mapping of solar farms is an active area of research, and this deep learning capacity evaluation pipeline is the first of its kind. Prior work in using satellite and aerial imagery has estimated the solar farm size, but not its estimated annual energy production capacity.

In summary:

1. We present a deep learning model capable of solar farm detection that achieves highly competitive performance metrics, including a mean accuracy of 96.87%, and a Jaccard Index (intersection over union of classified pixels) score of 95.5%.
2. Subjectively, our model was found to detect spaces between panels and pathways between panel rows producing a segmentation output that is better than human labeling. This has resulted in some of the most accurate detections in comparison with the existing literature.
3. We share the original, pixel-wise labeled dataset of solar farms comprising 23,000 images (256×256 pixels each) on which the model was trained.
4. Finally, we propose an original capacity evaluation model—extracting panel count, panel area, energy generation estimates, etc., of the detected solar energy facilities that were validated against publicly available data to within an average 4.5% error.

2. Materials and Methods

The capacity evaluation pipeline proposed in this paper comprises dataset creation, the deep learning model, and the capacity evaluation model. Our deep learning model was trained on an original dataset created by collecting the satellite imagery of several major solar farms in the US, and tested on images of farms unseen by the model. Data augmentation and ablation studies were performed to check the model's robustness to complex backgrounds and edge cases. This computationally intensive task of training was carried out with the help of the MIT Supercloud using a minimum of 2056 processors. Finally, the output polygons detected by the model were fed into the capacity evaluation model for further analysis.

2.1. Dataset

Seen in Figure 2 is what a typical solar farm looks like from space. The imperial county solar farm in Southeast California, close to the Mexico border, was all farmland in 2012, and has seen progressive development over the following years. Each of these images is a mosaic of geotiff tiles and serves as our source of data. Note that while it appears to be encroaching on farmland (one of the major criticisms of solar energy), the facility is actually in the middle of the arid Mojave desert and encroaching on highly irrigation intensive farms. The tradeoff in land use between farming and energy is an interesting use case but is beyond the scope of this paper.

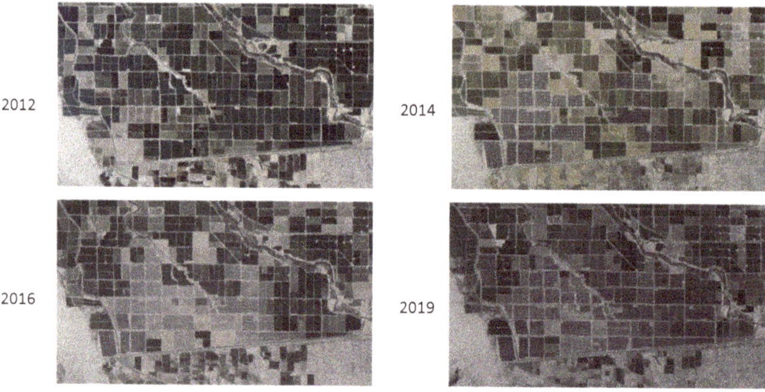

Figure 2. Growth of Mount Signal Solar in Imperial Valley, California, into one of the world's largest solar farms, over the last decade. Satellite imagery allows for a qualitative and quantitative "big picture" view of solar farms.

The first step in the process was to evaluate the sources of satellite imagery suitable for building a dataset of labeled images on which to train and test a deep learning model. In order to create our own dataset of imagery for this purpose, a number of satellite imagery sources were explored, with the criteria being resolution and availability across geographies. Sources range from freely accessible satellite imagery, low-resolution imagery from publicly owned assets (such as NASA's Landsat series of satellites), etc., to higher-resolution images from commercial resources like Planet, DigitalGlobe's WorldView, or ArcGIS. For the needs of this project, the USDA NAIP repository [8] (0.6 m GSD) sourced via USGS Earth Explorer [9] was chosen for analysis and dataset creation because it satisfied both the criteria of adequate resolution and uniform availability across the US.

Overhead imagery was collected, and detailed annotation was carried out on 10 major solar farms across the US (the annotated imagery of a few solar farms is included in Appendix A for reference). Solar farm areas were manually labeled to be used as ground truth (this is machine learning terminology, not the remote sensing definition) This is known as annotation. Annotation encompasses the negative labeling of nearby agricultural, semi-urban, and topographical relief systems. This was achieved using an open source tool called QGIS that helps build on geotiff files and creates masks that were then used as ground truth. A visual representation of the labeling process that was involved in dataset creation is seen in Figure 3.

Figure 3. The dataset annotation process.

Next, the large geotiff imagery was patchified into 256 × 256 patches, forming the basis of our novel dataset (one of the contributions of this paper) of about 23,500 labeled images in total for training, validating, and testing. Certain solar farms were set aside in entirety for testing so that the model could be evaluated on solar farms previously unseen by the model. Table 1 gives an overview of the composition of the dataset.

Table 1. Composition of the dataset. Some labeled and unlabeled solar farms are reserved exclusively for testing as an additional check for robustness and generalizability. The labeled portion of the dataset consists of 23,500 images, along with their corresponding labels/masks ready for training.

Solar Farm	Location	Capacity (mW)	Train/Test	Images	Labels
Mount Signal	Imperial County, CA, 32°40'24''N, 115°38'23''W	1165	Train	4000	4000
Techren Solar	Boulder, NV, 35°47'N, 114°59'W	700	Train	2500	2500
Topaz Solar	San Luis Obispo, CA, 35°23'N, 120°4'W	550	Train	6000	6000
Copper Mountain Solar	El Dorado, NV 35°47'N, 114°59'W	298	Train	2500	2500
Desert Sunlight	Desert Center, CA, 33°49'33''N, 115°24'08''W	1287	Test	4500	4500
Agua Caliente	Yuma County, AZ, 32°57.2'N, 113°29.4'W	740	Test	4500	4000
Solar Star	Rosamond, CA, 34°49'50''N, 118°23'53''W	831	Test	4000	-
Springbok	Kern county , CA, 35.25°N, 117.96°W	717	Test	4500	-
Great Valley Solar	Fresno County, CA, 36°34'52''N, 120°22'46''W	200	Test	4000	-
Mesquite	Maricopa County, AZ, 33°20'N, 112°55'W	400	Test	2000	-

2.2. Dataset Augmentation

Ideally, a robust convolutional neural network (CNN) should be able to classify objects even when they are positioned in different orientations or translations. However, CNNs are not architecturally invariant to translation, size, or illumination. In fact, several studies have found that these networks systematically fail to recognize new objects in untrained locations or orientations [10].

This is where data augmentation becomes essential. We account for the amount and diversity of data by training a neural network with additional synthetically modified data without actually collecting or labeling new data. This means applying minor alterations and changes to our existing dataset so that variations of the training set images are more likely to be seen by the model, dramatically improving subsequent generalization.

In this study, we augmented our dataset using contrast matching to bring out subtle differences in shade and to create a higher contrast image, as well as some commonly used morphological transformations in image processing, such as random rotations in 45 and 90 degree increments, and flipping the image horizontally and vertically with a 50% probability.

It is observed that augmentation techniques play a positive role in precise detection. Qualitative effects of image augmentation can be observed in the figure in Section 3.3.

2.3. Deep Learning Model Architecture

The structure of this problem calls for the use of a pixel-wise classifier, otherwise known as a semantic segmentation convolutional neural network (CNN). Semantic segmentation enables the precise localization of solar panel areas from satellite imagery for the most accurate estimate of the deployment area. This is because the output is a mask, rather than just a classification or bounding box. A standard CNN can classify a full image as containing a certain object. A bounding box level classifier will localize the detected object to within a square or rectangular box. A pixel-to-pixel classifier, however, can identify which pixel(s) of the image contains the object of interest, thus resulting in an output polygon of arbitrary shape. Since we are interested in the exact panel area of facilities, a pixel-level classifier can give us the most accurate area estimate. Similar problems have been addressed in [11,12] that used semantic segmentation convolutional neural networks for various purposes. The architecture of a CNN for semantic segmentation differs from the classification/bounding box CNNs, in that the output is at the pixel level. The choice to use such a CNN comes with the additional burden of requiring pixel-to-pixel labels for the datset. A semantic CNN also needs less data to train because the training labels specify exactly what to look for in the imagery.

An established CNN used as a benchmark semantic segmentation model is known as the "UNet", which is a traditional patch classification method first proposed in 2014. It gets its name from its architecture (U shaped) that contains two paths. The first path is the contraction path (also known as the encoder), which is used to capture the context in the image. The second path is the symmetric expanding path (also known as the decoder), which is used to enable precise localization. This is how U-Net combines low-level detail information and high-level semantic information. This architecture produces a prediction for each pixel, while retaining the spatial information in the original input image. The key to doing this is to change the last step of a CNN, making it fully convolutional instead of fully connected. This is why the UNet is an FCN (fully convolutional network), not a CNN (convolutional neural network).

Figure 4 visualizes the generalized architecture of UNet. It is similar to a CNN at every layer, except the final step, which is a 1×1 convolution used to map the channels to the desired number of classes retaining the pixel-to-pixel structure in the output. For comparison, a convolutional neural network (CNN) adopts the fully connected layer to obtain fixed-length feature vectors for classification. Instead of this, the deconvolution layer of FCN performs the feature map of the last volume-based layer. The UNet architecture that stems from FCN is used as a baseline model, and the network architecture is illustrated in Figure 4.

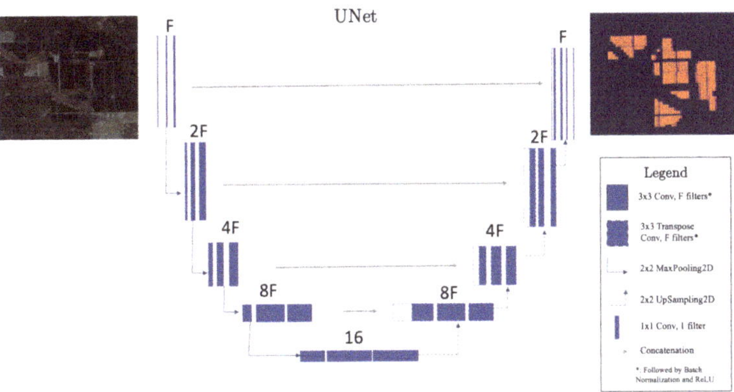

Figure 4. UNet architecture for solar farm detection. F = 64.

For this research, a deep learning model was developed using the open-source PyTorch library running in Python 3.7. We chose a UNet architecture with F = 64, which gives us a model with 1,940,000 trainable parameters—F was initially chosen based on the literature, and the parameters were finetuned until the best metrics were achieved. All FCN architectures explored were common in their utilization of normalized CMYK satellite images as input.

2.4. Model Evaluation

In order to properly train and test the proposed segmentation method, training images are generated by cropping the large original image tiles into patches of "digestible size", and these are fed into the network to learn the parameters. For deployment on larger images during the testing phase, the output masks can be stitched together as depicted in Figure 5, to conform with the input image, no matter the size. No data augmentation was used during initial training. The model was trained with an empirically optimal minibatch size of 10. The learning rate was initially set to 0.001 and then reduced to 0.1. The network converged in roughly 20–30 epochs.

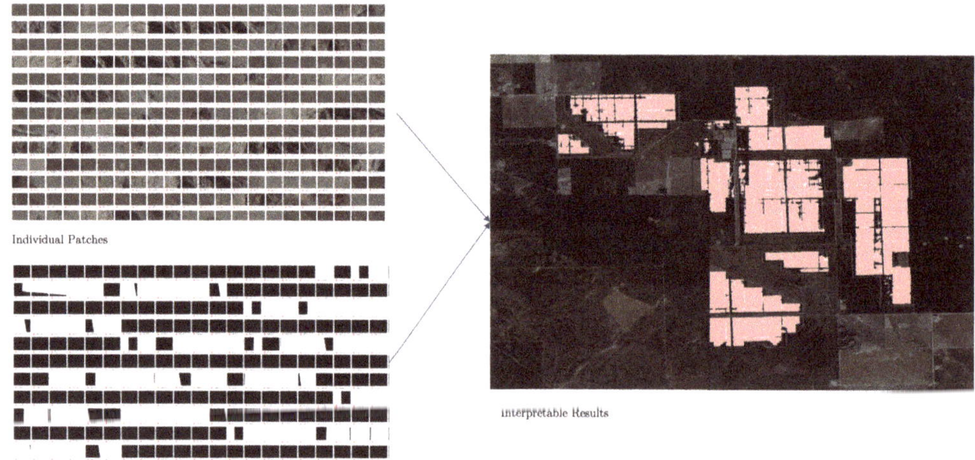

Figure 5. Postprocessing—hundreds of individual images were stitched together to visualize detected solar farm areas.

The metrics for evaluating any semantic segmentation model differ slightly from those of a CNN used for classification problems. Rather than precision and recall (completeness and correctness), insight is gleaned from metrics called pAcc (pixel accuracy), mAcc (mean accuracy), and the Intersection over Union (IoU)/Jaccard index.

Pixel accuracy is a metric that denotes the percent of pixels that are accurately classified in the image. This metric calculates the ratio between the amount of adequately classified pixels and the total number of pixels in the image as

$$pAcc = \frac{correctly\ classified\ pixels}{total\ pixels}$$

The mean accuracy is a metric that denotes the percent of images that are accurately classified in the dataset. This metric calculates the ratio between the amount of adequately classified images and the total number of images in the image as

$$mAcc = \frac{correctly\ classified\ images}{total\ images}$$

In semantic segmentation, a correctly classified image is hard to define. It is typically a threshold, say, more than half of the image is correctly segmented. As a consequence, poor detections can pass through this metric, making it more generous and less informative than what is needed. The most exacting metric is the intersection over union score, also known as the Jaccard similarity coefficient, a statistic that is used for gauging the similarity of the detected shape against its label.

$$IoU = \frac{A \cap B}{A \cup B}$$

2.5. Capacity Evaluation Model

In order to maximize utility to stakeholders, the final step in the proposed pipeline—the capacity evaluation model—explores the extraction of further information from remotely sensed solar energy facilities. Beyond verifying the existence and geographic location of the farms, can we estimate or count the number of panels? How do we predict how much solar energy is generated annually? These values can be used to triangulate self reported information, validate capacity figures, and even identify real-world inefficiencies.

The capacity evaluation model proposed in this section is a compound model of three independent steps—the accurate estimation of deployment area, the estimate of the number of panels, and finally, the evaluation of the energy production capacity of the facility. As depicted in Figure 6, the polygons detected via a deep learning pipeline are used to estimate the "convex hull" area of the facility. Next, the area estimate is distilled down to an estimate of panel area, and consequently, the number of panels. Finally, the energy production capacity is evaluated using a standard formula that includes efficiency, location (weather effects), and/or capacity factor. The area estimate hinges on model accuracy and quality of imagery; the estimate of the number of panels depends on panel dimensions, packing density, and axis system type. Finally, the energy production number depends on capacity factor, which in turn is governed by location, weather effects, panel efficiency, and so on.

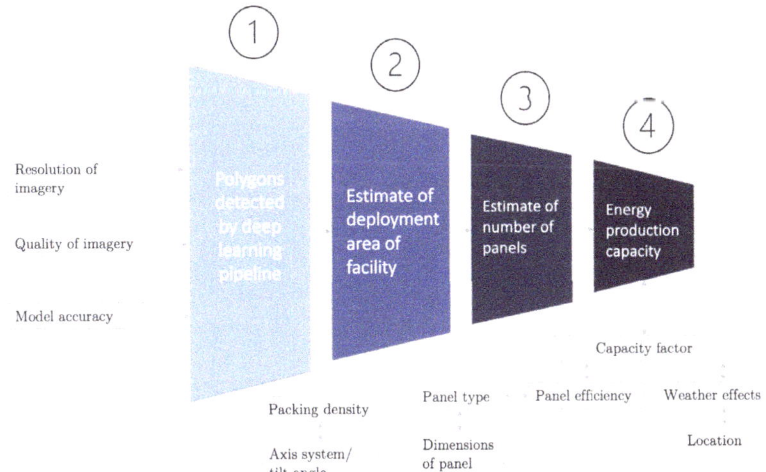

Figure 6. Depiction of the capacity evaluation model for solar farms. The goal is to estimate annual energy generation from polygons detected by the deep learning model on remotely sensed imagery.

Two approaches are explored to arrive at a capacity estimate. The first is a formula that uses a published capacity factor for a given geographic location, or the farm itself, if the number is available. The second, more complex method independent of assumptions, is based on NREL's PySAM model [13].

The capacity factor (CF), used in the first method, is defined as the ratio of actual energy delivered over a period of time over the maximum possible as per the rated capacity of a power plant operating non-stop. The typical capacity factors of most farms in the world range between 30 and 40%, while those in the Mohave desert are more specifically clustered at around 33–37%. CF depends on the geographic location and varies based on the actual weather events for a particular year.

$$Capacity\ Factor = \frac{Annual\ Energy\ Production\ (kWh/year)}{System\ Rated\ Capacity\ (kWh/h) \times 24\ (h/day) \times 365\ (days/year)} \tag{1}$$

Alternately, a more complex route may be taken that is independent of assumptions, and that is based on NREL's PySAM model. NREL's PySAM model uses a large number of criteria, including actual hourly meteorological data (horizontal irradiance, normal irradiance, diffuse irradiance, dew point, surface albedo, temperature, relative humidity, solar zenith angle...) to arrive at the energy generated by a panel on a given day. This estimate can then be fed into the model to calculate the actual annual production instead of the capacity factor method, which is an extrapolation from day to year.

The methodology is visualized in Figure 6, and step-by-step calculations and results are elaborated in the tables in Section 3.4. The method is as follows: first, the polygons detected by the deep learning pipeline are used to estimate the "convex hull" area of the facility, which is brought down to panel area using a packing density. The model accuracy, resolution of imagery available, and quality of imagery directly affect this number. Next, the area estimate is distilled down to an estimate of the panel area, and consequently, the number of panels.

$$Number\ of\ Panels = \frac{Total\ Panel\ Area}{Area\ per\ Panel} \tag{2}$$

$$\implies Number\ of\ Panels = \frac{Number\ of\ Pixels \times (Area/Pixel) \times Packing\ Density}{Area\ of\ Panel} \tag{3}$$

Ultimately, the energy production capacity is evaluated using Equation (1) as:

$$Annual\ Capacity\ (kWh/year) = CF \times System\ Rated\ Capacity\ (kW) \times 24 \times 365 \tag{4}$$

where,

$$System\ Rated\ Capacity\ (kW) = Panel\ Rated\ Capacity\ (kW) \times Number\ of\ Panels \tag{5}$$

3. Results

Summarized in Table 2 are the performance metrics achieved by our best model. Our best performing model produced a semantic segmentation output that is better than human labeling, and the patches can be seen in Figure 7. The segmentation performance on various full solar farms can be seen in Figure 8.

Table 2. Results—key performance metrics of the CNN.

Metric	Description	Result
pAcc (Pixel Accuracy)	Correctly classified pixels/total pixels	99.19%
mAcc (Mean Accuracy)	Mean accuracy considering optimal threshold	96.87%
mIoU (Mean IoU/Jaccard Index)	Overlap between mask and prediction	95.5%
fIoU (Frequency corrected IoU)	IoU reported for each class and weighted	97%

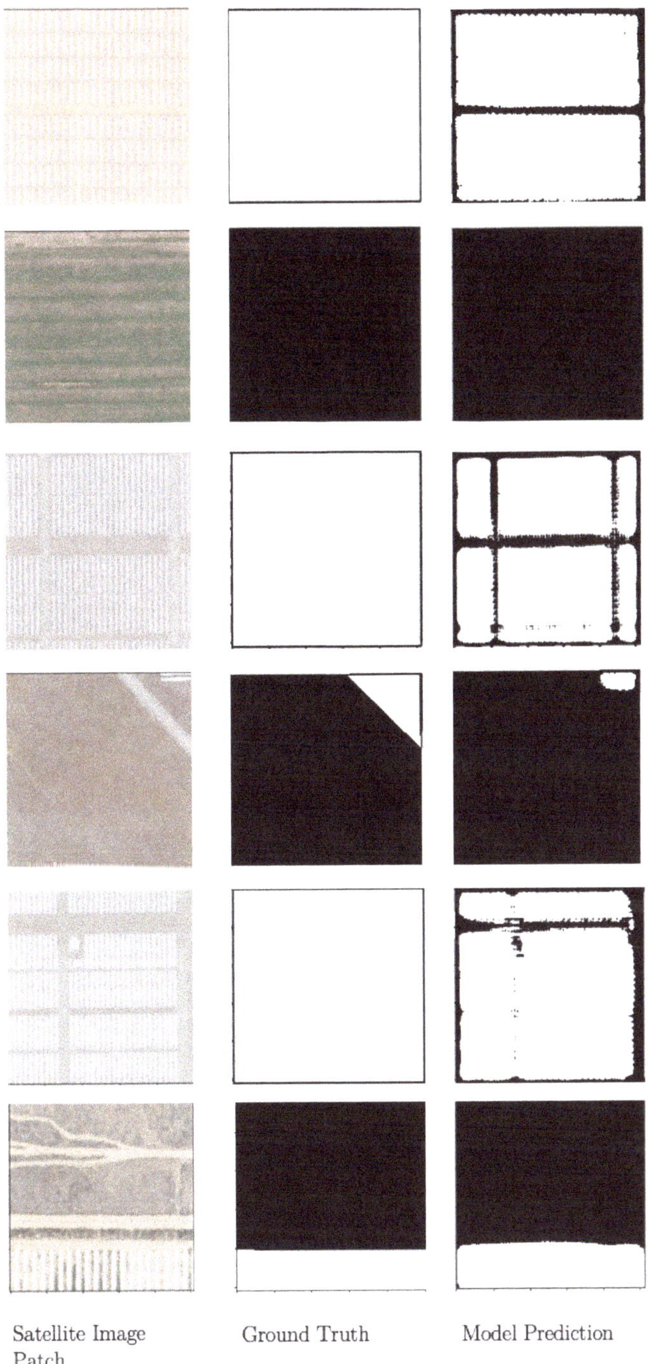

Figure 7. Predictions on individual patches (before postprocessing) show clearer outputs than human labeling (ground truth).

Figure 8. Segmentation performance on various test solar farms. Comparison of the confidence masks between teacher confidence (in black and white) and the student confidence (in color) shows that the model produces an output with better veracity than human labeling.

3.1. Performance Metrics

The best performing model achieved a mean accuracy of 96.87% and an mIoU of 95.5%. For comparison, solarNet achieved an mIoU score of 94.2%. The high IoU score is supported by Figures 7 and 8, which illustrate how the model is able to identify nuances within the solar farm at a sub-farm level, such as spaces between panel rows, pathways, and maintenance blocks.

3.2. Effect of Confidence Threshold

The IoU threshold is the confidence value at which a pixel is considered to be classified as containing photovoltaics. In standard practice, >0.5 confidence is considered as a positive prediction. A classification threshold is analogous to saying that there are higher/lower standards for accepting a pixel as yes/no. Seen in Figure 9 is the variation of the IoU score with the IoU threshold. As expected, there is a decline as the cutoff is made tighter. This can be interpreted in two ways. One, that the model is confident in its predictions, as IoU score only drops fast as the cutoff approaches 1. The best IoU score, 95.5%, was achieved with a cutoff at 0.4, which means the model is balanced but slightly more confident of negative predictions.

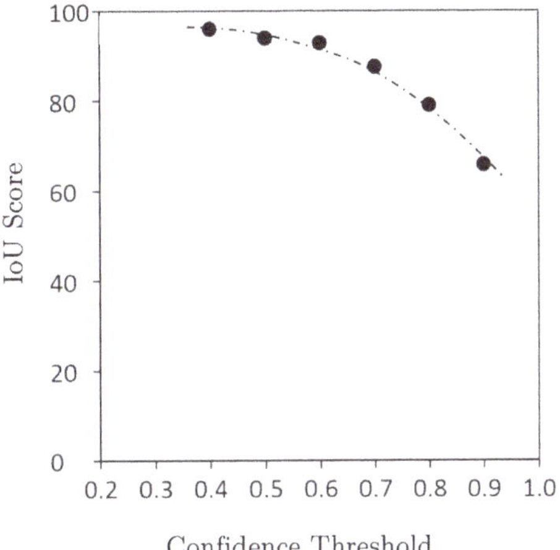

Figure 9. The variation of IoU score with confidence cutoff/threshold.

3.3. Effect of Image Augmentation

It can be qualitatively observed from Figure 10 that the augmentation techniques play a positive role in precise detection. As seen in Figure 10, the model detections are noisy before training the model on the augmented images. The results after applying the augmentation techniques (elucidated in the Methodology section) show that augmentation not only reduced the amount of noise, but was able to progressively help the model learn the essence of a solar farm.

Figure 10. Visualization of early improvement in the model with image and dataset augmentation.

3.4. Capacity Evaluation

The previous sections have affirmed our ability to input satellite imagery and extract detected polygon areas containing solar farms using a deep-learning-based pipeline. The capacity evaluation model developed in this paper comprises three pieces—the accurate estimation of the deployment area, the estimate of the number of panels, and finally, the evaluation of the energy production capacity of the facility. These steps are visualized in the model diagram in Figure 6, and are enumerated and presented in Tables 3–5.

Illustrated in the following tables are some case studies of the model applied to US solar farms. The results are presented step-by-step and are compared with reported numbers. Table 3 depicts the area detection, the first step in the capacity evaluation pipeline. The pixel count is multiplied by the square of the resolution (0.36 (m^2)) to arrive at the area estimate. Note that the differences between the detected area and the reported area are accounted for by the fact that the detected area comes from purely panel outlines detected by the CNN, whereas the reported area is a number from a commercial point of view—the area operated by farm management—and therefore, includes peripheral area, ongoing work, pathways, etc.

Table 3. Area Detection—the first step in capacity evaluation. Note that detected area is purely panel outlines while reported area includes peripheral area.

Solar Farm	Pixels Counted *Mil*	Area Detected (km^2)	Area Reported (km^2)	Panel Area (km^2)
Mount Signal	34.27	12.34	15.9	4.93
Agua Caliente	21.65	7.79	9.7	3.12
Desert Sunlight	38.53	13.87	16	5.55
Solar Star	25.33	9.12	13	3.65
Springbok	18.33	5.52	5.7	2.21

Table 4 depicts the estimate of the number of panels. The panel area is converted into panel count by taking into account the types of panels in the farm and their corresponding dimensions. This is because "number of panels" itself is not as relevant as total photovoltaic area. The difference in numbers is likely also caused in part due to somewhat incorrect data itself—the precise outlines of farms are dynamic, and reporting nomenclature can change as they are influenced by financial factors, taxation, timing, ownership change, etc.

Table 4. Estimate of the number of panels. The panel area is converted into panel count, taking into account the type of panels in the farm and their corresponding dimensions.

Solar Farm	Panel Type	Panel Area (km^2)	# Panels Counted (*Million*)	# Panels Reported (*Million*)	Error (%)
Mount Signal	FS 3&4	4.93	6.85	6.8	<1%
Agua Caliente	FS S4	3.12	4.33	4.8	9.7%
Desert Sunlight	FS S4	5.55	7.71	8.0	3.6%
Solar Star	Sunpower	3.65	1.55	1.7	8.8%
Springbok	FS S4	2.21	3.07	3.0	2.3%

Finally, Table 5 shows the capacity calculation results using capacity factors that are relevant to the geographical location of the farm. While the model ultimately gives fairly close estimates overall (all within 10%), there is notable variation between farms. There is a case of the capacity evaluation error percentage being low, despite panel estimates not being as precise (Agua Caliente), and vice versa (Springbok). Hence, the maximum of the two errors is also reported. This variation in numbers could be attributed to temporal factors—solar farms are dynamic and changing, whereas the reported figures are true for a point in time. Time changes, weather variations, and nuances have not been considered in our model, whether in panel count, capacity factor, or annual generation.

Table 5. Capacity calculation results. We report the average of the two error values for each solar farm, which lies in the range 2–7%.

Solar Farm	# Panels Counted	# Panels Reported	Annual Capacity Calculated (GWh)	Annual Capacity Reported (GWh)	Capacity Evaluation Error (%)	Max (Errors) (%)
Mount Signal	6.85	6.8	1165.1	1197	2.7%	2.7%
Agua Caliente	4.33	4.8	736.0	740	<1%	9.7%
Desert Sunlight	7.71	8.0	1309.9	1287	1.8%	3.6%
Solar Star	1.45	1.7	861.2	831	3.7%	8.8%
Springbok	3.07	3.0	623.2	717	13.1%	13.1%

4. Conclusions

The intersection of remote sensing and deep learning presents an exciting opportunity for geographically quantifying photovoltaic system deployment, essentially giving us the ability to draw insights on insofar lumped data. Insights from the remote quantification of photovoltaic deployment could have outcomes such as strategic decision-making, the cross-verification of reported data, and the incentivization of renewables targeting underserved territories.

This work explored several independent elements of a capacity pipeline that goes from raw overhead imagery to annual energy generation estimates by creating a dataset, labeling it, choosing a neural network, and training, testing, and optimizing the model for performance, and finally, by using results from the deep learning model to extract panel count, panel area, and capacity predictions of the detected solar energy facilities. Some of the key takeaways of this study are:

1. A semantic segmentation model that achieved strong performance metrics including a mean accuracy of 96.87%, a Jaccard Index of 95.5% (compared to SolarNet's 94.2%), and that is capable of highly precise and detailed detections. This has resulted in arguably some of the most precise/accurate solar farm detection imagery in the literature.
2. An original, pixel-wise labeled dataset of solar farms that was sourced, annotated, and built for this problem, comprising 23,000 256 × 256 images on which the model was trained.
3. A capacity evaluation model to extract panel count, panel area, energy generation estimates, etc., of the detected solar energy facilities that were validated against publicly available data to within 10% error, and an average error of 4.5%.

Future Work

There is plenty of scope for future work on this problem, as well as to the broader problem of applying remote sensing to renewable energy technology. This work fits into a longer-term goal of creating a granular global database of solar energy capacity production that could serve as a single source of truth for industries and policymakers to identify underserved areas and to inform decision-making. In the future, a highly refined version of this model could even be used as a replacement for conventional sources of knowledge, or as a secondary source of intelligence for the cross-validation of reported figures. We identify certain directions that future efforts at extending this research could take. They can be segmented as follows.

1. Exploring newer neural net architectures and conducting a more detailed optimization study.
2. Exploring other data sources, including hyperspectral imagery.

3. Testing the performance of the CNN on data from other countries, incorporating additional training data if necessary. What remains to be conducted is automatic deployment on large geographical areas such as states and countries.

4. Improving the accuracy and robustness of the capacity model. We were able to arrive at reasonably close estimates of solar farm areas, numbers of panels, and even the annual energy generated, but they are inconsistent. We enumerated some of the possible reasons for inconsistency that had to do with temporal changes, reporting, and data collection. With cleaner and more reliable data to compare to, the parameters/constants in the model, such as packing factor, can be updated with a least squares fit.

5. Identifying trends and consequently underserved areas with high solar energy potential. The CNN can be deployed on the imagery of various regions to assess the deployment of commercial PV over time, and garner insights regarding the impacts of historical political, social, and economic factors on the deployment of solar renewable energy technology at scale.

6. Identifying solar panel defects such as cracked solar cells, broken glass, and dust/sand build-up: defects in solar panels are unlikely to be detectable with imagery at a resolution of 0.4–0.7 m, so this will have to be completed with drone imagery.

Author Contributions: Conceptualization, R.R.; methodology, R.R. and E.A.; software, R.R. and E.A.; validation, R.R., O.L.d.W. and E.A.; formal analysis, R.R. and E.A.; investigation, R.R. and E.A.; resources, R.R. and E.A.; data curation, R.R.; writing—original draft preparation, R.R. and E.A.; writing—review and editing, R.R., E.A. and O.L.d.W.; visualization, R.R. and E.A.; supervision, O.L.d.W.; project administration, O.L.d.W. and A.H.; funding acquisition, O.L.d.W. All authors have read and agreed to the published version of the manuscript.

Funding: This research was funded by KACST under project number 6945909 (MIT cost object).

Data Availability Statement: Overhead dataset available.

Conflicts of Interest: The authors declare no conflict of interest.

Abbreviations

The following abbreviations are used in this manuscript:

CF	Capacity factor
CNN	Convolutional neural network
FCN	Fully connected network
FN	False Negative
FP	False Positive
GIS	Geographic Information System
GSD	Ground Sampling Distance
IoU	Intersection over Union
mAcc	mean Accuracy
mIoU	mean IoU
NAIP	National Agriculture Imagery Program
NREL	National Renewable Energy Labs
PV	Photovoltaics
PySAM	NREL Python System Advisor Model
QGIS	Quantum GIS
ReLu	Rectified Linear Unit
TN	True Negative
TP	True Positive
UNet	"U" Network
USDA	United States Department of Agriculture
USGS	United States Geological Survey

Appendix A

Figure A1. Mount Signal Solar (32°40′24″N, 115°38′23″W) in Imperial county, California, along with its corresponding hand labeled "ground truth".

Figure A2. Agua Caliente (32°57.2′N, 113°29.4′W), California, along with its corresponding label.

Figure A3. Solar Star (34°49′50″N, 118°23′53″W), the world's largest solar farm, along with its corresponding hand-labeled annotation.

Figure A4. Topaz Solar (35°23′N, 120°4′W) , along with its corresponding label.

Figure A5. Copper Mountain Solar (35°47′N, 114°59′W) , along with its corresponding label.

References

1. Chu, S.; Majumdar, A. Opportunities and challenges for a sustainable energy future. *Nature* **2012**, *488*, 294–303. [CrossRef] [PubMed]
2. BP Statistical Review of World Energy 2018: Two Steps Forward, One Step Back | News and Insights | Home. Available online: https://www.bp.com/en/global/corporate/news-and-insights/press-releases/bp-statistical-review-of-world-energy-2018.html (accessed on 10 October 2022).
3. Kruitwagen, L.; Story, K.T.; Friedrich, J.; Byers, L.; Skillman, S.; Hepburn, C. A global inventory of photovoltaic solar energy generating units. *Nature* **2021**, *598*, 604–610. . [CrossRef] [PubMed]
4. International Solar Alliance. Available online: https://newsroom.unfccc.int/news/international-solar-alliance (accessed on 10 October 2022).
5. Yu, J.; Wang, Z.; Majumdar, A.; Rajagopal, R. DeepSolar: A Machine Learning Framework to Efficiently Construct a Solar Deployment Database in the United States. *Joule* **2018**, *2*, 2605–2617. [CrossRef]
6. Hou, X.; Wang, B.; Hu, W.; Yin, L.; Wu, H. SolarNet: A Deep Learning Framework to Map Solar Power Plants In China From Satellite Imagery. *arXiv* **2019**, arXiv:1912.03685.
7. Malof, J.M.; Bradbury, K.; Collins, L.M.; Newell, R.G. Automatic detection of solar photovoltaic arrays in high resolution aerial imagery. *Appl. Energy* **2016**, *183*, 229–240. [CrossRef]
8. National Agriculture Imagery Program (NAIP). Available online: https://naip-usdaonline.hub.arcgis.com/ (accessed on 10 October 2022).
9. Science for a Changing World. Available online: https://www.usgs.gov/ (accessed on 10 October 2022).
10. Biscione, V.; Bowers, J.S. Convolutional Neural Networks Are Not Invariant to Translation, but They Can Learn to Be. *arXiv* **2021**, arXiv:2110.05861.
11. Agnew, S.; Dargusch, P. Effect of residential solar and storage on centralized electricity supply systems. *Nat. Clim. Chang.* **2015**, *5*, 315–318. [CrossRef]

12. Ekim, B.; Sertel, E. A Multi-Task Deep Learning Framework for Building Footprint Segmentation. In Proceedings of the 2021 IEEE International Geoscience and Remote Sensing Symposium IGARSS, Brussels, Belgium, 11–16 July 2021. [CrossRef]
13. NREL-PySAM—NREL-PySAM 3.0.0 Documentation. Available online: https://nrel-pysam.readthedocs.io/en/latest/version_changes/3.0.0.html (accessed on 10 October 2022).

Article

A Review of Image Super-Resolution Approaches Based on Deep Learning and Applications in Remote Sensing

Xuan Wang [1,†], Jinglei Yi [1,†], Jian Guo [1], Yongchao Song [1], Jun Lyu [1], Jindong Xu [1], Weiqing Yan [1], Jindong Zhao [1], Qing Cai [2,3] and Haigen Min [4,5,*]

[1] School of Computer and Control Engineering, Yantai University, Yantai 264005, China; xuanwang91@ytu.edu.cn (X.W.); jingleiyi@s.ytu.edu.cn (J.Y.); 1245467032@s.ytu.edu.cn (J.G.); ycsong@ytu.edu.cn (Y.S.); ljdream0710@pku.edu.cn (J.L.); xujindong@ytu.edu.cn (J.X.); wqyan@tju.edu.cn (W.Y.); zhjdong@ytu.edu.cn (J.Z.)

[2] School of Data Science, The Chinese University of Hong Kong, Shenzhen 518172, China; caiqing@cuhk.edu.cn

[3] School of Information Science and Technology, University of Science and Technology of China, Hefei 230026, China

[4] School of Information Engineering, Chang'an University, Xi'an 710064, China

[5] The Joint Laboratory for Internet of Vehicles, Ministry of Education-China Mobile Communications Corporation, Xi'an 710064, China

* Correspondence: hgmin@chd.edu.cn

† These authors contributed equally to this work.

Citation: Wang, X.; Yi, J.; Guo, J.; Song, Y.; Lyu, J.; Xu, J.; Yan, W.; Zhao, J.; Cai, Q.; Min, H. A Review of Image Super-Resolution Approaches Based on Deep Learning and Applications in Remote Sensing. *Remote Sens.* **2022**, *14*, 5423. https://doi.org/10.3390/rs14215423

Academic Editor: Gwanggil Jeon

Received: 29 September 2022
Accepted: 26 October 2022
Published: 28 October 2022

Abstract: At present, with the advance of satellite image processing technology, remote sensing images are becoming more widely used in real scenes. However, due to the limitations of current remote sensing imaging technology and the influence of the external environment, the resolution of remote sensing images often struggles to meet application requirements. In order to obtain high-resolution remote sensing images, image super-resolution methods are gradually being applied to the recovery and reconstruction of remote sensing images. The use of image super-resolution methods can overcome the current limitations of remote sensing image acquisition systems and acquisition environments, solving the problems of poor-quality remote sensing images, blurred regions of interest, and the requirement for high-efficiency image reconstruction, a research topic that is of significant relevance to image processing. In recent years, there has been tremendous progress made in image super-resolution methods, driven by the continuous development of deep learning algorithms. In this paper, we provide a comprehensive overview and analysis of deep-learning-based image super-resolution methods. Specifically, we first introduce the research background and details of image super-resolution techniques. Second, we present some important works on remote sensing image super-resolution, such as training and testing datasets, image quality and model performance evaluation methods, model design principles, related applications, etc. Finally, we point out some existing problems and future directions in the field of remote sensing image super-resolution.

Keywords: image super-resolution; deep learning; remote sensing; model design; evaluation methods

1. Introduction

Agriculture, meteorology, geography, the military, and other fields have benefited from remote sensing imaging technology. In application scenarios such as pest and disease monitoring, climate change prediction, geological survey, and military target identification, remote sensing images are indispensable. Therefore, in order to realize remote sensing image applications and analyses, high-resolution remote sensing images are essential. Despite this, factors such as sensor noise, optical distortion, and environmental interference can adversely affect the quality of remote sensing images and make it difficult to acquire high-resolution remote sensing images. Image super-resolution aims to reconstruct high-resolution (HR) images from low-resolution (LR) images (as shown in Figure 1), which is a typical computer vision task to mitigate the effects of acquisition equipment and environmental factors on remote sensing imaging results and improve the resolution of remote sensing images. However, the SR problem assumes that low-pass-filtered (blurred) LR data are a downsampled and noisy version of HR data. The loss of high-frequency information during the irreversible low-pass filtering and secondary sampling operations causes SR to be an ill-posed problem. In addition, the super-resolution (SR) operation is a pair of multiple mappings from LR to HR space, which can have multiple solution spaces for any LR input, so it is essential to determine the correct solution from it. Many methods have been proposed to solve such an inverse problem, which can be broadly classified into early interpolation-based methods [1–3], reconstruction-based methods [4–6], and learning-based methods [7–14]. Since interpolation-based methods, such as the bicubic interpolation method [15], typically upsample LR images to obtain HR images, although they are simple and fast, some high-frequency information is destroyed in the process, leading to a decrease in model accuracy. The reconstruction-based methods are implemented based on adding the prior knowledge of the image as a constraint to the super-resolution reconstruction process of the image. Based on the SoftCuts metric, [16] proposes an adaptive SR technique based on prior edge smoothing. Although this overcomes the problem of uncomfortable image super-resolution, it also has the disadvantages of slow convergence speed and high computational cost. To achieve super-resolution reconstruction, the learning-based method relies on a large number of LR and HR images as a priori information. In [17], local feature blocks are learned between LR and HR images using the neighbor embedding method. Nonetheless, if learning becomes difficult (for example, when super-resolution magnification damages detailed features in the image), the learning-based method will perform less well. Therefore, the currently popular super-resolution is based on deep learning, which learns the mapping between LR and HR image spaces to predict the missing high-frequency information in low-resolution images in a time-saving and efficient manner.

Figure 1. SR aims to reconstruct a high-resolution (HR) image from its degraded low-resolution (LR) counterpart.

The field of deep learning is continually developing. In recent years, many SR models based on deep learning have been proposed and have achieved significant results on benchmark test datasets of SR. Furthermore, the application of SR models to super-resolution tasks on remote sensing images has become an increasingly popular topic in the field of SR. Many attempts have been made by researchers to improve the performance of SR models on remote sensing images. In particular, Dong et al. first designed a model with three CNN layers, i.e., SRCNN [18]. Subsequently, Kim et al. increased the network depth to 20 in DRCN [19], and the experimental results were significantly improved compared with those of SRCNN [18]. On this basis, Liebel et al. [20] retrained SRCNN [18] using remote sensing image datasets to adapt the model to the multispectral nature of remote sensing data. VDSR [21] introduced residual learning and gradient cropping while increasing the number of network layers and solved the problem of processing multi-scale images in a single framework. LGCnet [22] is a combined local–global network based on VDSR proposed by Lei et al. The problem of missing local details in remote sensing images is solved by combining shallow and deep features through a branching structure, which makes full use of both local and global information. To solve the discomfort problem of the super-resolution of images, Guo et al. developed a dual regression model, DRN [23]. This model learns mappings directly from LR images without relying on HR images. Overall, in order to achieve better results, most SR methods are improved in terms of the following aspects: network architecture design, selection of the loss function, development of the learning strategy, etc.

Due to their superior performance, the exploration of deep-learning-based SR methods is growing deeper. Several survey articles on SR have been published. However, most of these reports highlight various evaluation metrics for the reconstruction results of SR algorithms. In this paper, instead of simply summarizing available survey works [24–28], we provided a comprehensive overview of SR methods, focusing on the principles and processes of deep learning to demonstrate their performance, innovation, strength and weakness, relevance, and challenges, while focusing on their application to remote sensing images. Figure 2 shows the hierarchically structured classification of SR in this paper.

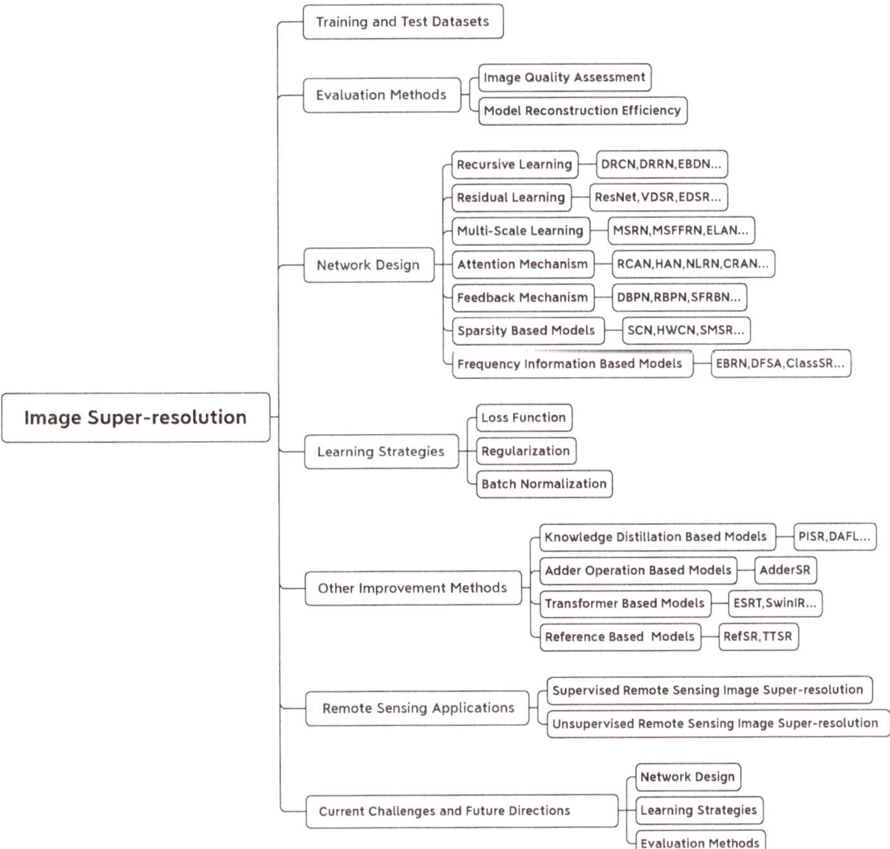

Figure 2. Hierarchically structured classification of SR in this paper.

The main contributions of this paper are as follows:

- We provide a comprehensive introduction to the deep-learning-based super-resolution process, including problem definitions, datasets, learning strategies, and evaluation methods, to give this review a detailed background.
- We classify the SR algorithms according to their design principles. In addition, we analyze the effectiveness of several performance metrics of representative SR algorithms on benchmark datasets, and some remote sensing image super-resolution methods proposed in recent years are also introduced. The visual effects of classical SR methods on remote sensing images are shown and discussed.
- We analyze the current issues and challenges of super-resolution remote sensing images from multiple perspectives and present valuable suggestions, in addition to clarifying future trends and directions for development.

The remaining sections of this review are arranged as follows. In Section 2, we briefly discuss what deep-learning-based SR is, the commonly used datasets, and the evaluation metrics. Section 3 describes in detail representative deep neural network architectures for SR tasks. In Section 4, several evaluation metrics are used to compare the performance of the SR methods mentioned in Section 3 and their application to remote sensing images. The applications of SR in remote domains are presented in Section 5. In Section 6, we discuss the current challenges and potential directions of SR. Finally, the work is concluded in Section 7.

2. Background

2.1. Deep-Learning-Based Super-Resolution

With advances in computing power, deep learning [29] in super-resolution has developed rapidly in recent years. Deep learning is a concept developed based on artificial neural networks [30], which is an extension of machine learning. Artificial neural networks imitate the way the human brain thinks, with artificial neurons as the computational units; the artificial neural network structure reflects the way these neurons are connected. The objective of deep learning is to determine the feature distribution of data by learning a hierarchical representation [31] of the underlying features. Specifically, deep learning continuously optimizes the super-resolution algorithm process through a series of learning strategies, such as deep network architecture, optimizer, and loss function design, while alleviating the ill-posed problem of super-resolution. Typically, the LR image I_x is modeled as the output of the following degradation:

$$I_x = \left(I_y \otimes k \right) \downarrow_s + n, \tag{1}$$

where $I_y \otimes k$ denotes the convolution operation between the HR image I_y and the degenerate blur kernel k (e.g., double cubic blur kernel, Gaussian blur kernel, etc.), $\downarrow_s$ is the downsampling operation with scale factor s, and n is the usually additive Gaussian white noise.

Deep learning differs from traditional algorithms because it can transfer the above processes into an end-to-end framework, saving time and efficiency. This is represented by the network structure of SRCNN [18], as shown in Figure 3. The image super-resolution process is roughly divided into three steps: feature extraction and representation, non-linear mapping, and image reconstruction. Specifically, first, the feature blocks are extracted from the low-resolution image by 9×9 convolution, and each feature block is represented as a high-dimensional vector. Then, each high-dimensional vector is non-linearly mapped to another high-dimensional vector by 5×5 convolution, where each mapped vector is a high-resolution patch. Finally, the final high-resolution image is generated by aggregating the above high-resolution patches by 5×5 convolution.

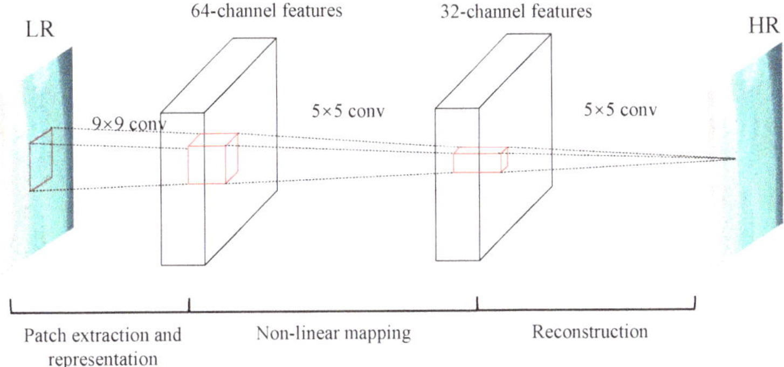

Figure 3. The network structure of SRCNN [18].

In comparison with natural images, remote sensing images differ in that (1) remote sensing images are captured from a distance of several hundred kilometers from the surface of the earth and are usually acquired by the use of aerial photography, land and ocean satellites, etc.; (2) most of the scenes in remote sensing images are forests, rivers, industrial areas, and airports, etc., which are typically large in scope, contain small objects, and have varied distribution forms; and (3) remote sensing images acquired under different weather conditions differ as well, due to factors such as lighting conditions on sensors, and clouds and fog that obscure the ground. The reconstruction of super-resolution remote sensing images, therefore, requires special considerations. For remote sensing images collected from forests and grasslands, the colors of the objects in the scene are very similar. It is difficult to classify the scene contents by color features alone. By referring to the texture features of these images, the "rough" forest and the "smooth" grass can be easily distinguished by the super-resolution method for these images.

2.2. Training and Test Datasets

Deep learning is a method of learning from data, and the goodness of the data plays an important role in the learning outcome of the model, with high-quality data being able to facilitate the improvement of the reconstruction performance of the deep learning SR-based model. Many diverse datasets for training and testing SR tasks have previously been proposed. Datasets commonly used for SR model training include BSDS300 [32], BSDS500 [33], DIV2K [34], etc. Similarly, BSD100 [32], Set5 [35], Set14 [36], Urban100 [37], etc. can be used to effectively test model performance. In particular, remote sensing image datasets such as AID [38], RSSCN7 [39], and WHU-RS19 [40] have been widely used in remote sensing image super-resolution tasks. In Table 1, we list some datasets that are commonly used in SR tasks (including the super-resolution of remote sensing images) and specify their image counts, image formats, resolutions, and content descriptions. Among them, the representative training dataset is the DIV2K [34] dataset, on which most SR models are trained. The DIV2K [34] dataset has three components: 800 training images, 100 validation images, and 100 test images. Set5 and Set14 are the classic test datasets for SR tasks, and they can accurately reflect the model performance. The OutdoorScene [41] dataset contains plants, animals, landscapes, reservoirs, etc., in outdoor scenes. AID [38] was originally used for the object detection task of remote sensing images, which contains 10,000 remote sensing images of 600 × 600 pixels, with scenes including airports, beaches, deserts, etc. RSSCN7 [39] contains 2800 remote sensing images from different seasons, arranged at four different scales, showing scenes such as farmland, parking lots, residential areas, and industrial areas. The WHU-RS19 [40] dataset comprises remote sensing images from 19 scenes, of which 50 images are included in each category. UC Merced [42] contains remote sensing images of 21 categories of scenes, 100 images per category, and each image size is 256 × 256 pixels. NWHU-RESISC45 [43] is published by Northwestern Polytechnic

University. The images represent 45 different categories of scenes, with 700 images per scene. RSC11 [44] is derived from Google Earth and contains 11 categories of scene, with each category having 100 images.

Table 1. Common datasets for image super-resolution (SR) and some remote sensing image datasets.

Dataset	Amount	Resolution	Format	Short Description
BSD300 [32]	300	(435, 367)	JPG	animal, scenery, decoration, plant, etc.
BSD500 [33]	500	(432, 370)	JPG	animal, scenery, decoration, plant, etc.
DIV2K [34]	1000	(1972, 1437)	PNG	people, scenery, animal, decoration, etc.
Set5 [35]	5	(313, 336)	PNG	baby, butterfly, bird, head, woman
Set14 [36]	14	(492, 446)	PNG	baboon, bridge, coastguard, foreman, etc.
T91 [45]	91	(264, 204)	PNG	flower, face, fruit, people, etc.
BSD100 [32]	100	(481,321)	JPG	animal, scenery, decoration, plant, etc.
Urban100 [37]	100	(984, 797)	PNG	construction, architecture, scenery, etc.
Manga109 [46]	109	(826, 1169)	PNG	comics
PIRM [47]	200	(617, 482)	PNG	animal, people, scenery, decoration, etc.
City100 [48]	100	(840,600)	RAW	city scene
OutdoorScene [41]	10624	(553, 440)	PNG	scenes outside
AID [38]	10000	(600, 600)	JPG	airport, bare land, beach, desert, etc.
RSSCN7 [39]	2800	(400, 400)	JPG	farmlands, parking lots, residential areas, lakes etc.
WHU-RS19 [40]	1005	(600, 600)	TIF	bridge, forest, pond, port, etc.
UC Merced [42]	2100	(256, 256)	PNG	farmland, bushes, highways, overpasses, etc.
NWHU-RESISC45 [43]	31,500	(256, 256)	PNG	airports, basketball courts, residential areas, ports, etc.
RSC11 [44]	1232	(512, 512)	TIF	grasslands, overpasses, roads, residential areas, etc.

In addition to the datasets introduced above, datasets such as ImageNet [49], VOC2012 [50], and CelebA [51] for other image processing tasks were also introduced into the SR task.

2.3. Evaluation Methods

The evaluation index of image reconstruction quality can reflect the reconstruction accuracy of an SR model. Meanwhile, the number of parameters, running time, and computation of a model reflect the performance of an SR model. In this section, the evaluation methods of image reconstruction quality and reconstruction efficiency are introduced.

2.3.1. Image Quality Assessment

Due to the widespread use of image super-resolution techniques, evaluating the quality of reconstructed images has become increasingly important. Image quality refers to the visual properties of an image, and the methods of image quality evaluation, distinguished from the point of view of human involvement, include two branches: subjective and objective evaluation. Using subjective evaluation, we can determine the quality of an image (whether it appears realistic or natural) based on statistical analysis and with a human being as the observer. This type of method can truly reflect human perception. The objective evaluation of an organization is usually conducted based on numerical calculations utilizing some mathematical algorithm that can automatically calculate the results. In general, the former is a straightforward approach and more relevant to practical needs, but these methods are difficult to implement and inefficient. Therefore, objective evaluation methods are more widely used in image quality assessment, especially complete reference methods, and several commonly used methods for image quality assessment are described below.

Peak Signal-to-Noise Ratio (PSNR)

PSNR [52] is one of the most popular objective image evaluation metrics in SR. Given a ground truth image I_y with N pixels and a reconstructed image I_{SR}, the PSNR can be defined by using the mean square error (MSE) as

$$PSNR = 10 \cdot \log_{10}\left(\frac{L^2}{MSE}\right), \tag{2}$$

$$MSE = \frac{1}{N}\sum_{i=1}^{N}\left(I_y - I_{SR}\right)^2, \tag{3}$$

where L refers to the peak signal, i.e., $L = 255$ for an 8-bit grayscale image. Although PSNR is relatively simple in its computational form and has a clear physical meaning, it essentially does not introduce human visual system characteristics into the image quality evaluation because it only considers MSE at the pixel level. Only the differences are analyzed purely from a mathematical perspective, resulting in the inability of PNSR to capture the differences in visual perception. However, it is more important to evaluate the constructive quality of the reconstructed image, so PNSR remains an accepted evaluation metric.

Structural Similarity (SSIM)

SSIM [52] is another popular image evaluation metric in the SR field. Unlike PSNR, which measures absolute error, SSIM belongs to the perceptual model and can measure the degree of distortion of a picture, as well as the degree of similarity between two pictures. As a full-reference objective image evaluation metric, SSIM is more in line with the intuition of the human eye. Specifically, SSIM is a comprehensive measure of similarity between images from three aspects, including structure, brightness, and contrast, which is defined as

$$SSIM = \left(l\left(I_{SR}, I_y\right)^\alpha \cdot c\left(I_{SR}, I_y\right)^\beta \cdot s\left(I_{SR}, I_y\right)^\gamma\right), \tag{4}$$

$$l\left(I_{SR}, I_y\right) = \frac{\left(2\mu_{I_{SR}}\mu_{I_y} + C_1\right)}{\mu_{I_{SR}}^2 + \mu_{I_y}^2 + C_1}, \tag{5}$$

$$c\left(I_{SR}, I_y\right) = \frac{\left(2\sigma_{I_{SR}}\sigma_{I_y} + C_2\right)}{\sigma_{I_{SR}}^2 + \sigma_{I_y}^2 C_2}, \tag{6}$$

$$s\left(I_{SR}, I_y\right) = \frac{\left(\sigma_{I_{SR}I_y} + C_3\right)}{\sigma_{I_{SR}}\sigma_{I_y} + C_3}, \tag{7}$$

where C_1, C_2, and C_3 are constants and α, β, and γ are weighting parameters. In order to avoid the case that the denominator is 0, $C_1 = (k_1 L)^2$, $C_2 = (k_2 L)^2$, $C_3 = \frac{C_2}{2}$, and $k_1 \ll 1, k_2 \ll 1$. SSIM takes values in the range of [0,1], and the larger the value, the higher the similarity between two images. MS-SSIM is a variant of SSIM, and due to the multivariate observation conditions, it takes into account the similarity between images at different scales and makes the image evaluation more flexible.

Mean Opinion Score (MOS)

MOS is a subjective evaluation method, usually using the two-stimulus method [53]. An observer directly rates the perception of image quality, and this result is mapped to a numerical value and averaged over all ratings, i.e., MOS. Many personal factors, such as emotion, professional background, motivation, etc., can impact the evaluation results when the observer performs the evaluation, which will cause the evaluation results to become unstable and not accurate enough to ensure fairness. Moreover, MOS is a time-consuming and expensive evaluation method because it requires the participation of the observer.

In addition to the above image evaluation metrics, there are many other image evaluation methods [54], such as the Natural Image Quality Evaluator (NIQE) [55], which is an entirely blind metric that does not rely on human opinion scores and expects a priori information to extract "quality-aware" features from images to predict their quality. The algorithmic process of NIQE is more accessible to implement than MOS. Learned Perceptual Image Patch Similarity (LPIPS) [56] is also known as "perceptual loss". Specifically, when evaluating the quality of super-resolution reconstructed images, it pays more attention to the depth features of the images and learns the inverse mapping from the reconstructed images to a high resolution, and then calculates the L2 distance between them. Compared with the traditional PNSR and SSIM methods, LPIPS is more in line with human perception.

2.3.2. Model Reconstruction Efficiency
Storage Efficiency (Params)

When evaluating an SR model, the quality of the reconstructed images it outputs is essential. Still, the complexity and performance of the model need to be paid attention to as well in order to promote the development and application of image super-resolution in other fields while considering the output results of the losing model. The number of parameters, running time, and computational efficiency of an SR model are important indicators reflecting the efficiency of model reconstruction.

Execution Time

The running time of a model is a direct reflection of its computational power. The current popular lightweight networks not only have a relatively low number of parameters but also have short running times. If an SR model adds complex operations, such as attention mechanisms, this can lead to an increase in running time and affect the performance of the model. Therefore, the running time is also an essential factor in determining the performance of the model.

Computational Efficiency (Mult & Adds)

Since the algorithmic process in convolutional neural networks is primarily dependent upon multiplication and addition operations, the multiplicative addition operands are used to assess the computational volume of the model, as well as to indirectly reflect its computational efficiency. The size of the model and the running time are the influencing factors of the multiplicative–additive operands.

To conclude, when evaluating the SR model, it is also important to take into account the complexity and performance of the model.

3. Deep Architectures for Super-Resolution
3.1. Network Design

Network design is a key part of the deep learning process, and this section will introduce and analyze some mainstream design principles and network models in the super-resolution domain, as well as explain and illustrate some deep learning strategies. Finally, some design methods that deserve further exploration will be discussed.

3.1.1. Recursive Learning

Increasing the depth and width of the model is a common means to improve the performance of the network, but this brings with it a large number of computational parameters, as shown in Figure 4. Recursive learning is proposed to control the number of model parameters and to achieve the sharing of parameters in recursive modules. In simple terms, recursive learning means reusing the same module multiple times. DRCN [19] applies recursive learning to super-resolution problems by using a single convolutional layer as the recursive unit and setting 16 recursions to increase the perceptual field to 41×41 without introducing too many parameters. However, the superimposed use of recursive modules also poses some problems: gradient explosion or disappearance. Therefore, in DRRN [57], global and local residual learning is introduced to solve the gradient problem, i.e., ResBlock is used as the recursive unit to reduce the training difficulty. Ahn et al. [58] made improvements to the recursive application of ResBlock. They proposed a global and local cascade connection structure to further speed up the network training and make the transfer of information more efficient. In addition, the EBRN presented by Qiu et al. [59] uses recursive learning to achieve the differentiation of information with a different frequency, i.e., low-frequency information is processed with shallow modules in the network, and high-frequency information is processed with deep modules. Recursive learning has also been widely used in some recent studies [60–62]. For example, in the SRRFN proposed by Li et al. [60], the recursive fractal module consists of a series of fractal modules with shared weights, which enables the reuse of model parameters.

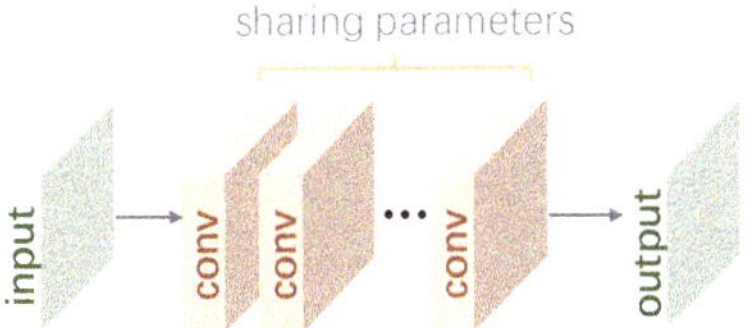

Figure 4. The structure of recursive learning.

3.1.2. Residual Learning

While recursive learning enables models to achieve a higher performance with as few parameters as possible, it also introduces the problem of exploding or vanishing gradients, and residual learning is a more popular approach to alleviate these problems. He et al. [63] proposed the use of residual learning in ResNet. It aims to mitigate the problem of exploding or disappearing gradients by constructing constant mappings using layer-hopping connections so that gradients in back-propagation can be passed directly to the network front-end through shortcuts, as shown in Figure 5. In image super-resolution tasks, low-resolution input images and high-resolution reconstructed images have most of the relevant information in terms of features, so only the residuals between them need to be learned to recover the lost information. In such a context, many residual learning based models [58,64–66] were proposed. Kim et al. proposed a profound super-resolution residual network VDSR [21] based on VGG-16, which has 20 layers, and takes the low-resolution image with bi-trivial interpolation as the input image. The residual information learned by the network is summed with the original input image as the output. Generally, the composition of the residual branch includes 3×3 convolutional layers, BN layers, and the relu activation function; some other ways this can be set up are mentioned in [67]. However, it is mentioned in EDSR [68] that the BN layer in the residual module is not suitable for super-resolution tasks because the distribution of colors of any image is normalized after BN layer processing. The original contrast information of the image is destroyed, which affects the quality of the output image of the network. Therefore, the BN layer is often chosen to be removed when designing residual modules in super-resolution tasks. RDN [64] proposes a residual dense block (RDB) that can fully preserve the features of the output of each convolutional layer. Nowadays, residual learning is a common strategy for super-resolution network design and has been applied in many models [69–73].

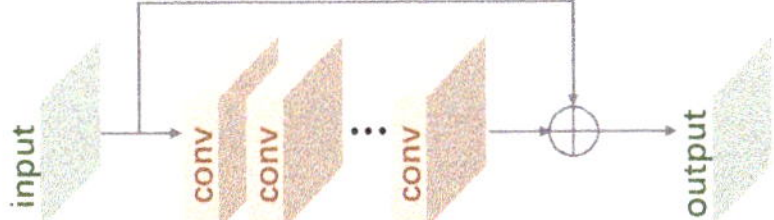

Figure 5. The structure of residual learning.

Although this global residual learning strategy achieves good results, global residual learning refers to the jump connection established from the input to the output. As the network levels deepen, global residual learning alone cannot recover a large amount of lost information, so researchers propose local residual learning, which is located in every few stacked layers and helps to preserve image details. A combination of global and local residual learning is applied in models such as VDSR [21], and EDSR [68].

3.1.3. Multi-Scale Learning

It has been pointed out [74,75] that images at different scales possess different features, and these rich features will help to generate high-quality reconstructed images. Therefore, multi-scale learning has been proposed to enable models to fully utilize features at different scales, while being applied to many SR models [76]. Li et al. [65] concluded that previous models were less robust to scale and less scalable, so multi-scale learning was applied to the SR task. He proposed a multi-scale residual module (MSRB) that used a 1×1 convolution kernel combined with 3×3 and 5×5 convolution kernels to obtain information at different scales (as shown in Figure 6), while local residual learning further improves the network training efficiency. In [77], the authors combined residual learning with multi-scale learning and proposed a multi-scale feature fusion residual block (MSFFRB) to extract and fuse image features of different scales. The multi-scale feature extraction and attention module (MSFEAAB) in [78] used convolutional kernels containing different sizes within the same layer to extract information of different frequencies. Among them, small-sized convolutional kernels primarily extract low-frequency components, while large-sized convolutional kernels extract high-frequency components. Not only are the rich image features obtained, but the computational complexity is not increased. Recently, more SR network models have adopted multi-scale learning to improve model performance. In ELAN [79], the authors proposed the grouped multi-scale self-attention (GMSA) module, in which self-attention is computed using windows of different sizes on a non-overlapping set of feature maps, as a way to establish long-range dependencies.

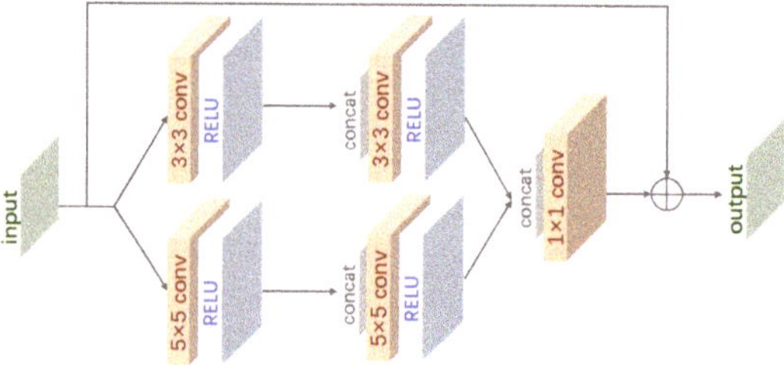

Figure 6. The structure of multi-scale residual block (MSRB) [65].

3.1.4. Attention Mechanism

The attention mechanism was proposed due to the fact that convolutional neural networks focus more on local information and ignore global features. The attention mechanism is widely used in various computer vision tasks, often inserted into the backbone network as a component, and its main purpose is to allocate computational resources to more important tasks with limited computational power. In short, the attention mechanism helps the network to ignore irrelevant information and focus on important details. Many works have previously been proposed to facilitate the development of attention mechanisms. For example, Hu et al. [80] proposed a novel "squeeze and excite" (SE) block, which adaptively adjusts channel feature responses according to the interdependencies between channels, as shown in Figure 7. With the continuous progress of the attention mechanism and the advancement of previous research work, the attention mechanism has begun to be applied to image super-resolution tasks.

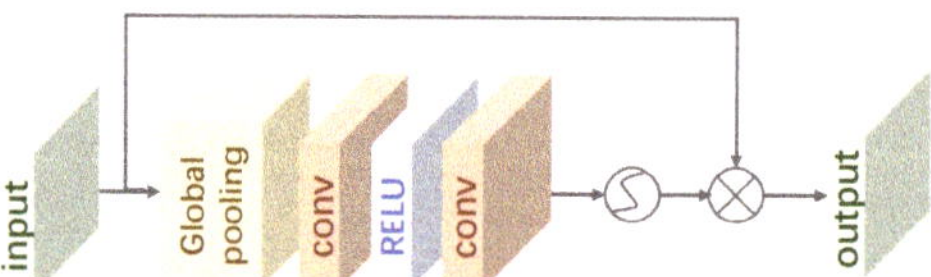

Figure 7. The structure of channel attention mechanism [80].

Channel Attention

In RCAN [81], Zhang et al. proposed a residual channel attention block (RCAB) to achieve higher accuracy by learning the correlation between channels to adjust channel features. To make the network pay more attention to the vital spatial features in the residual features, Liu et al. [66] proposed an enhanced spatial attention (ESA) block, which used a 1×1 convolution to reduce the number of channels to be light enough to be inserted into each residual block. Furthermore, three 3×3 convolution combinations are used in order to expand the perceptual field. Since channel attention treats each convolutional layer as a separate process and ignores the correlation between different layers, the use of this algorithm will lead to the loss of some intermediate features during the image reconstruction. Therefore, Niu et al. [82] proposed a holistic attention network (HAN) consisting of a layer attention module (LAM) and a channel space attention module (CSAM). The LAM can assign different attention weights to features in different layers by obtaining the dependencies between features of different depths, and then use the CSAM to learn the correlations at different locations in each feature map, so as to capture global features more efficiently. Similarly, the second-order channel attention (SOCA) module in SAN [83] learns the inter-channel feature correlations by using the second-order statistics of the features. The matrix multispectral channel attention (MMCA) module [84] first transformed the image features to the frequency domain by DCT and then learned the channel attention to achieve reconstruction accuracy in the SOTA results.

Non-Local Attention

Due to the limited perceptual field size, most image super-resolution networks are only good at extracting local features in images, ignoring the correlation between long-range features in images. However, this may provide critical information for reconstructing images. Therefore, some studies have been proposed for non-local feature correlation. For example, the purpose of the region non-local RL-NL module in SAN [83] is to divide the input image into specific sizes and perform non-local operations on each region. Liu et al. [85] proposed a non-local recurrent network (NLRN) to introduce non-local operations into recurrent neural networks (RNN) for image recovery tasks to obtain the correlation of deep features at each location with their neighboring features. Regarding non-local attention, a cross-scale non-local (CS-NL) attention module was proposed in CSNLN [86], which computes the similarity between LR feature blocks and HR target feature blocks in an image and improves the performance of the SR model.

Other Attention

In addition to the common attention mechanisms mentioned above, there are also some attention mechanisms designed from a specific perspective. For example, the contextual reasoning attention network [87] generates attention masks based on global contextual information, thus dynamically adjusting the convolutional kernel size to accommodate image feature changes. Zhang et al. [79] argued that the transformer's self-attention computation is too large and certain operations are redundant for super-resolution tasks, so the grouped multi-scale self-attention (GMSA) module was proposed, which computes attention within windows of different sizes while sharing attention to accelerate the computation. Mei et al. [88] introduced sparse representation to non-local self-attention to improve the performance of the attention mechanism and reduce the number of operations.

3.1.5. Feedback Mechanism

The feedback mechanism differs from the input-to-target object mapping by introducing a self-correcting phase to the learning process of the model, i.e., sending the output from the back end to the front end. The feedback mechanism is close to the recursive learning structure, but the difference is that the parameters of the feedback-based model are self-correcting, while the parameters of the recursive learning-based model are shared between modules. In recent years, feedback mechanisms have been gradually applied to computer vision tasks [89,90]. Feedback mechanisms are also widely used in SR models due to their ability to transfer deep information to the front end of the network to help further the processing of shallow information, which facilitates the reconstruction process from LR images to HR images. Haris et al. [91] proposed a depth inverse projection network for super-resolution, using an alternating upsampling and downsampling stage structure to achieve each stage of error feedback. RBPN [92] was proposed based on DBPN [91] for video super-resolution tasks, which also introduces a feedback mechanism, with the difference that RBPN integrates single-frame input and multi-frame input into one, using an encoder–decoder mechanism to integrate image details. In SFRBN [93], a feedback module (FB) is proposed, where the output of the previous module is fed back to the next module as part of its input, enabling the further refinement of low-level information.

3.1.6. Frequency Information-Based Models

In addition to improvements in model volume (increasing width and depth), some scholars have found that many current models for SR have a common problem: feature extraction or processing tends to ignore high-frequency information. The SR task is precisely a process of texture detail reconstruction for LR images, and such a problem can seriously affect the reconstruction results. Therefore, some SR methods that focus on image frequency information have been proposed. Qiu et al. [59] proposed an embedded block residual network in EBRN, which used a recursive approach to the hierarchical processing of features with different frequencies, with low-frequency information processed by a shallow module and high-frequency information processed by a deep module, as a way to achieve better results. Xie et al. [94] proposed a discrete cosine transform-based predictor that partitions the coefficients of the input image in terms of frequency information, thus implementing operations of different complexity for different regions, reducing computational effort and computational complexity. Magid et al. [84] proposed a dynamic high-pass filtering module (HPF) that dynamically adjusts the convolution kernel weights for different spatial locations, thus preserving high-frequency information. A matrix multispectral channel attention (MMCA) module was also proposed, which learns channel attention by transforming features to the frequency domain through DCT. Kong et al. [95] proposed ClassSR consisting of Class-Module and SR-Module to classify and super-resolve the input image based on frequency information. Specifically, the Class Module first decomposes the image into small sub-images and classifies their complexity, i.e., smooth regions are more accessible to reconstruct than textured regions. Then, these small sub-images are fed to different SR-Module branches for further processing according to different complexity levels.

3.1.7. Sparsity-Based Models

In addition to the recovery of high-frequency information, introducing image sparsity into CNN also leads to better performance. In SRN [96], an SR model incorporating sparse coding design was proposed with better performance than SRCNN. Gao et al. [97] presented a hybrid wavelet convolutional network (HWCN) to obtain scattered feature maps by predefined scattering convolution and then the sparse coding of these feature maps, used as the input to the SR model. Wang et al. [98] developed a sparse mask SR (SMSR) network to improve network inference efficiency by teaching sparse masks to cull redundant computations. In SMSR [98], "important" and "unimportant" regions are jointly distinguished by spatial and channel masks, thus skipping unnecessary computations.

3.2. Learning Strategies

Common problems in the training process of SR models based on deep learning include slow convergence and over-fitting. Solutions to these problems are closely related to deep learning strategies, such as selecting the loss function, including regularization, or performing batch normalization. These are critical steps in the training of deep learning models. The purpose of this section is to introduce common learning strategies and optimization algorithms used in deep learning.

3.2.1. Loss Function

Loss functions are used to calculate the error between reconstructed images and ground truth. The loss function is a crucial factor in determining the performance of the model and plays a role in guiding the model learning during the training process. A reasonable choice of the loss function can make the model converge faster on the dataset. The smaller the value of the loss function, the better the robustness of the model. In order to better reflect the reconstruction of images, researchers try to use a combination of multiple loss functions (e.g., pixel loss, texture loss, etc.). In this section, we will study several commonly used loss functions.

Pixel Loss

Pixel loss is the most popular loss function in image super-resolution tasks, which is used to calculate the difference between the reconstructed image and ground truth pixels to make the training process as close to convergence as possible. L1 loss, L2 loss, and Charbonnier loss are among the pixel-level loss functions:

$$L_{L1}(I_{SR}, I_y) = \frac{1}{hwc} \sum_{i,j,k} |I_{SR}^{i,j,k} - I_y^{i,j,k}|, \tag{8}$$

$$L_{L2}(I_{SR}, I_y) = \frac{1}{hwc} \sum_{i,j,k} \left(I_{SR}^{i,j,k} - I_y^{i,j,k}\right)^2, \tag{9}$$

$$L_{Cha}(I_{SR}, I_y) = \frac{1}{hwc} \sum_{i,j,k} \sqrt{\left(I_{SR}^{i,j,k} - I_y^{i,j,k}\right)^2 + \epsilon^2}, \tag{10}$$

where h, w, and c are the height, width, and number of channels of the image, respectively. ϵ is a constant (usually set to 10^{-3}) to ensure stable values. In image super-resolution tasks, many image evaluation metrics involve inter-pixel differences, such as PSNR, so pixel loss has been a popular loss function in super-resolution. However, pixel loss does not consider the perceptual quality and texture of the reconstructed image, which can lead to a lack of reconstructed images with lost high-frequency details; therefore, high-quality reconstructed images cannot be obtained.

Perceptual Loss

Perceptual loss is commonly used in GAN networks. In order to obtain reconstructed images with rich high-frequency features, researchers proposed perceptual loss in place of the L2 loss used previously to calculate inter-pixel differences. Specifically, perceptual loss is often used to compare two images that look similar but are different, because perceptual loss compares the perceptual quality and semantic differences between the reconstructed image and ground truth:

$$L_{perceptual} = \frac{1}{h_j w_j c_j} \left\|\varnothing_l(I_{SR}) - \varnothing_l(I_y)\right\|_2^2, \tag{11}$$

where h_l, w_l, and c_l denote the height, width, and number of channels of the l-th layer features, respectively. $\varnothing$ denotes the pre-trained network, and $\varnothing_{(l)}(I)$ denotes the high-level features extracted from a certain l-th layer of the network.

Content Loss

Content loss was applied early in the field of style migration, and is similar to perceptual loss, using the semantic difference between the generated and content images compared with the trained classification network, i.e., L2 distance:

$$L_{Content}(I_{SR}, I_y, \varnothing, l) = \frac{1}{h_l w_l c_l} \sum_{i,j,k} \left(\varnothing_{(l)}^{i,j,k}(I_{SR}) - \varnothing_{(l)}^{i,j,k}(I_y)\right), \tag{12}$$

where h_l, w_l, and c_l denote the height, width, and number of channels of the l-th layer features, respectively. $\varnothing$ denotes the pre-trained classification network, and $\varnothing_{(l)}(I)$ denotes the high-level features extracted from the l-th layer of the network.

Texture Loss

By obtaining the spatial correlation between the feature maps in the pre-trained network, texture loss is a modification of perceptual loss as introduced by Gatys et al. to the field of style migration. Since the reconstructed images possess the same style as ground truth, texture loss can also be applied in the field of super-resolution [14,99–101]. Texture loss is mainly achieved by computing the Gram matrix:

$$G_{(l)}^{ij}(I) = \text{vec}\left(\varnothing_{(l)}^{i}(I)\right) \cdot \text{vec}\left(\varnothing_{(l)}^{j}(I)\right), \tag{13}$$

where $G_{(l)}^{ij}(I)$ is the inner product between vectorized feature maps i and j at the l-th layer, which captures the tendency of features to appear simultaneously in different parts of the image. vec() denotes the vectorization operation and $\varnothing_{(l)}^{i}(I)$ denotes the i-th channel of the feature map on the l-th layer of image I. Then, the texture loss is defined as follows:

$$L_{\text{texture}}\left(I_{SR}, I_y, \varnothing, l\right) = \frac{1}{c_l^2}\sqrt{\sum_{i,j}\left(G_{(l)}^{ij}(I_{SR}) - G_{(l)}^{ij}(I_y)\right)^2}. \tag{14}$$

Adversarial Loss

Recent research has demonstrated that generative adversarial networks (GANs) perform well on image super-resolution tasks. GANs are gradually being applied to computer vision tasks. A generative adversarial network (GAN) consists of two core parts: generator and discriminator. It is the generator's responsibility to create data that do not exist, while the discriminator is responsible for determining whether the generated data are accurate or false. After iterative training, the ultimate goal of the generator is to generate data that look naturally real and are as close to the original data as possible, so that the discriminator cannot determine the authenticity. The task of the discriminative model is to identify the fake data as accurately as possible, and the application of GAN in the field of super-resolution takes the form of SRGAN [102]. The design of the SRGAN loss function is based on the cross-entropy in pixel loss, which is defined as follows:

$$L_{\text{Adversarial}}\left(I_x, G, D\right) = \sum_{n=1}^{N} - \log D(G(I_x)), \tag{15}$$

where $G(I_x)$ is the reconstructed SR image, and G and D represent the generator and discriminator, respectively. Some MOS tests have shown that SR models trained by a combination of content loss and adversarial loss perform better in terms of the perceptual quality of images than SR models that undergo only pixel loss. Still, with reduced PSNR, research continues on how to integrate GAN into SR models and stabilize the trained GAN.

3.2.2. Regularization

The SR model training process is prone to the over-fitting phenomenon; that is, the model overlearns the training dataset and has poor generalization ability, resulting in a high evaluation index of the reconstructed image output on the training set and poor performance on the test set. The reasons for over-fitting include (1) the small size of the training dataset, and (2) the high complexity of the model and numerous parameters. Therefore, the most direct way to avoid overfitting is to increase the size of the training dataset so that the training set samples are as close as possible to the ground-truth data distribution. However, this approach does not guarantee the effect and is time-consuming and laborious. In deep learning, a common learning strategy used to prevent overfitting is regularization.

The essence of regularization is to preserve the original features, make the input dataset smaller than the number of model parameters, and avoid training, in order to obtain parameters that improve the generalization ability of the model and prevent overfitting. The common regularization methods [103,104] in deep learning include L1\L2 regularization, dropout [105–107], early stopping, and data augmentation.

L1\L2 Regularization

L1\L2 regularization is the most commonly used regularization method. It essentially adds regular terms of L1\L2 parameterization to the loss function to reduce the number of parameters, acting as a parameter penalty to reduce the complexity of the model and limit its learning ability.

L1 regularization is defined as follows:

$$cost = Loss + \gamma \sum \|w\|, \tag{16}$$

L2 regularization is defined as follows:

$$cost = Loss + \gamma \sum \|w\|^2, \tag{17}$$

where γ is the hyperparameter that controls the proportion of the control loss term and the regularization term, and w is the model weight.

Since many parameter vectors in the L1 regularization term are sparse vectors, resulting in many parameters being zero after model regularization, L1 regularization is used when compressing the model in deep learning, while L2 regularization is commonly used in other cases.

Dropout

Hinton et al. [108] proposed that when the dataset is small, and the neural network model is large and complex, over-fitting tends to occur during training. To prevent over-fitting, some of the feature detectors can be stopped in each training batch so that the model does not rely too much on certain local features, thus improving the generalization ability and performance of the model. Compared with other regularization methods, dropout [109] is simpler to implement, has essentially no restrictions on the model structure, and has good performance on feedforward neural networks, probabilistic models, and recurrent neural networks, with a wide range of applicability. There are two typical dropout implementations: vanilla dropout and inverted dropout.

The process of vanilla dropout includes the model being trained by randomly dropping some neurons with a certain probability p. Then, forward propagation is performed, the loss is calculated, and backward propagation and gradient update are performed. Finally, the step of randomly dropping neurons is repeated. However, the selection of neurons is random for each dropout, and vanilla dropout requires scaling (i.e., multiplying by $(1 - p)$) the trained parameters at test time, which leads to different results for each test with the same input, making the model performance unstable and the operation of balancing expectations too cumbersome. Therefore, vanilla dropout is not widely used.

Inverted dropout is an improved version of vanilla dropout. It is based on the principle of dropping a portion of neurons with random probability p during the model training process. Unlike vanilla dropout, it does not process the parameters during the test stage. Inverted dropout scales the output values by a factor of $\frac{1}{1-p}$ during forward propagation, balancing the expectation values and keeping the training and testing processes consistent.

Early Stopping

As the number of training iterations increases, the training error of the model gradually decreases but the error in the test set increases again. The strategy of stopping the algorithm when the error on the test set does not improve further within a pre-specified number of cycles, at which point the parameters of the model are stored, and the parameter that minimizes the error on the test set is returned, is called early termination [110,111]. The early termination method hardly changes the model training parameters and optimization objectives and does not disrupt the learning process of the model. Due to its outstanding effectiveness and simplicity, the early termination method is the more commonly used regularization method.

Data Augmentation

Training with a larger number of datasets is the most direct way to improve model generalization and prevent over-fitting. Furthermore, data augmentation [112,113] is an important method to meet the demand of deep learning models for large amounts of data. In general, the size of a dataset is fixed, and data augmentation increases the amount of data by manually generating new data. For images, a single image can be flipped, rotated, cropped, or even Gaussian blurred to generate other forms of images.

3.2.3. Batch Normalization

To address the problem of the data distribution within a deep network changing during training, Sergey et al. [114] proposed batch normalization (BN) to avoid covariance shifts within parameters during training. Batch normalization is introduced into the deep learning network framework as a layer, commonly used after the convolution layer, to readjust the data distribution. The BN layer divides the input data into batches, a batch being the number of samples optimized each time, in order to calculate the mean and variance of the groups, and then normalizes them, since each group determines the gradient and reduces randomness when descending. Finally, scaling and offset operations are performed on the data to achieve a constant transformation, and the data recover their original distribution.

Batch normalization can prevent over-fitting from appearing to a certain extent, which is similar to the effect of dropout and improves the generalization ability of the model. Meanwhile, because batch normalization normalizes the mean and variance of parameters in each layer, it solves the problem of gradient disappearance. It supports the use of a larger learning rate, which increases the magnitude of gradient dropout and increases the training speed.

3.3. Other Improvement Methods

In addition to the network design strategies mentioned in Section 3.1, this section will add other design approaches that have further research value.

3.3.1. Knowledge-Distillation-Based Models

Hinton et al. [115] first introduced the concept of knowledge distillation, a model compression algorithm based on a "teacher–student network", where the critical problem is how to transfer the knowledge transformed from a large model (teacher model) to a small model (student model). Lee et al.[116] proposed a distillation structure for SR, which was the first time that knowledge distillation was introduced into the super-resolution domain. Knowledge distillation has been widely used in various computer vision tasks, and its advantages of saving computational and storage costs have been shown. In [116], features from the decoder of the teacher network are transferred to the student network in the form of feature distillation, which enables the student network to learn richer detailed information. Zhang et al. [117] proposed a network distillation method DAFL applicable to cell phones and smart cameras in the absence of raw data, using a GAN network to simulate the raw training data in the teacher network, and using a progressive distillation strategy to distill more information from the teacher network and better train the student network.

3.3.2. Adder-Operation-Based Models

Nowadays, the convolution operation is a common step in deep learning, and the primary purpose of convolution is to calculate the correlation between the input features and the filter, which will result in a large number of floating-point-valued multiplication operations. To reduce the computational cost, Chen et al. [118] proposed to use additive operations instead of multiplication operations in convolutional neural networks, i.e., L1 distance is used instead of convolution to calculate correlation, while L1-norm is used to calculate variance, and an adaptive learning rate scale change strategy is developed to speed up model convergence. Due to the superior results produced by AdderNet, Chen et al. [119] applied the additive operation to the image super-resolution task. In AdderSR [119], the relationship between adder operation and constant mapping is analyzed, and a shortcut is inserted to stabilize the performance. In addition, a learnable power activation is proposed to emphasize high-frequency information.

3.3.3. Transformer-Based Models

In recent years, the excellence of transformer in the field of natural language processing has driven its application in computer vision tasks. Many transformer-based image processing methods have been proposed one after another, e.g., image classification [120,121], image segmentation [122,123], etc. The advantage of transformer is that self-attention can model long-term dependencies in images [124] and obtain high-frequency information, which helps to recover the texture details of images. Yang et al. [101] proposed a texture transformer network for image super-resolution, where the texture transformer of the method extracts texture information based on the reference image and transfers it to the high-resolution image while fusing different levels of features in a cross-scale manner, obtaining better results compared with the latest methods. Chen et al. [125] proposed a hybrid attention transformer that improves the ability to explore pixel information by introducing channel attention into the transformer while proposing an overlapping cross-attention module (OCAB) to better fuse features from different windows. Lu et al. [126] proposed an efficient and lightweight super-resolution CNN combined with transformer (ESRT), where, on the one hand, the feature map is dynamically resized by the CNN part to extract deep features. On the other hand, the long-term dependencies between similar patches in an image are captured by the efficient transformer (ET) and efficient multi-headed attention (EMHA) mechanisms to save computational resources while improving model performance. The transformer combined with CNN for SwinIR [127] can be used for super-resolution reconstruction to learn the long-term dependencies of images using a shifted window mechanism. Cai et al. [128] proposed a hierarchical patch transformer, which is a hierarchical partitioning of the patches of an image for different regions, for example, using smaller patches for texture-rich regions of the image, to gradually reconstruct high-resolution images.

Transformer-based SR methods are quickly evolving and are being widely adopted due to their superior results, but their large number of parameters and the required amount of computational effort are still problems to be solved.

3.3.4. Reference-Based Models

The proposed reference-based SR method alleviates the inherent pathological problem of SR, i.e., an LR image can be obtained by degrading multiple HR images. RefSR used external images from various sources (e.g., cameras, video frames, and network images) as a reference to improve data diversity while conveying reference features and providing complementary information for the reconstruction of LR images. Zhang et al. [99] proposed that the previous RefSR suffers from the problem that the reference image is required to have similar content to the LR image, otherwise it will affect the reconstruction results. To solve the above problems, SRNTT [99] borrowed the idea of neural texture migration for semantically related features after matching the features of the LR image and reference image. In TTSR [101], Yang et al. proposed a texture transformer based on the reference image to extract the texture information of the reference image and transfer it to the high-resolution image.

4. Analyses and Comparisons of Various Models

4.1. Details of the Representative Models

To describe the performance of the SR models mentioned in Section 3 more intuitively, 16 of these representative models are listed in Table 2, including their PSNR and SSIM metrics on Set5 [35], Set14 [36], BSD100 [32], Urban100 [37], and Manga109 [46] datasets, training datasets, and the number of parameters (i.e., model size).

Table 2. PSNR/SSIM comparison on Set5 [35], Set14 [36], BSD100 [32], Urban100 [37], and Manga109 [46]. In addition, the number of training datasets and the parameters of the model are provided.

Models	Scale	Set5 PSNR/SSIM	Set14 PSNR/SSIM	BSD100 PSNR/SSIM	Urban100 PSNR/SSIM	Manga109 PSNR/SSIM	Train Data	Param.
SRCNN [18]	×2	36.66/0.9542	32.45/0.9067	31.36 0.8879	29.50/0.8946	35.60/0.9663	T91 + ImageNet	57 K
VDSR [21]	×2	37.53/0.9587	33.03/0.9124	31.90/0.8960	30.76/0.9140	-/-	BSD + T91	665 K
DRCN [19]	×2	37.63/0.9588	33.04/0.9118	31.85/0.8942	30.75/0.9133	-/-	T91	1.8 M
DRRN [57]	×2	37.74/0.9591	33.23/0.9136	32.05/0.8973	31.23/0.9188	-/-	BSD + T91	297 K
CARN [58]	×2	37.76/0.9590	33.52/0.9166	32.09/0.8978	31.92/0.9256	-/-	BSD + T91 + DIV2K	1.6 M
EDSR [68]	×2	38.11/0.9601	33.92/0.9195	32.32/0.9013	32.93/0.9351	-/-	DIV2K	43 M
ELAN [79]	×2	38.17/0.9611	33.94/0.9207	32.30/0.9012	32.76/0.9340	39.11/0.9782	DIV2K	8.3 M
MSRN [65]	×2	38.08/0.9605	33.74/0.9170	32.23/0.9013	32.22/0.9326	38.82/0.9868	DIV2K	6.5 M
RCAN [81]	×2	38.27/0.9614	34.12/0.9216	32.41/0.9027	33.34/0.9384	39.44/0.9786	DIV2K	16 M
HAN [82]	×2	38.27/0.9614	34.16/0.9217	32.41/0.9027	33.35/0.9385	39.46/0.9785	DIV2K	16.1 M
RDN [64]	×2	38.30/0.9616	34.10/0.9218	32.40/0.9022	33.09/0.9368	39.38/0.9784	DIV2K	22.6 M
NLSN [88]	×2	38.34/0.9618	34.08/0.9231	32.43/0.9027	33.42/0.9394	39.59/0.9789	DIV2K	16.1 M
RFANet [66]	×2	38.26/0.9615	34.16/0.9220	32.41/0.9026	33.33/0.9389	39.44/0.9783	DIV2K	11 M
SAN [83]	×2	38.31/0.9620	34.07/0.9213	32.42/0.9028	33.10/0.9370	39.32/0.9792	DIV2K	15.7 M
SMSR [98]	×2	38.00/0.9601	33.64/0.9179	32.17/0.8990	32.19/0.9284	38.76/0.9771	DIV2K	1 M
ESRT [126]	×2	-/-	-/-	-/-	-/-	-/-	DIV2K	751 K
TDPN [14]	×2	38.31/0.9621	34.16/0.9225	32.52/0.9045	33.36/0.9386	39.57/0.9795	DIV2K	12.8 M
SwinIR [127]	×2	38.42/0.9623	34.46/0.9250	32.53/0.9041	33.81/0.9427	39.92/0.9797	DIV2K + Flickr2K	12 M
SRCNN [18]	×3	32.75/0.9090	29.30/0.8215	28.41/0.7863	26.24/0.7989	30.48/0.9117	T91 + ImageNet	57 K
VDSR [21]	×3	33.66/0.9213	29.77/0.8314	28.82/0.7976	27.14/0.8279	-/-	BSD + T91	665 K
DRCN [19]	×3	33.82/0.9226	29.76/0.8311	28.80/0.7963	27.15/0.8276	-/-	T91	1.8 M
DRRN [57]	×3	34.03/0.9244	29.96/0.8349	28.95/0.8004	27.53/0.8378	-/-	BSD + T91	297 K
CARN [58]	×3	34.29/0.9255	30.29/0.8407	29.06/0.8034	28.06/0.8493	-/-	BSD + T91 + DIV2K	1.6 M
EDSR [68]	×3	34.65/0.9282	30.52/0.8462	27.71/0.7420	29.25/0.8093	-/-	DIV2K	43 M
ELAN [79]	×3	34.61/0.9288	30.55/0.8463	29.21/0.8081	28.69/0.8624	34.00/0.9478	DIV2K	8.3 M
MSRN [65]	×3	34.38/0.9262	30.34/0.8395	29.08/0.8041	28.08/0.8554	33.44/0.9477	DIV2K	6.5 M
RCAN [81]	×3	34.74/0.9299	30.65/0.8482	29.32/0.8111	29.09/0.8702	34.44/0.9499	DIV2K	16 M
HAN [82]	×3	34.75/0.9299	30.67/0.8483	29.32/0.8110	29.10/0.8705	34.48/0.9500	DIV2K	16.1 M
RDN [64]	×3	34.78/0.9300	30.67/0.8482	29.33/0.8105	29.00/0.8683	34.43/0.9498	DIV2K	22.6 M
NLSN [88]	×3	34.85/0.9306	30.70/0.8485	29.34/0.8117	29.25/0.8726	34.57 0.9508	DIV2K	16.1 M
RFANet [66]	×3	34.79/0.9300	30.67/0.8487	29.34/0.8115	29.15/0.8720	34.59/0.9506	DIV2K	11 M
SAN [83]	×3	34.75/0.9300	30.59/0.8476	30.59/0.8476	28.93/0.8671	34.30/0.9494	DIV2K	15.7 M
SMSR [98]	×3	34.40/0.9270	30.33/0.8412	29.10/0.8050	28.25/0.8536	33.68/0.9445	DIV2K	1 M
ESRT [126]	×3	34.42/0.9268	30.43/0.8433	29.15/0.8063	28.46/0.8574	33.95/0.9455	DIV2K	751 K
TDPN [14]	×3	34.86/0.9312	30.79/0.8501	29.45/0.8126	29.26/0.8724	34.48/0.9508	DIV2K+Flickr2K	12.8 M
SwinIR [127]	×3	34.97/0.9318	30.93/0.8534	29.46/0.8145	29.75/0.8826	35.12/0.9537	DIV2K + Flickr2K	12 M
SRCNN [18]	×4	30.48/0.8628	27.50/0.7513	26.90/0.7101	24.52/0.7221	27.58/0.8555	T91 + ImageNet	57 K
VDSR [21]	×4	31.35/0.8838	28.01/0.7674	27.29/0.7260	25.18/0.7524	-/-	BSD + T91	665 K
DRCN [19]	×4	31.53/0.8854	28.02/0.7670	27.23/0.7233	25.14/0.7510	-/-	T91	1.8 M
DRRN [57]	×4	31.68/0.8888	28.21/0.7720	27.38/0.7284	25.44/0.7638	-/-	BSD + T91	297 K
CARN [58]	×4	32.13/0.8937	28.60/0.7806	27.58/0.7349	26.07/0.7837	-/-	BSD + T91 + DIV2K	1.6 M
EDSR [68]	×4	32.46/0.8968	28.80/0.7876	27.71/0.7420	26.6 /0.8033	-/-	DIV2K	43M
ELAN [79]	×4	32.43/0.8975	28.78/0.7858	27.69/0.7406	26.54/0.7982	30.92/0.9150	DIV2K	8.3 M
MSRN [65]	×4	32.07/0.8903	28.60/0.7751	27.52/0.7273	26.04/0.7896	30.17/0.9034	DIV2K	6.5 M
RCAN [81]	×4	32.63/0.9002	28.87/0.7889	27.77/0.7436	26.82/0.8087	31.22/0.9173	DIV2K	16 M
HAN [82]	×4	32.64/0.9002	28.90/0.7890	27.80/0.7442	26.85/0.8094	31.42/0.9177	DIV2K	16.1 M
RDN [64]	×4	32.61/0.9003	28.92/0.7893	27.80/0.7434	26.82/0.8069	31.39/0.9184	DIV2K	22.6 M
NLSN [88]	×4	32.59 0.9000	28.87 0.7891	27.78 0.7444	26.96 0.8109	31.27 0.9184	DIV2K	16.1 M
RFANet [66]	×4	32.66/0.9004	28.88/0.7894	27.79/0.7442	26.92/0.8112	31.41/0.9187	DIV2K	11 M
SAN [83]	×4	32.64/0.9003	28.92/0.7888	27.78/0.7436	26.79/0.8068	31.18/0.9169	DIV2K	15.7 M
SMSR [98]	×4	32.12/0.8932	28.55/0.7808	27.55/0.7351	26.11/0.7868	30.54/0.9085	DIV2K	1 M
ESRT [126]	×4	32.19/0.8947	28.69/0.7833	27.69/0.7379	26.39/0.7962	30.75/0.9100	DIV2K	751 K
TDPN [14]	×4	32.69/0.9005	29.01/0.7943	27.93/0.7460	27.24/0.8171	31.58/0.9218	DIV2K	12.8 M
SwinIR [127]	×4	32.92/0.9044	29.09/0.7950	27.92/0.7489	27.45/0.8254	32.03/0.9260	DIV2K + Flickr2K	12 M

By comparing them, the following conclusions can be drawn: (1) To better visualize the performance differences between these models, we selected the number of parameters and the PSNR metrics of these models on the Set5 dataset and plotted a line graph, as shown in Figure 8. Usually, the larger the number of parameters, the better the reconstruction results, but this does not show that increasing the model size will improve the model performance, which is inaccurate. (2) Without considering the model size, the image super-resolution used for the transformer models tends to perform well. (3) Lightweight (that is, the number of parameters is less than 1000 K) and efficient models are in the minority in the field of image super-resolution, but in the future will become the mainstream direction of research.

Additionally, we list some classical methods, datasets, and evaluation metrics of remote sensing image super-resolution models in Table 3, sorted by year of publication. In analyzing the data, we can observe that, on the one hand, research methods in RSISR are gradually diversifying and have improved in terms of their performance in recent years. On the other hand, less attention is being paid to research on large-scale remote sensing super-resolution methods, which represents an area in which research will be challenging.

Table 3. PSNR/SSIM of some representative methods for remote sensing image super-resolution.

Models	Method	Scale	Dataset	PSNR/SSIM
LGCnet [22]	combination of local and global Information	×2 ×3 ×4	UC Merced	33.48/0.9235 29.28/0.8238 27.02/0.7333
RS-RCAN [129]	residual channel attention	×2 ×3 ×4	UC Merced	34.37/0.9296 30.26/0.8507 27.88/0.7707
WTCRR [130]	wavelet transform, recursive learning and residual learning	×2 ×3 ×4	NWPU-RESISC45	35.47/0.9586 31.80/0.9051 29.68/0.8497
CSAE [131]	sparse representation and coupled sparse autoencoder	×2 ×3	NWPU-RESISC45	29.070/0.9343 25.850/0.8155
DRGAN [132]	a dense residual generative adversarial	×2 ×3 ×4	NWPU-RESISC45	35.56/0.9631 31.92/0.9102 29.76/0.8544
MPSR [133]	enhanced residual block (ERB) and residual channel attention group(RCAG)	×2 ×3 ×4	UC Merced	39.78/0.9709 33.93/0.9199 30.34/0.8584
RDBPN [134]	residual dense backprojection network	×4 ×8	UC Merced	25.48/0.8027 21.63/0.5863
EBPN [135]	enhanced back-projection network(EBPN)	×2 ×4 ×8	UC Merced	39.84/0.9711 30.31/0.8588 24.13/0.6571
CARS [136]	channel attention	×4	Pleiades1A	30.8971/0.9489
FeNet [137]	a lightweight feature enhancement network)	×2 ×3 ×4	UC Merced	34.22/0.9337 29.80/0.8481 27.45/0.7672

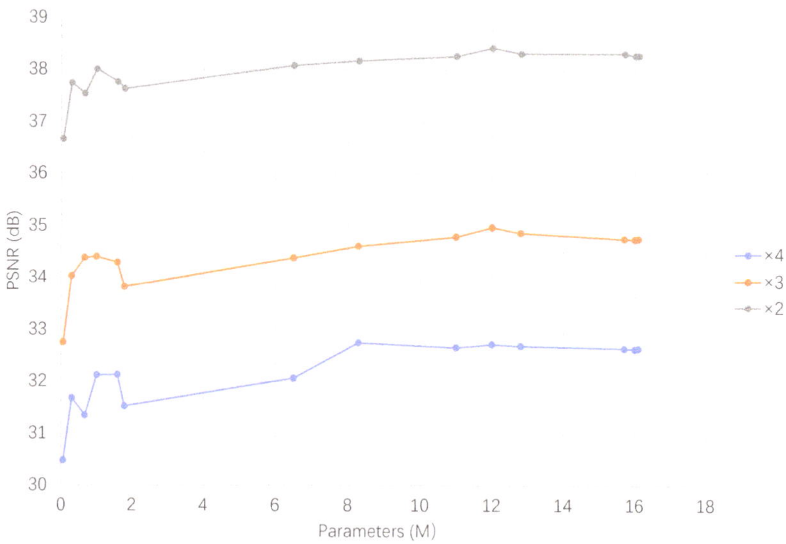

Figure 8. Variation of PSNR with the number of parameters.

4.2. Results and Discussion

To visualize the results of our experiments on remote sensing image datasets, we select classical SR models and present the visualization results (Figure 9) to visually and comprehensively illustrate their application on remote sensing images. In particular, we retrain these models and test them based on the WHU-RS19 [40] and RSC11 [44] datasets.

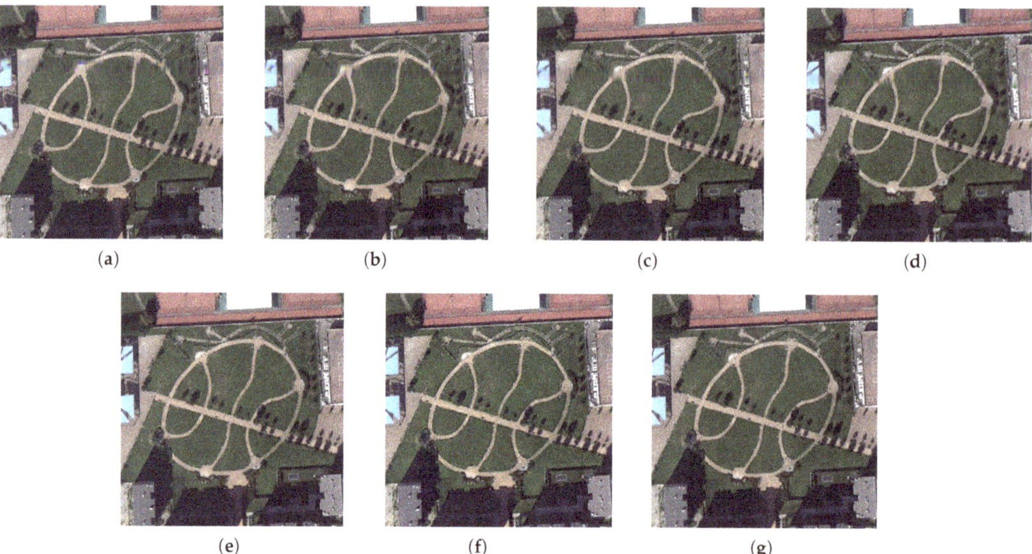

Figure 9. Comparison of visual results of different SR methods with ×2 super-resolution on the WHU-RS19 [40] dataset (square scene). (**a**) HR. (**b**) Bicubic. (**c**) EDSR [68]. (**d**) RCAN [81]. (**e**) RDN [64]. (**f**) SAN [83]. (**g**) NLSN [88].

Figure 9 illustrates the comparison of different SR methods for the super-resolution reconstruction of the WHU-RS19 [40] dataset from remote sensing images. When compared with the HR images, the results obtained by bicubic interpolation and EDSR [68] all exhibit a loss of detail and a smoothing effect. NLSN [88] appears to retain high-frequency information better, with the texture details of the reconstructed images being close to those of HR images, and the contours of the structures in the images being more clearly defined.

Figure 10 shows the results of the SR method for ×2 super-resolution reconstruction on a parking lot image in the WHU-RS19 [40] dataset. There are a variety of car colors present in the scene. Color shifts are observed using both the bicubic interpolation and RCAN [81] methods. RDN [64] with dense residual blocks provides accurate color results. The results of all other reconstruction methods are blurry.

The results of the SR method for ×2 super-resolution reconstruction on the WHU-RS19 [40] dataset from forests are given in Figure 11. Except for SAN [83] and RCAN [81], all other methods show high color similarity to the HR image. The results of several attention-based methods are also acceptable in terms of texture features, and the edge details of the forest are relatively well-defined.

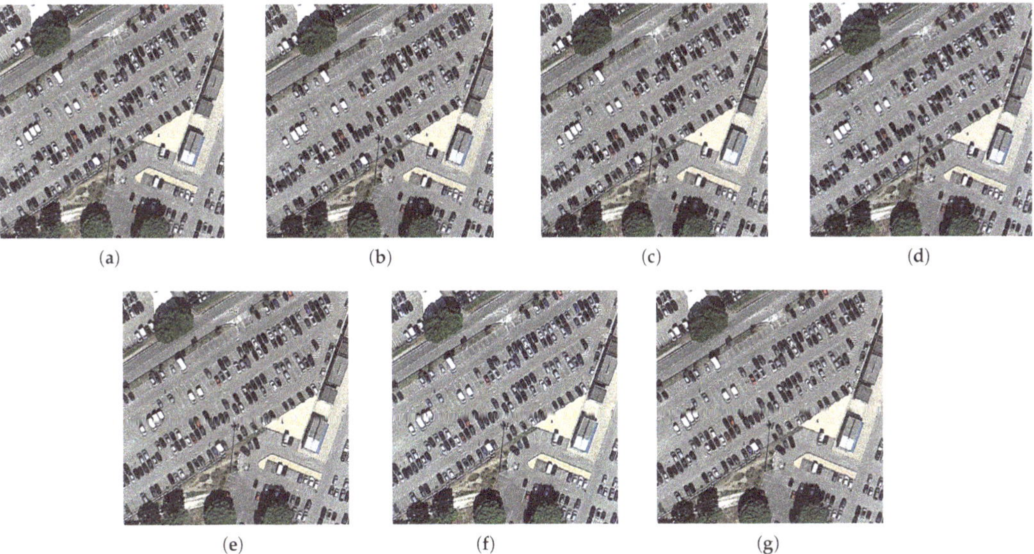

Figure 10. Comparison of visual results of different SR methods with ×2 super-resolution on the WHU-RS19 [40] dataset (parking lot scene). (**a**) HR. (**b**) Bicubic.(**c**) EDSR [68]. (**d**) RCAN [81]. (**e**) RDN [64]. (**f**) SAN [83]. (**g**) NLSN [88].

Figure 12 shows the results of the SR method for ×2 super-resolution reconstruction on the port images in the RSC11 [44] dataset. SAN [83] and RDN [64] methods provide better visual results both in terms of spatial and spectral characteristics. It is easier to identify objects such as boats in the scene based on the reconstruction results. EDSR [68] and bicubic interpolation results are blurrier around the edges.

Figure 13 shows the effect of the SR method on the ×2 super-resolution reconstruction of the residential area images in the RSC [44] dataset. In the reconstruction results of the CNN-based SR method, some exterior contours of the buildings can be observed, and useful geometric features are retained. The result of the bicubic interpolation process is blurrier and lacks some spatial detail features.

Figure 14 shows the results of the SR method for ×2 super-resolution reconstruction on sparse forest images in the RSC11 [44] dataset. The result generated by NLSN [88] is closer to the color characteristics of HR and better preserves the color of the plain land. RDN [64] retains more texture features and can observe detailed information such as branches and trunks of trees.

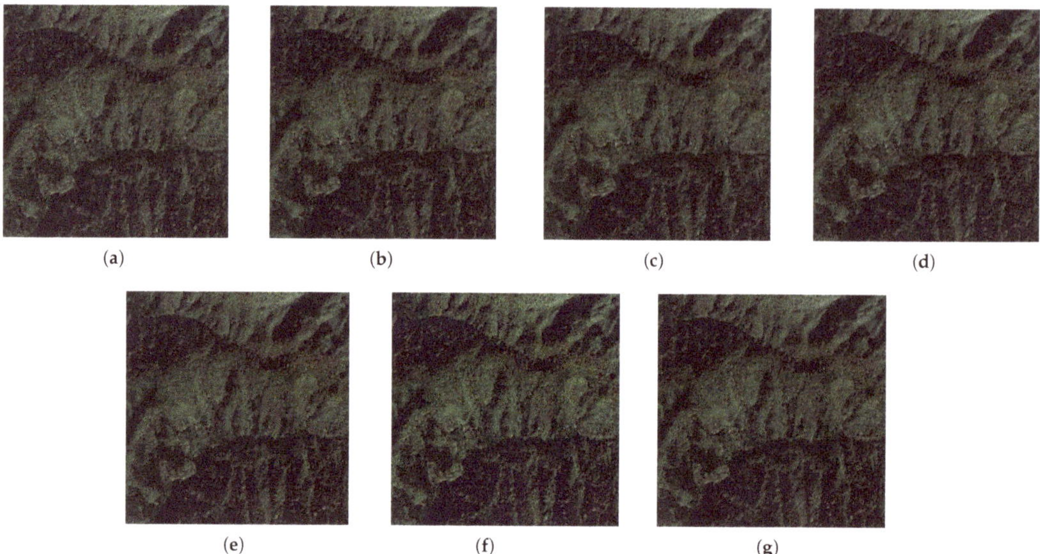

Figure 11. Comparison of visual results of different SR methods with ×2 super-resolution on the WHU-RS19 [40] dataset (forest scene). (**a**) HR. (**b**) Bicubic. (**c**) EDSR [68]. (**d**) RCAN [81]. (**e**) RDN [64]. (**f**) SAN [83]. (**g**) NLSN [88].

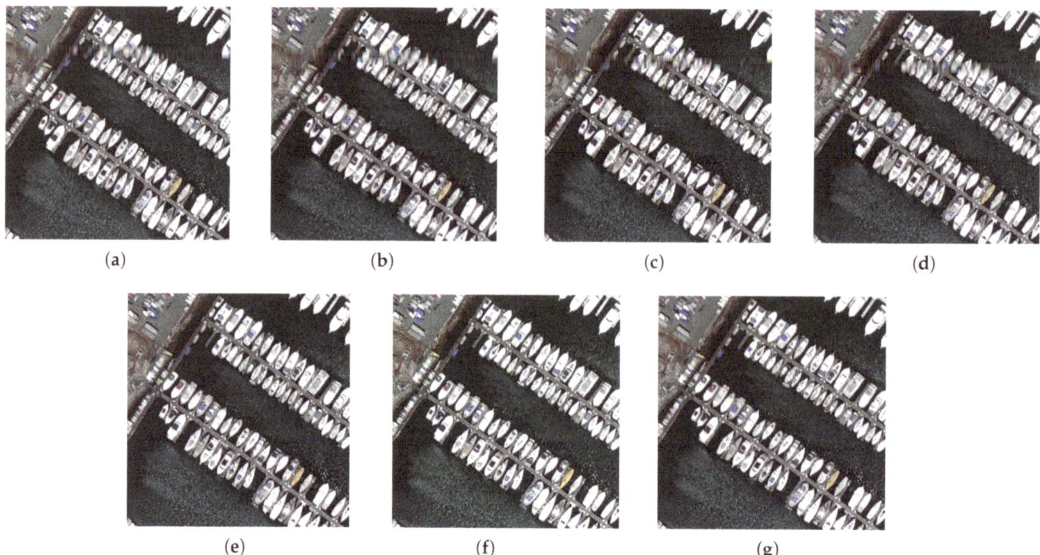

Figure 12. Comparison of visual results of different SR methods with ×2 super-resolution on the RSC11 [44] dataset (port scene). (**a**) HR. (**b**) Bicubic. (**c**) EDSR [68]. (**d**) RCAN [81]. (**e**) RDN [64]. (**f**) SAN [83]. (**g**) NLSN [88].

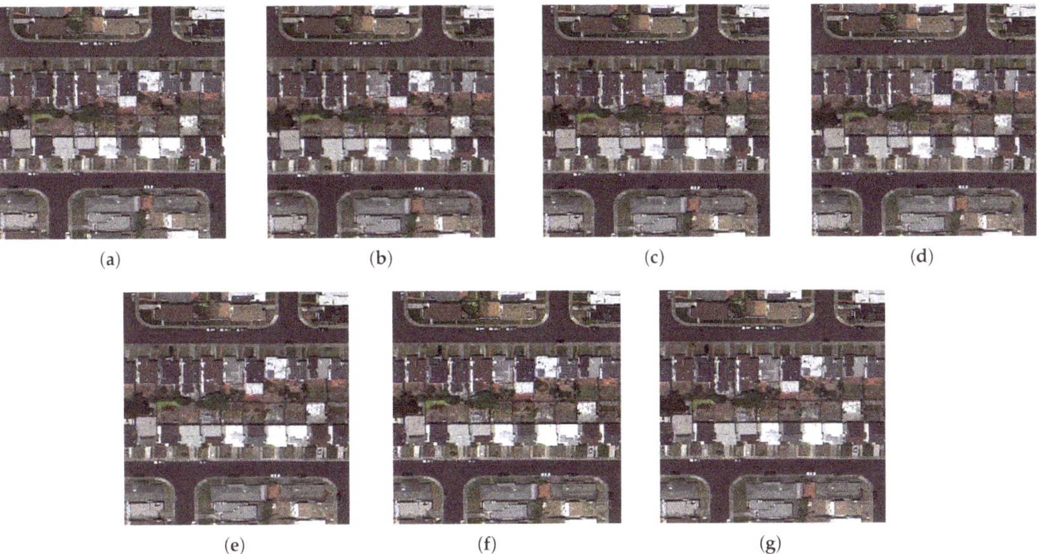

Figure 13. Comparison of visual results of different SR methods with ×2 super-resolution on the RSC11 [44] dataset (residential area scene). (**a**) HR. (**b**) Bicubic. (**c**) EDSR [68]. (**d**) RCAN [81]. (**e**) RDN [64]. (**f**) SAN [83]. (**g**) NLSN [88].

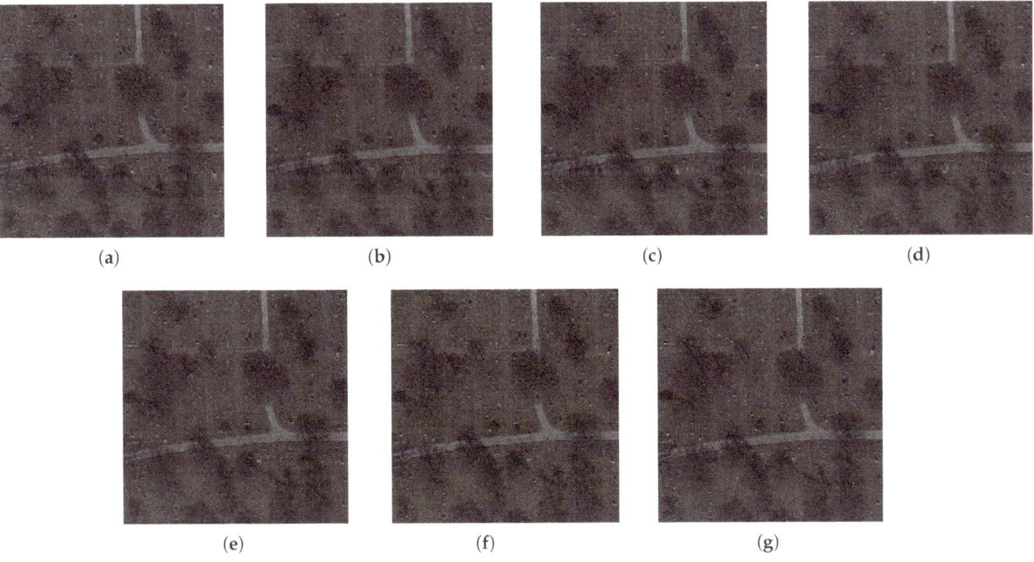

Figure 14. Comparison of visual results of different SR methods with ×2 super-resolution on the RSC11 [44] dataset (sparse forest scene). (**a**) HR. (**b**) Bicubic. (**c**) EDSR [68]. (**d**) RCAN [81]. (**e**) RDN [64]. (**f**) SAN [83]. (**g**) NLSN [88].

5. Remote Sensing Applications

Among the most critical factors for success in remote sensing applications, such as target detection and scene recognition, are high-resolution remote sensing images with rich detail. Thus, methods of super-resolution that can be used for remote sensing have received more attention from researchers. The characteristics of remote sensing images have been addressed by many researchers in recent years by proposing super-resolution methods [138–142]. In this section, these methods are divided into two categories: supervised remote sensing image super-resolution and unsupervised remote sensing image super-resolution, and their characteristics are summarized.

5.1. Supervised Remote Sensing Image Super-Resolution

Most current remote sensing image super-resolution methods use supervised learning, i.e., LR–HR remote sensing image pairs are used to train models to learn the mapping from low-resolution remote sensing images to high-resolution remote sensing images.

In [143], a multiscale convolutional network MSCNN is proposed to accomplish remote sensing image feature extraction using convolutional kernels of different sizes to obtain richer, deeper features. Inspired by DBPN [91] and ResNet [63], Pan et al. proposed the residual dense inverse projection network (RDBPN) [134], which consists of projection units with dense residual connections added to obtain local and global residuals, while achieving feature reuse to provide more comprehensive features for large-scale remote sensing image super-resolution. Lei et al. [144] focused on remote sensing images containing more flat regions (i.e., more low-frequency features), and proposed coupled-discriminate GAN (CDGAN). In CDGAN, the discriminator receives inputs from both real HR images and SR images to enhance the network's ability to discriminate low-frequency regions of remote sensing images, and a coupled adversarial loss function is introduced to further optimize the network. In [145], a hybrid higher-order attention network (MHAN) is proposed, including two parts: a feature extraction network and feature refinement network. Among them, the higher-order attention mechanism (HOA) is used to reconstruct the high-frequency features of remote sensing images while introducing frequency awareness to make full use of the layered features. E-DBPN (Enhanced-DBPN) [144] is a generator network constructed based on DBPN. Enhanced residual channel attention module (ERCAM) is added to E-DBPN, which has the advantage of not only preserving the input image original features but also allowing the network to concentrate on the most significant portions of the remote sensing images, thus extracting features that are more helpful for super-resolution. Meanwhile, a sequential feature fusion module (SFFM) is proposed in E-DBPN to process the feature output from different projection units in a progressive manner. Usually, remote sensing images have a wide range of scene scales and large differences in object sizes in the scene. To address this characteristic of remote sensing images, Zhang et al. [146] proposed the multi-scale attention network (MSAN), which uses a multi-level activation feature fusion module (MAFB) to extract features at different scales and further fuse them. In addition, a scene adaptive training strategy is proposed to make the model better adapt to remote sensing images from different scenes. In [147], a deep recurrent network is proposed. First, the encoder extracts the remote sensing image features, a gating-based recurrent unit (GRU) is responsible for feature fusion, and finally the decoder outputs the super-resolution results. To reduce the computation and network parameters, Wang et al. [148] proposed a lightweight context transformation network (CTN) for remote sensing images. The context transformation layer (CTL) in this network is a lightweight convolutional layer, which can maintain the network performance while saving computational resources. In addition, the context conversion block (CTB) composed of CTL and the context enhancement module (CEM) jointly complete the extraction and enhancement of the contextual features of remote sensing images. Finally, the feature representation is processed by the context aggregation module to obtain the reconstruction results of remote sensing images. The U-shaped attention connectivity network (US-ACN) for the super-resolution of remote sensing images proposed by Jiang et al. [149] solves the problem of the performance degradation of previous super-resolution models on real images by learning the commonality of the internal features of remote sensing images. Meanwhile, a 3D attention module is designed to calculate 3D weights by learning channels and spatial attention, which is more helpful for the learning of internal features. In addition, a U-shaped connection is added between the attention modules, which is more helpful for the learning of attention weights and the full utilization of contextual information. In [141], self-attention is used to improve the generative adversarial network and its texture enhancement function is used to solve the problems of edge blurring and artifacts in remote sensing images. The improved generator based on weight normalization mainly consists of dense residual blocks and a self-attentive mechanism for feature extraction, while stabilizing the training process to recover the edge details of remotely sensed images. In addition, a loss function is constructed by combining L1 parametric, perceptual, and texture losses, thus optimizing the network and removing remote sensing image artifacts. In [139], fuzzy kernel and noise are used to simulate the degradation patterns of real remote sensing images. The discriminator of Unet architecture is used to stabilize the training, while the residual balanced attention network (RBAN) is proposed to reconstruct the real texture of remote sensing images.

5.2. Unsupervised Remote Sensing Image Super-Resolution

Despite the fact that the super-resolution method with supervised learning has produced some results, there are still challenges associated with the pairing of LR–HR remote sensing images. On the one hand, the current remote sensing imaging technology and the influence of the external environment cannot meet the demand for high-resolution remote sensing images; on the other hand, the acquired high-resolution remote sensing images are processed with ideal degradation (such as double triple downsampling, Gaussian blur, etc.), and such degradation modes cannot approach the degradation of realistic low-resolution remote sensing images.

In [150], the generated random noise is first projected to the target resolution to ensure the reconstruction constraint on the LR input image, and the image is reconstructed using a generator network to obtain high-resolution remote sensing images by iterative iterations. In [151], a CycleGAN-based remote sensing super-resolution network is proposed. The training process uses the output of the degradation network as the input of the super-resolution network and the output of the super-resolution network as the input of the degradation network, so as to construct a cyclic loss function and thus improve the network performance. In [152], the unsupervised network UGAN is proposed. The network feeds low-resolution remote sensing images directly to the generator network and extracts features using convolutional kernels of different sizes to provide more information for the unsupervised super-resolution process. In [153], after training with a large amount of synthetic data, the most similar model to real degradation is developed, and then a loss function is derived from the difference between the original low-resolution image of the remote sensing network and the degraded image of the model.

6. Current Challenges and Future Directions

The models that have achieved excellent results in the field of image super-resolution in the past are presented in Section 3 and 4. The results of the application of these models on remotely sensed images show that they have driven the development of image super-resolution as well as remote sensing image processing techniques. The description of the methods for the super-resolution of remote sensing images in Section 5 also proves that this is a promising research topic. However, there are still many unresolved issues and challenges in the field of image super-resolution. Especially in the direction of the super-resolution of remote sensing images, on the one hand, remote sensing images, compared with natural images, are characterized by diverse application scenarios, a large number of targets, and complex types; on the other hand, external environments such as lighting and atmospheric conditions can affect the quality of remote sensing images. In this section, we will discuss these issues and introduce some popular and promising directions for future research. Remote sensing super-resolution can break through the limitations of technical level and environmental conditions, contributing to studies of resource development and utilization, disaster prediction, etc. We believe that these directions will encourage excellent work to emerge on the topic of image super-resolution, and further explore the application of super-resolution methods to remote sensing images, contributing to the advancement of remote sensing.

6.1. Network Design

A proper network architecture design not only has high evaluation metrics but also enables efficient learning by reducing the running time and computational resources required, resulting in an excellent performance. Some promising future directions for network design are described below.

(1) More Lightweight and Efficient Architecture. Although the proposed deep network models have shown excellent results on several benchmark datasets and better results on various evaluation methods, the good performance of the models is determined by multiple factors, such as the number of model parameters and the resources required for computation, which determine whether the image super-resolution methods can be applied in realistic scenarios (e.g., smartphones and cameras, etc.). Therefore, it is necessary to develop lighter and more efficient image super-resolution network architectures to achieve higher research value. For example, compressing the model size using techniques such as network binarization and network quantization is a desirable approach. In the future, achieving a lightweight and efficient network architecture will be a popular trend in the field of image super-resolution. In the meantime, the application of the network architecture to the super-resolution of remote sensing images not only improves the reconstruction efficiency but also speeds up the corresponding remote sensing image processing.

(2) Combination of Local and Global Information. For image super-resolution tasks, the integrity of local information makes the image texture more realistic, and the integrity of global information makes the image content more contextually relevant. Especially for remote sensing images, the feature details are more severely corrupted compared with natural images. Therefore, the combination of local and global information will provide richer features for image super-resolution, which helps in the generation of complete high-resolution reconstructed images. In the practical application of remote sensing images, feature-rich high-resolution images play an invaluable role. For example, when using remote sensing technology for geological exploration, the observation and analysis of the spectral characteristics of remote sensing images enables the timely acquisition of the surface conditions for accurate judgment.

(3) Combination of High-frequency and Low-frequency Information. Usually, convolutional networks are good at extracting low-frequency information, and high-frequency information (such as image texture, edge details, etc.) is easily lost in the feature transfer process. Due to the limitation of the imaging principle of the sensor, the acquired remote sensing images also occasionally have the problem of blurred edges and artifacts. Improving the network structure by designing a frequency domain information filtering mechanism, combining it with a transformer, etc., to retain the high-frequency information in the image by as much as possible will help in the reconstruction of high-resolution images. When remote sensing technology is applied to vegetation monitoring, the complete spectral and textural features in remote sensing images will help improve the classification accuracy for vegetation.

(4) Real-world Remote Sensing Image Super-resolution. In the process of remote sensing image acquisition, realistic training samples of LR–HR remote sensing images are often not obtained due to atmospheric influence and imaging system limitations. On the one hand, the LR remote sensing images obtained by most methods using ideal degradation modes (such as double triple downsampling, Gaussian fuzzy kernel, and noise) still have some differences from the spatial, positional, and spectral information of the real remote sensing images. Therefore, the methods used to generate images that are closer to the real degraded remote sensing images are of important research value. On the other hand, unsupervised super-resolution methods can learn the degradation process of LR remote sensing images

and reconstruct them in super-resolution without pairwise training samples. Therefore, research on unsupervised remote sensing image super-resolution methods should receive more attention so as to cope with some real scenarios of remote sensing image super-resolution tasks.

(5) Remote Sensing Image Super-resolution across Multiple Scales and Scenes. The scenes of remote sensing images often involve multiple landscapes, and the target objects in the same scene vary greatly in size, which presents some challenges to the learning and adaptive ability of the model. Meanwhile, most current remote sensing image super-resolution methods use $\times 2$, $\times 3$, and $\times 4$ scale factors. As a consequence, the model should be trained to learn how to map relationships between LR–HR remote sensing images from multiple scenes. For the characteristics of target objects in remote sensing images, more attention should be paid to the research of super-resolution methods with $\times 8$ and larger scale factors, so as to provide more useful information for remote sensing image processing tasks.

6.2. Learning Strategies

In addition to the network architecture design, a reasonable deep learning strategy is also an important factor in determining the network performance. Some promising learning strategy design solutions are presented here.

(1) Loss Function. Most of the previous network models choose MSE loss or L2 loss or use a weighted combination of loss functions. Most suitable loss functions for image super-resolution tasks are still to be investigated. Although some new loss functions have been proposed from other perspectives, such as perceptual loss, content loss, and texture loss, they have yet to produce satisfactory results regarding their applications in image super-resolution tasks. Therefore, it is necessary to further explore the balance between image super-resolution accuracy and perceptual quality to find more accurate loss functions.

(2) Batch Normalization. Batch normalization speeds up model training and has been widely used in various computer vision tasks. Although it solves the gradient disappearance problem, it is unsatisfactory for image super-resolution in some studies. Therefore, the normalization techniques suitable for super-resolution tasks need further research.

6.3. Evaluation Methods

Image quality evaluation, as an essential procedure in the process of image super-resolution based on deep learning, also faces certain challenges. How to propose an evaluation metric with simple implementation and accurate results still needs to be continuously explored. Some promising development directions to solve the current problem are presented below.

(1) More Precise Metrics. PSNR and SSIM, as currently popular evaluation metrics, also have some drawbacks. Although PSNR is a simple algorithm that can be implemented quickly, because it is a purely objective evaluation method, the calculated results sometimes differ greatly from those obtained by human vision. SSIM measures the quality of reconstructed images in terms of brightness, contrast, and structure. However, there are some limitations on the evaluation objects, and for images that have undergone non-structural distortion (e.g., displacement, rotation, etc.), SSIM cannot evaluate them properly. Therefore, it is necessary to propose a more accurate image evaluation index.

(2) More Diverse Metrics. As image super-resolution technology continues to advance, it is used in more fields. In this case, it is inaccurate to use only mainstream evaluation metrics such as PSNR or SSIM to evaluate reconstruction results. For example, reconstructed images applied in the medical field tend to focus more on the recovery of detailed areas, and it is necessary to refer to evaluation criteria that focus on the high-frequency information of the image. MOS, as a subjective evaluation method, evaluates the results in a manner that is closer to the visual perception of the human eye, but in practice, it is difficult to implement this method because it requires a large number of people to participate. There is a need to propose more targeted evaluation indices for certain characteristics of remote sensing images in particular. The spatial resolution and spectral resolution of remote sensing

images play a vital role in practical applications, such as weather forecasting, forestry, and geological surveying, etc. Thus, to evaluate the quality of reconstructed remote sensing images, one should consider whether the reconstruction results can optimize a particular property of these images. In general, the diversification of image evaluation metrics is also a popular development direction.

7. Conclusions

This paper provides a comprehensive summary of deep-learning-based image super-resolution methods, including common datasets, image quality evaluation methods, model reconstruction efficiency, deep learning strategies, and some techniques to optimize network metrics. In addition, the applications of image super-resolution methods in remote sensing images are comprehensively presented. Finally, although the research on image super-resolution methods, especially for remote sensing image super-resolution reconstruction, has made great progress in recent years, significant challenges remain, such as low model inference efficiency, the unsatisfactory reconstruction of real-world images, and a single approach to measuring the quality of images. Thus, we point out some promising development directions, such as more lightweight and effective model design strategies, remote sensing image super-resolution methods that are more adaptable to realistic scenes, and more accurate and diversified image evaluation metrics. We believe this review can help researchers to gain a deeper understanding of image super-resolution techniques and the application of super-resolution methods in the field of remote sensing image processing, thus promoting progress and development.

Author Contributions: Conceptualization, X.W., Y.S. and W.Y.; software, J.Y. and J.G.; investigation, J.Y., J.X. and J.L.; formal analysis, Q.C.; writing—original draft preparation, J.Y., X.W. and H.M.; writing—review and editing, Q.C., W.Y. and Y.S.; supervision, J.Z., J.X., W.Y. and J.L.; funding acquisition, X.W., J.X., J.Z., Q.C. and H.M. All authors have read and agreed to the published version of the manuscript.

Funding: This research was funded by the Natural Science Foundation of Shandong Province (ZR2020QF108, ZR2022QF037, ZR2020MF148, ZR2020QF031, ZR2020QF046, ZR2022MF238), and the National Natural Science Foundation of China (62272405, 62072391, 62066013, 62172351, 62102338, 62273290, 62103350), and in part by the China Postdoctoral Science Foundation under Grant 2021M693078, and Shaanxi Key R & D Program (2021GY-290), and the Youth Innovation Science and Technology Support Program of Shandong Province under Grant 2021KJ080, Yantai Science and Technology Innovation Development Plan Project under Grant 2021YT06000645, the Open Foundation of State key Laboratory of Networking and Switching Technology (Beijing University of Posts and Telecommunications) under Grant SKLNST-2022-1-12.

Institutional Review Board Statement: Not applicable.

Informed Consent Statement: Not applicable.

Data Availability Statement: The datasets are available on Github at https://github.com/Leilei111 11/DOWNLOADLINK, accessed on 28 September 2022.

Acknowledgments: We would like to thank the anonymous reviewers for their supportive comments to improve our manuscript.

Conflicts of Interest: The authors declare no conflict of interest.

References

1. Jo, Y.; Kim, S.J. Practical single-image super-resolution using look-up table. In Proceedings of the IEEE/CVF Conference on Computer Vision and Pattern Recognition, Nashville, TN, USA, 20–25 June 2021; pp. 691–700.
2. Loghmani, G.B.; Zaini, A.M.E.; Latif, A. Image Zooming Using Barycentric Rational Interpolation. *J. Math. Ext.* **2018**, *12*, 67–86.
3. Cherifi, T.; Hamami-Metiche, L.; Kerrouchi, S. Comparative study between super-resolution based on polynomial interpolations and Whittaker filtering interpolations. In Proceedings of the 2020 1st International Conference on Communications, Control Systems and Signal Processing (CCSSP), El-Oued , Algeria, 16–17 March 2020; pp. 235–241.
4. Xu, Y.; Li, J.; Song, H.; Du, L. Single-Image Super-Resolution Using Panchromatic Gradient Prior and Variational Model. *Math. Probl. Eng.* **2021**, *2021*, 9944385. [CrossRef]

5. Huang, Y.; Li, J.; Gao, X.; He, L.; Lu, W. Single image super-resolution via multiple mixture prior models. *IEEE Trans. Image Process.* **2018**, *27*, 5904–5917. [CrossRef] [PubMed]

6. Yang, Q.; Zhang, Y.; Zhao, T.; Chen, Y.Q. Single image super-resolution using self-optimizing mask via fractional-order gradient interpolation and reconstruction. *ISA Trans.* **2018**, *82*, 163–171. [CrossRef]

7. Xiong, M.; Song, Y.; Xiang, Y.; Xie, B.; Deng, Z. Anchor neighborhood embedding based single-image super-resolution reconstruction with similarity threshold adjustment. In Proceedings of the 2021 2nd International Conference on Artificial Intelligence and Information Systems, Chongqing, China, 28–30 May 2021; pp. 1–8.

8. Hardiansyah, B.; Lu, Y. Single image super-resolution via multiple linear mapping anchored neighborhood regression. *Multimed. Tools Appl.* **2021**, *80*, 28713–28730. [CrossRef]

9. Liu, J.; Liu, Y.; Wu, H.; Wang, J.; Li, X.; Zhang, C. Single image super-resolution using feature adaptive learning and global structure sparsity. *Signal Process.* **2021**, *188*, 108184. [CrossRef]

10. Yang, B.; Wu, G. Efficient Single Image Super-Resolution Using Dual Path Connections with Multiple Scale Learning. *arXiv* **2021**, arXiv:2112.15386.

11. Yang, Q.; Zhang, Y.; Zhao, T. Example-based image super-resolution via blur kernel estimation and variational reconstruction. *Pattern Recognit. Lett.* **2019**, *117*, 83–89. [CrossRef]

12. Wang, L.; Du, J.; Gholipour, A.; Zhu, H.; He, Z.; Jia, Y. 3D dense convolutional neural network for fast and accurate single MR image super-resolution. *Comput. Med. Imaging Graph. Off. J. Comput. Med. Imaging Soc.* **2021**, *93*, 101973. [CrossRef]

13. Yutani, T.; Yono, O.; Kuwatani, T.; Matsuoka, D.; Kaneko, J.; Hidaka, M.; Kasaya, T.; Kido, Y.; Ishikawa, Y.; Ueki, T.; et al. Super-Resolution and Feature Extraction for Ocean Bathymetric Maps Using Sparse Coding. *Sensors* **2022**, *22*, 3198. [CrossRef]

14. Cai, Q.; Li, J.; Li, H.; Yang, Y.H.; Wu, F.; Zhang, D. TDPN: Texture and Detail-Preserving Network for Single Image Super-Resolution. *IEEE Trans. Image Process.* **2022**, *31*, 2375–2389. [CrossRef] [PubMed]

15. Keys, R. Cubic convolution interpolation for digital image processing. *IEEE Trans. Acoust. Speech, Signal Process.* **1981**, *29*, 1153–1160. [CrossRef]

16. Dai, S.; Han, M.; Xu, W.; Wu, Y.; Gong, Y.; Katsaggelos, A.K. SoftCuts: A Soft Edge Smoothness Prior for Color Image Super-Resolution. *IEEE Trans. Image Process.* **2009**, *18*, 969–981. [PubMed]

17. Chang, H.; Yeung, D.Y.; Xiong, Y. Super-resolution through neighbor embedding. In Proceedings of the 2004 IEEE Computer Society Conference on Computer Vision and Pattern Recognition, Washington, DC, USA, 27 June–2 July 2004; Volume 1, p. 1.

18. Dong, C.; Loy, C.C.; He, K.; Tang, X. Image super-resolution using deep convolutional networks. *IEEE Trans. Pattern Anal. Mach. Intell.* **2015**, *38*, 295–307. [CrossRef]

19. Kim, J.; Lee, J.K.; Lee, K.M. Deeply-recursive convolutional network for image super-resolution. In Proceedings of the IEEE Conference on Computer Vision and Pattern Recognition, Las Vegas, NV, USA, 26 June–1 July 2016; pp. 1637–1645.

20. Liebel, L.; Körner, M. Single-image super resolution for multispectral remote sensing data using convolutional neural networks. *ISPRS-Int. Arch. Photogramm. Remote Sens. Spat. Inf. Sci.* **2016**, *41*, 883–890. [CrossRef]

21. Kim, J.; Lee, J.K.; Lee, K.M. Accurate image super-resolution using very deep convolutional networks. In Proceedings of the IEEE Conference on Computer Vision and Pattern Recognition, Las Vegas, NV, USA, 26 June–1 July 2016; pp. 1646–1654.

22. Lei, S.; Shi, Z.; Zou, Z. Super-resolution for remote sensing images via local–global combined network. *IEEE Geosci. Remote Sens. Lett.* **2017**, *14*, 1243–1247. [CrossRef]

23. Guo, Y.; Chen, J.; Wang, J.; Chen, Q.; Cao, J.; Deng, Z.; Xu, Y.; Tan, M. Closed-loop matters: Dual regression networks for single image super-resolution. In Proceedings of the IEEE/CVF Conference on Computer Vision and Pattern Recognition, Seattle, WA, USA, 13–19 June 2020; pp. 5407–5416.

24. Wang, Z.; Chen, J.; Hoi, S.C. Deep learning for image super-resolution: A survey. *IEEE Trans. Pattern Anal. Mach. Intell.* **2020**, *43*, 3365–3387. [CrossRef]

25. Tian, C.; Zhang, X.; Lin, J.C.W.; Zuo, W.; Zhang, Y. Generative Adversarial Networks for Image Super-Resolution: A Survey. *arXiv* **2022**, arXiv:2204.13620.

26. Chen, H.; He, X.; Qing, L.; Wu, Y.; Ren, C.; Sheriff, R.E.; Zhu, C. Real-world single image super-resolution: A brief review. *Inf. Fusion* **2022**, *79*, 124–145. [CrossRef]

27. Liu, H.; Ruan, Z.; Zhao, P.; Dong, C.; Shang, F.; Liu, Y.; Yang, L.; Timofte, R. Video super-resolution based on deep learning: A comprehensive survey. *Artif. Intell. Rev.* **2022**, 1–55. [CrossRef]

28. Yan, B.; Bare, B.; Ma, C.; Li, K.; Tan, W. Deep Objective Quality Assessment Driven Single Image Super-Resolution. *IEEE Trans. Multimed.* **2019**, *21*, 2957–2971. [CrossRef]

29. LeCun, Y.; Bengio, Y.; Hinton, G. Deep learning. *Nature* **2015**, *521*, 436–444. [CrossRef] [PubMed]

30. Abiodun, O.I.; Jantan, A.; Omolara, A.E.; Dada, K.V.; Mohamed, N.A.; Arshad, H. State-of-the-art in artificial neural network applications: A survey. *Heliyon* **2018**, *4*, e00938. [CrossRef]

31. Bengio, Y.; Courville, A.C.; Vincent, P. Representation Learning: A Review and New Perspectives. *IEEE Trans. Pattern Anal. Mach. Intell.* **2013**, *35*, 1798–1828. [CrossRef]

32. Martin, D.; Fowlkes, C.; Tal, D.; Malik, J. A database of human segmented natural images and its application to evaluating segmentation algorithms and measuring ecological statistics. In Proceedings of the Eighth IEEE International Conference on Computer Vision (ICCV 2001), Vancouver, BC, Canada, 9–12 July 2001; Volume 2, pp. 416–423.

33. Arbelaez, P.; Maire, M.; Fowlkes, C.; Malik, J. Contour detection and hierarchical image segmentation. *IEEE Trans. Pattern Anal. Mach. Intell.* **2010**, *33*, 898–916. [CrossRef] [PubMed]

34. Agustsson, E.; Timofte, R. Ntire 2017 challenge on single image super-resolution: Dataset and study. In Proceedings of the IEEE Conference on Computer Vision and Pattern Recognition Workshops, Honolulu, HI, USA, 21–16 July 2017; pp. 126–135.

35. Bevilacqua, M.; Roumy, A.; Guillemot, C.; Alberi-Morel, M.L. Low-complexity single-image super-resolution based on nonnegative neighbor embedding. In Proceedings of the 23rd British Machine Vision Conference (BMVC), Surrey, UK, 7–10 September 2012; pp. 135.1–135.10.

36. Zeyde, R.; Elad, M.; Protter, M. On single image scale-up using sparse-representations. In Proceedings of the International Conference on Curves and Surfaces, Avignon, France, 24–30 June 2010; Springer: Berlin/Heidelberg, Germany, 2010; pp. 711–730.

37. Huang, J.B.; Singh, A.; Ahuja, N. Single image super-resolution from transformed self-exemplars. In Proceedings of the IEEE Conference on Computer Vision and Pattern Recognition, Boston, MA, USA, 7–12 June 2015; pp. 5197–5206.

38. Xia, G.S.; Hu, J.; Hu, F.; Shi, B.; Bai, X.; Zhong, Y.; Zhang, L.; Lu, X. AID: A benchmark data set for performance evaluation of aerial scene classification. *IEEE Trans. Geosci. Remote Sens.* **2017**, *55*, 3965–3981. [CrossRef]

39. Zou, Q.; Ni, L.; Zhang, T.; Wang, Q. Deep learning based feature selection for remote sensing scene classification. *IEEE Geosci. Remote Sens. Lett.* **2015**, *12*, 2321–2325. [CrossRef]

40. Dai, D.; Yang, W. Satellite image classification via two-layer sparse coding with biased image representation. *IEEE Geosci. Remote Sens. Lett.* **2010**, *8*, 173–176. [CrossRef]

41. Wang, X.; Yu, K.; Dong, C.; Loy, C.C. Recovering realistic texture in image super-resolution by deep spatial feature transform. In Proceedings of the IEEE Conference on Computer Vision and Pattern Recognition, Salt Lake City, UT, USA, 18–23 June 2018; pp. 606–615.

42. Yang, Y.; Newsam, S. Bag-of-visual-words and spatial extensions for land-use classification. In Proceedings of the 18th SIGSPATIAL International Conference on Advances in Geographic Information Systems, San Jose, CA, USA, 3–5 November 2010; pp. 270–279.

43. Cheng, G.; Han, J.; Lu, X. Remote Sensing Image Scene Classification: Benchmark and State of the Art. *Proc. IEEE* **2017**, *105*, 1865–1883. [CrossRef]

44. Zhao, L.; Tang, P.; Huo, L. Feature significance-based multibag-of-visual-words model for remote sensing image scene classification. *J. Appl. Remote Sens.* **2016**, *10*, 035004. [CrossRef]

45. Yang, J.; Wright, J.; Huang, T.S.; Ma, Y. Image super-resolution via sparse representation. *IEEE Trans. Image Process.* **2010**, *19*, 2861–2873. [CrossRef] [PubMed]

46. Fujimoto, A.; Ogawa, T.; Yamamoto, K.; Matsui, Y.; Yamasaki, T.; Aizawa, K. Manga109 dataset and creation of metadata. In Proceedings of the 1st International Workshop on Comics Analysis, Processing and Understanding, Cancun, Mexico, 4 December 2016; pp. 1–5.

47. Blau, Y.; Mechrez, R.; Timofte, R.; Michaeli, T.; Zelnik-Manor, L. The 2018 PIRM challenge on perceptual image super-resolution. In Proceedings of the European Conference on Computer Vision (ECCV) Workshops, Munich, Germany, 8–14 September 2018.

48. Chen, C.; Xiong, Z.; Tian, X.; Zha, Z.J.; Wu, F. Camera lens super-resolution. In Proceedings of the IEEE/CVF Conference on Computer Vision and Pattern Recognition, Long Beach, CA, USA, 16–17 June 2019; pp. 1652–1660.

49. Deng, J.; Dong, W.; Socher, R.; Li, L.J.; Li, K.; Fei-Fei, L. Imagenet: A large-scale hierarchical image database. In Proceedings of the 2009 IEEE Conference on Computer Vision and Pattern Recognition, Miami, FL, USA, 20–25 June 2009; pp. 248–255.

50. Everingham, M.; Eslami, S.; Van Gool, L.; Williams, C.K.; Winn, J.; Zisserman, A. The pascal visual object classes challenge: A retrospective. *Int. J. Comput. Vis.* **2015**, *111*, 98–136. [CrossRef]

51. Liu, Z.; Luo, P.; Wang, X.; Tang, X. Deep learning face attributes in the wild. In Proceedings of the IEEE International Conference on Computer Vision, Santiago, Chile, 7–13 December 2015; pp. 3730–3738.

52. Wang, Z.; Bovik, A.C.; Sheikh, H.R.; Simoncelli, E.P. Image quality assessment: From error visibility to structural similarity. *IEEE Trans. Image Process.* **2004**, *13*, 600–612. [CrossRef] [PubMed]

53. Mittal, A.; Moorthy, A.K.; Bovik, A.C. No-reference image quality assessment in the spatial domain. *IEEE Trans. Image Process.* **2012**, *21*, 4695–4708. [CrossRef] [PubMed]

54. Zhang, K.; Zhao, T.; Chen, W.; Niu, Y.; Hu, J.F. SPQE: Structure-and-Perception-Based Quality Evaluation for Image Super-Resolution. *arXiv* **2022**, arXiv:2205.03584.

55. Mittal, A.; Soundararajan, R.; Bovik, A.C. Making a "Completely Blind" Image Quality Analyzer. *IEEE Signal Process. Lett.* **2013**, *20*, 209–212. [CrossRef]

56. Zhang, R.; Isola, P.; Efros, A.A.; Shechtman, E.; Wang, O. The Unreasonable Effectiveness of Deep Features as a Perceptual Metric. In Proceedings of the 2018 IEEE/CVF Conference on Computer Vision and Pattern Recognition, Salt Lake City, UT, USA, 18–23 June 2018; pp. 586–595.

57. Tai, Y.; Yang, J.; Liu, X. Image super-resolution via deep recursive residual network. In Proceedings of the IEEE Conference on Computer Vision and Pattern Recognition, Honolulu, HI, USA, 21–26 July 2017; pp. 3147–3155.

58. Ahn, N.; Kang, B.; Sohn, K.A. Fast, accurate, and lightweight super-resolution with cascading residual network. In Proceedings of the European Conference on Computer Vision (ECCV), Munich, Germany, 8–14 September 2018; pp. 252–268.

59. Qiu, Y.; Wang, R.; Tao, D.; Cheng, J. Embedded block residual network: A recursive restoration model for single-image super-resolution. In Proceedings of the IEEE/CVF International Conference on Computer Vision, Seoul, Korea, 27 October–2 November 2019; pp. 4180–4189.

60. Li, J.; Yuan, Y.; Mei, K.; Fang, F. Lightweight and Accurate Recursive Fractal Network for Image Super-Resolution. In Proceedings of the 2019 IEEE/CVF International Conference on Computer Vision Workshop (ICCVW), Seoul, Korea, 27 October–2 November 2019; pp. 3814–3823.

61. Luo, Z.; Huang, Y.; Li, S.; Wang, L.; Tan, T. Efficient Super Resolution by Recursive Aggregation. In Proceedings of the 2020 25th International Conference on Pattern Recognition (ICPR), Milan, Italy, 10–15 January 2021; pp. 8592–8599.

62. Gao, G.; Wang, Z.; Li, J.; Li, W.; Yu, Y.; Zeng, T. Lightweight Bimodal Network for Single-Image Super-Resolution via Symmetric CNN and Recursive Transformer. *arXiv* **2022**, arXiv:2204.13286.

63. He, K.; Zhang, X.; Ren, S.; Sun, J. Deep residual learning for image recognition. In Proceedings of the IEEE Conference on Computer Vision and Pattern Recognition, Las Vegas, NV, USA, 27–30 June 2016; pp. 770–778.

64. Zhang, Y.; Tian, Y.; Kong, Y.; Zhong, B.; Fu, Y. Residual Dense Network for Image Super-Resolution. In Proceedings of the 2018 IEEE/CVF Conference on Computer Vision and Pattern Recognition, Salt Lake City, UT, USA, 18–23 June 2018; pp. 2472–2481.

65. Li, J.; Fang, F.; Mei, K.; Zhang, G. Multi-scale Residual Network for Image Super-Resolution. In Proceedings of the European Conference on Computer Vision (ECCV), Munich, Germany, 8–14 September 2018; pp. 517–532.

66. Liu, J.; Zhang, W.; Tang, Y.; Tang, J.; Wu, G. Residual feature aggregation network for image super-resolution. In Proceedings of the IEEE/CVF Conference on Computer Vision and Pattern Recognition, Seattle, WA, USA, 13–19 June 2020; pp. 2359–2368.

67. He, K.; Zhang, X.; Ren, S.; Sun, J. Identity mappings in deep residual networks. In Proceedings of the European Conference on Computer Vision, Amsterdam, The Netherlands, 11–14 October 2016; Springer: Berlin/Heidelberg, Germany, 2016; pp. 630–645.

68. Lim, B.; Son, S.; Kim, H.; Nah, S.; Mu Lee, K. Enhanced deep residual networks for single image super-resolution. In Proceedings of the IEEE Conference on Computer Vision and Pattern Recognition Workshops, Honolulu, HI, USA, 21–26 July 2017; pp. 136–144.

69. Qin, J.; He, Z.; Yan, B.; Jeon, G.; Yang, X. Multi-Residual Feature Fusion Network for lightweight Single Image Super-Resolution. In Proceedings of the 2021 Asia-Pacific Signal and Information Processing Association Annual Summit and Conference (APSIPA ASC), Tokyo, Japan, 14–17 December 2021; pp. 1511–1518.

70. Park, K.; Soh, J.W.; Cho, N.I. A Dynamic Residual Self-Attention Network for Lightweight Single Image Super-Resolution. *IEEE Trans. Multimed.* **2021**. [CrossRef]

71. Sun, L.; Liu, Z.; Sun, X.; Liu, L.; Lan, R.; Luo, X. Lightweight Image Super-Resolution via Weighted Multi-Scale Residual Network. *IEEE/CAA J. Autom. Sin.* **2021**, *8*, 1271–1280. [CrossRef]

72. Liu, D.; Li, J.; Yuan, Q. A Spectral Grouping and Attention-Driven Residual Dense Network for Hyperspectral Image Super-Resolution. *IEEE Trans. Geosci. Remote Sens.* **2021**, *59*, 7711–7725. [CrossRef]

73. Liu, J.; Tang, J.; Wu, G. Residual Feature Distillation Network for Lightweight Image Super-Resolution. In Proceedings of the European conference on computer vision (ECCV), Glasgow, UK, 23–28 August 2020; Springer: Berlin/Heidelberg, Germany, 2020; pp. 41–55.

74. Lai, W.S.; Huang, J.B.; Ahuja, N.; Yang, M.H. Deep Laplacian Pyramid Networks for Fast and Accurate Super-Resolution. In Proceedings of the 2017 IEEE Conference on Computer Vision and Pattern Recognition (CVPR), Honolulu, HI, USA, 21–26 July 2017; pp. 5835–5843.

75. Chollet, F. Xception: Deep Learning with Depthwise Separable Convolutions. In Proceedings of the 2017 IEEE Conference on Computer Vision and Pattern Recognition (CVPR), Honolulu, HI, USA, 21–26 July 2017; pp. 1800–1807.

76. Chen, X.; Sun, C. Multiscale Recursive Feedback Network for Image Super-Resolution. *IEEE Access* **2022**, *10*, 6393–6406. [CrossRef]

77. Qin, J.; Huang, Y.; Wen, W. Multi-scale feature fusion residual network for Single Image Super-Resolution. *Neurocomputing* **2020**, *379*, 334–342. [CrossRef]

78. Pandey, G.; Ghanekar, U. Single image super-resolution using multi-scale feature enhancement attention residual network. *Optik* **2021**, *231*, 166359. [CrossRef]

79. Zhang, X.; Zeng, H.; Guo, S.; Zhang, L. Efficient Long-Range Attention Network for Image Super-resolution. *arXiv* **2022**, arXiv:2203.06697.

80. Hu, J.; Shen, L.; Sun, G. Squeeze-and-excitation networks. In Proceedings of the IEEE Conference on Computer Vision and Pattern Recognition, Salt Lake City, UT, USA, 18–23 June 2018; pp. 7132–7141.

81. Zhang, Y.; Li, K.; Li, K.; Wang, L.; Zhong, B.; Fu, Y. Image super-resolution using very deep residual channel attention networks. In Proceedings of the European Conference on Computer Vision (ECCV), Munich, Germany, 8–14 September 2018; pp. 286–301.

82. Niu, B.; Wen, W.; Ren, W.; Zhang, X.; Yang, L.; Wang, S.; Zhang, K.; Cao, X.; Shen, H. Single image super-resolution via a holistic attention network. In Proceedings of the European Conference on Computer Vision, Glasgow, UK, 23–28 August 2020; Springer: Berlin/Heidelberg, Germany, 2020; pp. 191–207.

83. Dai, T.; Cai, J.; Zhang, Y.; Xia, S.T.; Zhang, L. Second-order attention network for single image super-resolution. In Proceedings of the IEEE/CVF Conference on Computer Vision and Pattern Recognition, Long Beach, CA, USA, 15–20 June 2019; pp. 11065–11074.

84. Magid, S.A.; Zhang, Y.; Wei, D.; Jang, W.D.; Lin, Z.; Fu, Y.; Pfister, H. Dynamic high-pass filtering and multi-spectral attention for image super-resolution. In Proceedings of the IEEE/CVF International Conference on Computer Vision, Montreal, QC, Canada, 10–17 October 2021; pp. 4288–4297.
85. Liu, D.; Wen, B.; Fan, Y.; Loy, C.C.; Huang, T.S. Non-local recurrent network for image restoration. *Adv. Neural Inf. Process. Syst.* **2018**, *31*. Available online: http://s.dic.cool/S/tamTpxhq (accessed on 28 September 2022).
86. Mei, Y.; Fan, Y.; Zhou, Y.; Huang, L.; Huang, T.S.; Shi, H. Image super-resolution with cross-scale non-local attention and exhaustive self-exemplars mining. In Proceedings of the IEEE/CVF Conference on Computer Vision and Pattern Recognition, Seattle, WA, USA, 13–19 June 2020; pp. 5690–5699.
87. Zhang, Y.; Wei, D.; Qin, C.; Wang, H.; Pfister, H.; Fu, Y. Context reasoning attention network for image super-resolution. In Proceedings of the IEEE/CVF International Conference on Computer Vision, Montreal, QC, Canada, 10–17 October 2021; pp. 4278–4287.
88. Mei, Y.; Fan, Y.; Zhou, Y. Image super-resolution with non-local sparse attention. In Proceedings of the IEEE/CVF Conference on Computer Vision and Pattern Recognition, Nashville, TN, USA, 20–25 June 2021; pp. 3517–3526.
89. Li, K.; Hariharan, B.; Malik, J. Iterative instance segmentation. In Proceedings of the IEEE Conference on Computer Vision and Pattern Recognition, Las Vegas, NV, USA, 27–30 June 2016; pp. 3659–3667.
90. Carreira, J.; Agrawal, P.; Fragkiadaki, K.; Malik, J. Human Pose Estimation with Iterative Error Feedback. In Proceedings of the 2016 IEEE Conference on Computer Vision and Pattern Recognition (CVPR),Las Vegas, NV, USA, 27–30 June 2016; pp. 4733–4742.
91. Haris, M.; Shakhnarovich, G.; Ukita, N. Deep back-projection networks for super-resolution. In Proceedings of the IEEE Conference on Computer Vision and Pattern Recognition, Salt Lake City, UT, USA, 18–21 June 2018; pp. 1664–1673.
92. Haris, M.; Shakhnarovich, G.; Ukita, N. Recurrent back-projection network for video super-resolution. In Proceedings of the IEEE/CVF Conference on Computer Vision and Pattern Recognition, Long Beach, CA, USA, 15–20 June 2019; pp. 3897–3906.
93. Li, Z.; Yang, J.; Liu, Z.; Yang, X.; Jeon, G.; Wu, W. Feedback Network for Image Super-Resolution. In Proceedings of the 2019 IEEE/CVF Conference on Computer Vision and Pattern Recognition (CVPR), Long Beach, CA, USA, 15–20 June 2019; pp. 3862–3871.
94. Xie, W.; Song, D.; Xu, C.; Xu, C.; Zhang, H.; Wang, Y. Learning frequency-aware dynamic network for efficient super-resolution. In Proceedings of the IEEE/CVF International Conference on Computer Vision, Montreal, QC, Canada, 10–17 October 2021; pp. 4308–4317.
95. Kong, X.; Zhao, H.; Qiao, Y.; Dong, C. Classsr: A general framework to accelerate super-resolution networks by data characteristic. In Proceedings of the IEEE/CVF Conference on Computer Vision and Pattern Recognition, Nashville, TN, USA, 20–25 June 2021; pp. 12016–12025.
96. Liu, D.; Wang, Z.; Wen, B.; Yang, J.; Han, W.; Huang, T.S. Robust single image super-resolution via deep networks with sparse prior. *IEEE Trans. Image Process.* **2016**, *25*, 3194–3207. [CrossRef]
97. Gao, X.; Xiong, H. A hybrid wavelet convolution network with sparse-coding for image super-resolution. In Proceedings of the 2016 IEEE International Conference on Image Processing (ICIP), Phoenix, AZ, USA, 25–28 September 2016; pp. 1439–1443.
98. Wang, L.; Dong, X.; Wang, Y.; Ying, X.; Lin, Z.; An, W.; Guo, Y. Exploring sparsity in image super-resolution for efficient inference. In Proceedings of the IEEE/CVF Conference on Computer Vision and Pattern Recognition, Nashville, TN, USA, 20–25 June 2021; pp. 4917–4926.
99. Zhang, Z.; Wang, Z.; Lin, Z.L.; Qi, H. Image Super-Resolution by Neural Texture Transfer. In Proceedings of the 2019 IEEE/CVF Conference on Computer Vision and Pattern Recognition (CVPR), Long Beach, CA, USA, 15–20 June 2019; pp. 7974–7983.
100. MadhuMithraK, K.; Ramanarayanan, S.; Ram, K.; Sivaprakasam, M. Reference-Based Texture Transfer For Single Image Super-Resolution Of Magnetic Resonance Images. In Proceedings of the 2021 IEEE 18th International Symposium on Biomedical Imaging (ISBI), Nice, France, 13–16 April 2021; pp. 579–583.
101. Yang, F.; Yang, H.; Fu, J.; Lu, H.; Guo, B. Learning Texture Transformer Network for Image Super-Resolution. In Proceedings of the 2020 IEEE/CVF Conference on Computer Vision and Pattern Recognition (CVPR), Seattle, WA, USA, 13–19 June 2020; pp. 5790–5799.
102. Ledig, C.; Theis, L.; Huszár, F.; Caballero, J.; Aitken, A.P.; Tejani, A.; Totz, J.; Wang, Z.; Shi, W. Photo-Realistic Single Image Super-Resolution Using a Generative Adversarial Network. In Proceedings of the 2017 IEEE Conference on Computer Vision and Pattern Recognition (CVPR), Honolulu, HI, USA, 21–26 July 2017; pp. 105–114.
103. Moradi, R.; Berangi, R.; Minaei, B. A survey of regularization strategies for deep models. *Artif. Intell. Rev.* **2019**, *53*, 3947–3986. [CrossRef]
104. Kukačka, J.; Golkov, V.; Cremers, D. Regularization for Deep Learning: A Taxonomy. *arXiv* **2017**, arXiv:1710.10686.
105. Srivastava, N.; Hinton, G.E.; Krizhevsky, A.; Sutskever, I.; Salakhutdinov, R. Dropout: A simple way to prevent neural networks from overfitting. *J. Mach. Learn. Res.* **2014**, *15*, 1929–1958.
106. Srivastava, N. Improving neural networks with dropout. *Univ. Tor.* **2013**, *182*, 7.
107. Konda, K.R.; Bouthillier, X.; Memisevic, R.; Vincent, P. Dropout as data augmentation. *arXiv* **2015**, arXiv:1506.08700.
108. Hinton, G.E.; Srivastava, N.; Krizhevsky, A.; Sutskever, I.; Salakhutdinov, R.R. Improving neural networks by preventing co-adaptation of feature detectors. *arXiv* **2012**, arXiv:1207.0580.
109. Krizhevsky, A.; Sutskever, I.; Hinton, G.E. ImageNet classification with deep convolutional neural networks. *Commun. ACM* **2012**, *60*, 84–90. [CrossRef]

110. Li, M.; Soltanolkotabi, M.; Oymak, S. Gradient Descent with Early Stopping is Provably Robust to Label Noise for Overparameterized Neural Networks. *arXiv* **2020**, arXiv:1903.11680.

111. Prechelt, L. Early stopping-but when? In *Neural Networks: Tricks of the Trade*; Springer: Berlin/Heidelberg, Germany, 1998; pp. 55–69.

112. Zoph, B.; Cubuk, E.D.; Ghiasi, G.; Lin, T.Y.; Shlens, J.; Le, Q.V. Learning data augmentation strategies for object detection. In Proceedings of the European Conference on Computer Vision, Glasgow, UK, 23–28 August 2020; Springer: Berlin/Heidelberg, Germany, 2020; pp. 566–583.

113. Zhong, Z.; Zheng, L.; Kang, G.; Li, S.; Yang, Y. Random erasing data augmentation. In Proceedings of the AAAI Conference on Artificial Intelligence, New York, NY, USA, 7–8 February 2020; Volume 34, pp. 13001–13008.

114. Ioffe, S.; Szegedy, C. Batch normalization: Accelerating deep network training by reducing internal covariate shift. In Proceedings of the International Conference on Machine Learning (PMLR), Lille, France, 7–9 July 2015; pp. 448–456.

115. Hinton, G.E.; Vinyals, O.; Dean, J. Distilling the Knowledge in a Neural Network. *arXiv* **2015**, arXiv:1503.02531.

116. Lee, W.; Lee, J.; Kim, D.; Ham, B. Learning with Privileged Information for Efficient Image Super-Resolution. *arXiv* **2020**, arXiv:2007.07524.

117. Zhang, Y.; Chen, H.; Chen, X.; Deng, Y.; Xu, C.; Wang, Y. Data-free knowledge distillation for image super-resolution. In Proceedings of the IEEE/CVF Conference on Computer Vision and Pattern Recognition, Nashville, TN, USA, 20–25 June 2021; pp. 7852–7861.

118. Chen, H.; Wang, Y.; Xu, C.; Shi, B.; Xu, C.; Tian, Q.; Xu, C. AdderNet: Do we really need multiplications in deep learning? In Proceedings of the IEEE/CVF Conference on Computer Vision and Pattern Recognition, Seattle, WA, USA, 13–19 June 2020; pp. 1468–1477.

119. Song, D.; Wang, Y.; Chen, H.; Xu, C.; Xu, C.; Tao, D. Addersr: Towards energy efficient image super-resolution. In Proceedings of the IEEE/CVF Conference on Computer Vision and Pattern Recognition, Nashville, TN, USA, 20–25 June 2021; pp. 15648–15657.

120. Vaswani, A.; Ramachandran, P.; Srinivas, A.; Parmar, N.; Hechtman, B.; Shlens, J. Scaling local self-attention for parameter efficient visual backbones. In Proceedings of the IEEE/CVF Conference on Computer Vision and Pattern Recognition, Nashville, TN, USA, 20–25 June 2021; pp. 12894–12904.

121. Liu, Z.; Lin, Y.; Cao, Y.; Hu, H.; Wei, Y.; Zhang, Z.; Lin, S.; Guo, B. Swin transformer: Hierarchical vision transformer using shifted windows. In Proceedings of the IEEE/CVF International Conference on Computer Vision, Montreal, QC, Canada, 10–17 October 2021; pp. 10012–10022.

122. Touvron, H.; Cord, M.; Douze, M.; Massa, F.; Sablayrolles, A.; Jégou, H. Training data-efficient image transformers & distillation through attention. In Proceedings of the International Conference on Machine Learning (PMLR), Virtual, 18–24 July 2021; pp. 10347–10357.

123. Chu, X.; Tian, Z.; Wang, Y.; Zhang, B.; Ren, H.; Wei, X.; Xia, H.; Shen, C. Twins: Revisiting the design of spatial attention in vision transformers. *Adv. Neural Inf. Process. Syst.* **2021**, *34*, 9355–9366.

124. Raghu, M.; Unterthiner, T.; Kornblith, S.; Zhang, C.; Dosovitskiy, A. Do vision transformers see like convolutional neural networks? *Adv. Neural Inf. Process. Syst.* **2021**, *34*, 12116–12128.

125. Chen, X.; Wang, X.; Zhou, J.; Dong, C. Activating More Pixels in Image Super-Resolution Transformer. *arXiv* **2022**, arXiv:2205.04437.

126. Lu, Z.; Li, J.; Liu, H.; Huang, C.; Zhang, L.; Zeng, T. Transformer for single image super-resolution. In Proceedings of the IEEE/CVF Conference on Computer Vision and Pattern Recognition, New Orleans, LA, USA, 19–24 June 2022; pp. 457–466.

127. Liang, J.; Cao, J.; Sun, G.; Zhang, K.; Van Gool, L.; Timofte, R. Swinir: Image restoration using swin transformer. In Proceedings of the IEEE/CVF International Conference on Computer Vision, Montreal, QC, Canada, 10–17 October 2021; pp. 1833–1844.

128. Cai, Q.; Qian, Y.; Li, J.; Lv, J.; Yang, Y.H.; Wu, F.; Zhang, D. HIPA: Hierarchical Patch Transformer for Single Image Super Resolution. *arXiv* **2022**, arXiv:2203.10247.

129. Haut, J.M.; Fernández-Beltran, R.; Paoletti, M.E.; Plaza, J.; Plaza, A.J. Remote Sensing Image Superresolution Using Deep Residual Channel Attention. *IEEE Trans. Geosci. Remote Sens.* **2019**, *57*, 9277–9289. [CrossRef]

130. Ma, W.; Pan, Z.; Guo, J.; Lei, B. Achieving Super-Resolution Remote Sensing Images via the Wavelet Transform Combined With the Recursive Res-Net. *IEEE Trans. Geosci. Remote Sens.* **2019**, *57*, 3512–3527.

131. Shao, Z.; Wang, L.; Wang, Z.; Deng, J. Remote sensing image super-resolution using sparse representation and coupled sparse autoencoder. *IEEE J. Sel. Top. Appl. Earth Obs. Remote Sens.* **2019**, *12*, 2663–2674. [CrossRef]

132. Ma, W.; Pan, Z.; Yuan, F.; Lei, B. Super-resolution of remote sensing images via a dense residual generative adversarial network. *Remote Sens.* **2019**, *11*, 2578. [CrossRef]

133. Dong, X.; Xi, Z.; Sun, X.; Gao, L. Transferred Multi-Perception Attention Networks for Remote Sensing Image Super-Resolution. *Remote Sens.* **2019**, *11*, 2857. [CrossRef]

134. Pan, Z.; Ma, W.; Guo, J.; Lei, B. Super-Resolution of Single Remote Sensing Image Based on Residual Dense Backprojection Networks. *IEEE Trans. Geosci. Remote Sens.* **2019**, *57*, 7918–7933. [CrossRef]

135. Dong, X.; Xi, Z.; Sun, X.; Yang, L. Remote Sensing Image Super-Resolution via Enhanced Back-Projection Networks. In Proceedings of the IGARSS 2020 - 2020 IEEE International Geoscience and Remote Sensing Symposium, Waikoloa, HI, USA, 26 September–2 October 2020; pp. 1480–1483.

136. Wang, P.; Bayram, B.; Sertel, E. Super-resolution of remotely sensed data using channel attention based deep learning approach. *Int. J. Remote Sens.* **2021**, *42*, 6048–6065. [CrossRef]
137. Wang, Z.; Li, L.; Xue, Y.; Jiang, C.; Wang, J.; Sun, K.; Ma, H. FeNet: Feature Enhancement Network for Lightweight Remote-Sensing Image Super-Resolution. *IEEE Trans. Geosci. Remote Sens.* **2022**, *60*, 1–12. [CrossRef]
138. Huang, B.; Guo, Z.; Wu, L.; He, B.; Li, X.; Lin, Y. Pyramid Information Distillation Attention Network for Super-Resolution Reconstruction of Remote Sensing Images. *Remote Sens.* **2021**, *13*, 5143. [CrossRef]
139. Zhang, J.; Xu, T.; Li, J.; Jiang, S.; Zhang, Y. Single-Image Super Resolution of Remote Sensing Images with Real-World Degradation Modeling. *Remote Sens.* **2022**, *14*, 2895. [CrossRef]
140. Yue, X.; Chen, X.; Zhang, W.; Ma, H.; Wang, L.; Zhang, J.; Wang, M.; Jiang, B. Super-Resolution Network for Remote Sensing Images via Preclassification and Deep–Shallow Features Fusion. *Remote Sens.* **2022**, *14*, 925. [CrossRef]
141. Xu, Y.; Luo, W.; Hu, A.; Xie, Z.; Xie, X.; Tao, L. TE-SAGAN: An Improved Generative Adversarial Network for Remote Sensing Super-Resolution Images. *Remote Sens.* **2022**, *14*, 2425. [CrossRef]
142. Guo, M.; Zhang, Z.; Liu, H.; Huang, Y. NDSRGAN: A Novel Dense Generative Adversarial Network for Real Aerial Imagery Super-Resolution Reconstruction. *Remote Sens.* **2022**, *14*, 1574. [CrossRef]
143. Qin, X.; Gao, X.; Yue, K. Remote Sensing Image Super-Resolution using Multi-Scale Convolutional Neural Network. In Proceedings of the 2018 11th UK-Europe-China Workshop on Millimeter Waves and Terahertz Technologies (UCMMT), Hangzhou, China, 5–7 September 2018; Volume 1, pp. 1–3.
144. Yu, Y.; Li, X.; Liu, F. E-DBPN: Enhanced deep back-projection networks for remote sensing scene image superresolution. *IEEE Trans. Geosci. Remote Sens.* **2020**, *58*, 5503–5515. [CrossRef]
145. Zhang, D.; Shao, J.; Li, X.; Shen, H.T. Remote sensing image super-resolution via mixed high-order attention network. *IEEE Trans. Geosci. Remote Sens.* **2020**, *59*, 5183–5196. [CrossRef]
146. Zhang, S.; Yuan, Q.; Li, J.; Sun, J.; Zhang, X. Scene-adaptive remote sensing image super-resolution using a multiscale attention network. *IEEE Trans. Geosci. Remote Sens.* **2020**, *58*, 4764–4779. [CrossRef]
147. Arefin, M.R.; Michalski, V.; St-Charles, P.L.; Kalaitzis, A.; Kim, S.; Kahou, S.E.; Bengio, Y. Multi-image super-resolution for remote sensing using deep recurrent networks. In Proceedings of the IEEE/CVF Conference on Computer Vision and Pattern Recognition Workshops, Seattle, WA, USA, 14–19 June 2020; pp. 206–207.
148. Wang, S.; Zhou, T.; Lu, Y.; Di, H. Contextual Transformation Network for Lightweight Remote-Sensing Image Super-Resolution. *IEEE Trans. Geosci. Remote Sens.* **2021**, *60*, 1–13. [CrossRef]
149. Jiang, W.; Zhao, L.; Wang, Y.J.; Liu, W.; Liu, B.D. U-Shaped Attention Connection Network for Remote-Sensing Image Super-Resolution. *IEEE Geosci. Remote Sens. Lett.* **2021**, *19*, 1–5. [CrossRef]
150. Haut, J.M.; Fernandez-Beltran, R.; Paoletti, M.E.; Plaza, J.; Plaza, A.; Pla, F. A new deep generative network for unsupervised remote sensing single-image super-resolution. *IEEE Trans. Geosci. Remote Sens.* **2018**, *56*, 6792–6810. [CrossRef]
151. Wang, P.; Zhang, H.; Zhou, F.; Jiang, Z. Unsupervised remote sensing image super-resolution using cycle CNN. In Proceedings of the IGARSS 2019—2019 IEEE International Geoscience and Remote Sensing Symposium, Yokohama, Japan, 28 July–2 August 2019; pp. 3117–3120.
152. Zhang, N.; Wang, Y.; Zhang, X.; Xu, D.; Wang, X. An unsupervised remote sensing single-image super-resolution method based on generative adversarial network. *IEEE Access* **2020**, *8*, 29027–29039. [CrossRef]
153. Zhang, N.; Wang, Y.; Zhang, X.; Xu, D.; Wang, X.; Ben, G.; Zhao, Z.; Li, Z. A multi-degradation aided method for unsupervised remote sensing image super resolution with convolution neural networks. *IEEE Trans. Geosci. Remote Sens.* **2020**, *60*, 1–14. [CrossRef]

Article

Deep-Learning-Based Feature Extraction Approach for Significant Wave Height Prediction in SAR Mode Altimeter Data

Ghada Atteia [1], Michael J. Collins [2], Abeer D. Algarni [1] and Nagwan Abdel Samee [1,*]

[1] Department of Information Technology, College of Computer and Information Sciences, Princess Nourah bint Abdulrahman University, P.O. Box 84428, Riyadh 11671, Saudi Arabia
[2] Department of Geomatics Engineering, University of Calgary, Calgary, AB T2N 1N4, Canada
* Correspondence: nmabdelsamee@pnu.edu.sa

Abstract: Predicting sea wave parameters such as significant wave height (SWH) has recently been identified as a critical requirement for maritime security and economy. Earth observation satellite missions have resulted in a massive rise in marine data volume and dimensionality. Deep learning technologies have proven their capabilities to process large amounts of data, draw useful insights, and assist in environmental decision making. In this study, a new deep-learning-based hybrid feature selection approach is proposed for SWH prediction using satellite Synthetic Aperture Radar (SAR) mode altimeter data. The introduced approach integrates the power of autoencoder deep neural networks in mapping input features into representative latent-space features with the feature selection power of the principal component analysis (PCA) algorithm to create significant features from altimeter observations. Several hybrid feature sets were generated using the proposed approach and utilized for modeling SWH using Gaussian Process Regression (GPR) and Neural Network Regression (NNR). SAR mode altimeter data from the Sentinel-3A mission calibrated by in situ buoy data was used for training and evaluating the SWH models. The significance of the autoencoder-based feature sets in improving the prediction performance of SWH models is investigated against original, traditionally selected, and hybrid features. The autoencoder–PCA hybrid feature set generated by the proposed approach recorded the lowest average RMSE values of 0.11069 for GPR models, which outperforms the state-of-the-art results. The findings of this study reveal the superiority of the autoencoder deep learning network in generating latent features that aid in improving the prediction performance of SWH models over traditional feature extraction methods.

Keywords: significant wave height; deep learning; autoencoder; principal component analysis; SAR; altimeter; Gaussian process regression

Citation: Atteia, G.; Collins, M.J.; Algarni, A.D.; Samee, N.A. Deep-Learning-Based Feature Extraction Approach for Significant Wave Height Prediction in SAR Mode Altimeter Data. *Remote Sens.* **2022**, *14*, 5569. https://doi.org/10.3390/rs14215569

Academic Editor: Gwanggil Jeon

Received: 28 September 2022
Accepted: 1 November 2022
Published: 4 November 2022

1. Introduction

Wave conditions are important parameters in coastal engineering and the research of maritime processes. Wave conditions such as wave height and wind speed may assist in optimizing shipping routes and harvesting times of aquaculture farms. Wave height plays a crucial influence in energy extraction from waves, sediment movement, harbor design, and soil erosion. For any practical applications, long-term observed data are necessary. Methods for determining wave heights include field measurements, theoretical research, and numerical simulation. In most of these instances, however, there will be no long-term measurements, making wave height prediction vital.

Recently, satellite-based remote sensing systems including electro-optical, microwave radiometers, Synthetic Aperture Radar, and altimeters have been providing tremendous amounts of data about earth. Satellite data collection and processing is being used to significantly help to make operational decisions in many challenging environmental problems. For ocean observation from space, satellite imaging systems have demonstrated their capability to provide ocean wave spectra at high spatial resolution [1–3]. The Wave Mode (WM) has been specifically adopted by Envisat, ERS-1/2, and Sentinel-1A/B SARs to provide information on ocean waves in open ocean [4–7].

Traditionally, SWH retrieval schemes in satellite imagery can be classified into three categories as described in this section. The first group of algorithms depends on integrating the directional ocean wave spectrum estimated from the SAR spectrum. These methods require wind information or a first guess for the wave spectra [8–11]. Given that the relation between the wave spectrum and the SAR spectrum is nonlinear [12], and that it is not possible to predict the wave height below a certain frequency, estimation of wave height using this scheme is incomplete [12].

The second group includes empirical algorithms that have emerged since the 2000s. Empirical models can estimate SWH directly from features computed from SAR images and/or SAR spectra and do not require prior wave/wind information as in the first scheme. An example of these models that estimates significant wave height is the C-band WAVE algorithm called CWAVE. The original CWAVE algorithm has two versions, one that uses a mean and variance of image intensity (the base model) and one that adds 20 variables calculated from the image spectrum (the full-spectrum model) [12]. Many versions were developed for the CWAVE models such as the CWAVE_ERS for ERS-2 wave mode [13], CWAVE_ENV for Envisat wave mode [14], and other empirical Hs retrieval attempts for SAR data provided by Sentinel-1A [15,16], Radarsat-2 [17,18], and TerraSAR-X [19].

In the third category, various machine learning (ML) algorithms are employed for the purpose of wave parameters estimation. Machine and deep learning techniques have proven high prediction performance in several life fields. For instance, machine learning has been used for the medical diagnosis of many diseases [20–23], cyberbullying detection [24], environmental monitoring [25], augmentation of turbulence models [26], management of vegetated water resources [27,28], and in other applications. In oceanography and Earth sciences, ML has a diverse range of real-time applications. The primary applications of machine learning in oceanography include ocean weather and climate prediction, wave modeling, SWH, and wind speed predictions in regular sea state conditions [29,30] and in complex sea state conditions [29,31,32]. For instance, the study in [29] developed an ensemble of neural networks for the prediction of significant wave height from satellite images in an offshore region of a wind farm. The study by Stefanakos [31] integrated the Fuzzy Inference System with the Adaptive Network-based Fuzzy Inference System to predict wind and SWH parameters from a nonstationary wave parameters time series. Classical ML algorithms were used for wave height/wind speed estimation in the study by Stopa and Mouche [33], in which they implemented the CWAVE using a shallow feed-forward neural network using SAR images. They tested the full-spectrum model and the base model, and experimented with a few other parameters in the base model [33]. Collins et al. in [18] implemented the base and full-spectrum CWAVE models as neural networks and used Radarsat-2 Fine Quad data. They trained and tested the networks using buoy observations and investigated as well the effects of incidence angle and polarization. The common conclusion among the aforementioned studies is that neural networks extend the ability of retrieving the wave parameters using SAR images under a large range of environmental conditions in which SWH estimation is challenging. Although the results of the aforementioned study are promising, the approach of predicting SWH from satellite imagery itself is complicated and tedious.

For more than 30 years, satellite radar altimeters have provided comprehensive coverage of wind speed and significant wave height [34]. Numerous applications have made use of these data, such as offshore engineering design, numerical model validation, wind and wave climatology, and the analysis of long-term trends in oceanographic wind speed and wave height. However, the use of altimeter data for modeling SWH received little attention in the literature. Altimeter data provide several SWH and wind-speed-related parameters. The significance of these parameters for the prediction of SWH has not yet been investigated in the literature. Nevertheless, a single study has been found to utilize some altimeter features in the context of SWH prediction. The study of Quach et al. [35] integrated features from satellite altimeter data with a number of features that were derived from the modulation spectra of SAR images and developed a deep-learning-based prediction model for SWH. Their results show an improved prediction performance using their proposed method. Studies in the literature used other dataset types for predicting SWH. The majority of studies used buoy measurements for modeling SWH [32,36,37], while some recent studies used satellite imagery and extracted image features and used them for SWH prediction [12,14,29]. Only few papers have utilized altimeter data features for SWH forecasting [35]. The investigation of the significance of the entire set of features in altimeter data for SWH prediction is considered a gap in the literature. Motivated to fill this research gap, in this study, we propose a new framework to investigate the significance of altimeter data features in modeling SWH. Within this framework, a deep-learning-based feature extraction approach is introduced to extract significant features from SAR mode satellite altimeter data. The autoencoder deep learning neural network is utilized to extract latent features from the altimeter data. The autoencoder network has the capability to map the original input feature into an abstract set of significant latent features. Two traditional feature extraction approaches are utilized as well to extract extra features: the Pearson Correlation Coefficient (PCC) Analysis and the PCA. Several hybrid feature sets are then formed by fusing traditionally extracted and deep-learning-derived features. The feature sets are used for modeling SWH individually. This study proposes a novel hybrid approach for extracting significant features from altimeter data for SWH prediction. The deep learning autoencoder neural network was utilized, separately, and hybridized with other traditional feature extraction methods uniquely in this study for the prediction of significant wave height. To the best of our knowledge, no research has used autoencoders for SWH prediction in satellite data. Moreover, the hybrid combination of the (autoencoder–PCA) has not been presented in the literature for wave parameter prediction to data. The main contributions of the present study are listed as follows:

1. Proposal of a new hybrid deep-learning-based approach for extracting features from SAR mode satellite altimeter data.
2. Proposal of a new framework to investigate the significance of altimeter data-driven features for SWH prediction.
3. Utilization of autoencoder deep learning neural network to extract latent features from the altimeter data.
4. Generation of several feature sets composed of the original data features, traditionally extracted features, deep learning-derived features, and hybrid combinations from them.
5. Utilization of the generated feature sets to model SWH using the Gaussian Process Regression and Neural Network Regression algorithms and evaluate the prediction performance.
6. Comparing the prediction performance of the SWH models trained using the basic and hybrid feature sets.
7. Evaluation of the significance of the proposed features using hypothesis testing.

The paper is structured as follows: Section 2 describes the dataset used in this work, Section 3 presents the used methods, Section 4 discusses the obtained results, and Section 5 concludes the work.

2. Dataset

The used dataset is satellite records of significant wave height and wind speed measured by the SENTINEL-3A altimeter. Sentinel-3A is an Earth observation satellite specialized to oceanography. It is the first of four Sentinel-3 satellites planned as part of the Copernicus Program. On 16 February 2016, the European Space Agency launched the Sentinel-3A satellite to measure sea surface topography, temperature, and color with high accuracy and dependability to support ocean forecasting systems, as well as environmental and climate monitoring [38]. SAR Radar Altimeter (SRAL) of SENTINEL-3A SLAR is a new-generation altimeter that operates in Synthetic Aperture Radar (SAR) mode at all times [39]. SAR mode is the optimum mode for data recording over open ocean surface since it is designed to achieve high along-track resolution over generally flat surfaces [39]. A summary of Sentinel-3A altimeter operating characteristics is provided in Table 1. Altimetry instrument, exact repeat mission period, orbit parameters such as inclination and altitude, antenna properties such as frequency and frequency band, latitude coverage, and operational time for Sentinel-3A are depicted in Table 1.

Table 1. Summary of Sentinel-3A altimeter operating characteristics [40].

Altimetry Instrument	Revisit Time	Inclination	Frequency	Frequency Band	Altitude	Latitude Coverage	Life Time
SRAL	27 days	98.650	13.575 GHz 5.41 GHz	KU C	814.5 km	−78 to 81	2016–ongoing

The dataset used in this study is a subset of the IMOS (Integrated Marine Observing System, Battery Point, Australia) Surface Waves Sub-Facility Altimeter Wave/Wind database publicly available through the Australian Ocean Data Network portal (AODN: https://portal.aodn.org.au/, accessed on 15 August 2022). The IMOS dataset is a large archive of global significant wave height and wind speed records measured by 13 satellite altimeters over 33 years from 1985 to 2018 [34]. The altimeters of GEOSAT, ERS-1, TOPEX, ERS-2, GFO, JASON-1, ENVISAT, JASON-2, CRYOSAT-2, HY-2A, SARAL, JASON-3, and SENTINEL-3A were used to collect the SWH and wind speed measurements. Values of significant wave height and wind speed are derived from high-frequency altimeter data by fitting a functional form to the radar return from the ocean surface through the waveform retracking process. Altimeter data in this database were calibrated using a long-term high-quality wind speed and wave height database measured by in situ buoys from the National Oceanographic Data Center (NODC). Due to land and ice contamination, and the quality of the altimeter waveform received by the satellite, altimeter-generated Geophysical Data Records may contain data spikes. Therefore, quality flags were used to specify the goodness level of the data and aid in quality controlling it. The archive data contains a series of data flags defined as 1, 2, 3, 4, and 9; these flags represent Good data, Probably good data, Hardware error, Bad data, and Missing data, respectively [34]. In this study, only good quality and probably good data are used.

Data of two geographical positions were selected for this study; throughout the paper, the first position is referred to as P0, while the second location is referred to as P1. Position P0 is located at 0° latitude and 0° longitude (0°N 0°E), which is a point in the Atlantic Ocean. This point is called the Null Island and is located where the prime meridian meets the equator. The Null Island lies in international waters in the Atlantic Ocean, about 600 km off the coast of West Africa in the Gulf of Guinea [41]. Position P1 is located at 0° latitude and 1° longitude (0°N 1°E), which is located as well in the Atlantic Ocean. For P0, data records were acquired for the period from 26 March 2016 at 09:57:02 Z' to 11 July 2018 at 09:57:30 Z'. The data file for P0 contains 1008 records. The data of position P1 contain 1033 entries and were acquired from 3 March 2016 at 09:53:25 Z' to 15 July 2018 at 09:53:46 Z'. For each position, the data file contains 26 variables, as depicted in Table 2. The records are binned into bins of 1° by 1°. Full data resolution is provided within each bin for the corresponding latitude and longitude of every 1 Hz measurement [34].

Table 2. Data variables names and their definitions [34].

Feature Name	Feature Description
TIME	Time of data acquisition provided as a number referenced to 1985-01-01, 00:00:00 UTC.
LATITUDE	The angle that is created when a vector that is perpendicular to an ellipsoidal surface is drawn from a point on the surface.
LONGITUDE	A type of geographic coordinate that indicates the position of a point on the surface of the Earth with relation to the east–west axis.
BOT_DEPTH	Ocean floor depths underwater.
DIST2COAST	Distance from the coast.
SIG0_C	Backscatter coefficient for C-band altimetry.
SIG0_C_quality_control	Backscatter coefficients quality flags in C-band altimetry.
SIG0_C_num_obs	The number of valid C-band altimetry backscatter coefficient measurements at 20 Hz that make up the 1 Hz measurement.
SIG0_C_std_dev	The 1 Hz measurement is comprised of the standard deviation of the data that make up the 20 Hz C-band altimetry backscatter coefficient.
SIG0_KU	Coefficient of backscatter for Ku band altimetry.
SIG0_KU_quality_control	Quality flags of backscatter coefficient in Ku-band altimetry.
SIG0_KU_num_obs	Amount of all valid 20 Hz Ku-band altimetry backscatter coefficient data used to calculate the 1 Hz value.
SIG0_KU_std_dev	The 1 Hz measurement is based on the standard deviation of the data for the 20 Hz Ku-band altimetry backscatter coefficient.
SWH_C	The height of a significant wave, as measured by uncalibrated C-band altimetry.
SWH_C_quality_control	Significant wave height quality flag for C-band altimetry.
SWH_C_num_obs	Significant wave height values taken at 20 Hz by C-band altimetry and converted to a 1 Hz scale.
SWH_C_std_dev	Standard deviation of significant wave height measured at 1 Hz using C-band altimetry, based on data collected at 20 Hz.
SWH_KU	Significant wave height as measured by uncalibrated Ku-band altimetry.
SWH_KU_CAL	The significant wave height was calibrated using the Ku-band altimetry.
SWH_KU_quality_control	Flag indicating the quality of the Ku-band altimetry significant wave height data.
SWH_KU_num_obs	The number of valid Ku-band altimetry readings of significant wave height that were used to construct the 1 Hz measurement.
SWH_KU_std_dev	The standard deviation of the significant wave height data collected at 20 Hz by Ku-band altimetry and used to construct the 1 Hz measurement.
UWND	Modeling zonal wind speed using ECMWF.
VWND	Modeling meridional wind speed using ECMWF.
WSPD	Wind speed derived from wind function alone and not calibrated.
WSPD_CAL	The wind speed was calibrated based on the wind function.

3. Methods

In this section, the proposed framework and methods used for feature extraction are presented. Regression algorithms used for SWH modeling and performance evaluation methods are also provided.

3.1. Proposed Framework

In this study, the proposed framework introduces a hybrid approach for extracting the significant features for the prediction of SWH from altimeter data. This hybrid approach combines the features generated by three feature extraction techniques. The Pearson Correlation Analysis, Principal Component Analysis algorithms, and Sparse Autoencoder deep neural network are utilized to extract the most significant attributes from the input features. Multiple hybrid feature combinations are introduced and examined for modeling SWH using Gaussian Process Regression and Neural Network Regression. The proposed framework is composed of four phases: the data preprocessing phase, feature sets formation phase, SWR modeling phase, and model evaluation and testing phase. In the data preprocessing phase, multiple preprocessing steps are conducted to prepare the data for the feature sets formation phase. In the feature sets formation phase, a number of basic and hybrid feature sets are created from the input data. Basic sets include the

ALL-Set, PCC-Set, PCA-Set, and AUT-Set-N. The ALL-Set is composed of all features in the dataset excluding the response variable to be predicted, namely SWH. Pearson Correlation Coefficients between input features and the response variable are thresholded to select the features encompassed in the PCC-Set. Features in the PCA-Set are generated by the PCA algorithm with 95% variance. Autoencoder-driven features are generated by training a sparse autoencoder neural network by all input features and extracting a specified number of latent space features from the encoder. Up to three latent features are derived by the autoencoder network and formed three autoencoder-driven feature sets, namely AUT-Set-1, AUT-Set-2, and AUT-Set-3. Multiple hybrid feature sets are further formed using various combinations of the PCC, PCA, and AUT feature sets. Hybrid sets include the HAT-N and HCAT-N sets. The composition of theses sets is elaborated in the Results section. In the SWH modeling phase, the training dataset is used for training a number of Gaussian Process regression and Neural Network regression models. The regression models are validated using a 5-cross validation scheme and tested on a holdout test set in the final model evaluation and testing phase. The prediction performance of the SWH models trained on the hybrid feature sets are compared with that trained by the basic PCC, PCA, and autoencoder feature sets, as well as all input features set. The proposed framework is presented in Figure 1.

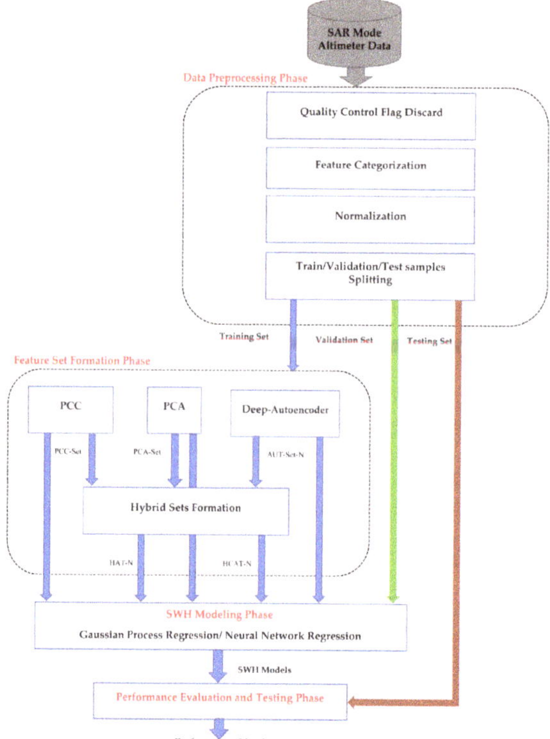

Figure 1. Proposed framework of the current study.

3.2. Data Preprocessing

In this phase, multiple data preprocessing steps are conducted to prepare the data for the feature sets formation phase, as shown in Figure 2. The target/response variable to be predicted in this work is SWH. The remaining variables are preprocessed to prepare the input features that will be used for predicting the target. In this study, quality control flags for the SWH and SIG0 are discarded, and the remaining features are divided into four categories: observing condition features, site related features, wind speed features, and measured features. The features under each category are depicted in Table 3. The measured features are further categorized according to the frequency band used for data acquisition into KU-band-related features and C-band features. The SRAL altimeter on Sentinel 3A uses the KU-band (13.575 GHz, bandwidth 350 MHz) for range measurements. However, it uses the C-band (5.41 GHz, bandwidth 320 MHz) for ionospheric correction [42]. This is achieved in the SAR acquisition mode by using bursts of 64 KU-band pulses surrounded by two C-band pulses [42]. Therefore, in this study, the SWH modeling was conducted using only the KU-band-measured features along with the other site and observing condition features.

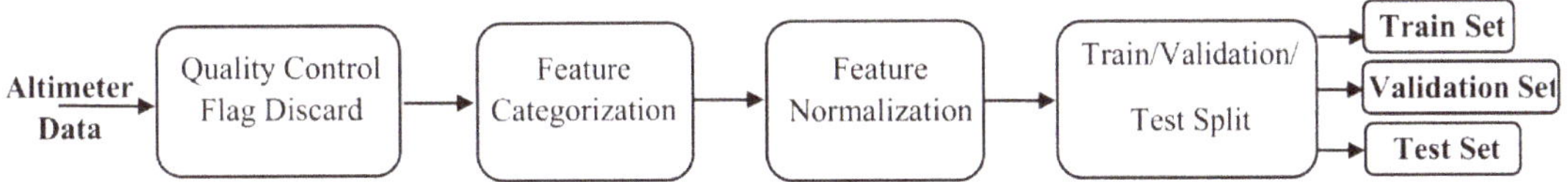

Figure 2. Preprocessing phase of the proposed framework.

Table 3. Categorization of input features.

Observing Condition Features	Site Features	Measured Features		Wind Speed Features
		KU-Band Features	**C-Band Features**	
TIME	DIST2COAST	SIG0_KU	SIG0_C	VWND
LATITUDE	BOT_DEPTH	SIG0_KU_std_dev	SIG0_C_std_dev	WSPD
LONGITUDE		SIG0_KU_num_obs	SIG0_C_num_obs	UWND
		SWH_KU_num_obs	SWH_C_num_obs	WSPD_CAL
		SWH_KU_std_dev	SWH_C_std_dev	

In order to maintain close ranges of the input variables, the features are normalized to have unit standard deviation and zero mean, with the following exceptions. The latitude and longitude features are replaced by their sine and cosine values after converting them into angles in the range $[0, 2\pi]$ rad. Features containing the number of observations are converted to discrete values in the range [0–3] by subtracting each entry by the features maximum value. After data normalization, the dataset is subdivided into training and testing sets with 90:10 training to testing ratio.

3.3. Feature Sets Formation

In this phase, a number of basic and hybrid feature sets are generated and used to model the SWH. A number of feature extraction and reduction approaches were used to extract significant features from the input data. The Pearson Correlation Analysis, Principal Component Analysis, and the autoencoder deep neural network are used for feature extraction and selection. Three basic feature sets are formed using features extracted from the all-features set (ALL-Set) by these algorithms: PCC-Set, PCA-Set, and AUT-Sets. The feature formation phase is depicted in Figure 3.

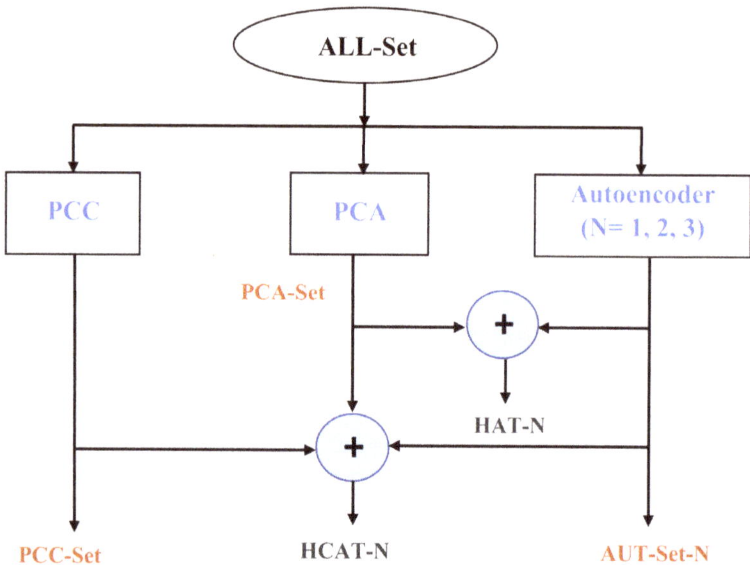

Figure 3. The feature formation phase of the proposed framework; + represents the fusion between feature sets.

3.3.1. Pearson Correlation Analysis

Pearson Correlation Analysis is an approach to find the linear correlation between two random variables. The Pearson correlation coefficient is considered a measure of dependency between two vectors. PCC between a pair of variables X and Y can be evaluated using Equation (1). PCC can take values in the range $[-1, 1]$. Absolute PCC values near 1 mean high linear dependency between variables, while values close to zero show low dependency.

$$PCC = \frac{cov(X, Y)}{\sqrt[2]{\sigma(X)\sigma(Y)}} \tag{1}$$

where, $\sigma(X)$, $\sigma(Y)$ are the variance of X and Y, respectively, and $cov(X, Y)$ is the covariance matrix between X and Y.

In this study, Person Correlation Coefficients between input features and the response variable are computed and thresholded to select the features encompassed in the PCC-Set. The selection of the threshold value is data-dependent, as discussed in the Results section.

3.3.2. Principal Component Analysis

Principal component analysis, or PCA, is traditional data analysis approach that generates a series of the best linear approximations for a given dataset. It is considered the most widely used method for dimensionality reduction with minimum information loss [22,43,44]. In this research, the PCA is employed to extract a sequence of uncorrelated features, or principal components (PCs), from the altimeter observational data. The new PC features represent linear combinations of the input variables and comprise the major information contained in the original data. For data matrix Z with m number of variables and n number of samples given as $Z = \begin{pmatrix} v_{11} & v_{21} & \cdots & v_{m1} \\ \vdots & & \ddots & \vdots \\ v_{1n} & v_{2n} & \cdots & v_{mn} \end{pmatrix}$, the PCA algorithm could generate k uncorrelated features using linear combinations of the input variables. The principal components denoted as u_1, u_2, u_3, ..., u_k are given in Equation (2), where l_{ij} is the linear combinations coefficient [44].

$$
\left\{
\begin{aligned}
u_1 &= l_{11}\, v_1 + l_{12}\, v_2 + l_{13}\, v_3 + \ldots + l_{1m}\, v_m = \sum_{i=1}^{m} l_{1i}\, v_i \\
u_2 &= l_{21}\, v_1 + l_{22}\, v_2 + l_{23}\, v_3 + \ldots + l_{2m}\, v_m = \sum_{i=1}^{m} l_{2i}\, v_i \\
&\quad\quad\quad\quad\quad \cdot \\
&\quad\quad\quad\quad\quad \cdot \\
&\quad\quad\quad\quad\quad \cdot \\
u_k &= l_{k1}\, v_1 + l_{k2}\, v_2 + l_{k3}\, v_3 + \ldots + l_{km}\, v_m = \sum_{i=1}^{m} l_{ki}\, v_i
\end{aligned}
\right.
\tag{2}
$$

The principal components satisfy two conditions; the retrieved features (u_1, u_2, u_3, ..., u_k) are uncorrelated, and the first principal component, u_1, has the highest variance followed by u_2, etc. The number of extracted features, PCs, is determined based on the Cumulative Percent Variance (CPV). CPV is used as a threshold to determine the k number of PCs that covers the required percent of information in the original data. The level of CPV is decided in advance. In this work, the PCA-Set includes the principal components generated by the PCA algorithm with a CPV of 95%.

3.3.3. Autoencoder Neural Network

An autoencoder is a deep learning neural network composed of an encoder–decoder structure, as shown in Figure 4, that learns a compressed version of input data [45]. Basically, autoencoder networks are used for the reconstruction of input data. The encoder converts the input to a compressed representation, while the decoder attempts to reverse the mapping in order to reconstruct the input. The ability of autoencoder network to learn a compacted representation of the input and deliver it at the encoder end makes it an effective tool for feature extraction and dimensionality reduction. Autoencoders can map input information into abstract latent space features, which are more informative and smaller in size. In this study, an autoencoder is used to generate a set of compact latent features that capture the most important attributes from the input data. These features are then used as predictors for the SWH model. Unsupervised sparse autoencoder training is performed in this study to generate the latent features. The autoencoder objective function is the mean squared error function with weight regularization, Ω_w, and sparsity regularization, Ω_{sp}, provided in Equation (3) [46]. Sparsity and weight regularization were included in the objective function to enable the autoencoder to learn representations from a small number of the training samples. The coefficients β and λ in Equation (3) control the effect of the sparsity and weight regularizers on the objective function, respectively.

$$
E = \underbrace{\frac{1}{S} \sum_{a=1}^{S} \sum_{b=1}^{V} (x_{ba} - \hat{x}_{ba})^2}_{mean\ squared\ error} + \lambda \times \Omega_w + \beta \times \Omega_{sp}
\tag{3}
$$

where x is a training example, $\hat{x}$ is the estimate of the training example, and S and V are the number of samples and the number of variables in the data, respectively. Ω_{sp} and Ω_w are calculated using the Equations (4) and (5) [46]

$$
\Omega_w = \frac{1}{S} \sum_{l}^{L} \sum_{j}^{S} \sum_{i}^{V} \left(w_{ji}^{(l)} \right)^2
\tag{4}
$$

$$
\Omega_{sp} = \sum_{i=1}^{D^{(1)}} KL\left(\rho \| \hat{\rho}_i \right) = \sum_{i=1}^{D^{(1)}} \rho\, log\left(\frac{\rho}{\hat{\rho}_i} \right) + (1 - \rho)\, log\left(\frac{1 - \rho}{1 - \hat{\rho}_i} \right)
\tag{5}
$$

where L is the number of network layers, and w is the weight of a network neuron located according to the indices i, j, l. $\hat{\rho}_i$ is the average activation of the ith network neuron, ρ is the average of the first layer ($D^{(1)}$) neurons, and $KL\left(\rho\|\hat{\rho}_i\right)$ is the Kullback–Leibler divergence between ρ and $\hat{\rho}_i$ [46].

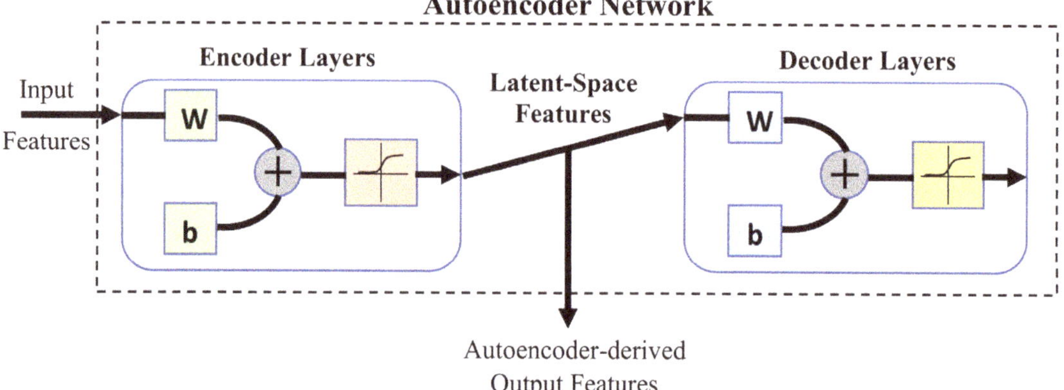

Figure 4. Structure of the autoencoder deep learning neural network; W and b are the weight and bias of the network neurons.

In this work, to generate the latent features at the encoder end, the autoencoder was fed by all input features and trained in an unsupervised fashion using the scaled conjugate gradient algorithm (SCGA) [47]. The training process ends when either the gradient reaches a minimum of 1×10^{-6} or the number of epochs approaches 5000. The weight and sparsity regularizer coefficients were set to $\lambda = 0.001$ and $\beta = 0.01$, respectively, and the Logistic Sigmoid function was used as the transfer function of both the encoder and encoder. These values were selected by experiment as they provide the best autoencoder performance. After training, the latent features are extracted from the encoder, and the decoder is discarded. These latent features form the AUT-Sets are used for modeling the SWH. Up to three latent features are derived by the autoencoder network, forming three autoencoder-driven feature sets, namely AUT-Set-1, AUT-Set-2, and AUT-Set-3.

3.3.4. Hybrid Feature Set Generation

After generating the PCC-Set, PCA-Set, and AUT-Sets, several hybrid feature sets were composed by merging features from these basic sets. Hybrid sets that were composed by fusing the features of the PCC-Set, PCA-Set, and AUT-Set-N are demoted throughout the paper as HCAT–N, where N is the number of autoencoder output features. Another group of hybrid feature sets is formed by merging the features in the PCA-Set with that of the AUT-Set-N. These sets are denoted herein as HAT-N. In this study, N takes the values 1, 2, and 3. Therefore, there are three HCAT sets and three HAT sets: HCAT-1, HCAT-2, HCAT-3, HAT-1, HAT-2, and HAT-3. The number of features in each hybrid set is dependent on the number of features in the basic sets which, itself, is data-dependent.

3.4. SWH Modeling

An accurate prediction of SWH is challenging due to its strong intermittency and instability [48]. Traditional regression models such as regression trees and K-nearest Neighbor (KNN) are insufficient for an accurate prediction of SWH due to the complexity of the data [29]. On the other, more sophisticated regression algorithms such as the artificial neural networks and kernel-based models could offer better fits to this problem. The Gaussian Processes is a kernel-based algorithm that provides flexible models that could work well with such data due to its capability of defining distributions over functions [49].

Therefore, the Gaussian Process Regression and neural network regression are utilized for modeling SWH using altimeter data. Multiple GPR models with various kernel functions were trained using the training set associated with each of the basic and hybrid sets. Kernels utilized for the GPR models include the Exponential, Squared Exponential, Rational Quadratic, and Matern functions.

3.4.1. Gaussian Process Regression

Gaussian Process Regression is a Bayesian approach to regression that is nonparametric. GPR computes the probability distribution for all admissible data-fitting functions [50]. Using the training data, the posterior probability is obtained, and then the predictive posterior distribution on the points of interest is computed. In GPR, we begin by assuming a Gaussian process prior, f(x), which may be characterized by a mean function, m(x), and covariance function, k(x, x′), for every input x. Expressions of m and k are given by Equations (6)–(8) [50].

$$f(x) \sim GP(m, k) \tag{6}$$

$$m(x) = \frac{1}{4} x^2 \tag{7}$$

$$k(x, x') = e^{\left(-\frac{1}{2}(x-x')^2\right)} \tag{8}$$

Specifically, a Gaussian process is comparable to an infinite-dimensional multivariate Gaussian distribution in which all sets of dataset labels are jointly Gaussian distributed. By selecting the mean and covariance functions, we can include previous knowledge about the space of functions into this GP prior. During model selection, the shape of the mean function and covariance kernel function in the GP prior are chosen and tweaked. The mean function can be zero or equals the mean of the training dataset. There are numerous alternatives for the covariance kernel function. In this work, multiple kernel functions are used for modeling the SWH using each feature set. The Exponential, Squared Exponential, Matern, Quadratic, and Rational Quadratic kernel functions are used.

3.4.2. Neural Network Regression

The neural network used for the SWH regression in this study is a narrow feedforward NN with one hidden fully connected layer and one fully connected output layer. This architecture was selected to accommodate the limited number of input features and data size. The hidden layer contains 10 neurons and is followed by a ReLu activation function. The first hidden layer is connected to the training data (the input feature matrix). Each input is multiplied with a weight and then added to a bias at each neuron in the fully connected layer. The output from this layer passes to the activation function and then to the final fully connected layer, which produces the predicted response as the NN output.

3.5. Model Evaluation and Testing

After the GPR and NNR models are trained using the training set associated with each of the feature sets individually, the models are evaluated in a 5-fold cross-validation scheme to reduce potential overfitting. The trained models are then assessed using a hold-out set. The prediction performance of the trained models is assessed using the root mean square error (RMSE) and the coefficient of determination R^2.

The RMSE is a measure of how far the predicted values and the true values in a dataset differ from one another. The mathematical expression of the RMSE is given by Equation (9).

$$RMSE = \sqrt{\frac{\sum_{i=1}^{n}(PV_i - TV_i)^2}{n}} \tag{9}$$

where PV_i and TV_i represent the predicted and true values of the ith observation of n samples.

The coefficient of determination is a measure of the amount of variation in the dependent variable that can be accounted for by the predictors in a regression analysis. R^2 is an indicator of how well a model fits a dataset. The value of R^2 can be anywhere from zero to one. R^2 can be calculated using the formula of Equation (10);

$$R^2 = 1 - RSS/TSS \tag{10}$$

where $RSS = \sum_{i=1}^{n} (y_i - \hat{y}_i)^2$ represent the sum of squares of residuals, and $TSS = \sum_{i=1}^{n} (y_i - \overline{y})^2$ is the total sum of squares, respectively. The true target of the ith sample is denoted as y, the true observations mean is $\overline{y}$, and $\hat{y}_i$ is the predicted value of the target.

4. Results and Discussion

According to the proposed framework, the input features were preprocessed, and multiple feature selection techniques were used to generate several combination sets of features. In the experiment conducted within the proposed framework, the SWH is modeled using the KU-band features. The KU-band-measured features along with the observing condition features, site-related features, and wind speed features are used to form the feature sets. These sets are used individually to model the SWH measured by the altimeters KU frequency band. Table 4 illustrates the KU-band-based features used in this study.

Table 4. KU-band-based features used for modeling KU-based SWH.

KU-Band-Based Features
TIME
LATITUDE (sine and cosine): LATSINE, LATCOSINE
LONGITUDE (sine and cosine): LONGSINE, LONGCOSINE
DIST2COAST
BOT_DEPTH
SIG0_KU
SIG0_KU_std_dev
SIG0_KU_num_obs
SWH_KU_num_obs
SWH_KU_std_dev
VWND
WSPD
UWND
WSPD_CAL

4.1. Feature Sets Formation

In this work, the calibrated SWH measured using the altimeters KU frequency band, SWH_KU_CAL, is considered the response variable. The KU-based features depicted in Table 4 are used to form the basic and hybrid feature sets in this experiment. The ALL-Set is composed of 16 features which represent all KU-based features except the target variable and the noncalibrated version of it. To create the PCC-Set, Pearson correlation coefficients between the input features and the target variable were calculated. Table 5 depicts the absolute values of the PCC for each input feature. Normally, SWH is highly correlated with itself and its noncalibrated version. However, the recorded |PCC| values for the other predictors are less than 0.6. For both positions P0 and P1, the calibrated and noncalibrated wind speed based on the wind function predictors, WSPD_CAL and WSPD, record the highest correlation with the target, followed by the VWND, and then the KU-altimeter backscatter coefficient, SIG0_KU. It was noticed that the correlation between the target and the rest of the predictors is low (less than 0.1); therefore, the absolute correlation coefficients between the input features and the SWH_CAL were thresholded with a value of 0.1. Thus, the PCC-Set is formulated from the features that satisfy the criterion $|PCC| \geq CCt$. The features included in the PCC-Set for P0 and P1 and their correlation values are highlighted in gray in Table 5. The SIG0_KU, VWND, WSPD_CAL, SWH_KU_std_dev, SIG0_KU_std_dev, and WSPD are included in the PCC-Set of both positions P0 and P1. However, it was noticed that for P1, the TIME variable achieved a PCC of 0.1, and therefore, it was included in the PCC-Set of this position.

Table 5. Absolute values of Pearson correlation coefficients between SWH_KU_CAL and the KU-based features for positions P0 and P1; The features included in the PCC-Set and their correlation values are highlighted in gray.

Position P0		Position P1	
Feature	|PCC|	Feature	|PCC|
SWH_KU_CAL	1	SWH_KU_CAL	1
TIME	0.0163990439235463	TIME	0.102288865532865
SWH	0.999999732870120	SWH	0.999999542438490
SIG0_KU	0.336577351930497	SIG0_KU	0.445150343609593
UWND	0.082578003464963	UWND	0.0603089207682942
VWND	0.395006371045506	VWND	0.451649290833774
WSPD_CAL	0.455941025835115	WSPD_CAL	0.579889225755999
SWH_KU_std_dev	0.173924815008291	SWH_KU_std_dev	0.371314384227738
SIG0_KU_std_dev	0.188570572892636	SIG0_KU_std_dev	0.209765508152799
DIS2COAST	0.0120031039897664	DIS2COAST	0.0702488522142944
BOT_DEPTH	0.00854422341173639	BOT_DEPTH	0.0184796601456242
WSPD	0.457591377033189	WSPD	0.579140739443493
LATSINE	0.00510651233370158	LATSINE	0.000559112294287260
LATCOSINE	0.000914880701032926	LATCOSINE	0.0223517634977151
LONGSINE	0.00209464350081526	LONGSINE	0.0108063236584730
LONGCOSINE	0.0185392818790396	LONGCOSINE	0.00620089102115493
SWH_KU_num_obs	0.0204934366118391	SWH_KU_num_obs	0.00454090475465734
SIG0_KU_num_obs	0.0204934366118391	SIG0_KU_num_obs	0.00454090475465734

As the TIME feature records different PCC values for P0 and P1, we further investigate the correlation behavior between the TIME feature and the target variable for seven geographical positions. Table 6 presents the \|PCC\| values for the TIME feature for the tested positions, the number of observations, and the time period over which the records were collected for each position. It is observable from Table 6 that the TIME feature generally records low correlation with the SWH. For, P1, P3, and P4, the correlation coefficient equals roughly 0.1. Therefore, for the PCC threshold used in this work, the TIME feature is included in the PCC-Set of these positions. However, the PCC values for P0, P2, P5, and P6 are 10 times lower than the other positions, and thus the TIME feature is discarded from the corresponding PCC-Set.

Table 6. Absolute values of Pearson correlation coefficients between SWH_KU_CAL and the TIME feature for seven geographical positions; # DP is the number of data points (observations).

Position	P0	P1	P2	P3	P4	P5	P6
Location	(0°N 0°E)	(0°N 0°E)	(0°N 2°E)	(0°N 3°E)	(0°N 4°E)	(0°N 5°E)	(0°N 6°E)
Period of Acquisition	26 March 2016–11 July 2018	3 March 2016–15 July 2018	7 March 2016–5 July 2018	11 March 2016–9 July 2018	1 March 2016–13 July 2018	19 March 2016–4 July 2018	9 March 2016–8 July 2018
# DP	1008	1033	1006	999	1034	1017	1089
|PCC|	0.01639	0.10228	0.06201	0.12204	0.12279	0.08812	0.02018

To generate the PCA features, the PCA algorithm was fed with the ALL-Set, and the CPV was set to 95%. The PCA-Set contains the principal components that explain 95% of the variance. It was found that for both positions P0 and P1, the PCA-Set contains the first principle component only, which captures 95% of the variance contained in the data.

The autoencoder-derived feature sets were generated through feeding a sparse autoencoder by the ALL-Set. By setting the number of latent features output from the encoder end into a number less than the number of features in the ALL-Set, the autoencoder network was utilized as a latent-feature generator and a dimensionality reduction tool. The number of latent features output from the autoencoder, N, was set to 1, 2, and 3. Therefore, three autoencoder sets are generated: AUT-Set-1, AUT-Set-2, and AUT-Set-3. The autoencoder was trained in an unsupervised manner over 5000 epochs with the settings depicted previously in the Methods section. The performance of the autoencoder is measured using the mean squared error with weight and sparsity regularizers (MSE-WSR). Table 7 shows the starting and stopping values of the gradient and the MSE-WSR values for positions P0 and P1 when N equals 1, 2, and 3. It is observable from Table 8 that the MSE-WSR decreases with increasing the number of output features. Increasing the number of output features helps including more details from the original data, which aids in reducing the output cost. However, increasing the number of latent features would not guarantee better prediction performance of the regression model. Therefore, the maximum number of output features from the encoder was selected to be 3. This setting helped reduce the computational load and time, and it was proved by experiment to be sufficient to enhance the regression model performance. It is also noticed that the values of the gradient and MSE are the highest at the beginning of the training process and the lowest at the stopping, which is a normal result of algorithm learning. The behavior of the autoencoder performance against the training epochs is depicted in Figure 5, which shows sample plots of the autoencoder performance in Experiment 1 for N = 1 at P0 and P1.

Table 7. Autoencoder performance in generating latent features from original input features for positions P0 and P1; # denotes the number of features.

Position	# Output Features	MSE-WSR		Gradient	
		Initial	Stopped	Initial	Stopped
P0	1	3.71×10^3	40	138	0.076
	2	3.71×10^3	3.22	178	0.031
	3	3.71×10^3	1.66	200	1.16
P1	1	3.79×10^3	34.7	182	0.14
	2	3.79×10^3	3.07	100	0.035
	3	3.79×10^3	2.03	246	0.079

Table 8. Feature sets used for modelling SWH_KU_CAL using Positions P0 and P1 data. # F denotes the number of features included in the feature set.

Position P0			Position P1		
Feature Set	# F	Included Features	Feature Set	# F	Included Features
ALL-Set	16	TIME, SIG0_KU, UWND, VWND, WSPD_CAL, SWH_KU_std_dev, SIG0_KU_std_dev, DIS2COAST, BOT_DEPTH, WSPD, LATSINE, LATCOSINE, LONGSINE, LONGCOSINE, SWH_KU_num_obs, SIG0_KU_num_obs.	ALL-Set	16	TIME, SIG0_KU, UWND, VWND, WSPD_CAL, SWH_KU_std_dev, SIG0_KU_std_dev, DIS2COAST, BOT_DEPTH, WSPD, LATSINE, LATCOSINE, LONGSINE, LONGCOSINE, SWH_KU_num_obs, SIG0_KU_num_obs.
PCC-Set	6	SIG0_KU, VWND, WSPD_CAL, SWH_KU_std_dev, SIG0_KU_std_dev	PCC-Set	7	TIME, SIG0_KU, VWND, WSPD_CAL, SWH_KU_std_dev, SIG0_KU_std_dev.
PCA-Set	1	First principal component explaining 95% of data variance.	PCA-Set	1	First principal component explaining 95% of data variance.
AUT-Set-1	1	Single latent feature output from the encoder	AUT-Set-1	1	Single latent feature output from the encoder

Table 8. *Cont.*

Position P0			Position P1		
Feature Set	**# F**	**Included Features**	**Feature Set**	**# F**	**Included Features**
AUT-Set-2	2	Two latent features output from the encoder	AUT-Set-2	2	Two latent features output from the encoder
AUT-Set-3	3	Three latent features output from the encoder	AUT-Set-3	3	Three latent features output from the encoder
HCAT-1	8	Hybrid set composed by fusing the features in PCC-Set, PCA-Set, and AUT-Set-1	HCAT-1	9	Hybrid set composed by fusing the features in PCC-Set, PCA-Set, and AUT-Set-1
HCAT-2	9	Hybrid set composed by fusing the features in PCC-Set, PCA-Set, and AUT-Set-2	HCAT-2	10	Hybrid set composed by fusing the features in PCC-Set, PCA-Set, and AUT-Set-2
HCAT-3	10	Hybrid set composed by fusing the features in PCC-Set, PCA-Set, and AUT-Set-3	HCAT-3	11	Hybrid set composed by fusing the features in PCC-Set, PCA-Set, and AUT-Set-3
HAT-1	2	Hybrid set composed by fusing the features in PCA-Set and AUT-Set-1	HAT-1	2	Hybrid set composed by fusing the features in PCA-Set and AUT-Set-1
HAT-2	3	Hybrid set composed by fusing the features in PCA-Set and AUT-Set-2	HAT-2	3	Hybrid set composed by fusing the features in PCA-Set and AUT-Set-2
HAT-3	4	Hybrid set composed by fusing the features in PCA-Set and AUT-Set-3	HAT-3	4	Hybrid set composed by fusing the features in PCA-Set and AUT-Set-3

Hybrid feature sets were formed by merging features from the basic feature sets. Table 8 depicts the features in the basic and hybrid feature sets and their number of features used for SWH_KU_CAL modeling for Positions P0 and P1.

The performance of the GPR and NNR models trained individually by the basic and hybrid sets for modeling SWH_KU_CAL is depicted in Tables 9 and 10. Table 9 shows the RMSE and R2 values for the regressors trained on position P0 data, while Table 10 presents the regression performance for position P1. For position P0, the results show that GPR models recorded higher prediction performance than the NNR models for all feature sets. It was noticed that the basic feature sets generally yielded lower regression performance than the hybrid sets. It is noticeable that GPR models trained by the HAT sets recorded higher performance than the other hybrid sets. The best GPR model records the highest R2 value of 0.92 and an RMSE value of 0.11724. This model has a Rational Quadratic kernel and was trained by the HAT-2 set. The second-best GPR model recorded an R2 value of 0.91 and was trained by the hybrid set HAT-1. On the other hand, the NNR model trained on the AUT-Set-2 set recorded the highest performance, followed by the HAT-2-based model over the other NNR models. The best models are highlighted in dark gray, and the second-best performance regressor is highlighted in light gray in Tables 9 and 10.

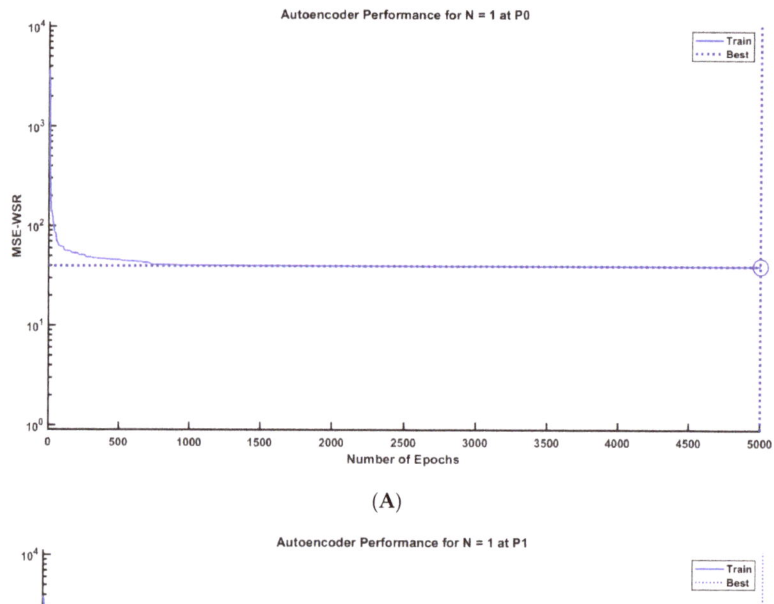

(**A**)

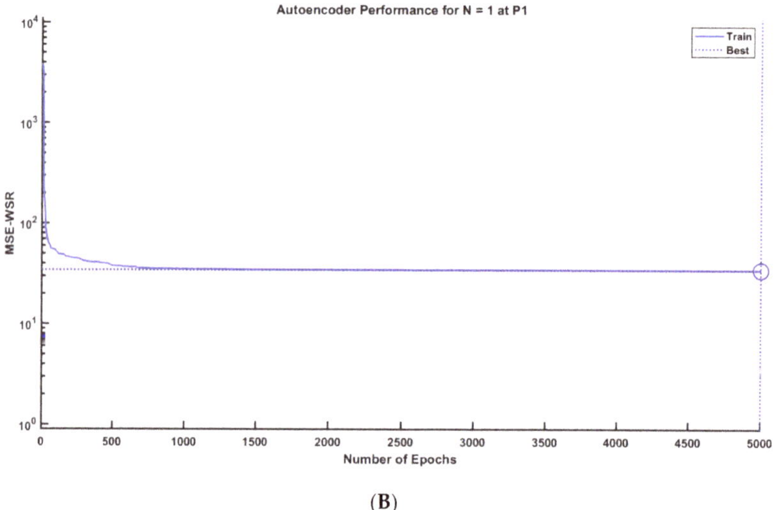

(**B**)

Figure 5. Autoencoder performance during the training process with N = 1 for (**A**) P0, (**B**) P1.

Table 9. SWH_KU_CAL prediction performance of GPR and NNR models trained on KU-based feature combination sets for Position P0; Best models are highlighted in dark gray, and the second-best performance regressor is highlighted in light gray; # denotes the number of features.

Position: P0		GPR			NNR	
Feature Set	**# F**	**RMSE**	**R^2**	**Kernel Function**	**RMSE**	**R^2**
ALL-Set	16	0.29262	0.41	Rational Quadratic	0.31634	0.32
PCA-Set	1	0.12792	0.87	Rational Quadratic	0.29677	0.31
PCC-Set	6	0.31881	0.36	Matern 5/2	0.33156	0.31
AUT-Set-1	1	0.20963	0.73	Squared Exponential	0.31853	0.37
HAT-1	2	0.12188	0.91	Squared Exponential	0.33354	0.36
HCAT-1	8	0.29877	0.38	Rational Quadratic	0.2625	0.52

Table 9. *Cont.*

Position: P0		GPR			NNR	
Feature Set	**# F**	**RMSE**	**R^2**	**Kernel Function**	**RMSE**	**R^2**
AUT-Set-2	2	0.12985	0.9	Rational Quadratic	0.24259	0.64
HAT-2	3	0.11724	0.92	Rational Quadratic	0.2601	0.6
HCAT-2	9	0.32058	0.32	Rational Quadratic	0.29551	0.4
AUT-Set-3	3	0.14791	0.89	Squared Exponential	0.31347	0.49
HAT-3	4	0.13112	0.8	Rational Quadratic	0.27078	0.13
HCAT-3	10	0.39404	0.23	Squared Exponential	0.33302	0.45

Table 10. SWH_KU_CAL prediction performance using GPR and NNR trained by various feature combinations for Position P1; # denotes the number of features.

Position: P1		GPR			NNR	
Feature Set	**# F**	**RMSE**	**R^2**	**Kernel Function**	**RMSE**	**R^2**
ALL-Set	16	0.25525	0.44	Exponential	0.27248	0.36
PCA-Set	1	0.11961	0.86	Rational Quadratic	0.22502	0.49
PCC-Set	7	0.24238	0.4	Rational Quadratic	0.24024	0.41
AUT-Set-1	1	0.17635	0.64	Rational Quadratic	0.2258	0.33
HAT-1	2	0.10414	0.89	Squared Exponential	0.20098	0.6
HCAT-1	9	0.23234	0.47	Exponential	0.22728	0.49
AUT-Set-2	2	0.1046	0.87	Exponential	0.19511	0.54
HAT-2	3	0.1113	0.85	Matern 5/2	0.1889	0.58
HCAT-2	10	0.23529	0.31	Exponential	0.2236	0.38
AUT-Set-3	3	o.12272	0.84	Exponential	0.18277	0.65
HAT-3	4	0.13351	0.82	Matern 5/2	0.19522	0.61
HCAT-3	11	0.2367	0.37	Exponential	0.22549	0.41

Figure 6 illustrates the goodness of fit of the SWH predictions generated for the test set by the best GPR and NNR models trained on P0 data. The plots of Figure 6 show the predicted versus true values of the response, SWH_KU_CAL, and the residuals for the best GPR and NNR models highlighted in dark gray in Table 9. It is clear that the GPR model predictions are closer to the diagonal line, which represents the perfect prediction, than those predicted by the NNR. This observation is consistent with the high R2 value of the GPR model and is confirmed by the residual plot. The residuals of the GPR predictions range between $[-0.3, 0.3]$, while it ranges from $[-0.8$ to $0.7]$ for the NNR predictions.

For position P1, it is clear from Table 9 that the GPR model trained on the HAT-1 set achieved the highest performance compared with the NNR based on the highest R2. The second-best performance is recorded by the AUT-Set-2-based GPR model with an exponential kernel. On the other hand, the best NNR model recorded 0.65 for the coefficient of determination and was trained by the AUT-Set-3. The second-best performer was the HAT-3-based NNR model. Similarly to P0, the GPR models achieved higher performance that the NNR. It is observed that the regressors trained on the PCC-Set, and the hybrid features based on it, the HCAT sets, suffered from poor performance. This could be interpreted as a result of the low correlation between the predictors in the PCC-Set and the target, which hindered the improvement of the model performance, even after fusing the PCC, PCA, and AUT features together. It was also noticed that the HAT sets provides better regression performance than the PCA-Set and the AUT sets. This indicates the improving impact of the autoencoder features on the prediction performance when added to the PCA features.

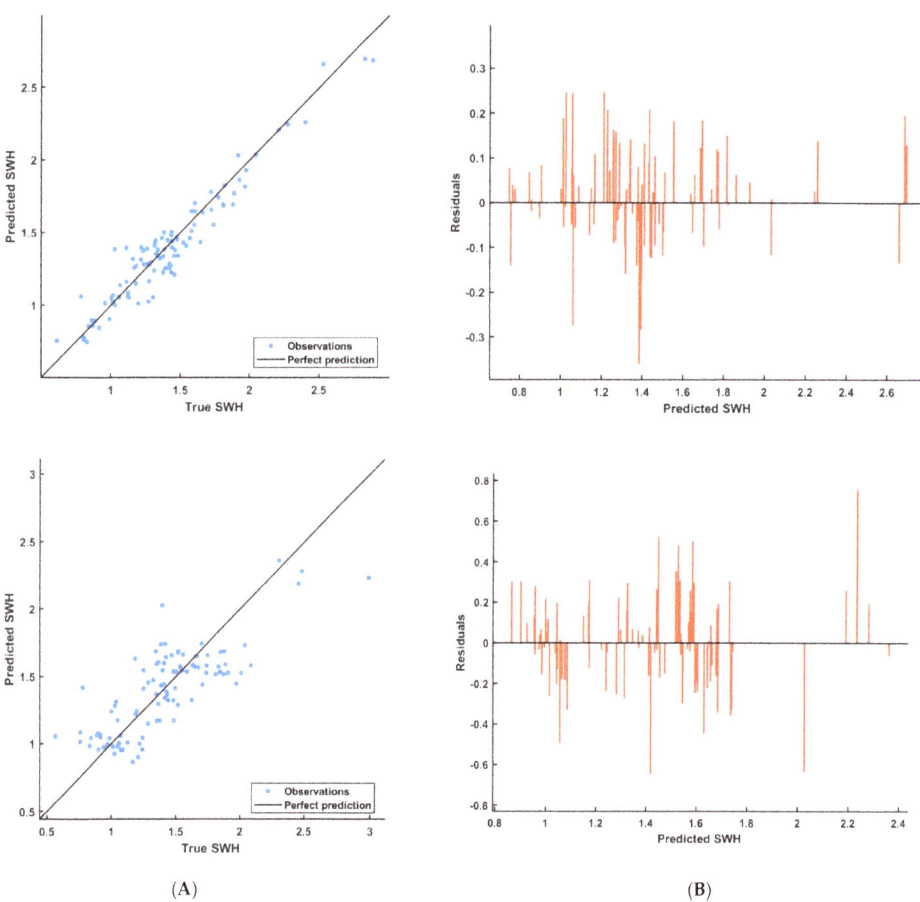

(**A**)   (**B**)

Figure 6. Testing prediction performance of the best performing regressors trained on the KU-based features of P0 data. (**A**) Predicted versus true SWH; (**B**) residuals versus predicted SWH; upper row: GPR model based on HAT-2; lower row: NNR model based on AUT-Set-2.

Figure 7 presents the goodness of fit of the SWH predictions generated by the best GPR and NNR models trained on P1 data. The plots of Figure 7 illustrate the predicted versus true values of the response, SWH_KU_CAL, and the residuals for the best GPR and NNR models on the test set. It is clear that the predictions are scattered roughly symmetrically around the diagonal line for both GPR and NNR. The predictions of the GPR model are closer to the diagonal line than the NNR predictions. This observation is reflected in the residual plots, which show the difference between the true and predicted target. The error in the predictions with respect to the SWH true values ranges between $[-0.3, 0.4]$ for the GPR model and $[-0.6, 0.5]$ for the NNR model. The performance plots of Figure 5 reveal the superiority of the GPR model over the NNR.

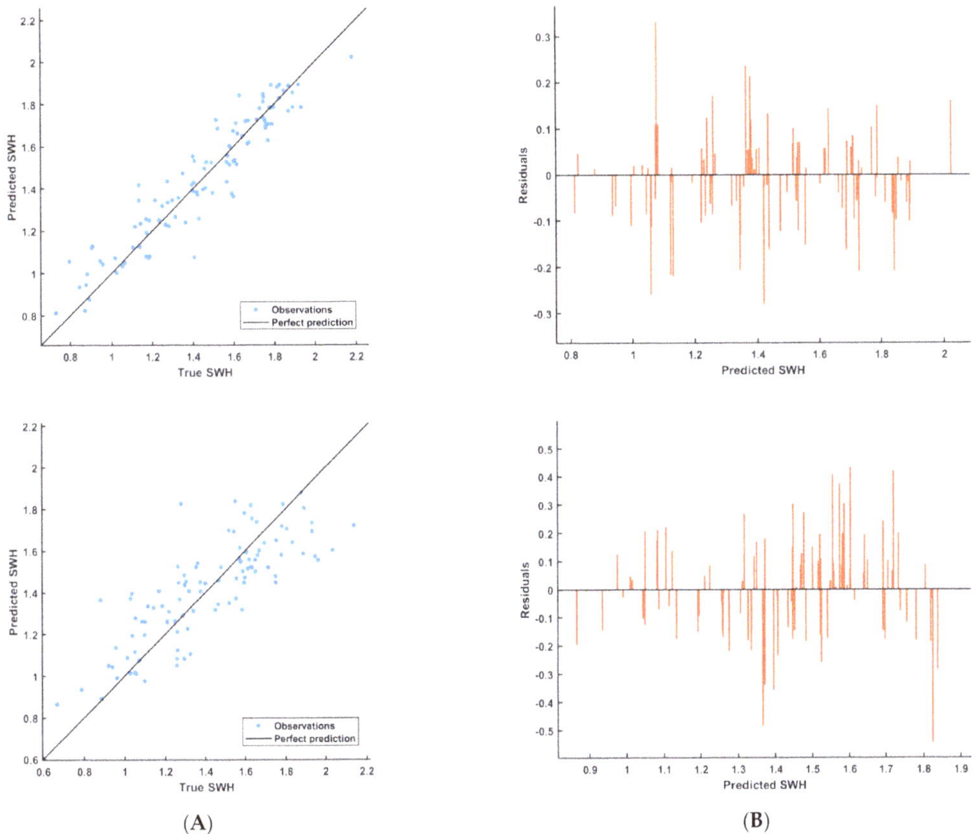

(A)   **(B)**

Figure 7. Testing prediction performance of the best performing regressors trained on the KU- based features of P1 data. (**A**) Predicted versus true SWH; (**B**) residuals versus predicted SWH; upper row: GPR model based on HAT-1; lower row: NNR model based on AUT-Set-3.

To summarize the findings of the current research, the prediction performance of the first and second-best regressors recorded by the GPR and NNR models for position P0 and P1 is presented in Table 11. The highest average RMSEs obtained over the two positions are 0.11069 and 0.21268 for the GPR and NNR models, respectively. It was noticed that The GPR models provides better prediction performance than the NNR models in terms of RMSE and R^2 metrics for both positions. This observation was further proved by the residual plots of the regression models. It was noticed that the HAT feature sets boosted the GPR model performance over that trained by the basic PCA or AUT feature sets individually. In contrast, pure autoencoder features yielded better performance of the NNR models over that of NNR models trained individually by the basic as well as the hybrid sets. Moreover, it was observed that the HCAT sets yielded lower prediction performance than the AUT sets and HAT sets for both the GPR and NNR. This observation could be referred to the low correlation of the original predictors in the PCC-Set with the response variable. Adding such features to the PCA and autoencoder-derived features hindered the significant improvement of the model performance. It was shown that the autoencoder-derived features aid in providing improved prediction performance of the GPR and NNR models over the basic feature sets.

Table 11. Summary of the performance of best SWH regression models for position P0 and P1; # denotes the number of features.

Position	Rank	GPR				NNR			
		Features Set	# F	RMSE	R^2	Feature Set	# F	RMSE	R^2
P0	1	HAT-2	3	0.11724	0.92	AUT-Set-2	2	0.24259	0.64
	2	HAT-1	2	0.12188	0.91	HAT-2	3	0.2601	0.6
P1	1	HAT-1	2	0.10414	0.89	AUT-Set-3	3	0.18277	0.65
	2	AUT-Set-2	2	0.1046	0.87	HAT-3	4	0.19522	0.61

To discuss the results from the sea area (site) perspective, the PCC analysis showed that the DIST2COAST, BOT_DEPTH, LONGITUDE, and LATITUDE-related features are not significant with respect to SWH from the correlation perspective for both positions P0 and P1 (these features recorded very low PCC values). The observation that could be made here is that these site-related features do not contribute significantly to SWH measurements. However, the measured features showed generally higher PCC values than the site-related features, and thus could effectively affect to SWH measurements. The measured features, especially the wind speed, are characterized by their intermittent and stochastic nature. Moreover, the data of the two used positions were collected over different times, and the two positions are approximately 69 miles apart to the east, which means that the two sites had different sea states at the time of data acquisition. Such variations would interpret the difference in the best feature sets of the two positions (HAT-2 for P0 versus HAT-1 for P1 for the GPR and AUT_Set-2 versus AUT-Set-3 for the NNR). Nonetheless, the best feature sets for both sites were based on the autoencoder-derived features, which reveal the effectiveness of this technique in extracting significant features from the original data features. The autoencoder-derived features even improved the prediction performance when combined with the PCA features (in the HAT feature set).

4.2. Hypothesis Testing for Feature Significance

In order to reinforce the findings of the current study, the significance of the features included in the feature sets that yielded the highest prediction performance of the GPR and NNR is examined using hypothesis testing. In the present study, the ANOVA F-statistics test was utilized to identify the significance of the features included in the HAT-2 and AUT-Set-2 feature sets of P0 data as well as the features of HAT-1 and AUT-Set-3 features of P1. In this test, the input features are used to model the response variable using a linear regression model and determine the significance of the predicted model coefficients through statistical metrics, namely the F-value and p-value. The null hypothesis of the test, H0, assumes that there is no relationship between the response variable, SWH, and the input features i.e., all dependent variable coefficients are zero. On the other hand, the alternative hypothesis, H1, implies that the model is accurate if there is at least one instance where any of the dependent variable coefficients are nonzero. The outcomes of the ANOVA Test of the significance of the aforementioned four feature sets in predicting the SWH are depicted in Table 12. The significance level is considered 0.05 for the p-value. The values obtained for both the F-value and the p-value indicate that there is a significant association between the response variable, SWH, and the input predictors for all feature sets. Therefore, the Null hypothesis can be rejected, and the significance of the examined autoencoder-derived features and hybrid features is confirmed.

Table 12. Outcomes of ANOVA Test of Autoencoder-based features significance in predicting the SWH using P0 and P1 data.

Position	Feature Set	Feature Symbol	Test F-Value	Test *p*-Value
P0	HAT-2	F1	27.9	8.19×10^{-7}
		F2	30.70119	4.08×10^{-8}
		F3	49.92715	3.46×10^{-12}
	AUT-Set-2	F1	11.57716	0.0007
		F2	43.37607	8.15×10^{-11}
P1	HAT-1	F1	7.9	0.004
		F2	7.3	0.006
	AUT-Set-3	F1	464.4652	4.75×10^{-82}
		F2	17.44876	3.27×10^{-5}
		F3	23.47258	1.5×10^{-6}

The prediction performance of the SWH regression model trained on the feature sets generated using the proposed deep-learning-based approach is further evaluated against the state of the art. Numerous research studies have addressed the problem of SWH prediction from satellite data from different perspectives and using various types of satellite data. In order to have a meaningful benchmarking, only studies that tackled the problem of SWH prediction using the IMOS Surface Waves Sub-Facility dataset are considered for comparison. The IMOS Surface Waves Sub-Facility dataset is a recent dataset that was published in 2019 and has received slight coverage in the literature. Only a single recent study was found to use the IMOS dataset for the prediction of SWH. The study by Quach et al. [35] investigated the use of deep learning to predict significant wave height from a dataset created from collocations between the Sentinel-1SAR and altimeter satellites observations from the IMOS dataset. Quach et al. integrated features from the IMOS altimeter data with a number of CWAVE features that were derived from the SAR image modulation spectra and developed a deep-learning-based regression model for SWH prediction. The results of that study show an improved RMSE of the deep learning model of 0.26. In our study, we employed the autoencoder deep learning network to generate significant features from the altimeter observations for the prediction of SWH using GPR and NNR. The proposed deep-learning-based feature generation method yielded average RMSE values of 0.11069 and 0.21268 for the GPR and NNR models, respectively. Therefore, the deep-learning-based SWH modeling approach proposed in the present study provides improved prediction performance over the state of the art.

5. Conclusions

In this research, we introduced a framework to extract features from SAR mode altimeter data using a hybrid deep-learning-based approach for the prediction of SWH. The proposed approach is based on the proficiency of the autoencoder neural network in representing input features in the latent space. The proposed framework is composed of four phases: data preprocessing, feature sets formation, SWR modeling, and model evaluation and testing. After the data were preprocessed, a number of basic feature sets were created from the input data. The basic sets include the ALL-Set, PCC-Set, PCA-Set, and AUT-Set-N. Multiple hybrid feature sets were further formed using various combinations of the PCC, PCA, and AUT feature sets, as well as the HAT, and HCAT sets. These sets were used for modeling SWH using the GPR and NNR. The regression models were validated using a 5-cross validation scheme and tested on a holdout test set. The prediction performance of the SWH models trained on the hybrid feature sets are compared with that trained by the basic PCC, PCA, and autoencoder-driven feature sets as well as all input features set. The results show that hybridizing the PCA and AUT feature sets yielded improved prediction performance for the GPR models, while pure autoencoder-derived features boasted the performance of the NNR models. The significance of the autoencoder-based pure and hybrid feature sets was proven through hypothesis testing. The presented results reveal the significance of the autoencoder-derived features in improving the performance of SWH prediction from altimeter data. In general, the findings of this study reveal the superiority of the autoencoder deep learning network in generating latent features that aid in improving SWH prediction performance over traditional feature extraction methods.

Author Contributions: Conceptualization, G.A., N.A.S. and M.J.C.; methodology, G.A., N.A.S. and M.J.C.; software, G.A. and N.A.S.; validation, G.A.; formal analysis, N.A.S. and G.A.; investigation, G.A., N.A.S. and M.J.C.; resources, G.A. and N.A.S.; data curation, G.A.; writing—original draft preparation, G.A. and N.A.S.; writing—review and editing, G.A., N.A.S., A.D.A. and M.J.C.; visualization, G.A. and N.A.S.; supervision, M.J.C.; project administration, G.A.; funding acquisition, G.A., N.A.S., A.D.A. and M.J.C. All authors have read and agreed to the published version of the manuscript.

Funding: Princess Nourah bint Abdulrahman University Researchers Supporting Project number (PNURSP2022R51), Princess Nourah bint Abdulrahman University, Riyadh, Saudi Arabia.

Institutional Review Board Statement: Not applicable.

Informed Consent Statement: Not applicable.

Data Availability Statement: Not applicable.

Acknowledgments: This project was funded by Princess Nourah bint Abdulrahman University Researchers Supporting Project number (PNURSP2022R51), Princess Nourah bint Abdulrahman University, Riyadh, Saudi Arabia.

Conflicts of Interest: The authors declare no conflict of interest.

References

1. Hasselmann, K.; Hasselmann, S. On the nonlinear mapping of an ocean wave spectrum into a synthetic aperture radar image spectrum and its inversion. *J. Geophys. Res.* **2008**, *96*, 10713–10729. [CrossRef]
2. Pugliese Carratelli, E.; Dentale, F.; Reale, F. *Numerical PSEUDO—Random Simulation of SAR Sea and Wind Response*; Special Publication; European Space Agency (ESA): Paris, France, 2006.
3. Carratelli, E.P.; Dentale, F.; Reale, F. *Reconstruction of SAR Wave Image Effects through Pseudo Random Simulation*; Special Publication; European Space Agency (ESA): Paris, France, 2007.
4. Hasselmann, K.; Chapron, B.; Aouf, L.; Ardhuin, F.; Collard, F.; Engen, G.; Hasselmann, S.; Heimbach, P.; Janssen, P.; Johnsen, H.; et al. *The ERS SAR Wave Mode: A Breakthrough in Global Ocean Wave Observations*; Special Publication; European Space Agency (ESA): Paris, France, 2013.
5. Collard, F.; Ardhuin, F.; Chapron, B. Monitoring and analysis of ocean swell fields from space: New methods for routine observations. *J. Geophys. Res. Ocean.* **2009**, *114*, C07023. [CrossRef]
6. Ardhuin, F.; Chapron, B.; Collard, F. Observation of swell dissipation across oceans. *Geophys. Res. Lett.* **2009**, *36*, L06607. [CrossRef]
7. Ardhuin, F.; Collard, F.; Chapron, B.; Girard-Ardhuin, F.; Guitton, G.; Mouche, A.; Stopa, J.E. Estimates of ocean wave heights and attenuation in sea ice using the SAR wave mode on Sentinel-1A. *Geophys. Res. Lett.* **2015**, *42*, 2317–2325. [CrossRef]
8. Hasselmann, S.; Brüning, C.; Hasselmann, K.; Heimbach, P. An improved algorithm for the retrieval of ocean wave spectra from synthetic aperture radar image spectra. *J. Geophys. Res. C Ocean.* **1996**, *101*, 16615–16629. [CrossRef]
9. Sun, J.; Kawamura, H. Retrieval of surface wave parameters from sar images and their validation in the coastal seas around Japan. *J. Oceanogr.* **2009**, *65*, 567–577. [CrossRef]
10. Zhang, B.; Li, X.; Perrie, W.; He, Y. Synergistic measurements of ocean winds and waves from SAR. *J. Geophys. Res. Ocean.* **2015**, *120*, 6164–6184. [CrossRef]
11. Schulz-Stellenfleth, J.; Lehner, S.; Hoja, D. A parametric scheme for the retrieval of two-dimensional ocean wave spectra from synthetic aperture radar look cross spectra. *J. Geophys. Res. C Ocean.* **2005**, *110*, C05004. [CrossRef]
12. Collins, M.J.; Ma, M.; Dabboor, M. On the Effect of Polarization and Incidence Angle on the Estimation of Significant Wave Height From SAR Data. *IEEE Trans. Geosci. Remote Sens.* **2019**, *57*, 4529–4543. [CrossRef]
13. Schulz-Stellenfleth, J.; König, T.; Lehner, S. An empirical approach for the retrieval of ocean wave parameters from synthetic aperture radar data. In Proceedings of the International Geoscience and Remote Sensing Symposium (IGARSS), Denver, CO, USA, 31 July–4 August 2006.
14. Li, X.M.; Lehner, S.; Bruns, T. Ocean wave integral parameter measurements using envisat ASAR wave mode data. *IEEE Trans. Geosci. Remote Sens.* **2011**, *49*, 155–174. [CrossRef]
15. Grieco, G.; Lin, W.; Migliaccio, M.; Nirchio, F.; Portabella, M. Dependency of the Sentinel-1 azimuth wavelength cut-off on significant wave height and wind speed. *Int. J. Remote Sens.* **2016**, *37*, 5086–5104. [CrossRef]

16. Shao, W.; Zhang, Z.; Li, X.; Li, H. Ocean wave parameters retrieval from Sentinel-1 SAR imagery. *Remote Sens.* **2016**, *8*, 707. [CrossRef]

17. Romeiser, R.; Graber, H.C.; Caruso, M.J.; Jensen, R.E.; Walker, D.T.; Cox, A.T. A new approach to ocean wave parameter estimates from C-band ScanSAR images. *IEEE Trans. Geosci. Remote Sens.* **2014**, *53*, 1320–1345. [CrossRef]

18. Ren, L.; Yang, J.; Zheng, G.; Wang, J. Significant wave height estimation using azimuth cutoff of C-band RADARSAT-2 single-polarization SAR images. *Acta Oceanol. Sin.* **2015**, *34*, 93–101. [CrossRef]

19. Shao, W.; Wang, J.; Li, X.; Sun, J. An empirical algorithm for wave retrieval from co-polarization X-band SAR imagery. *Remote Sens.* **2017**, *9*, 711. [CrossRef]

20. Atteia, G.E. Latent Space Representational Learning of Deep Features for Acute Lymphoblastic Leukemia Diagnosis. *Comput. Syst. Sci. Eng.* **2022**, *45*, 361–376. [CrossRef]

21. Atteia, G.; Abdel Samee, N.; El-Kenawy, E.S.M.; Ibrahim, A. CNN-Hyperparameter Optimization for Diabetic Maculopathy Diagnosis in Optical Coherence Tomography and Fundus Retinography. *Mathematics* **2022**, *10*, 3274. [CrossRef]

22. Samee, N.A.; Alhussan, A.A.; Ghoneim, V.F.; Atteia, G.; Alkanhel, R.; Al-antari, M.A.; Kadah, Y.M. A Hybrid Deep Transfer Learning of CNN-Based LR-PCA for Breast Lesion Diagnosis via Medical Breast Mammograms. *Sensors* **2022**, *22*, 4938. [CrossRef]

23. Atteia, G.; Alhussan, A.A.; Samee, N.A. BO-ALLCNN: Bayesian-Based Optimized CNN for Acute Lymphoblastic Leukemia Detection in Microscopic Blood Smear Images. *Sensors* **2022**, *22*, 5520. [CrossRef]

24. Khan, U.; Khan, S.; Rizwan, A.; Atteia, G.; Jamjoom, M.M.; Samee, N.A. Aggression Detection in Social Media from Textual Data Using Deep Learning Models. *Appl. Sci.* **2022**, *12*, 5083. [CrossRef]

25. Zhong, S.; Zhang, K.; Bagheri, M.; Burken, J.G.; Gu, A.; Li, B.; Ma, X.; Marrone, B.L.; Ren, Z.J.; Schrier, J.; et al. Machine Learning: New Ideas and Tools in Environmental Science and Engineering. *Environ. Sci. Technol.* **2021**, *55*, 12741–12754. [CrossRef] [PubMed]

26. Wu, J.L.; Xiao, H.; Paterson, E. Physics-informed machine learning approach for augmenting turbulence models: A comprehensive framework. *Phys. Rev. Fluids* **2018**, *7*, 074602. [CrossRef]

27. Lama, G.F.C.; Errico, A.; Pasquino, V.; Mirzaei, S.; Preti, F.; Chirico, G.B. Velocity uncertainty quantification based on Riparian vegetation indices in open channels colonized by *Phragmites australis*. *J. Ecohydraulics* **2021**, *7*, 71–76. [CrossRef]

28. Hardy, A.; Ettritch, G.; Cross, D.E.; Bunting, P.; Liywalii, F.; Sakala, J.; Silumesii, A.; Singini, D.; Smith, M.; Willis, T.; et al. Automatic Detection of Open and Vegetated Water Bodies Using Sentinel 1 to Map African Malaria Vector Mosquito Breeding Habitats. *Remote Sens.* **2019**, *11*, 593. [CrossRef]

29. Tapoglou, E.; Forster, R.M.; Dorrell, R.M.; Parsons, D. Machine learning for satellite-based sea-state prediction in an offshore windfarm. *Ocean Eng.* **2021**, *235*, 109280. [CrossRef]

30. Dhiman, H.S.; Deb, D.; Guerrero, J.M. Hybrid Machine Intelligent SVR Variants for Wind Forecasting and Ramp Events. *Renew. Sustain. Energy Rev.* **2019**, *108*, 369–379. [CrossRef]

31. Stefanakos, C. Fuzzy time series forecasting of nonstationary wind and wave data. *Ocean Eng.* **2016**, *121*, 1–12. [CrossRef]

32. Feng, Z.; Hu, P.; Li, S.; Mo, D. Prediction of Significant Wave Height in Offshore China Based on the Machine Learning Method. *J. Mar. Sci. Eng.* **2022**, *10*, 836. [CrossRef]

33. Stopa, J.E.; Mouche, A. Significant wave heights from Sentinel-1 SAR: Validation and applications. *J. Geophys. Res. Ocean.* **2017**, *122*, 1827–1848. [CrossRef]

34. Ribal, A.; Young, I.R. 33 years of globally calibrated wave height and wind speed data based on altimeter observations. *Sci. Data* **2019**, *6*, 77. [CrossRef]

35. Quach, B.; Glaser, Y.; Stopa, J.E.; Mouche, A.A.; Sadowski, P. Deep Learning for Predicting Significant Wave Height from Synthetic Aperture Radar. *IEEE Trans. Geosci. Remote Sens.* **2021**, *59*, 1859–1867. [CrossRef]

36. Zhang, X.; Dai, H. Significant Wave Height Prediction with the CRBM-DBN Model. *J. Atmos. Ocean. Technol.* **2019**, *36*, 333–351. [CrossRef]

37. Fan, S.; Xiao, N.; Dong, S. A novel model to predict significant wave height based on long short-term memory network. *Ocean Eng.* **2020**, *205*, 107298. [CrossRef]

38. Sentinel-3—Sentinel Online. Available online: https://sentinels.copernicus.eu/web/sentinel/missions/sentinel-3 (accessed on 27 September 2022).

39. User Guides—Sentinel-3 Altimetry—Operating Modes—Sentinel Online. Available online: https://sentinels.copernicus.eu/web/sentinel/user-guides/sentinel-3-altimetry/overview/modes (accessed on 27 September 2022).

40. User Guides—Sentinel-3 Altimetry—Heritage and Future—Sentinel Online. Available online: https://sentinels.copernicus.eu/web/sentinel/user-guides/sentinel-3-altimetry/overview/heritage-and-future (accessed on 27 September 2022).

41. The Geographical Oddity of Null Island. Worlds Revealed: Geography & Maps at The Library Of Congress. Available online: https://blogs.loc.gov/maps/2016/04/the-geographical-oddity-of-null-island/ (accessed on 6 September 2022).

42. SRAL Instrument—Sentinel-3 Altimetry Technical Guide—Sentinel Online. Available online: https://sentinels.copernicus.eu/web/sentinel/technical-guides/sentinel-3-altimetry/instrument/sral (accessed on 26 September 2022).

43. Ma, J.; Yuan, Y. Dimension reduction of image deep feature using PCA. *J. Vis. Commun. Image Represent.* **2019**, *63*, 102578. [CrossRef]

44. Jolliffe, I.T.; Cadima, J. Principal component analysis: A review and recent developments. *Philos. Trans. R. Soc. A Math. Phys. Eng. Sci.* **2016**, *374*, 20150202. [CrossRef]
45. Goodfellow, I.; Bengio, Y.; Courville, A. *Deep Learning—Adaptive Computation and Machine Learning*; MIT Press: Cambridge, MA, USA, 2017; Volume 1, ISBN 978-0-262-03561-3.
46. Olshausen, B.A.; Field, D.J. Sparse coding with an overcomplete basis set: A strategy employed by V1? *Vis. Res.* **1997**, *37*, 3311–3325. [CrossRef]
47. Møller, M.F. A scaled conjugate gradient algorithm for fast supervised learning. *Neural Netw.* **1993**, *6*, 525–533. [CrossRef]
48. Li, M.; Liu, K. Probabilistic Prediction of Significant Wave Height Using Dynamic Bayesian Network and Information Flow. *Water* **2020**, *12*, 2075. [CrossRef]
49. MacKay, D.J.C.; MacKay, D.J.C. *Gaussian Processes—A Replacement for Supervised Neural Networks?* Cambridge University: Cambridge, UK, 1997.
50. Rasmussen, C.E. Gaussian Processes in machine learning. In *Advanced Lectures on Machine Learning*; Lecture Notes in Computer Science; Springer: Berlin/Heidelberg, Germany, 2004; Volume 3176, pp. 63–71. [CrossRef]

Article

Deep-Separation Guided Progressive Reconstruction Network for Semantic Segmentation of Remote Sensing Images

Jiabao Ma [1], Wujie Zhou [1,2,*], Xiaohong Qian [1] and Lu Yu [2]

1 School of Information and Electronic Engineering, Zhejiang University of Science and Technology, Hangzhou 310023, China
2 College of Information Science and Electronic Engineering, Zhejiang University, Hangzhou 310027, China
* Correspondence: 109029@zust.edu.cn

Abstract: The success of deep learning and the segmentation of remote sensing images (RSIs) has improved semantic segmentation in recent years. However, existing RSI segmentation methods have two inherent problems: (1) detecting objects of various scales in RSIs of complex scenes is challenging, and (2) feature reconstruction for accurate segmentation is difficult. To solve these problems, we propose a deep-separation-guided progressive reconstruction network that achieves accurate RSI segmentation. First, we design a decoder comprising progressive reconstruction blocks capturing detailed features at various resolutions through multi-scale features obtained from various receptive fields to preserve accuracy during reconstruction. Subsequently, we propose a deep separation module that distinguishes various classes based on semantic features to use deep features to detect objects of different scales. Moreover, adjacent middle features are complemented during decoding to improve the segmentation performance. Extensive experimental results on two optical RSI datasets show that the proposed network outperforms 11 state-of-the-art methods.

Keywords: digital surface model; multimodal; multi-scale supervision; feature separation; reconstruction refinement

Citation: Ma, J.; Zhou, W.; Qian, X.; Yu, L. Deep-Separation Guided Progressive Reconstruction Network for Semantic Segmentation of Remote Sensing Images. *Remote Sens.* **2022**, *11*, 5510. https://doi.org/10.3390/rs14215510

Academic Editor: Gwanggil Jeon

Received: 15 September 2022
Accepted: 30 October 2022
Published: 1 November 2022

Publisher's Note: MDPI stays neutral with regard to jurisdictional claims in published maps and institutional affiliations.

1. Introduction

Semantic segmentation aims to semantically classify the pixels in an image [1]. In remote sensing, semantic segmentation is crucial in several applications, such as scene understanding [2], land cover classification [3], and urban planning [4]. Owing to the success of deep learning (DL) and the promising results obtained on multiple semantic segmentation benchmarks containing natural images [5–7], semantic segmentation of remote sensing images (RSIs) increasingly adopts DL approaches [8–10]. However, an RSI is substantially larger than a typical natural image for computer vision applications; it contains objects of different sizes and shows complex scenes. Moreover, during data acquisition, the tilted perspective of RSIs can lead to scale variations in objects captured at different distances [11,12], exacerbating problems related to multi-scale changes.

With the continuous development of DL, convolutional neural networks have ushered in a new era of computer vision. The full convolution was proposed by Long et al. [13] to replace a fully connected layer with a convolutional layer in a classification network. However, decoding relies on the deep semantic features obtained from upsampling to obtain an output prediction map. Accordingly, using U-shaped architectures, Ronneberger et al. [14] and Vijay et al. [15] proposed UNet and SegNet, respectively, which use upsampling and continuous convolutions to complete decoding; each layer splices features from the encoding stage. The method of supplementing the features extracted from the encoder to the decoder is also often used in the later semantic segmentation methods, enhancing the complementarity of features in a different phase. Inspired by the above methods, Jiang et al. [16] proposed RedNet in 2018 with the same decoding approach, obtaining intermediate prediction maps at each stage to supervise the network at different resolutions.

Moreover, Chen et al. [17] used the splicing of deep semantic features with shallow features and upsampling for the prediction map. This study proposed the idea of atrous convolution to expand the receptive field in the convolution process. Meanwhile, Chen et al. [18] proposed the feature separation and aggregation models for fusing multimodal features and fully exploring the characteristics of different stages. However, the architecture of the decoder makes it simple to fully use the features extracted by the encoder. In addition, Yu et al. [19] used deep features from two encoder branches to construct prediction maps during decoding for fast inference and the real-time performance of the proposed method. Xu et al. [20] used an attention mechanism and multi-branch parallel architecture to build lightweight networks for real-time segmentation. Such structures helped obtain representations of objects at different scales. For the multimodal data, Zhou et al. [21] and Seichter et al. [22] used RGB and depth multimodal data to complete the semantic segmentation task of indoor scenes. The proposed network followed the encoder–decoder architecture combined with the last three high-level features of the encoder to construct the prediction map, which was simple in structure. However, this approach failed to make full use of shallow features. Some details need to be included in the refactoring process. Hu et al. [23] used five decoder blocks with the same encoding structure and applied upsampling to each block to restore the resolution of the prediction map. This is one of the most widely used architectures in semantic segmentation in recent years. Middle feature streams are deployed in multimodal data to deal with fused features, which can enhance the representation of multimodal features. In addition, researchers have widely favored multimodal data in different fields [24–26]. In particular, to handle quality variations across multimodality RSI datasets, Zheng et al. [27] use a DSM (Digital Surface Model) as auxiliary information to improve the segmentation performance of the model on single-modal data. Nevertheless, the method only applies self-attention to the deepest feature, and the structure of the decoder is relatively simple. Thus, it fails to detect the object in the complex scene. Similarly, Ma et al. [28] used powerful encoding features with a transformer to extract multimodality information. In this approach, the transformer is fully combined with CNN to deal with multi-scale features.

Most networks for tasks such as semantic segmentation have encoder–decoder architectures. Common encoders include VGG [29], ResNet [30] and, recently, the transformer [31]. However, the feature extraction ability of these encoders is limited to some extent. In particular, network performance improvements depend on how to handle the above features and, most importantly, how to reconstruct the features in the decoder. For developing decoders, different architectures have been devised; however, the bottom-up approach is typically used after feature extraction. A typical decoder is UNet [14], which is the basis for several subsequently developed networks. Various studies [32] have performed a fusion of features with different scales after extraction to improve feature reconstruction. In such methods, the decoder contains common convolutional and upsampling layers. Although its implementation is simple, this type of decoder lacks efficiency. Moreover, the features extracted by the encoder contain different levels of meaning at different resolutions. That is, current methods cannot take advantage of these features. How to reconstruct features efficiently and cooperate with each other is crucial in the design model.

To solve the abovementioned problems, we propose a deep-separation-guided progressive reconstruction network (DGPRNet) comprising a deep separation module (DSEM) for semantic segmentation of RSIs. In particular, to improve feature reconstruction, we design a progressive reconstruction block (PRB) based on atrous spatial pyramid pooling (ASPP) [33] with multiple convolutional layers combining various receptive fields for refactoring characteristics at each resolution. Unlike other methods based on upsampling to increase the resolution [34], the PRBs use deconvolution to adjust the resolution, increasing through each block until the input image is solved. Moreover, to enhance the forward guidance of deep semantic features to shallow layers, the proposed deep separation module (DSEM) processes semantic features such that pixels of the same class are clustered, whereas the separation between pixels from different classes is maximized. The prediction map is multi-supervised. Thus, the expression ability of deep semantic features is enhanced, and the PRB provides positive feedback.

The study's contributions are as follows:

1. A PRB based on ASPP [33] was embedded in the decoder to strengthen the feature reconstruction and reduce the error in this process. Five features with different resolutions were processed serially using atrous convolution layers with different ratios, and the feature resolution was expanded by deconvolution to obtain the decoding output of each block.
2. The proposed DSEM processed the last three semantic features from the decoder to emphasize semantic information to use deep semantic features. Intraclass separation was minimized, while interclass separation was maximized. Meanwhile, multi-supervision was applied to DGPRNet for segmentation, improving the reconstruction ability of each module.
3. Experiments on two RSI datasets showed that the proposed model outperforms 11 state-of-the-art methods, including current semantic segmentation methods.

2. Proposed DGPRNet

Figure 1 shows the architecture of the proposed DGPRNet. In particular, the architecture comprised symmetric ResNet-50 [30] backbones for feature extraction and the novel decoder consisting of PRBs and DSEM for processing semantic features. As seen in Figure 1, the DGPRNet adopts an encoder–decoder architecture. The two symmetric ResNet-50 backbones constituted the encoder processing input images by extracting features at five different resolutions from RGB (red–green–blue)/DSM (digital surface model) RSIs. According to the features extracted by the encoder, the adjacent modules from [35] were used between features F_2, F_3, and F_4 for feature aggregation. During decoding, inspired by ASPP [33], we used the proposed PRBs to reconstruct and combine the features at various resolutions, and each PRB provided a prediction map at the corresponding resolution. The DSEM classified the last three deep semantic features. Finally, we obtained the prediction map from five scales.

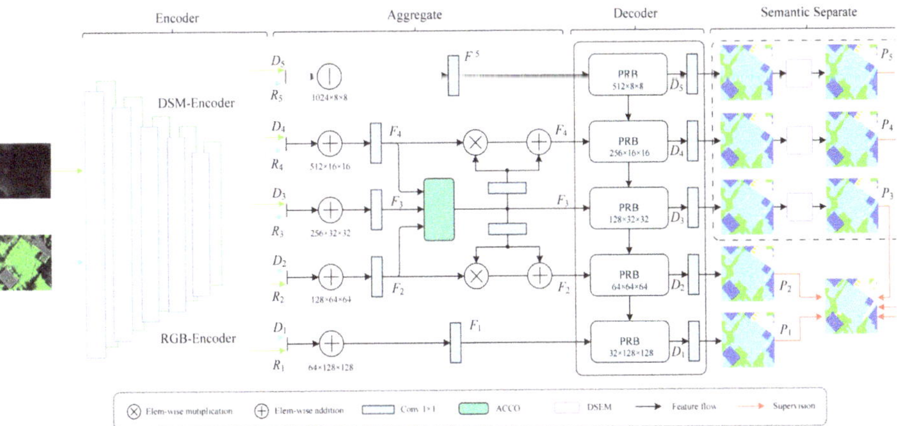

Figure 1. Overall architecture of the proposed DGPRNet. The network includes four stages: encoding, feature aggregation, decoding, and semantic separation.

2.1. Encoder

RGB and DSM images contained unreliable information and objects of different sizes due to the complexity of real scenes and the diversity of RSIs. In particular, feature extraction was essential in existing image semantic segmentation methods based on DL.

We used ResNet-50 as the encoder to obtain five features with different resolutions, R_i and D_i, $i \in \{1, 2, 3, 4, 5\}$, from the RGB and DSM images, respectively, and fused the features by simple pixel-wise addition [36–39], obtaining feature F_i. Shallow features contained details such as object boundaries, and deep features reflected semantic information such as the class and location of an object. RSIs contained various objects of different sizes. In particular, detecting these objects is crucial. Moreover, we used the module from [34] to aggregate multi-scale information. We retained the original information of the shallowest and deepest features and only used the features of the middle three resolutions to obtain the aggregated representation of adjacent features. The aggregated features were supplemented with features at the corresponding resolution. The encoder was formulated as follows:

$$F_i = Conv(R_i \oplus D_i),\tag{1}$$

$$\begin{cases} F_3 = ACCO(F_2, F_3, F_4) \\ F_2 = F_2 \otimes Conv(F_3) \oplus Conv(F_3) \\ F_4 = F_4 \otimes Conv(F_3) \oplus Conv(F_3), \end{cases}\tag{2}$$

where $\oplus$ denotes pixel-wise addition, $\otimes$ denotes pixel-wise multiplication, and *Conv* represents a convolutional layer with batch normalization and rectified linear unit activation. Subsequently, the aggregated representation of an object at different resolutions can be obtained. In this way, multi-scale objects can be accurately detected.

2.2. PRB

Universal networks work well on all datasets. Therefore, a critical problem in applying DL to computer vision is reconstructing the features extracted by the encoder according to the characteristics of a specific dataset, and finally providing an accurate prediction map. Therefore, we proposed the PRB (shown in Figure 2), where the encoder extracted features with different resolutions. Based on ASPP, we used dilated convolutions with different rates in series to enlarge the receptive field at each resolution. In each block, objects of different sizes and those with different dimensions were detected at different resolutions.

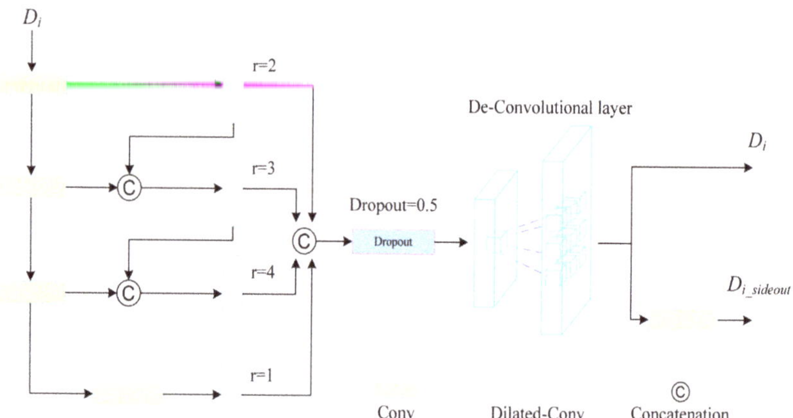

Figure 2. Architecture of the proposed PRB.

After the encoded features were obtained, the PRB reconstructed the features at each resolution. In [33], ASPP modules were deployed at the bottom of the network, acting on the deepest semantic features to expand the receptive field. However, the intended effect was limited. Based on ASPP, we expanded the receptive field in each layer during decoding to detect objects at different resolutions. Specifically, the PRB contained four convolutional layers with different dilation rates. Moreover, the features were serially transferred between convolutional layers. Then, they were concatenated in parallel for feature aggregation under different receptive fields. Moreover, upsampling enables us to increase the resolution of features [31]. Inspired by deconvolution, we merged upsampling and feature aggregation into one step, and a dropout layer was added to prevent overfitting. The PRB was formulated as follows:

$$D_i = Cat(F_i, D_{i+1}),\tag{3}$$

$$\begin{cases} D_{i,c1} = Conv(D_i) \\ D_{i,c2} = Conv(D_{i,c1}) \\ D_{i,c3} = Conv(D_{i,c2}), \end{cases}\tag{4}$$

$$\begin{cases} D_{i,d2} = DConv(D_i, 2) \\ D_{i,d3} = DConv(Cat(D_{i,d2}, D_{i,c2}), 3) \\ D_{i,d4} = DConv(Cat(D_{i,d3}, D_{i,c3}), 4), \end{cases}\tag{5}$$

$$D_i = DeConv(Dropout(Cat(D_{i,c3}, D_{i,d2}, D_{i,d3}, D_{i,d4}), 0.5)),\tag{6}$$

where Cat, $DConv$, and $DeConv$ denote concatenation, a dilated convolutional layer, and a deconvolutional layer, respectively, and $i \in \{1, 2, 3, 4, 5\}$, with D_{i+1} being omitted for $i = 5$ in Equation (3).

2.3. DSEM

Deep semantic features represent the mapping of an image onto a semantic space. Moreover, the feature representation of pixels belonging to a class in complex scenes showed high variability, and RSIs corresponded to complex scenes. Consequently, different objects might be classified into the same class in some cases. To increase the classification accuracy, we proposed the DSEM that modeled intraclass and interclass features to strengthen their distinguishability and reduce ambiguity. First, high-level semantic feature map D_i, $i \in \{3, 4, 5\}$ was processed by a 1×1 convolutional layer to obtain feature maps α, β, $\gamma \in R^{C \times H \times W}$. Then, the features were processed to obtain different expressions within and between classes. The DSEM was formulated as follows:

$$\begin{cases} intra = Softmax(R(\alpha) \times T(R(\beta))) \times R(\gamma) \\ intra = F(intra) + D_i, \end{cases}\tag{7}$$

$$\begin{cases} inter = Softmax(T(R(D_i)) \times R(D_i)) \\ inter = F(inter) \times D_i + D_i, \end{cases}\tag{8}$$

$$P_i = (inter + intra) \otimes D_i,\tag{9}$$

where R denotes a resizing function from $R^{C \times H \times W}$ to $R^{C \times HW}$, T is the transposition from $R^{C \times HW}$ to $R^{HW \times C}$, F denotes the inverse mapping of R, and $\times$ denotes matrix multiplication.

The original semantic features were combined with the weights for intraclass and interclass features to obtain a deep separation prediction with higher resolution and more detailed feature classification performance while reducing feature redundancy. We applied the DSEM to features of the last three resolutions obtained from decoding. The network simultaneously performed prediction at five resolutions during training and supervised the network. Hence, the reconstruction ability during decoding was strengthened by integrating the DSEM.

2.4. Loss Function

We used binary cross-entropy as the loss function between the prediction map and the segmentation ground truth. The obtained prediction maps at five resolutions were resized to the dimension of the ground truth to calculate the loss. Given the five prediction maps, the binary cross-entropy loss function was defined as follows:

$$Loss = \sum_{i=1}^{5} BCE(P_i, GT),$$ (10)

where GT and P denote the ground truth and a prediction map, respectively. During testing, P_1 is the segmentation result of DGPRNet.

3. Experiments and Results

3.1. Datasets and Performance Indicators

The Potsdam [40] and Vaihingen [41] RSI datasets were used in semantic segmentation experiments to verify the performance of the proposed DGPRNet. The Potsdam dataset contains 38 patches of 6000 × 6000 pixels, in our experiments, we considered 17 patches for training (2_10, 3_10, 3_11, 3_12, 4_11, 4_12, 5_10, 5_12, 6_8, 6_9, 6_10, 6_11, 6_12, 7_7, 7_9, 7_11, 7_12) and 7 patches for testing (2_11, 2_12, 4_10, 5_11, 6_7, 7_8, 7_10). Furthermore, the Vaihingen dataset comprised 33 images with pixels of 2494 × 2064. We split the 16 patches for training (1, 3, 5, 7, 13, 17, 21, 23, 26, 32, 37) and 5 patches for testing (11, 15, 28, 30, 34). For evaluation, we used the average pixel accuracy of each class and the average intersection over union as performance indicators. Moreover, the intersection over union was applied between the prediction and target regions to obtain the optimal segmentation weight. Therefore, the intersection over union was the main indicator for training and evaluating different methods on the two RSI datasets.

3.2. Implementation Details

The experimental platform was implemented in Ubuntu 20.04 using the PyTorch 1.9.1 environment, and the network was trained on a computer equipped with an NVIDIA Titan V graphics card and 12 GB of memory. Owing to the large size of the original RSIs, the patches of input RGB and DSM images were scaled to 256 × 256 pixels. Finally, we obtained 35,972 slices for training, 4032 slices for validation, and 252 slices for testing on Potsdam. Moreover, we obtained 17,656 slices for training, 412 slices for validation, and 111 slices for testing on Veihingen. The pretrained ResNet-50 was used as the backbone for feature extraction. Considering the dataset characteristics and training time, training proceeded over 300 epochs on the Vaihingen dataset and over 100 epochs on the Potsdam dataset. We used stochastic gradient descent with a momentum of 0.9, weight decay of 0.9, batch size of 10, and learning rate of 5×10^{-4} for optimization. Moreover, we used a poly strategy [42] to adjust the learning rate during training. The training process of the model on the two datasets took approximately 32 h, and the test time was 73 min and 6 min, respectively, on Potsdam and Vaihingen, including the model inference time and the concatenation from slices into high-resolution remote sensing images.

3.3. Comparison with State-of-the-Art Methods

3.3.1. Quantitative Evaluation

In particular, Tables 1 and 2 list the performance indicators obtained by applying various methods on the two RSI semantic segmentation datasets. Table 3 summarizes the comparison results of all models in terms of flops and parameters, including the method based on Transformer [43]. The proposed DGPRNet outperformed the comparison methods on the Potsdam and Vaihingen datasets, and the indicators verified the high detection performance of the proposed method. Moreover, the DGPRNet detection of the class car on the two datasets was remarkable, confirming correct object detection in challenging scenes. Compared with existing methods, DGPRNet showed outstanding results in three classes, namely impervious surfaces, buildings, and cars, on the Vaihingen dataset. In addition, both mAcc and mIoU outperformed the best indicators in the comparisons. In particular, the IoU indicator of DGPRNet in the impervious surface and building outperformed the participating methods by 0.52% (SA-Gate) and 0.96% (ACNet). Especially for the class car, DGPRNet reached 92.30% and 84.84% in Acc and IoU indicators, which exceeded 8.03% and 6.77% compared with SA-Gate. In addition, the overall indicators mAcc and mIoU reached 90.43% and 82.36%, respectively, increasing by 1.81% and 1.69% compared to SA-Gate. Furthermore, on the Potsdam dataset, the DGPRNet outperformed the comparison methods in almost all classes except for low vegetation, tree, and clutter on the classification accuracy. Similarly, IoU in car and cluster reached 92.46% and 47.02%, respectively, compared with more than 2.03% and 3.48% for ACNet and Deeplabv3+. The accuracy on the class car exceeded HRCNet by 2.09% and reached 96.03%. In terms of overall performance, the proposed DGPRNet achieved mAcc of 85.69% and mIoU of 77.69%, increasing by 1.27% and 1.79% compared to the SA-Gate and RedNet, respectively. The improvement in the overall indicators of the proposed method in the small category can be explained as follows: by complementing each other at different resolutions, the aggregate representation information of a specific category at multiple scales can be obtained, greatly improving the accuracy and IoU on the small objects. Therefore, the improvement in this category is particularly significant.

Table 1. Quantitative results of the proposed DGPRNet and 11 state-of-the-art methods on the Vaihingen dataset. The values in bold indicate the best scores in the evaluation matrix.

		FCN-8S [13]	U-Net [14]	SegNet [15]	DeepLabv3+ [17]	BiseNetV2 [19]	HRCNet [20]	RedNet [16]	ACNet [23]	SA-Gate [18]	TSNet [21]	ESANet [22]	DCSwin [43]	Ours
Imp.surf	Acc	89.66	91.68	89.88	90.06	90.56	91.62	91.49	91.95	90.99	87.93	92.09	91.40	91.55
	IoU	79.71	80.90	80.93	81.11	80.97	81.60	84.62	85.34	85.70	78.98	85.18	84.45	86.22
Building	Acc	93.22	89.84	90.88	87.04	91.24	91.72	94.81	95.45	93.85	95.81	94.93	95.29	95.80
	IoU	86.80	86.50	86.54	82.70	86.69	88.01	91.07	91.82	91.72	91.47	91.16	91.30	92.78
Low veg.	Acc	75.83	77.97	78.66	76.65	74.68	79.24	78.67	78.64	84.95	71.62	75.72	79.02	81.27
	IoU	64.33	65.91	64.07	64.44	63.66	67.38	66.59	66.87	68.68	57.03	65.48	66.26	68.62
Tree	Acc	89.22	91.30	88.96	88.60	91.66	90.55	91.41	91.20	89.06	94.26	92.35	89.85	91.22
	IoU	75.58	77.86	75.96	76.64	76.54	78.58	78.27	78.55	79.15	81.26	77.65	77.54	79.34
Car	Acc	45.12	75.80	43.93	42.51	63.75	70.69	59.77	83.12	84.27	67.63	75.92	81.51	92.30
	IoU	40.16	71.22	43.16	43.10	61.80	68.73	56.06	76.81	78.07	66.86	70.11	73.47	84.84
mAcc		78.61	79.75	78.46	76.97	82.38	84.76	83.23	88.07	88.62	83.54	86.20	87.41	90.43
mIoU		69.32	71.34	70.13	69.49	73.93	76.86	75.32	79.88	80.67	75.12	77.92	78.60	82.36

Table 2. Quantitative results of the proposed DGPRNet and 11 state-of-the-art methods on the Potsdam dataset.

		FCN-8S [13]	U-Net [14]	SegNet [15]	DeepLabv3+ [17]	BiseNetV2 [19]	HRCNet [20]	RedNet [16]	ACNet [23]	SA-Gate [18]	TSNet [21]	ESANet [22]	DCSwin [43]	Ours
Imp.surf	Acc	89.47	90.03	90.18	91.57	90.12	90.03	92.19	91.32	85.84	85.22	91.38	91.66	92.76
	IoU	79.77	80.27	80.46	82.49	80.58	81.68	82.83	82.74	80.64	76.85	82.92	82.28	83.33
Building	Acc	90.69	88.71	90.21	91.78	88.88	90.87	93.61	93.83	93.65	91.85	93.69	92.92	93.94
	IoU	83.60	82.92	84.18	87.59	83.70	85.75	90.13	90.06	88.51	86.65	89.82	89.12	91.26
Low veg.	Acc	85.13	85.82	85.88	87.36	87.68	88.17	87.00	86.16	86.46	88.52	87.10	87.31	87.12
	IoU	71.12	71.60	71.63	73.63	71.55	73.18	73.22	73.53	72.71	67.98	73.16	74.48	74.46
Tree	Acc	82.86	84.06	82.49	85.45	81.11	82.02	83.00	86.03	85.70	78.75	82.48	84.46	85.84
	IoU	71.23	72.05	70.68	73.32	71.55	71.32	71.77	72.87	72.89	67.49	70.81	73.23	73.80
Car	Acc	91.02	93.89	93.15	93.89	93.14	93.94	93.36	93.79	92.18	78.22	93.08	96.31	96.03
	IoU	81.53	90.24	89.72	90.04	89.17	89.82	90.08	90.43	89.39	76.85	88.53	90.12	92.46
Clutter	Acc	49.05	50.30	51.76	53.80	50.66	56.72	56.74	54.51	62.70	37.49	55.68	56.01	58.48
	IoU	36.49	36.26	37.21	43.54	36.35	40.03	43.51	41.65	40.59	30.85	43.38	43.37	47.02
mAcc		78.61	82.13	82.28	83.97	81.93	83.63	84.32	84.27	84.42	76.68	83.90	84.61	85.69
mIoU		69.32	72.22	72.31	75.10	71.86	73.63	75.26	75.21	74.12	67.78	74.77	75.43	77.05

3.3.2. Qualitative Evaluation

Figure 3 shows the segmentation results obtained using DGPRNet and 11 state-of-the-art methods. Examples of multiple scenes were included, such as scenes with objects of different scales (clutter), small objects (car), large objects (building), low contrast with the background, and blurred boundaries. In general, the qualitative results show that DGPRNet has improved scene adaptability and reconstruction accuracy compared to similar methods.

In the visual contrast result, the area marked by the red rectangle shows the place that differs the most. Some methods could not precisely locate the objects of clutter because the object was usually located in a complex scene with different sizes. As shown in the first to fifth lines of Figure 3, the clutter can be accurately located in many complex scenes compared with other models. As the classes with the largest proportion in the dataset, the detection results of the proposed method for buildings are accurate and the edges are smooth, as shown in the second, third, sixth, and seventh rows in Figure 3. Compared with other methods, there are fewer cases of incomplete detection. The key problem to be solved in this study is to accurately segment the small objects (car) in the dataset. As seen from the fifth to the ninth rows of Figure 3, the segmentation result of the class car is more precise than that of other models. The qualitative results showed that DGPRNet better adapted to different scenes and reconstructed features with higher accuracy than similar methods. Moreover, DGPRNet performed highly in various complex scenes and detected small objects and the object's edges better than the other evaluated methods.

Table 3. The comparison on flops and parameters in all methods.

	Flops (GMac)	Params (M)
FCN8s	74.55	134.29
UNet	55.93	26.36
SegNet	18.3	53.56
DeepLabv3+	32.45	59.33
BiseNet	3.23	3.63
HRCNet	30.28	62.71
RedNet	21.17	81.95
ACNet	26.41	116.6
SA-Gate	41.23	110.85
TSNet	34.27	41.8
ESANet	10.15	45.42
DCSwin	34.4	118.39
Ours	55.39	142.82

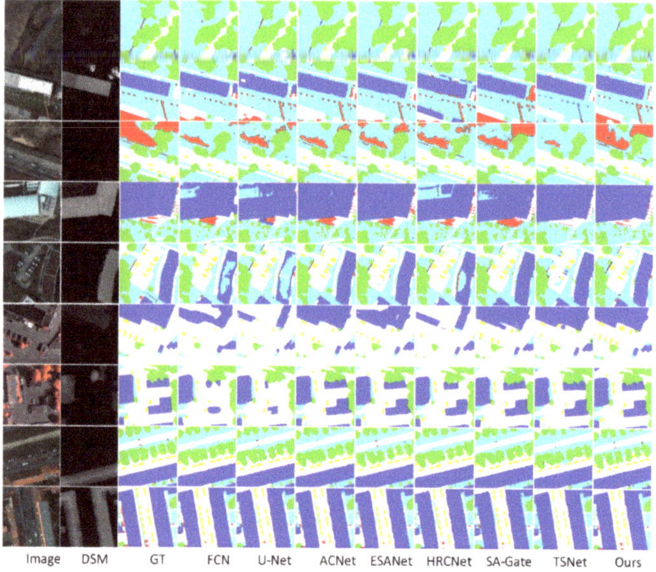

Figure 3. Comparison of segmentation results from different methods.

3.4. Ablation Study

To verify the effectiveness of the adopted modules, we conducted a comparative experiment on two datasets. Table 4 lists the comparison of the ablation indicators.

Table 4. Ablation study on the Vaihingen and Potsdam datasets.

	Vaihingen		Potsdam	
	mAcc	mIoU	mAcc	mIoU
Baseline	84.49	76.66	83.72	74.79
W/o DSM	89.67	81.07	84.89	75.88
W/o PRB	88.50	80.73	84.54	74.82
W/o DSEM	87.14	78.84	80.81	69.85
Ours	**90.43**	**82.36**	**85.69**	**77.05**

3.4.1. Effect of Modal DSM

The effectiveness of multimodal data was verified. In this regard, the DSM data were removed and represented as w/o DSM in Table 4. The ablation results show that the scores of the model in the single modal are slightly lower than those in the multimodal data. The results indicate that DSM modal data can indeed improve the performance of the model from another perspective.

3.4.2. Effects of Module PRB

The w/o PRB indicates the scheme implemented without the PRB module. In the decoding part, we replaced the PRB module with the convolution block combined with 3×3 convolutional layers + BN + ReLU to verify the effectiveness of the PRB module. Figure 4 shows the prediction map. The scheme without the PRB module performed lower than the full model. For example, the detection area of the building in the first, second, and fourth rows is discontinuous. Furthermore, a clutterer was present with incorrect classification in the third and fourth rows. The above comparison diagram also verifies that the PRB module reduces the feature reconstruction error in the process of network decoding and plays a crucial role in network inference. Compared with the full model, the ablation indicators mAcc and mIoU of PRB decreased by 1.93% and 1.63% on the Vaihingen dataset and 1.15% and 2.23% on the Potsdam dataset, respectively. The above results demonstrate the importance of the PRB module from both qualitative and quantitative perspectives.

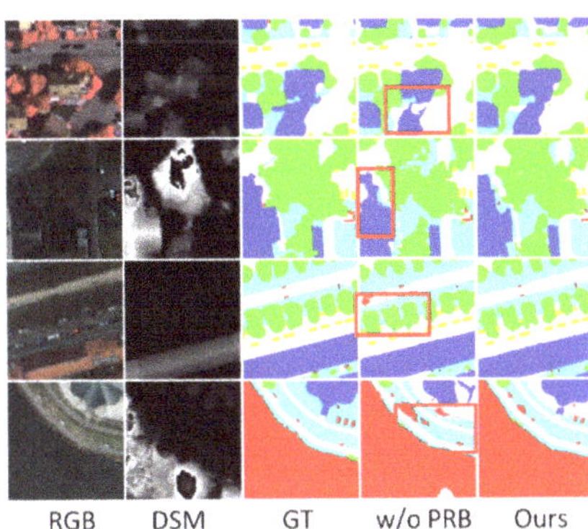

Figure 4. Ablation performance comparisons with the effect of PRB.

3.4.3. Effects of Module DSEM

Similarly, we verified the effectiveness of DSEM and the multi-supervision strategy. We removed the DSEM and multi-supervision strategy of the last three layers of semantic features and used the scheme w/o DSEM to represent it. The performance of the scheme w/o DSEM was low. As shown in Table 4, compared with the full model, the DSEM scheme decreased the two indicators by 3.29% and 3.52% in the Vaihingen dataset and 4.88% and 7.2% in the Potsdam dataset, respectively. Moreover, from the perspective of visualization, if no further exploration of the deep feature exists between classes, the first and second lines of Figure 5 are confused with building and clutter. Similarly, a category misclassification will be present in the third and fourth lines. Hence, we concluded that this module considerably facilitated the reconstruction ability of the network decoding layer. The module improved the specification effect on the deep semantic features and helped the decoding module enhance reconstruction under different resolutions.

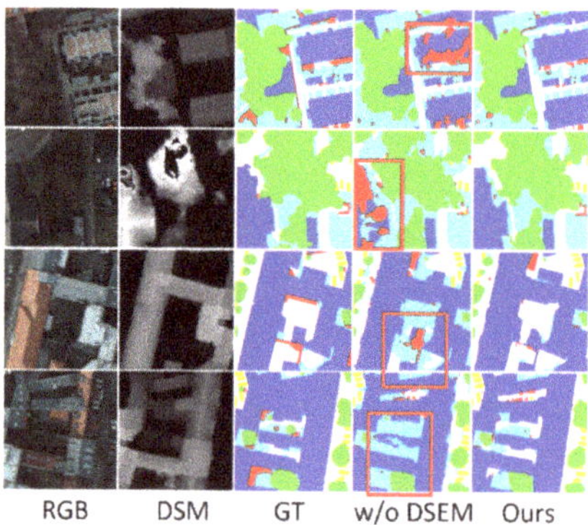

Figure 5. Ablation performance comparisons with the effect of DSEM.

4. Conclusions

This study proposed a novel network framework called DGPRNet for semantic segmentation of remote sensing images by exploring inter and intraclass relationships in deep features and decreasing feature reconstruction loss in the decoder. First, adjacent intermediate features were complemented before decoding to improve the expression of multi-scale features. Second, PRB was developed and deployed at five stages in the decoder to capture detailed features obtained from different receiving fields at multiple resolutions, reducing error and maintaining accuracy during reconstruction. Finally, the proposed DSEM distinguished and aggregated interclass and intraclass features based on semantic features to leverage deep features in detecting objects with different scales. Experimental results on two RSI datasets showed that DGPRNet outperformed 11 state-of-the-art methods.

Author Contributions: Conceptualization, J.M. and W.Z.; methodology, J.M., X.Q.; software, X.Q.; validation, L.Y.; writing—review and editing, W.Z., L.Y.; supervision, W.Z.; project administration, W.Z.; funding acquisition, X.Q. All authors have read and agreed to the published version of the manuscript.

Funding: This work was supported by National Natural Science Foundation of China (61502429, 61672337, 61972357); the Zhejiang Provincial Natural Science Foundation of China (LY18F020012, LY17F020011), and Zhejiang Key R & D Program (2019C03135).

Data Availability Statement: The code used and the datasets generated during the different steps of the analysis are available from the corresponding author on reasonable request.

Acknowledgments: The authors sincerely appreciate the helpful comments and constructive suggestions given by the academic editors and reviewers.

Conflicts of Interest: The authors declare no conflict of interest.

References

1. Neupane, B.; Horanont, T.; Aryal, J. Deep learning-based semantic segmentation of urban features in satellite images: A review and meta-analysis. *Remote Sens.* **2021**, *13*, 808. [CrossRef]
2. Zhou, W.; Liu, J.; Lei, J.; Hwang, J.-N.; Yu, L. GMNet: Graded-feature multilabel-Learning network for RGB-Thermal urban scene semantic segmentation. *IEEE Trans. Image Process.* **2021**, *30*, 7790–7802. [CrossRef]
3. Chen, Z.; Li, D.; Fan, W.; Guan, H.; Wang, C.; Li, J. Self-attention in reconstruction bias U-Net for semantic segmentation of building rooftops in optical remote sensing images. *Remote Sens.* **2021**, *13*, 2524. [CrossRef]
4. Wang, L.; Li, R.; Wang, D.; Duan, C.; Wang, T.; Meng, X. Transformer meets convolution: A bilateral awareness network for semantic segmentation of very fine resolution urban scene images. *Remote Sens.* **2021**, *13*, 3065. [CrossRef]
5. Zhou, W.; Yang, E.; Lei, J.; Wan, J.; Yu, L. PGDENet: Progressive Guided Fusion and Depth Enhancement Network for RGB-D Indoor Scene Parsing. *IEEE Trans. Multimed.* **2022**. [CrossRef]
6. Zhou, W.; Liu, W.; Lei, J.; Luo, T.; Yu, L. Deep binocular fixation prediction using hierarchical multimodal fusion network. *IEEE Trans. Cogn. Dev. Syst.* **2021**. [CrossRef]
7. Wu, J.; Zhou, W.; Luo, T.; Yu, L.; Lei, J. Multiscale multilevel context and multimodal fusion for RGB-D salient object detection. *Signal Process.* **2021**, *178*, 107766. [CrossRef]
8. Zhou, W.; Jin, J.; Lei, J.; Yu, L. CIMFNet: Cross-Layer Interaction and Multiscale Fusion Network for Semantic Segmentation of High-Resolution Remote Sensing Images. *IEEE J. Sel. Top. Signal Process.* **2022**, *16*, 666–676. [CrossRef]
9. Liu, X.; Jiao, L.; Zhao, J.; Zhao, J.; Zhang, D.; Liu, F.; Tang, X. Deep multiple instance learning-based spatial–spectral classification for PAN and MS imagery. *IEEE Trans. Geosci. Remote Sens.* **2017**, *56*, 461–473. [CrossRef]
10. Mou, L.; Hua, Y.; Zhu, X.X. A relation-augmented fully convolutional network for semantic segmentation in aerial scenes. In Proceedings of the IEEE/CVF Conference on Computer Vision and Pattern Recognition, Salt Lake City, UT, USA, 18–23 June 2018; pp. 12416–12425.
11. Zhou, W.; Dong, S.; Lei, J.; Yu, L. MTANet: Multitask-Aware Network with Hierarchical Multimodal Fusion for RGB-T Urban Scene Understanding. *IEEE Trans. Intell. Veh.* **2022**. [CrossRef]
12. Zhou, W.; Guo, Q.; Lei, J.; Yu, L.; Hwang, J.-N. IRFR-Net: Interactive recursive feature-reshaping network for detecting salient objects in RGB-D images. *IEEE Trans. Neural Netw. Learn. Syst.* **2021**. [CrossRef]
13. Long, J.; Shelhamer, E.; Darrell, T. Fully convolutional networks for semantic segmentation. In Proceedings of the IEEE Conference on Computer Vision and Pattern Recognition, Boston, MA, USA, 7–12 June 2015; pp. 3431–3440.
14. Ronneberger, O.; Fischer, P.; Brox, T. U-net: Convolutional networks for biomedical image segmentation. In Proceedings of the International Conference on Medical Image Computing and Computer-Assisted Intervention, Munich, Germany, 5–9 October 2015; pp. 234–241.
15. Vijay, B.; Alex, K.; Roberto, C. SegNet: A Deep Convolutional Encoder-Decoder Architecture for Image Segmentation. *IEEE Trans. Pattern Anal. Mach. Intell.* **2017**, *39*, 2481–2495.
16. Jiang, J.; Zheng, L.; Luo, F.; Zhang, Z. Rednet: Residual encoder-decoder network for indoor rgb-d semantic segmentation. *arXiv* **2018**, arXiv:1806.01054.
17. Chen, L.C.; Zhu, Y.; Papandreou, G.; Schroff, F.; Adam, H. Encoder-decoder with atrous separable convolution for semantic image segmentation. In Proceedings of the European Conference on Computer Vision, Munich, Germany, 8–14 September 2018; pp. 801–818.
18. Chen, X.; Lin, K.Y.; Wang, J.; Wu, W.; Qian, C.; Li, H.; Zeng, G. Bi-directional cross-modality feature propagation with separation-and-aggregation gate for RGB-D semantic segmentation. In *European Conference on Computer Vision*; Springer: Cham, Switzerland, 2020; pp. 561–577.
19. Yu, C.; Gao, C.; Wang, J.; Yu, G.; Shen, C.; Sang, N. Bisenet v2: Bilateral network with guided aggregation for real-time semantic segmentation. *Int. J. Comput. Vis.* **2021**, *129*, 3051–3068. [CrossRef]
20. Xu, Z.; Zhang, W.; Zhang, T.; Li, J. HRCNet: High-resolution context extraction network for semantic segmentation of remote sensing images. *Remote Sens.* **2020**, *13*, 71. [CrossRef]
21. Zhou, W.; Yuan, J.; Lei, J.; Luo, T. TSNet: Three-stream self-attention network for RGB-D indoor semantic segmentation. *IEEE Intell. Syst.* **2020**, *36*, 73–78. [CrossRef]
22. Seichter, D.; Köhler, M.; Lewandowski, B.; Wengefeld, T.; Gross, H.M. Efficient rgb-d semantic segmentation for indoor scene analysis. In Proceedings of the IEEE International Conference on Robotics and Automation, Xi'an, China, 30 May–5 June 2021; pp. 13525–13531.
23. Hu, X.; Yang, K.; Fei, L.; Wang, K. Acnet: Attention based network to exploit complementary features for rgbd semantic segmentation. In Proceedings of the IEEE International Conference on Image Processing, Taipei, Taiwan, 22–25 September 2019; pp. 1440–1444.
24. Zhou, W.; Yu, L.; Zhou, Y.; Qiu, W.; Wu, M.; Luo, T. Local and global feature learning for blind quality evaluation of screen content and natural scene images. *IEEE Trans. Image Process.* **2018**, *27*, 2086–2095. [CrossRef]
25. Zhou, W.; Yang, E.; Lei, J.; Yu, L. FRNet: Feature Reconstruction Network for RGB-D Indoor Scene Parsing. *IEEE J. Sel. Top. Signal Process.* **2022**, *16*, 677–687. [CrossRef]
26. Zhou, W.; Lin, X.; Lei, J.; Yu, L.; Hwang, J.-N. MFFENet: Multiscale feature fusion and enhancement network for RGB–Thermal urban road scene parsing. *IEEE Trans. Multimed.* **2022**, *24*, 2526–2538. [CrossRef]

27. Zheng, X.; Wu, X.; Huan, L.; He, W.; Zhang, H. A Gather-to-Guide Network for Remote Sensing Semantic Segmentation of RGB and Auxiliary Image. *IEEE Trans. Geosci. Remote Sens.* **2021**, *60*, 1–15. [CrossRef]
28. Ma, X.; Zhang, X.; Pun, M.O. A Crossmodal Multiscale Fusion Network for Semantic Segmentation of Remote Sensing Data. *IEEE J. Sel. Top. Appl. Earth Obs. Remote Sens.* **2022**, *15*, 3463–3474. [CrossRef]
29. Simonyan, K.; Zisserman, A. Very deep convolutional networks for large-scale image recognition. *arXiv* **2014**, arXiv:1409.1556.
30. He, K.; Zhang, X.; Ren, S.; Sun, J. Deep residual learning for image recognition. In Proceedings of the IEEE conference on Computer Vision and Pattern Recognition, Las Vegas, NE, USA, 27–30 June 2016; pp. 770–778.
31. Zhou, W.; Liu, C.; Lei, J.; Yu, L.; Luo, T. HFNet: Hierarchical feedback network with multilevel atrous spatial pyramid pooling for RGB-D saliency detection. *Neurocomputing* **2022**, *490*, 347–357. [CrossRef]
32. Zhou, W.; Lv, Y.; Lei, J.; Yu, L. Global and Local-Contrast Guides Content-Aware Fusion for RGB-D Saliency Prediction. *IEEE Trans. Syst. Man Cybern. Syst.* **2021**, *51*, 3641–3649. [CrossRef]
33. He, K.; Zhang, X.; Ren, S.; Sun, J. Spatial pyramid pooling in deep convolutional networks for visual recognition. *IEEE Trans. Pattern Anal. Mach. Intell.* **2015**, *37*, 1904–1916. [CrossRef]
34. Zhou, W.; Guo, Q.; Lei, J.; Yu, L.; Hwang, J.-N. ECFFNet: Effective and consistent feature fusion network for RGB-T salient object detection. *IEEE Trans. Circuits Syst. Video Technol.* **2022**, *32*, 1224–1235. [CrossRef]
35. Li, G.; Liu, Z.; Zeng, D.; Lin, W.; Ling, H. Adjacent context coordination network for salient object detection in optical remote sensing images. *IEEE Trans. Cybern.* **2022**. [CrossRef]
36. Zhou, W.; Liu, C.; Lei, J.; Yu, L. RLLNet: A lightweight remaking learning network for saliency redetection on RGB-D images. *Sci. China Inf. Sci.* **2022**, *65*, 160107. [CrossRef]
37. Gong, T.; Zhou, W.; Qian, X.; Lei, J.; Yu, L. Global contextually guided lightweight network for RGB-thermal urban scene understanding. *Eng. Appl. Artif. Intell.* **2023**, *117*, 105510. [CrossRef]
38. Zhou, W.; Wu, J.; Lei, J.; Hwang, J.-N.; Yu, L. Salient object detection in stereoscopic 3D images using a deep convolutional residual autoencoder. *IEEE Trans. Multimed.* **2021**, *23*, 3388–3399. [CrossRef]
39. Zhou, W.; Zhu, Y.; Lei, J.; Wan, J.; Yu, L. CCAFNet: Crossflow and cross-scale adaptive fusion network for detecting salient objects in RGB-D images. *IEEE Trans. Multimed.* **2022**, *24*, 2192–2204. [CrossRef]
40. International Society for Photogrammetry and Remote Sensing. 2D Semantic Labeling Contest-Potsdam. Available online: https://www.isprs.org/education/benchmarks/UrbanSemLab/2d-sem-label-potsdam.aspx (accessed on 1 January 2020).
41. International Society for Photogrammetry and Remote Sensing. 2D Semantic Labeling Contest-Vaihingen. Available online: https://www.isprs.org/education/benchmarks/UrbanSemLab/2d-sem-label-vaihingen.aspx (accessed on 1 January 2020).
42. Chen, L.C.; Papandreou, G.; Kokkinos, I.; Murphy, K.; Yuille, A.L. Deeplab: Semantic image segmentation with deep convolutional nets, atrous convolution, and fully connected crfs. *IEEE Trans. Pattern Anal. Mach. Intell.* **2017**, *40*, 834–848. [CrossRef]
43. Wang, L.; Li, R.; Duan, C.; Zhang, C.; Meng, X.; Fang, S. A novel transformer based semantic segmentation scheme for fine-resolution remote sensing images. *IEEE Geosci. Remote Sens. Lett.* **2022**, *19*, 1–5. [CrossRef]

remote sensing

Article

A Multi-Dimensional Deep-Learning-Based Evaporation Duct Height Prediction Model Derived from MAGIC Data

Cheng Yang [1,2,†], Jian Wang [1,2,3,*,†] and Yafei Shi [1,2]

[1] School of Microelectronics, Tianjin University, Tianjin 300072, China
[2] Qingdao Institute for Ocean Technology, Tianjin University, Qingdao 266200, China
[3] Shandong Technology Research Center of Marine Information Perception and Transmission Engineering, Qingdao 266200, China
* Correspondence: wangjian16@tju.edu.cn
† These authors contributed equally to this work.

Abstract: The evaporation duct height (EDH) can reflect the main characteristics of the near-surface meteorological environment, which is essential for designing a communication system under this propagation mechanism. This study proposes an EDH prediction network with multi-layer perception (MLP). Further, we construct a multi-dimensional EDH prediction model (multilayer-MLP-EDH) for the first time by adding spatial and temporal "extra data" derived from the meteorological measurements. The experimental results show that: (1) compared with the naval-postgraduate-school (NPS) model, the root-mean-square error (RMSE) of the meteorological-MLP-EDH model is reduced to 2.15 m, and the percentage improvement reached 54.00%; (2) spatial and temporal parameters can reduce the RMSE to 1.54 m with an improvement of 66.96%; (3) the multilayer-MLP-EDH model can match measurements well at both large and small scales by attaching meteorological parameters at extra height, the error is further reduced to 1.05 m, with 77.51% improvement compared with the NPS model. The proposed model can significantly improve the prediction accuracy of the EDH and has great potential to improve the communication quality, reliability, and efficiency of ducting in evaporation ducts.

Keywords: maritime communication; evaporation duct; deep learning; multi-dimensional prediction model

Citation: Yang, C.; Wang, J.; Shi, Y. A Multi-Dimensional Deep-Learning-Based Evaporation Duct Height Prediction Model Derived from MAGIC Data. *Remote Sens.* **2022**, *14*, 5484. https://doi.org/10.3390/rs14215484

Academic Editor: Gwanggil Jeon

Received: 3 September 2022
Accepted: 28 October 2022
Published: 31 October 2022

Publisher's Note: MDPI stays neutral with regard to jurisdictional claims in published maps and institutional affiliations.

1. Introduction

An atmospheric duct is a unique phenomenon in the lower atmosphere, and electromagnetic waves can experience less attenuation in the trapped layer by limiting the spread of the wavefront from spherical to cylindrical expansion, where the waves are bent by atmospheric refraction [1,2]. A long-range transmission in microwave bands can also be realized. Namely, microwave radio signals may refract in the lower layers of Earth's atmosphere and propagate far beyond the line of sight [3]. This feature may be appropriate for communications at sea [4], while the public land mobile network (PLMN) [5,6] is limited due to the special meteorological conditions and terrain features [7,8]. Moreover, communication systems using evaporation ducts are expected to become an important means of the sixth-generation communication system until we further understand its distribution characteristics.

As a strong negative vertical humidity gradient near the sea surface, the evaporation duct exists due to the moisture content rapidly decreasing with increasing altitude [9,10]. Dramatic effects may be applied to the microwave communication system while transmitting in the ducting layer, especially for frequencies above 1 GHz [11], which may meet the demand for large-bandwidth, high-speed, and long-range applications [12]. The evaporation duct frequently occurs over the ocean and the occurrence in the South China Sea (SCS) exceeds 75% [13]. However, the spatial and temporal refractivity variations significantly affect shipboard communication performances at sea and near shore [4,14,15]. As a result, communication effects may experience several disadvantages, including a specific time loss (about 25% in the SCS based on the above statistical characteristic of the evaporation duct) and an available low antenna height with minimal effects on any land–land or ship-based system, etc., constraining it from becoming a convenient and widely used maritime communication. Therefore, the prerequisite for communication using evaporation ducts is the accurate prediction of transmission effects, which relies on two critical components: the atmospheric refractive condition and the prediction in the given environment [16]. The evaporation duct height (EDH) is the characteristic parameter of the refractivity profile. The accurate prediction of the EDH has special significance in the practical design of the evaporation duct communication system and the instrument parameters. The EDH can be directly measured [17,18] and evaluated by numerous methods [19–21] and theoretical models [9,22–26]. The directly measured method may take lots of time and effort, and the latter two methods can calculate from meteorological detection at a certain height, but the accuracy needs further improvement. Therefore, a focus topic in the present and the future is to maximize the use of measurement datasets and to improve the prediction accuracy of EDH.

Presently, "artificial intelligent (AI) enhanced operation" has become one of the hotspot directions [14,27–30]. The combination of AI and the analysis of evaporation ducts has also boosted the accuracy of EDH prediction. In modeling construction, Yan et al., propose a numerical profiling method that adopts the artificial neural network and training data from the remote sensing data and the naval postgraduate school (NPS) model [14]. Zhao et al., propose a method based on a multi-layer perception (MLP) of five hidden layers to predict the EDH, and the applicability in different areas is analyzed [28]. In short-term prediction, Zhao et al., constructed an EDH prediction model based on a long short-term memory network [29]. In addition, Mai et al., introduced the Darwinian evolutionary algorithm and compared the accuracy with the neural network in EDH prediction [30].

In this paper, deep learning methods are utilized to improve the prediction accuracy of the EDH so that the communication system can be better designed and operated. Furthermore, we construct a multi-dimensional EDH prediction model for the first time by blending with spatial and temporal "extra data" during meteorological detection [31]. Section II describes the background and previous EDH prediction method and Section III describes the modeling process of the proposed model. Finally, predictions of the proposed model and the theoretical method are compared with the measurements; the effectiveness has also been verified.

2. Background and Methods

2.1. Evaporation duct Diagnosis

The refraction in the atmosphere refers to the bending characteristics while the electromagnetic wave propagates in the medium, and the degree can be measured by the refraction index n

$$n = \frac{c}{v} \tag{1}$$

where c and v are the propagation speed of the electromagnetic wave in free space and the medium, respectively.

Radio refractivity N (N-unit) is usually used in the troposphere to reflect the corresponding spatial structure characteristics. According to the ITU-R Recommendation P.453-14 [32]

$$N = (n-1) \times 10^6 = \frac{77.6}{T} \times \left(P + \frac{4810e}{T} \right) \tag{2}$$

where P is the atmospheric pressure (hPa), T is the absolute temperature (K), and e is the water vapor pressure (hPa).

For the convenience of considering the curvature of the Earth, the modified refractivity M (M-unit) is often utilized as [9,10]

$$M = N + \frac{z}{r_e} \times 10^6 \approx N + 0.157z \tag{3}$$

$$\frac{dM}{dz} = \frac{dN}{dz} + 0.157 \tag{4}$$

where z is the height above the ground (m) and r_e is Earth's radius (the average Earth radius is 6371 km).

Electromagnetic waves are bent towards the ground by atmospheric refraction, while the vertical gradient of modified refractivity becomes negative ($dM/dz < 0$). Signals can refract and propagate over the horizon with matching frequencies and angles.

2.2. Theoretical Models of EDH

Based on the Monin–Obukhov similarity theory [33,34], the vertical profile of mean wind speed U(m/s), potential temperature θ (K), and specific humidity q (kg/kg) in the surface layer can be calculated. At present, the extensively utilized numerical methods in the evaporation duct prediction include the Paulus–Jeske (PJ) model [22], the Musson–Gauthier–Bruth (MGB) model [23], the Babin–Young–Carton (BYC) model [9], the NPS model [24], and the surface heat budget of the arctic ocean experiment (SHEBA) model [26] are extensively utilized at present. An evaporation duct's modified refractivity (M-profile) can be defined by a limited number of meteorological factors, such as pressure and temperature at the sea surface, relative humidity, temperature, and wind speed at a certain altitude.

The comparison between the BYC model, the NPS model, and the SHEBA model are listed in Table 1. During the calculation, the scale parameters and the thermodynamically roughness height of the sea surface are defined by the COARE algorithm [33], and the profile stability functions calculate the wind speed and temperature under stable conditions.

Taking the NPS model as an example, the advanced air–sea flux algorithm COARE 3.0 is adopted, keeping good consistency with the measured results [35]. The input parameters are used to determine the modified refractivity profile, and the altitude with minimum value is the EDH. The vertical profile of air temperature T and specific humidity q at altitude z can be calculated as [24]:

$$T(z) = T_0 + \frac{\theta_*}{\kappa}\left[\ln\left(\frac{z}{z_{0t}}\right) - \psi_h\left(\frac{z}{L}\right)\right] - \Gamma_d z \tag{5}$$

$$q(z) = q_0 + \frac{q_*}{\kappa}\left[\ln\left(\frac{z}{z_{0t}}\right) - \psi_h\left(\frac{z}{L}\right)\right] \tag{6}$$

where θ_* and q_* are the characteristic scales of potential temperature and specific humidity, respectively; ψ_h is stability functions; z_{0t} is thermodynamic roughness height; Γ_d is the dry adiabatic decline rate; κ is Karman constant; L is the Monin–Obukhov length.

According to theoretical models, the EDH can be calculated with meteorological parameters at the sea surface and at a certain height. The calculation function can be expressed as

$$\text{EDH} = \mathcal{F}_{\text{Theoretical}}\left(T_{h_0}, P_{h_0}, T_{h_1}, U_{h_1}, RH_{h_1}, h_1\right) \tag{7}$$

where T_{h_0} and P_{h_0} are the pressure and temperature at the sea surface h_0, respectively, RH_{h_1}, T_{h_1}, and U_{h_1} are the relative humidity, temperature, and wind speed at the altitude h_1, respectively.

Table 1. The prediction methods of evaporation ducts.

Year	Models	Reference	Stability Functions in Stable Conditions	
			Form	Functions
1996	BYC model	[9]	Businger–Dyer	$\psi_m = \frac{\psi_{uk} + \xi^2 \psi_k}{1 + \xi^2}$ $\psi_h = \frac{\psi_{tk} + \xi^2 \psi_k}{1 + \xi^2}$ $\psi_{uk} = 2\ln\left(\frac{1+z_{pu}}{2}\right) + \ln\left(\frac{1+z_{pu}^2}{2}\right) - 2\arctan(z_{pu}) + \frac{\pi}{2}$ $\psi_k = 1.5\ln\left(\frac{z_{pg}^2 + z_{pg} + 1}{3}\right) - \sqrt{3}\arctan\left(\frac{2z_{pg}+1}{\sqrt{3}}\right) + \frac{\pi}{\sqrt{3}}$ $\psi_{tk} = 2\ln\left(\frac{1+z_{pt}}{2}\right)$ $z_{pu} = (1 - 16\xi)^{0.25}$ $z_{pt} = (1 - 16\xi)^{0.5}$ where $z_{pg} = (1 - 10\xi)^{0.333}$ for wind speed; $z_{pg} = (1 - 34\xi)^{0.333}$ for temperature.
2000	NPS model	[24] [25]	Beljaars and Holtslag (BH91)	$\psi_m = \psi_h = -\frac{b_h}{2}\ln(1 + c_h\xi + \xi^2)\left(-\frac{a_h}{B_h} + \frac{b_h c_h}{2B_h}\right) \times$ $\left(\ln\frac{2\xi + c_h B_h}{2\xi + c_h + B_h} - \ln\frac{c_h - B_h}{c_h + B_h}\right)$ where $a_h = b_h = 5$, $c_h = 3$, $B_h = \sqrt{5}$.
2007	SHEBA model	[26]	Grachev and Andreas (SHEBA07)	$\psi_m = -\frac{3a_m}{b_m}(x-1) + \frac{a_m B_m}{2b_m}\left[2\ln\frac{x-B_m}{1+B_m} - \ln\frac{x^2 - xB_m + B_m^2}{1 - B_m + B_m^2}\right]$ $\quad + 2\sqrt{3}\left(\arctan\frac{2x - B_m}{\sqrt{3}B_m} - \arctan\frac{2 - B_m}{\sqrt{3}B_m}\right), R_{iB} < 0.2$ $\psi_h = -\frac{b_h}{2}\ln(1 + c_h\xi + \xi^2)\left(-\frac{a_h}{B_h} + \frac{b_h c_h}{2B_h}\right) \times$ $\left(\ln\frac{2\xi + c_h B_h}{2\xi + c_h + B_h} - \ln\frac{c_h - B_h}{c_h + B_h}\right), R_{iB} < 0.2$ $a_m = 5$, $b_m = a_m/6.5$, $a_h = b_h = 5$, $c_h = 3$, $x = (1+\xi)^{1/3}$, $B_m = \left(\frac{1-b_m}{b_m}\right)^{1/3} > 0$, $B_h = \sqrt{c_h^2 - 4}$, R_{iB} is the bulk Richardson number.

where $\xi = z/L$ is the Monin–Obukhov parameter, used to express the atmospheric stability; z is the altitude; L is the similarity length.

2.3. Analysis of Transmission Effects

For long distance transmission in the microwave band, the main transmission mechanisms are normal propagation close to the Earth's surface and troposcatter propagation [36,37]. Anomalous propagation of transmission in the ducting layer may apply to communication systems over the ocean, signals may experience less attenuation under appropriate conditions.

The spatial and temporal distribution of meteorological parameters is uneven, leading to the changes of transmission effect changes with the variation of the evaporation duct. As a result, extra propagation loss among the designed communication system may be incurred. To analyze the influence of the variation of evaporation ducts quantitatively, a communication link was designed with the antenna heights fixed at 10 m. A specific position in the Pacific Ocean (30.0°N, 130.0°W) was selected, where the annual refractivity was 346.68 N-units [32]. The transmission loss diagram at 12 GHz with a distance of 0–500 km is shown in Figure 1. To perform the propagation curves corresponding to the typical EDH varied from 6 m to 18 m, the parabolic equation toolbox (PETOOL) [38,39] with the parabolic equation (PE) method [40,41] has been applied.

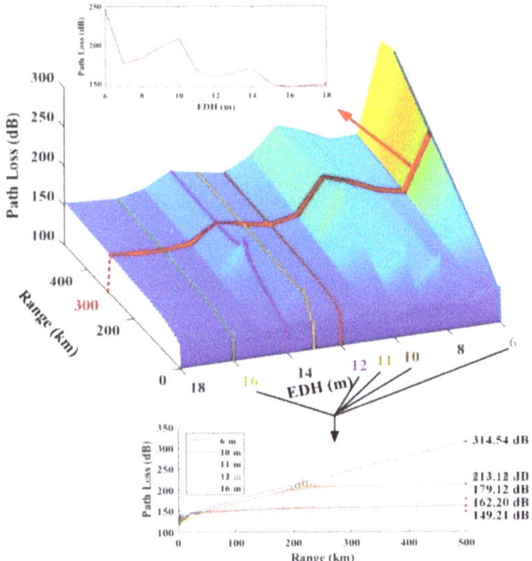

Figure 1. The transmission loss diagram at 12 GHz with a distance of 0–500 km.

In Figure 1, the path loss curves varied considerably with different EDH. Path losses increase with the EDH fixed at 6 m and lower than the antenna height, resulting in a poor communication effect. Significant improvement of channel conditions arises when the EDH is between 16 m and 18 m and the path loss fluctuates around 150 dB. Overall, the transmission loss fluctuates with the increase of the EDH. Especially in the range of 10–12 m, a 1 m variation in the height may lead to an increase in the path loss at 500 km of 16.92 dB to 34.00 dB, which brings much uncertainty to the operation of the communication system. This is roughly the same as the results of [38,42].

3. Datasets and Methodology

3.1. Modeling Data

3.1.1. MAGIC Datasets

To verify the prediction accuracy of existing models and to explore a better method, we use measured meteorological data from the ship-based marine ARM GCSS Pacific cross-section intercomparison (GPCI) investigation of clouds (MAGIC) field campaign. The ship-based MAGIC field campaign, with the marine-capable second ARM mobile facility (AMF2) deployed, lasted for nearly 200 days between Los Angeles, California and Honolulu, Hawaii, which provided high-resolution measured datasets of clouds, precipitation, and marine boundary layer (MBL) [43,44]. The ship completed 20 round trips from October 2012 to September 2013. Lots of instruments were deployed to measure meteorological parameters aboard the ship throughout the campaign: a Vaisala weather station, an inertial navigational location and attitude system, an infrared SST autonomous radiometer (ISAR), radiosondes, etc. [44].

The meteorological parameters of the MAGIC datasets collected in this paper mainly include temperature, pressure, relative humidity, wind speed, and direction measured by the marine meteorological system (MARMET) at approximately 27 m above sea level; the sea surface skin temperature (SSST) measured by the ISAR; the ship location by a navigation system. The time resolution of the MARMET and the ISAR devices is 1 min. The time distribution of the sounding data collected from MAGIC datasets is shown in Figure 2.

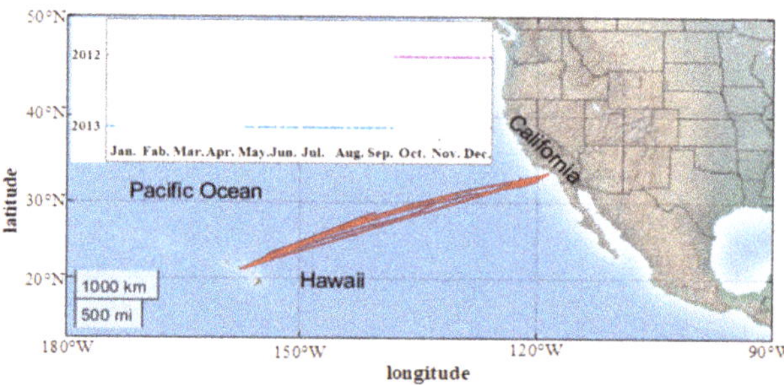

Figure 2. The spatial and temporal distribution of MAGIC measurements.

In addition, standard radiosondes (Vaisala model MW-31, SN E50401) were launched at 1 m to measure the vertical profiles of temperature, pressure, relative humidity, and wind speed and direction. Meteorological data at different altitudes were also collected using the Vaisala radiosonde (MW-31), with 0.5 Hz vertical resolution at fixed times [31]. As a result, 571 sets of radiosonde data were formed.

3.1.2. Data Processing

Using the boat measurements of the MARMET and ISAR devices at the sea surface and 27 m above sea level and the radiosonde-collected data during their ascent, we can obtain meteorological parameters for at least seven altitudes at each geographic location. The sea surface relative humidity (RH) was set at 98% [43]. As shown in Table 2, the measured datasets are preprocessed before modeling, with invalid datasets removed.

Table 2. The dataset preprocessing process.

Serial Number	Invalid Datasets	Number
1	NaN in the dataset	60 sets of data
2	0 in the dataset	18 sets of data
3	All measured altitudes fixed at 1 m	22 sets of data

In addition, we selected 476 effective datasets from 571 collections of radiosonde sounding data. Then, the modified refractivity index at seven layers can be obtained based on the effective meteorological datasets. A least-squares curve fit was applied to each of the 476 measurements. Furthermore, we got the M-refractivity profile by a log-linear function [9,10,45].

$$M = f_0 h - f_1 \ln(h + 0.001) + f_2 \tag{8}$$

where M is the modified refractivity (M-units) and h is the altitude (m). f_0, f_1, and f_2 are coefficients that can be calculated for a least-squares best fit. The constant 0.001 was used to prevent the curve from blowing up at zero altitudes [9,10].

Figure 3 shows the calculating process of the EDH, modified refractivity M for every 0.1 m between the surface and 40 m altitude based on MAGIC datasets, and the height at which the minimum M is achieved is the EDH.

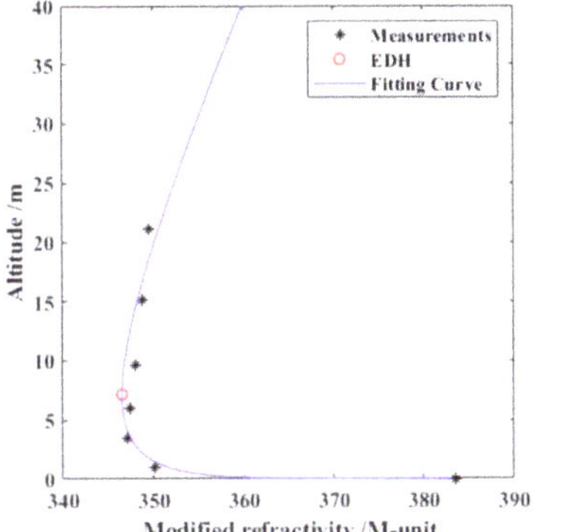

Figure 3. The calculating method of the EDH.

3.1.3. Reliability Assessment of Theoretical Models

Meteorological parameters, including the pressure and temperature at the surface, the relative humidity, the temperature, the wind speed in the air, and the altitude, will be considered in the calculation of the evaporation duct characteristics using theoretical models [24]. The statistical root-mean-square error (RMSE) of the EDH predictions of the BYC model, the NPS model, and the SHEBA model based on the MAGIC datasets are described in Table 3. The minimum RMSE is 4.52 m by the NPS model. Let x and y represent measured EDH and predicted EDH of the NPS model, and the fitting line is $y = 0.42x + 5.52$, far from the evaluation criteria $y = x$, as shown in Figure 4.

Table 3. The statistical RMSEs of three theoretical methods.

Models	BYC Model	NPS Model	SHEBA Model
RMSEs	4.72 m	4.52 m	4.79 m

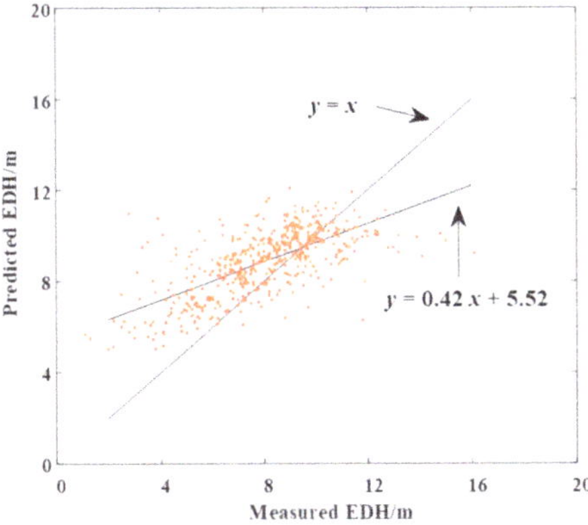

Figure 4. The EDH fitting line by the MAGIC datasets and inverted by the NPS model.

The EDH calculated by the NPS model does not match the measured data well. According to Figure 1, the RMSE exceeds 4 m, which may lead to a transmission loss error of more than 100 dB, which may bring significant deviation to the receiving effect of the transmission system.

3.2. Modeling Method

Theoretical models are generally based on the Monin–Obukhov similarity theory and are constrained by some basic physical boundary layer assumptions [3]. On the contrary, the neural network training prediction method can be derived entirely from original data. Therefore, it is more suitable for the natural atmospheric environment and will not be constrained by theoretical assumptions.

Considering that MAGIC has special characteristics from other similar experiments: (1) the experimental positions were spatially repeatable (the ship completed 20 round trips); (2) the radiosonde data were concentrated at several hours (it launched every 6 h); (3) the experiment instruments were set at multilayers, which have great data background both in time and space. Therefore, combine the experimental data with the neural network by adding the spatial data, such as latitude, longitude, and meteorological parameters in multilayers, and the temporal data, such as experiment time, to construct new datasets as training input for the prediction model.

Artificial intelligence originated in the 20th century and has been used in various industries, but it is seldom used in EDH prediction [27–30]. MLP is a kind of artificial neural network (ANN) with a forward structure [46–48] that maps a group of input vectors to a group of output vectors. The MLP consists of multiple layers and their neurons are fully connected to the next layer. It has a high nonlinear global function and powerful adaptive and self-learning ability, which is suitable for finding the characteristics of EDH prediction data in multi-dimensions. Here, the MLP model is considered to implement the construction of the training network.

3.2.1. Principle of the MLP

The MLP has universal approximation property [46]. Theoretically, an MLP network composed of a linear output layer and at least one hidden layer with activation functions can describe any function from a finite dimensional space to another with arbitrarily high precision with sufficient hidden neurons supplied. Each node in the MLP is the neuron with a nonlinear activation function, except the input node. MLP is an extension of perceptron, which overcomes the weakness of not recognizing linear non-fractional data.

Compared with the single-layer perceptron, the hidden layer of MLP changes from one to multiple. The training purpose of MLP is to make the network approximate the function that needs to be fitted. During the training process, the information is carried out from the input layer to the hidden layers and then to the output layer. The input layer is responsible for receiving the characteristics of the training data and is connected to the hidden layer with weight parameters. In contrast, the output layer is the target value that the training is expected to achieve through the hidden layers to realize the nonlinear mapping of the input space.

A typical MLP training process is as follows: (1) the weights are randomly allocated; (2) the neural network is activated by using all features of the training datasets from the input layer and then the output value is obtained through forwarding propagation; (3) the error is calculated between the output and the target value and the weight is updated by backpropagation; (4) the training is repeated until the output error is lower than the established standard. The trained MLP network can accept new input datasets at the end of this process.

3.2.2. Modeling

The essence of prediction is complex regression function construction and a multi-dimensional EDH prediction model can be constructed with the "extra data" in the experiment. The essence of MLP is also a nonlinear function mapping from input vector to output, similar to the model we tried to train. The advantages of MLP in learning and in processing nonlinear global data may solve the regression problem of meteorological characteristics. With a reasonable network structure and hyperparameters combined with enough training data, the performance of MLP can be excellent compared with the theoretical model.

While constructing the complete dataset of the MLP model, spatial data, temporal data, and meteorological data at multiple altitudes, including temperature, pressure, wind speed, and RH at the data measurement location of the MAGIC field campaign, were collected as a set of modeling data. To complete a comprehensively trained network for the validation of testing datasets and the generality of the method, we should select the training dataset that would cover the main features of the total dataset. A commonly accepted hold-out approach [49] is a 7:3 ratio between the training and the testing set. Namely, the training and the testing set proportion is 70% and 30% of the total dataset. Therefore, the first 334 groups of data in about 12 round trips were selected and randomly reordered as training datasets. The remaining 142 groups of data were used as a testing dataset.

By classifying and selecting the corresponding parameter information with the training datasets, meteorological-MLP-EDH, spatial-MLP-EDH, temporal-MLP-EDH, spatial–temporal-MLP-EDH, and multilayer-MLP-EDH models were constructed. The modeling process is shown in Figure 5.

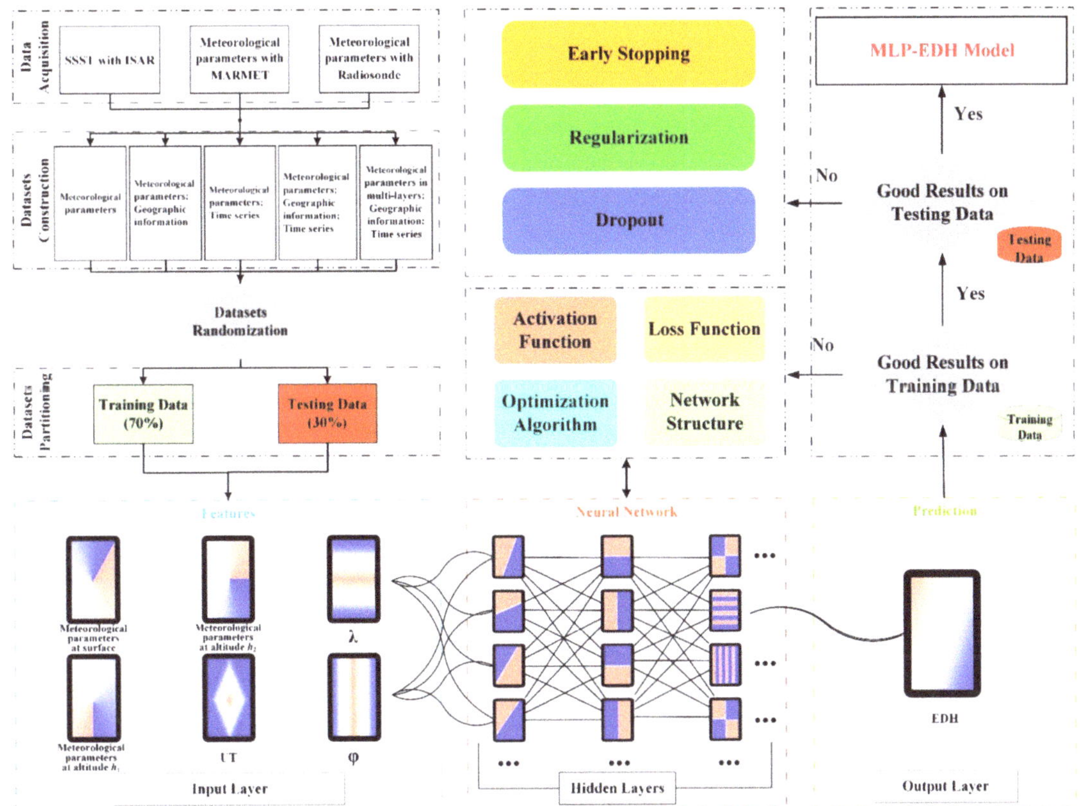

Figure 5. The modeling processes.

(1)   Meteorological-MLP-EDH

As a comparison with the theoretical models, the meteorological-MLP-EDH model takes the temperature (T) and the pressure (P) at the sea surface (h_0); temperature, wind speed (U), and RH in the air (h_1); measured altitudes in each training dataset as the input parameters. The mapping output is the corresponding EDH and the calculation function can be expressed as

$$EDH = \mathcal{F}_{\text{Meteorological}}(T_{h_0}, P_{h_0}, T_{h_1}, U_{h_1}, RH_{h_1}, h_1) \tag{9}$$

(2)   Spatial-MLP-EDH

The spatial data such as latitude and longitude may positively affect the prediction results of the experiment ship completing 20 round trips. The input vector of the spatial-MLP-EDH model takes the same parameters as the meteorological-MLP-EDH model. Furthermore, the spatial parameters of the experimental positions with latitude (λ) and longitude (φ) in the MAGIC campaign are also used as additional information to supply the feature of the selected meteorological parameters on a complete path.

$$EDH = \mathcal{F}_{\text{Spatial}}(\lambda, \varphi, T_{h_0}, P_{h_0}, T_{h_1}, U_{h_1}, RH_{h_1}, h_1) \tag{10}$$

(3)   Temporal-MLP-EDH

The radiosonde data were collected every 6 h in the MAGIC campaign. The temporal information may have a positive effect on prediction accuracy. With the measured time (UT), new datasets can be collected to construct the temporal-MLP-EDH model, implying the feature of the selected meteorological parameters at the specific time.

$$EDH = \mathcal{F}_{\text{Temporal}}(UT, T_{h_0}, P_{h_0}, T_{h_1}, U_{h_1}, RH_{h_1}, h_1) \tag{11}$$

(4)   Spatial–Temporal-MLP-EDH

The spatial–temporal-MLP-EDH model is a three-dimensional regression function consisting of spatial–temporal information and meteorological parameters at a single layer at sea surface and air. The input values supply the feature of selected meteorological parameters at a specific time on a complete path.

$$EDH = \mathcal{F}_{\text{Spatial–Temporal}}(\lambda, \varphi, UT, T_{h_0}, P_{h_0}, T_{h_1}, U_{h_1}, RH_{h_1}, h_1) \tag{12}$$

(5)   Multilayer-MLP-EDH

In addition, we constructed a four-dimensional regression function multilayer-MLP-EDH with meteorological parameters located on another layer. The sensitivity and accuracy of predicted results have been explored. The input vector mainly consists of spatial–temporal information and meteorological parameters at multiple layers with the altitudes of h_0, h_1, and h_2, which implies the feature of selected meteorological parameters over a wide vertical range at a specific time on a complete path.

$$EDH = \mathcal{F}_{\text{Multilayers}}(\lambda, \varphi, UT, T_{h_0}, P_{h_0}, T_{h_1}, U_{h_1}, RH_{h_1}, h_1, T_{h_2}, U_{h_2}, RH_{h_2}, h_2) \tag{13}$$

During the modeling process, the design of MLP and the selection of corresponding parameters will also greatly influence the prediction accuracy of the training data, so the related parameters need to be adjusted systematically. The section for MLP design mainly includes the activation function, loss function, optimization algorithm, and network structure [50].

(1)   Activation Function

In the hidden layer of MLP, the activation function is to introduce nonlinear changes to enhance the approximation ability of the neural network [28]. It uses differentiable functions and a back-propagation algorithm for effective learning. The most commonly used activation functions include rectified linear unit (ReLU), logistic sigmoid function, radial basis function (RBF), etc. In this paper, ReLU was used so that it can be tuned in a biomimetic way. The problem of gradient explosion and gradient disappearance is avoided by more efficient gradient descent and backpropagation [51]. ReLU function can be expressed as

$$\text{ReLU}(x) = \max(0, x) \tag{14}$$

where x is the input data that the neuron received.

(2)   Loss Function

In addition to the activation function, the loss function also needs to be defined to evaluate the difference between the output of the current network and the expected result. The network will update the weight parameter automatically according to the difference so that the whole network can fit the nonlinear mapping relationship as much as possible.

The general loss function mainly includes mean squared error (MSE), cross-entropy (CE), etc. The CE function is usually chosen when facing the problem of image classification and recognition. MSE is mainly used to deal with data prediction and inversion, as in this paper, and its calculation function is

$$\text{MSE} = \frac{1}{n}\sum_{i=1}^{n}\left(y_i - y_i^p\right)^2 \tag{15}$$

where n is the dataset number, y_i is the measured EDH, and y_i^p is the predicted EDH.

(3)   Optimization Algorithm

The original intention of the optimization algorithm is to define the parameters to be optimized, to create the objective function, to set the learning rate, etc. Then, the descent gradient is calculated and iterated according to the gradient.

Stochastic gradient descent with momentum (SGDM) and adaptive moment estimation (Adam) are the most commonly used optimization algorithms [52]. SGDM can reach the optimal global solution, but it has strict requirements on the learning rate and is easy to stop at the saddle point, which is suitable for reliable initialization parameters. Meanwhile, with the progress of training, the speed of the SGDM method will slow down and the learning rate needs to be manually adjusted. Sometimes, it will converge to the optimal local value and the training results will also be affected. Adam has the advantages of fast speed, small memory requirements, and adaptive learning rates for different parameters. It is good at handling sparse gradients and non-stationary objects and is more suitable for large datasets and high-dimensional spaces to be processed in this paper. Using the Adam function will eventually converge to the optimal global value by automatically adjusting the learning rate. Therefore, the Adam function is finally selected as the optimization algorithm given the inversion problem to be solved in this paper. The initial learning rate is set as 0.0001.

(4)   Network Structure

MLP introduces one-to-multiple hidden layers based on the single-layer neural network; the appropriate hidden layers can be selected according to the original intention.

For the data input module, the hold-out method [49] was used to randomly divide the 476 sets of measurements into fixed mutually exclusive datasets; the proportion is 70% in the training set and 30% in the testing set. To avoid the impact of deviations introduced in the partitioning process, we tried to maintain the spatial and temporal consistency of the training and the testing set. The first 334 groups of data in about 12 round trips were selected and randomly reordered as training datasets and the remaining 142 groups of data were used as a testing dataset.

The selection of hyperparameters is complicated and engineering work and network hyperparameters, including the hidden layers, the number of neurons in each layer, the batch size, and the number of training epochs, are introduced during the modeling process.

The number of hidden layers is essential to the hyperparameter in the MLP design, which is directly related to the function approximation capability of the network. However, excessive hidden layers may lead to overfitting by learning extra characteristics of the training datasets. Therefore, in the parameter adjustment experiment, we explored the parameter ranges during the parameter selection: the number of hidden layers (1-8) and the number of neurons per hidden layer (1-300). In the end, we selected an MLP with four hidden layers by a large number of computer experiments; the neurons in each layer were 100, 50, 20, and 5, respectively.

When constructing the EDH prediction model, it is necessary to consider that its design performs well on training data and can generalize on new input datasets. A deep learning model with too many parameters and few training datasets is easily overfitting during the training progress. The specific performance of overfitting is as follows: the loss function of the model is small in the training data and the prediction accuracy is high; however, the loss function of the testing data is relatively large and the prediction accuracy is low. In deep learning, regularization strategies are designed to reduce test errors, which may come at the expense of increasing training errors.

(1)   Early stopping

The regularization strategy most commonly used in deep learning is called early stopping. When the training has sufficient representation ability and even overfits the model, the training error will gradually decrease with time, but the verification error will rise as a consequence. The early stopping strategy means storing a copy of the model parameters after each validation error improvement. The algorithm terminates when the validation error does not improve further within a predetermined number of cycles.

(2)   L2 regularization

L2 regularization is one of the means to prevent overfitting. The model complexity is controlled by limiting the parameter range space, thus overfitting can be avoided. In this paper, L2 regularization is adopted for the convenience of derivation and optimization.

(3)   Dropout

Dropout can be a choice for training deep neural networks. The concept of dropout makes the model more generalized by stopping the activation of a particular neuron with a certain probability, thus it will not fully connect to local features. In addition, the interaction between neurons in the hidden layer can be reduced.

4. Results and Discussion

To evaluate the accuracy and improvement of the prediction model, we introduced three evaluating standards as follows: (1) bias, which reflects the deviation from the measurements; (2) variance, which reflects the stability and robustness of the prediction model; (3) improvement, which reflects the enhancement compared with the original model. Meanwhile, three performance indexes were also introduced to measure better the bias of multi-dimensional EDH prediction models: the RMSE, the mean absolute error (MAE), and the coefficient of determination (R^2). In addition, the variance of prediction error (Var) and the improvement (σ) are also used to assess the accuracy of predictions. The definitions and characteristics of these indexes are listed in Table 4.

Table 4. Equation and characteristics of the performance criteria.

Index	Definition	Characteristic
MAE	$\mathrm{MAE} = \frac{1}{n}\sum_{i=1}^{n}\left\lvert y_i - y_i^p \right\rvert$	Evaluate the absolute deviation between the predicted value y_i^p and measured value y_i, it is not susceptible to extreme values, where n is the number of samples.
R^2	$R^2 = 1 - \dfrac{\sum\limits_{i=1}^{n}\left(y_i - y_i^p\right)^2}{\sum\limits_{i=1}^{n}\left(y_i - \overline{y_i}\right)^2}$ $\overline{y_i} = \frac{1}{n}\sum_{i=1}^{n} y_i$	Evaluate the conformance of fitting the estimated regression equation, it indicates the degree of linear correlation between the predicted and measured value.
Var	$\mathrm{Var} = E(e - E(e))^2$ $e = y_i - y_i^p$	Evaluate the deviation of the prediction error e and the stability of the accuracy of the predictions.
σ	$\sigma = \left(\dfrac{\mathrm{RMSE_{NPS}} - \mathrm{RMSE_{MLP}}}{\mathrm{RMSE_{NPS}}}\right) \times 100\%$	Evaluate the percentage improvement of the MLP model compared with the prediction results of the NPS model, where $\mathrm{RMSE_{NPS}}$ and $\mathrm{RMSE_{MLP}}$ are the RMSE of the NPS model and the improved model based on MLP, respectively.

4.1. Generalization Performance of Spatial–Temporal Models Based on MLP

In order to better analyze the robustness of the trained model, testing datasets were used for prediction accuracy analysis. Meanwhile, the number of floating-point operations (FLOPs) is utilized to compare the computational load of the algorithm, considering that the number of input parameters used for models (9)–(13) is different [53]. The analysis results are shown in Figures 6 and 7 and Table 5.

$$\mathrm{FLOPs} = (2I - 1)O \tag{16}$$

where I and O are the input and output neuron numbers.

It can be seen that:

(1) In Figure 6a, the trained meteorological-MLP-EDH model with the same input parameters as the NPS model has a better-matched degree with the measured data. The RMSE decreases from 4.67 m to 2.15 m and the percentage improvement reaches 54.00%. In addition, the MAE and variance all improve, while the coefficient of determination R^2 remains at a low level with the promotion of the MLP. The RMSE of the meteorological-MLP-EDH model exceeds 2 m so that the maximum variation of transmission loss at 500 km could exceed 120 dB, according to Figure 1.

(2) The prediction curve of the model fits much closer to the measurements by continuously adding spatial information (such as latitude and longitude) and temporal information (such as UT). The blue bar in the diagram, which symbolizes absolute deviation, gradually decreases. While the RMSE in Figure 6d has been greatly improved, the RMSE of the spatial-MLP-EDH, the temporal-MLP-EDH, and the spatial–temporal-MLP-EDH is 1.84 m, 1.75 m, and 1.54 m, and the coefficient R^2 has also made furtherly progress. The corresponding percentage improvement reached 60.53%, 62.53%, and 66.96%, respectively. Notably, introducing spatial and temporal parameters has little effect on the variation results. In Figure 6 and Table 5, the spatial–temporal-MLP-EDH essentially agrees with the measured EDH, but it still fails to match the local maximum.

(3) The statistical results in Figure 7 show the deviation variation of the abovementioned models. The box of each frequency represents the upper and the lower quartiles of the deviations and the horizontal line in the middle of the box is the median of deviations. The black line connected with the colored box shows the confidence interval of the

deviations. Diamond symbols of corresponding colors represent outliers that deviate from the confidence range. The variation range of each model changes on a small scale, but the median value of deviation changes from -1.57 m of the meteorological-MLP-EDH model to 0.13 m of the spatial–temporal-MLP-EDH model, which is essentially in agreement with the measurements on a large scale.

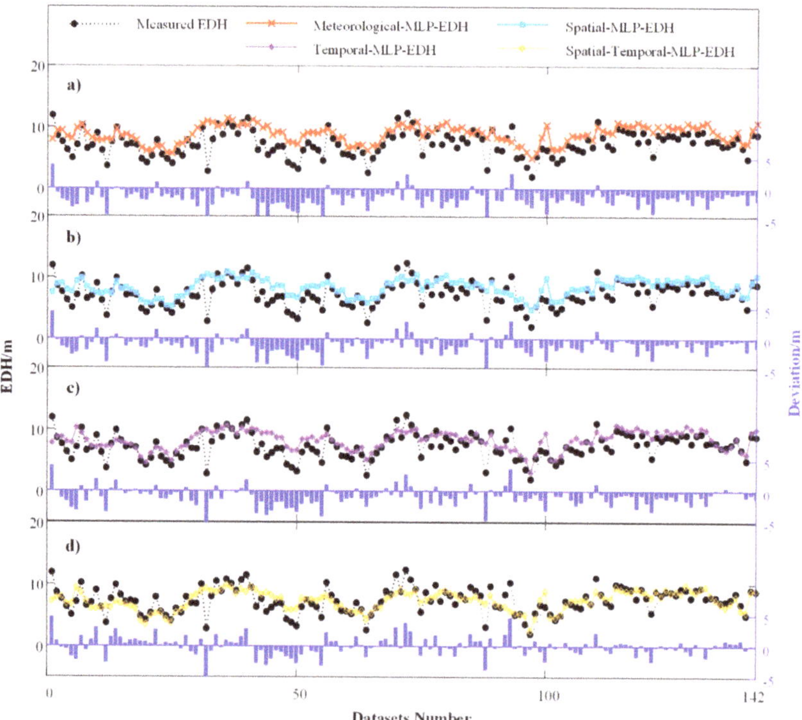

Figure 6. Generalization performance of the EDH prediction models based on MLP.

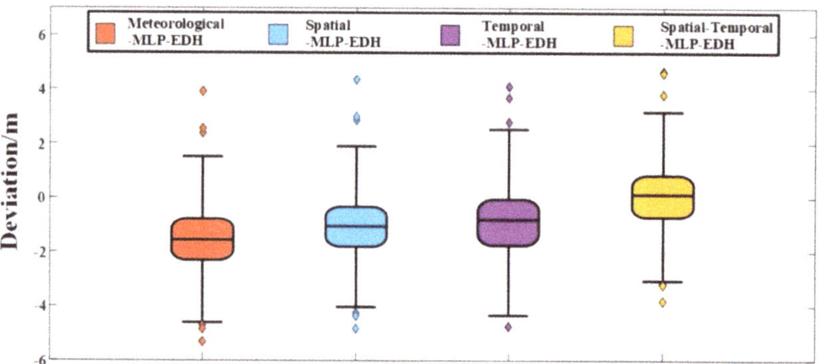

Figure 7. Statistical results of EDH prediction models based on MLP.

Table 5. Equation and characteristics of the performance criteria.

Method	Model	Training Data	FLOPs	Analysis of Testing Data				
				RMSE	MAE	R^2	Var	σ
Theoretical method	NPS -EDH	1. Temperature and pressure at the sea surface; 2. Temperature, wind speed, and relative humidity at layer h_0.	-	4.67	3.58	−4.18	12.37	-
MLP with a single layer	Meteorological-MLP-EDH	1. Temperature and pressure at the sea surface; 2. Temperature, wind speed, and relative humidity at layer h_0.	3.80×10^{13}	2.15	1.78	−0.09	2.24	54.00%
	Spatial-MLP-EDH	1. Temperature and pressure at the sea surface; 2. Temperature, wind speed, and relative humidity at layer h_0; 3. Latitude and longitude.	5.19×10^{13}	1.84	1.44	0.19	2.25	60.53%
	Temporal-MLP-EDH	1. Temperature and pressure at the sea surface; 2. Temperature, wind speed, and relative humidity at layer h_0; 3. UT.	4.49×10^{13}	1.75	1.34	0.26	2.29	62.53%
	Spatial–temporal-MLP-EDH	1. Temperature and pressure at the sea surface; 2. Temperature, wind speed, and relative humidity at layer h_0. 3. Latitude and longitude; 4. UT.	5.88×10^{13}	1.54	1.12	0.44	2.38	66.96%
MLP with multilayers	Multilayer-MLP-EDH	1. Temperature and pressure at the sea surface; 2. Temperature, wind speed, and relative humidity at layer h_0; 3. Temperature, wind speed, and relative humidity at layer h_1; 4. Latitude and longitude; 5. UT.	8.64×10^{13}	1.05	0.73	0.74	1.02	77.51%

Overall, the model has excellent generalization ability after training 70% of the original datasets and maintains good consistency in the testing datasets. Overall, the EDH prediction model based on MLP can maintain good consistency with the measurements at a large scale. However, a significant difference exists when predicting small-scale fluctuations, such as local maximum and minimum. Moreover, an optimization model with low bias and variance is always preferable based on MLP.

The training and testing datasets were collected from 20 repeated trips along one path and the experiment time was covered diurnal cycles. With the spatial parameters of the experimental positions with latitude and longitude and temporal parameters every 6 h in the MAGIC campaign introduced, the prediction accuracy of the model is gradually improved, indicating that the spatial and temporal variability is significant. By extracting much more "hidden information" from "extra data" in the training process, the spatial variability of the three-dimensional geographic information and the temporal variability of the diurnal cycles are repeatedly learned and memorized based on the MLP method. With the constructed multi-dimensional deep learning model, the geographic and time domain feature can be extracted, which supplies an improvement in EDH prediction.

4.2. Generalization Performance of Multilayer Model Based on MLP

The RMSE of the spatial–temporal-MLP-EDH model has improved to 1.54 m and the parameters as a coefficient of determination and variance of prediction error still have room for improvement. Adding the temperature, RH, and wind speed at an additional height, new datasets with the original parameters may improve prediction accuracy. The comparison results with measured data, EDH predicted by the NPS model, and the multilayer-MLP-EDH model are shown in Figure 8.

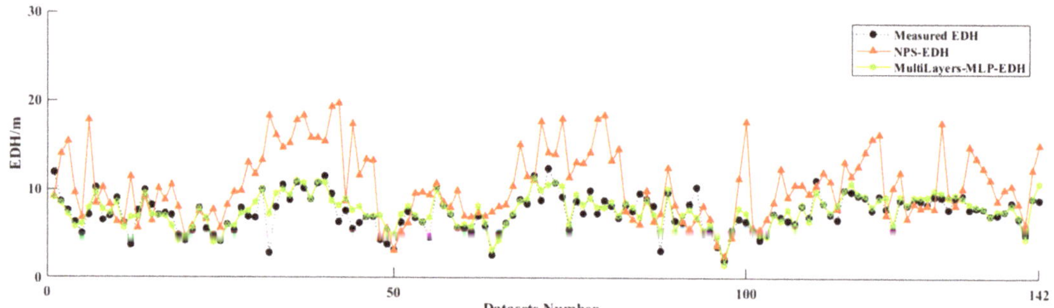

Figure 8. Generalization performance of the multilayer-MLP-EDH model.

The variance achieves another reduction with meteorological parameters in multilayers and decreases to 1.02 m. The trained model can match the trend of measurements at a large scale; meanwhile, the maximum and minimum values of the measurements at a small scale can also become significantly matched. According to statistical analysis, the RMSE of the multilayer-MLP-EDH method reached 1.05 m and the improvement percentage reached 77.51%, compared with the NPS model. Furthermore, the computational load of this algorithm (FLOPs) is 2.27 times as much as the meteorological-MLP-EDH model, which reached 8.64×10^{13}.

The overall trend of predicted EDH by NPS differs significantly from the measurements, mainly because air–sea coupling conditions limit the NPS model. The prediction accuracy is hard to maintain when the air–sea temperature difference (ASTD) is greater than 0 [14]. Table 6 shows the statistical RMSEs in stable and unstable conditions, and the prediction error of the NPS model increases when ASTD > 0. The multilayer-MLP-EDH method is maintained in RMSE, which reflects the consistency of the proposed method in dealing with different conditions.

Table 6. The statistical RMSEs in different conditions.

Models	ASTD > 0	ASTD < 0
NPS-EDH	5.31 m	4.60 m
Multilayer-MLP-EDH	1.07 m	1.05 m

As shown in Figure 9, the fitting line between the predicted and the measured data changes from $y = 0.42\,x + 5.52$ of the NPS model to $y = 0.93\,x + 0.99$, close to $y = x$. Therefore, this method has better operability by setting meteorological instruments at two different heights (the cabin and the deck, for instance). The EDH predicted error could reach nearly 1 m combined with the sea surface meteorological parameters.

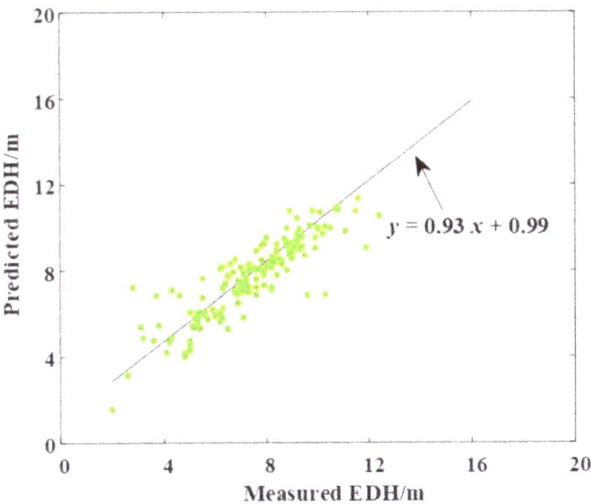

Figure 9. Scatter plot of measured EDH against modeled EDH using the multilayer-MLP-EDH model.

The predicted RMSE of EDH of the theoretical method is 4.67 m, which may lead to the uncertainty range of path loss exceeding 120 dB at the 500 km transmission range. For instance, as shown in Figure 1, the path loss can increase from the original design of 179.12 dB at a predicted EDH of 11 m to exceed 300 dB at a true EDH of nearly 6 m. This state will leave the transmission system in an unstable situation. However, a significant improvement arises when single-layer models based on MLP become involved. The predicted deviation of EDH decreases to 1.54–2.15 m, corresponding to a path loss variation from 162.20 to 213.12dB. The prediction accuracy of the evaporation duct channel continues to improve with multilayers. Furthermore, the uncertainty of path loss is reduced by 16.92 dB on the single-layer models. Therefore, the multilayer-MLP-EDH model can be essential in designing a communication system using the evaporation duct.

Table 7 provides a summary and comparison of the performance of EDH prediction, with the AI method introduced. From the comparison results, the four-dimensional regression function multilayer-MLP-EDH with meteorological parameters located on another layer proposed in this paper has the advantage of extracting the spatial–temporal information and the meteorological parameters at multiple altitudes in the training process. At the same time, a wider application range, higher precision, and model generalization are also achieved. Furthermore, the proposed model has great potential for enhancing the communication quality, reliability, and efficiency of ducting in evaporation ducts.

Table 7. Comparison of different models in EDH prediction.

Ref.	Modeling Category	Modeling Datasets	AI Method and Features	Network Structure	Prediction Results
[14]	Long-term	The calculated reults based on the NPS model and the remote sensing dataset	Artificial neural network	A 5-15-24 feedforward backpropagation network	1.91 m in the RMSE for air–sea temperature difference < 0, and 9.43 m for the difference > 0
[28]		Observation of experimental datasets in the northern hemisphere	MLP with rectified linear unit activation function	A five-hidden-layer network with neurons of 50, 30, 20, 10, and 5 in each layer	An enhancement between 80.82% and 93.77% compared with the PJ model
[29]	Short-term	Observation of experimental datasets in the northern hemisphere	Long short-term memory network	One hidden layer with 50 neurons	0.72 m in the average RMSE
[30]		High resolution meteorological sounding balloon data at a sea area near the equator	Darwinian evolutionary algorithm	The evolutionary process of selection based on A grid search method	0.2248 m in the RMSE
This work	Long-term	MAGIC datasets in the Pacific Ocean	MLP with spatial–temporal information and meteorological parameters at multiple altitudes introduced	A four-hidden-layer network with neurons of 100, 50, 20, and 5 in each layer	1.05 m in the RMSE

5. Conclusions

Low altitude atmospheric refractive conditions significantly affect the performance of shipboard communications at sea and near shore [12]. The accurate prediction of the EDH is thus crucial in the demonstration, design, development, operation, and maintenance management of the communication system under this mechanism. Based on the MLP deep-learning method, the multidimensional deep-learning model was proposed to improve the prediction accuracy of EDH. First, the meteorological-MLP-EDH model was designed, which improved the prediction accuracy by 54.00%, with the same input parameters as the NPS model. The spatial–temporal-MLP-EDH model has gone one step further by superimposing the spatial–temporal "extra data" in the experiment. As a result, it can be essentially in agreement with measurements at large scales and the predicted RMSE is 1.54 m, with a 66.96% percentage improvement compared with the NPS model. Lastly, the multilayer-MLP-EDH model with the temperature, RH, and wind speed at an additional height was trained, significantly matching measurements at large and small scales. According to statistical results, the predicted RMSE can reach 1.05 m and the percentage improvement reached 77.51%.

The proposed model in this paper can break through the limitations of theoretical models by extracting much more "hidden information" from "extra data" in the training process, significantly improving EDH prediction accuracy. As a result, the proposed model has great potential for enhancing the communication quality, reliability, and efficiency of ducting in evaporation ducts.

The models constructed in this paper are based on 476 sets of MAGIC data in the Pacific Ocean; the training and testing datasets are limited to a sea area of 21.2197°N, 33.6001°N, 118.3299°W, 157.7416°W at specific experimental time intervals. Future experiments should be performed to more completely validate the models. In addition, measurements should be made at comprehensive coverage, massive data acquisition, and high spatial and temporal resolution to improve the constructed model. Furthermore, the distribution of EDH in high precision, detailed resolution, and broad coverage with the improved proposed model would be valuable to the communication system using evaporation ducts over the ocean.

Author Contributions: C.Y.: conceptualization (equal); data curation (equal); investigation (equal); methodology (equal); software (equal); validation (equal); visualization (equal); writing—original draft (equal); writing—review and editing (equal). J.W.: conceptualization (equal); data curation (equal); investigation (equal); methodology (equal); software (equal); validation (equal); visualization (equal); writing—original draft (equal); writing—review and editing (equal). Y.S.: data curation (equal); investigation (equal); software (equal); validation (equal); visualization (equal); writing—initial draft (supporting). All authors have read and agreed to the published version of the manuscript.

Funding: This research was funded by the National Natural Science Foundation of China (No. 62031008) and the State Key Laboratory of Complex Electromagnetic Environment Effects on Electronics and Information Systems (No. CEMEE2022G0201, CEMEE-002-20220224).

Institutional Review Board Statement: Not applicable.

Informed Consent Statement: Not applicable.

Data Availability Statement: The MAGIC data used for the analyses were obtained from the Atmospheric Radiation Measurement (ARM) Program (https://www.arm.gov/research/campaigns/amf2012magic (accessed on 18 February 2022)).

Acknowledgments: The authors would like to acknowledge the reviewers for their constructive comments and suggestions to improve this manuscript.

Conflicts of Interest: The authors declare no conflict of interest.

References

1. Danklmayer, A.; Förster, J.; Fabbro, V.; Biegel, G.; Brehm, T.; Colditz, P.; Castanet, L.; Hurtaud, Y. Radar Propagation Experiment in the North Sea: The Sylt Campaign. *IEEE Trans. Geosci. Remote Sens.* **2018**, *56*, 835–846. [CrossRef]
2. Huang, L.F.; Liu, C.G.; Wang, H.G.; Zhu, Q.L.; Zhang, L.J.; Han, J.; Zhang, Y.S.; Wang, Q.N. Experimental Analysis of Atmospheric Ducts and Navigation Radar Over-the-Horizon Detection. *Remote Sens.* **2022**, *14*, 2588. [CrossRef]
3. Gilles, M.A.; Earls, S.; Bindel, D. A Subspace Pursuit Method to Infer Refractivity in the Marine Atmospheric Boundary Layer. *IEEE Trans. Geosci. Remote Sens.* **2019**, *57*, 5606–5617. [CrossRef]
4. Woods, G.S.; Ruxton, A.; Huddlestone-Holmes, C.; Gigan, G. High-Capacity, Long-Range, Over Ocean Microwave Link Using the Evaporation Duct. *IEEE J. Ocean. Eng.* **2009**, *34*, 323–330. [CrossRef]
5. Wang, J.; Yang, C.; Yan, N.N. Study on digital twin channel for the B5G and 6G communication. *Chin. J. Radio Sci.* **2021**, *36*, 340–348.
6. Kim, S.M.; Kim, J.; Han, C.; Min, S.S.; Kim, S.L. Opportunism in Spectrum Sharing for Beyond 5G With Sub-6 GHz: A Concept and Its Application to Duplexing. *IEEE Access* **2020**, *8*, 148877–148891. [CrossRef]
7. Wang, J.; Zhou, H.; Ye, L.; Qiang, S.; Chen, X. Wireless Channel Models for Maritime Communications. *IEEE Access* **2018**, *6*, 68070–68088. [CrossRef]
8. Zaidi, K.S.; Jeoti, V.; Awang, A. Wireless backhaul for broadband communication over Sea. In Proceedings of the 2013 IEEE 11th Malaysia International Conference on Communications (MICC), Kuala Lumpur, Malaysia, 26–28 November 2013; pp. 298–303.
9. Babin, S.M.; Young, G.S.; Carton, J.A. A new model of the oceanic evaporation duct. *J. Appl. Meteorol. Climatol.* **1997**, *36*, 193–204. [CrossRef]
10. Babin, S.M.; Dockery, G.D. LKB-Based Evaporation Duct Model Comparison with Buoy Data. *J. Appl. Meteorol. Climatol.* **2002**, *41*, 434–446. [CrossRef]
11. Shi, Y.; Yang, K.; Yang, Y.; Ma, Y. A new evaporation duct climatology over the South China Sea. *J. Meteorol. Res.* **2015**, *41*, 764–778. [CrossRef]
12. Yang, C.; Wang, J.; Ma, J.G. Exploration of X-band Communication for Maritime Applications in the South China Sea. *IEEE Antennas Wirel. Propag. Lett.* **2022**, *21*, 481–485. [CrossRef]
13. Zhao, X.F.; Wang, D.X.; Huang, S.X.; Huang, K.; Chen, J. Statistical estimations of atmospheric duct over the South China Sea and the Tropical Eastern Indian Ocean. *Chin. Sci. Bull.* **2013**, *58*, 2794–2797. [CrossRef]
14. Yan, X.; Yang, K.; Ma, Y. Calculation Method for Evaporation Duct Profiles Based on Artificial Neural Network. *IEEE Antennas Wirel. Propag. Lett.* **2018**, *17*, 2274–2278. [CrossRef]
15. Fountoulakis, V.; Earls, C. Inverting for Maritime Environments Using Proper Orthogonal Bases From Sparsely Sampled Electromagnetic Propagation Data. *IEEE Trans. Geosci. Remote Sens.* **2016**, *54*, 7166–7176. [CrossRef]

16. Wang, Q.; Burkholder, R.J.; Yardim, C.; Xu, L.Y.; Pozderac, J.; Fernando, H.J.S.; Alappattu, D.P.; Wang, Q. Range and Height Measurement of X-Band EM Propagation in the Marine Atmospheric Boundary Layer. *IEEE Trans. Antennas Propag.* **2019**, *67*, 2063–2073. [CrossRef]

17. Tian, B.; Han, L.; Kong, D.W.; Liu, C.G.; Zhou, M.; Yu, M.H. Study on Wireless Detection Using the Pseudo-Refractivity Model. In Proceedings of the 2012 8th International Conference on Wireless Communications, Networking and Mobile Computing, Shanghai, China, 21–23 September 2012; pp. 1–4.

18. Wang, H.; Wu, Z.; Kang, S.; Zhao, Z. Monitoring the Marine Atmospheric Refractivity Profiles by Ground-Based GPS Occultation. *IEEE Geosci. Remote Sens. Lett.* **2013**, *10*, 962–965. [CrossRef]

19. Karimian, A.; Yardim, C.; Gerstoft, P.; Hodgkiss, W.S.; Barrios, A.E. Refractivity estimation from sea clutter: An invited review. *Radio Sci.* **2011**, *46*, 1–16. [CrossRef]

20. Bussey, H.E.; Birnbaum, G. Measurement of variations in atmospheric refractive index with an airborne microwave refractometer. *J. Res. Nat. Bur. Stand.* **1953**, *51*, 171–178. [CrossRef]

21. Rowland, J.R.; Babin, S.M. Fine-scale measurements of microwave refractivity profiles with helicopter and low-cost rocket probes. *Johns Hopkins APL Tech. Dig* **1987**, *8*, 413–417.

22. Paulus, R.A. Practical application of an evaporation duct model. *Radio Sci.* **1985**, *20*, 887–896. [CrossRef]

23. Musson-Genon, L.; Gauthier, S.; Bruth, E. A simple method to determine evaporation duct height in the sea surface boundary layer. *Radio Sci.* **1992**, *27*, 635–644. [CrossRef]

24. Frederickson, P.A.; Davidson, K.L.; Goroch, A.K. *Operational Bulk Evaporation Duct Model for MORIAH Version 1.2*; Naval Postgraduate School: Monterey, CA, USA, 2000.

25. Beljaars, A.C.M.; Holtslag, A.A.M. Flux parameterization over land surfaces for atmospheric models. *J. App. Meteorol.* **1991**, *30*, 327–341. [CrossRef]

26. Grachev, A.A.; Andreas, E.L.; Fairall, C.W.; Guest, P.S.; Persson, P.O.G. SHEBA flux–profile relationships in the stable atmospheric boundary layer. *Bound.-Layer Meteor.* **2007**, *124*, 315–333. [CrossRef]

27. Han, J.; Wu, J.; Zhu, Q.; Wang, H.; Zhou, Y.; Jiang, M.; Zhang, S.; Wang, B. Evaporation Duct Height Nowcasting in China's Yellow Sea Based on Deep Learning. *Remote Sens.* **2021**, *13*, 1577. [CrossRef]

28. Zhu, X.; Li, J.; Zhu, M.; Jiang, Z.; Li, Y. An Evaporation Duct Height Prediction Method Based on Deep Learning. *IEEE Geosci. Remote Sens. Lett.* **2018**, *15*, 1307–1311. [CrossRef]

29. Zhao, W.; Zhao, J.; Li, J.; Zhao, D.; Huang, L.; Zhu, J.; Lu, J.; Wang, X. An evaporation duct height prediction model based on a Long Short-Term Memory Neural Network. *IEEE Trans. Antennas Propag.* **2021**, *69*, 7795–7804. [CrossRef]

30. Mai, Y.; Sheng, Z.; Shi, H.; Li, C.; Liao, Q.; Bao, J. A New Short-Term Prediction Method for Estimation of the Evaporation Duct Height. *IEEE Access* **2020**, *8*, 136036–136045. [CrossRef]

31. Vaisala MARWIN Sounding System MW32 Features. Available online: https://www.vaisala.com/en/products/weather-environmental-sensors/marwin-sounding-system-mw32 (accessed on 1 May 2021).

32. Recommendation ITU-R P.453-14 The Radio Refractive Index: Its Formula and Refractivity Data. Available online: https://www.itu.int/rec/R-REC-P.453-14-201908-I/en (accessed on 8 March 2020).

33. Fairall, C.W.; Bradley, E.F.; Hare, J.E.; Grachev, A.A.; Edson, J.B. Bulk parameterization of air-sea fluxes: Updates and verification for the COARE algorithm. *J. Clim.* **2003**, *16*, 571–591. [CrossRef]

34. Grachev, A.A.; Fairall, C.W. Dependence of the Monin–Obukhov Stability Parameter on the Bulk Richardson Number over the Ocean. *J. Appl. Meteorol. Climatol.* **1997**, *36*, 406–415. [CrossRef]

35. Shi, Y.; Zhang, Q.; Wang, S.; Yang, K.; Ma, Y. Impact of Typhoon on Evaporation Duct in the Northwest Pacific Ocean. *IEEE Access* **2019**, *7*, 109111–109119. [CrossRef]

36. Recommendation ITU-R P.525-4 Calculation of Free-Space Attenuation. Available online: https://www.itu.int/dms_pubrec/itu-r/rec/p/R-REC-P.525-4-201908-I!!PDF-E.pdf (accessed on 17 October 2020).

37. Recommendation ITU-R P.2001-4 A General Purpose Wide-Range Terrestrial Propagation Model in the Frequency Range 30 MHz to 50 GHz. Available online: https://www.itu.int/rec/R-REC-P.2001-4-202109-I/en (accessed on 4 January 2022).

38. Ozgun, O.; Apaydin, G.; Kuzuoglu, M.; Sevgi, L. PETOOL: MATLAB-Based One-Way and Two-Way Split-Step Parabolic Equation Tool for Radiowave Propagation over Variable Terrain. *Comput. Phys. Commun.* **2011**, *182*, 2638–2654. [CrossRef]

39. Ozgun, O.; Sahin, V.; Erguden, M.E.; Apaydin, G.; Yilmaz, A.E.; Kuzuoglu, M.; Sevgi, L. PETOOL v2.0: Parabolic Equation Toolbox with evaporation duct models and real environment data. *Comput. Phys. Commun.* **2020**, *256*, 107454. [CrossRef]

40. Dockery, G.D.; Kuttler, J.R. An improved impedance-boundary algorithm for Fourier split-step solutions of the parabolic wave equation. *IEEE Trans. Antennas Propag.* **1996**, *44*, 1592–1599. [CrossRef]

41. Hardin, R.; Tappert, F. Applications of the Split-Step Fourier Method to the Numerical Solution of Nonlinear and Variable Coefficient Wave Equations. *SIAM Rev.* **1973**, *15*, 423.

42. Shi, Y.; Zhang, Q.; Wang, S.W.; Yang, K.D.; Yang, Y.X.; Yan, X.D.; Ma, Y.L. A Comprehensive Study on Maximum Wavelength of Electromagnetic Propagation in Different Evaporation Ducts. *IEEE Access* **2019**, *7*, 82308–82319. [CrossRef]

43. Alappattu, D.P.; Wang, Q.; Kalogiros, J. Anomalous propagation conditions over eastern Pacific Ocean derived from MAGIC data. *Radio Sci.* **2016**, *51*, 1142–1156. [CrossRef]

44. Zhou, X.; Kollias, P.; Lewis, E.R. Clouds, Precipitation, and Marine Boundary Layer Structure during the MAGIC Field Campaign. *J. Climate.* **2015**, *28*, 2420–2442. [CrossRef]

45. Guo, X.M.; Zhao, D.L.; Zhang, L.J.; Wang, H.G.; Kang, S.F.; Lin, L.K. C band transhorizon signal characterisations in evaporation duct propagation environment over Bohai Sea of China. *IET Microw. Antennas Propag.* **2019**, *13*, 407–413. [CrossRef]
46. Ruck, D.W.; Rogers, S.K.; Kabrisky, M.; Oxley, M.E.; Suter, B.W. The multilayer perceptron as an approximation to a Bayes optimal discriminant function. *IEEE Trans. Neural Networks* **1990**, *1*, 296–298. [CrossRef]
47. Isaakidis, S.A.; Dimou, I.N.; Xenos, T.D.; Dris, N.A. An artificial neural network predictor for tropospheric surface duct phenomena. *Nonlinear Process. Geophys.* **2007**, *14*, 569–573. [CrossRef]
48. Sit, H.; Earls, C.J. Characterizing evaporation ducts within the marine atmospheric boundary layer using artificial neural networks. *Radio Sci.* **2019**, *54*, 1181–1191. [CrossRef]
49. Zhou, Z.H. *Machine Learning*; Tsinghua University Press: Beijing, China, 2016.
50. Minnis, P.; Mack, S.S.; Chen, Y.; Chang, F.; Yost, C.R. CERES MODIS Cloud Product Retrievals for Edition 4—Part I: Algorithm Changes. *IEEE Trans. Geosci. Remote Sens.* **2021**, *59*, 2744–2780. [CrossRef]
51. Vinod, N.; Hinton, G.E. Rectified Linear Units Improve Restricted Boltzmann Machines Vinod Nair. In Proceedings of the International Conference on Machine Learning Omnipress, Haifa, Israel, 21–24 June 2010; pp. 807–814.
52. Kingma, P.D.; Jimmy, B. Adam: A Method for Stochastic Optimization. In Proceedings of the 3rd International Conference for Learning Representations, San Diego, CA, USA, 7–9 May 2015.
53. Molchanov, P.; Tyree, S.; Karras, T.; Aila, T.; Kautz, J. Pruning convolutional neural networks for resource efficient transfer learning. In Proceedings of the 5th International Conference on Learning Representations (ICLR), Toulon, France, 24–26 April 2017.

Communication

Real-Time Vehicle Sound Detection System Based on Depthwise Separable Convolution Neural Network and Spectrogram Augmentation

Chaoyi Wang *,†, Yaozhe Song †, Haolong Liu †, Huawei Liu †, Jianpo Liu †, Baoqing Li † and Xiaobing Yuan †

Science and Technology on Micro-System Laboratory, Shanghai Institute of Microsystem and Information Technology, Chinese Academy of Sciences, Shanghai 201899, China
* Correspondence: chaoyiwang@mail.sim.ac.cn
† Current address: 1455 Pingcheng Road, Jiading District, Shanghai 201899, China.

Abstract: This paper proposes a lightweight model combined with data augmentation for vehicle detection in an intelligent sensor system. Vehicle detection can be considered as a binary classification problem, vehicle or non-vehicle. Deep neural networks have shown high accuracy in audio classification, and convolution neural networks are widely used for audio feature extraction and audio classification. However, the performance of deep neural networks is highly dependent on the availability of large quantities of training data. Recordings such as tracked vehicles are limited, and data augmentation techniques can be applied to improve the overall detection accuracy. In our case, spectrogram augmentation is applied on the mel spectrogram before extracting the Mel-scale Frequency Cepstral Coefficients (MFCC) features to improve the robustness of the system. Then depthwise separable convolution is applied to the CNN network for model compression and migrated to the hardware platform of the intelligent sensor system. The proposed approach is evaluated on a dataset recorded in the field using intelligent sensor systems with microphones. The final frame-level accuracy achieved was 94.64% for the test recordings and 34% of the parameters were reduced after compression.

Keywords: depthwise separable convolutional neural networks; spectrogram augmentation; sound detection; vehicle detection

Citation: Wang, C.; Song, Y.; Liu, H.; Liu, H.; Liu, J.; Li, B.; Yuan, X. Real-Time Vehicle Sound Detection System Based on Depthwise Separable Convolution Neural Network and Spectrogram Augmentation. *Remote Sens.* **2022**, *14*, 4848. https://doi.org/10.3390/rs14194848

Academic Editor: Gwanggil Jeon

Received: 24 August 2022
Accepted: 26 September 2022
Published: 28 September 2022

Publisher's Note: MDPI stays neutral with regard to jurisdictional claims in published maps and institutional affiliations.

1. Introduction

Vehicle detection and identification(VDI) systems are in growing demand as development of information and communication technology [1] increases, and the need for sophisticated signal processing and data analysis techniques is becoming increasingly apparent [2]. A growing number of novel applications such as smart navigation, traffic monitoring and transportation infrastructure monitoring have been accompanied by a corresponding improvement in overall system performance and efficiency [3]. Accurate and rapid detection of moving vehicles is fundamental in these applications.

Vehicle detection aims to detect a vehicle passing by a deployed sensor. Vehicle detection and classification systems are mainly based on ultrasonic sensors, acoustic sensors, infrared sensors, inductive loops, magnetic sensors, video sensors, laser sensors and microwave radars [4]. Currently, video sensors and image detection techniques are frequently adopted for vehicle detection [5,6]. However, these image-based methods require the camera to be placed directly towards the road, and the lens cannot be blocked. In our scenario, the sensors are mostly placed in the field or forests, where vehicles may come from all directions and objects such as weeds and trees are likely to disturb the view.

Acoustic communications are attractive because they do not require extra hardware on either transmitter and receiver sides, which facilitates numerous tasks in IoT and other applications [7]. Therefore, in our intelligent sensor system, the acoustic signals are collected using acoustic sensors and processed on the chips. The vehicle detection task can be solved as an acoustic event classification task. Vehicle detection and identification using features extracted from vehicle audio with supervised learning have been widely explored, such as support vector machine classifiers, k-nearest neighbor classifiers, Gaussian mixture models, hidden Markov models, etc. [3].

Recently, deep neural networks have shown promising results in many pattern recognition applications [8], such as acoustic event classification. The vehicle detection task can be considered as a binary acoustic event classification of a vehicle or a non-vehicle. Deep neural networks are powerful pattern classifiers which enable the networks to learn the highly nonlinear relationships between the input features and the output targets [9]. Convolutional neural networks (CNNs) have also been widely used for remote sensing recognition tasks [10–12] and acoustic event classification tasks [13], as CNNs have shared-weight architecture based on convolution kernels which is efficient in extracting acoustic features for acoustic classification.

Many feature extraction techniques have been studied for analyzing acoustic characteristics over decades, including temporal domain, frequency domain, cepstral domain, wavelet domain and time-frequency domain [14]. Mel frequency cepstral coefficients (MFCC), a kind of cepstral domain feature, are widely used for acoustic classification [15]. Recent works exploring CNN-based approaches have shown significant improvements over hand-crafted feature-based methods such as MFCC [16–21]. In our practical application, the locations of the sensors deployed are different, and therefore the distances between the sensors and road are uncertain. MFCCs are relatively independent of the absolute signal level [22]; thus, MFCCs are appropriate for vehicle detection in our case as the amplitudes of the vehicle signals vary with the distance between the sensors and roads.

However, the performance of deep neural networks is highly dependent on the availability of large quantities of training data in order to learn a nonlinear function from input to output that generalizes well and yields high classification accuracy on unseen data [23]. The recordings for vehicles of specific types are limited, such as an armored vehicle. To solve this problem, data augmentation is applied to the original recordings to generate more samples for training. Data augmentation is a common strategy adopted to increase the quantity of training data, avoid overfitting and improve robustness of the models [24]. Commonly used strategies for acoustic data augmentation are vocal tract length perturbation, tempo perturbation, speed perturbation [24], time shifting, pitch shifting, time streching [25] and spectrogram augmentation [26].

After a neural network for vehicle detection is trained, it has to be migrated to the hardware platform where the computation cost and battery life is limited. Typical approaches include linear quantization of network weights and inputs [27] and a reduction in the number of parameters [28]. Depthwise separable convolutions are a form of factorized convolutions which factorize a standard convolution into a depthwise convolution and a 1×1 convolution called a pointwise convolution [29]. The computational cost can be reduced using depthwise separable convolution with only a small reduction in accuracy.

This paper aims to solve a practical issue for vehicle detection by using a lightweight CNN model for acoustic classification. To summarize, the main contributions of this paper are as follows:

1. A spectrogram augmentation method is applied to the mel spectrogram of the acoustic signals to improve the robustness of the proposed model.
2. A CNN classification model is trained on the original data and the augmented samples to achieve a high classification accuracy of each frame.
3. Depthwise separable convolution is applied to the original CNN network for model compression. The lightweight model can be migrated to the chips of the intelligent sensor system and realize the task of real-time vehicle detection.

The paper is organized as follows: Section 2 describes the materials and methods including both hardware structure and algorithm implementation. Section 3 presents the detailed results of the experiments. Section 4 discusses the experiment results. Section 5 presents the conclusion of this paper.

2. Materials and Methods

This section describes the system hardware structure, data collection method, dataset description, feature extraction, data augmentation, two-stage detection method and experiment setup. The codes for the experiments including feature extraction, spectrogram augmentation and deep neural network structures are published in the Github website: https://github.com/chaoyiwang09/Vehicle-Detection-CNN.git (accessed on 23 August 2022).

2.1. System Hardware Structure

Our implemented system can be divided into the four modules according to their functions: microphone array (MA), preprocessing and sampling (P and S) module, real-time data processing and acquisition (P and A) module and transmission module [30]. Four microphone arrays are used to collect the acoustic signal in the deployed area. The collected acoustic signals are then sampled in the P and S module to obtain four simultaneous digital signals by the synchronized filters and amplifiers [31]. The detection algorithm is implemented on the digital signal processors (DSP) chip of the real-time P and A module. The detection results are finally transmitted to a terminal device through radio frequency. The diagram of the system hardware process is shown in Figure 1.

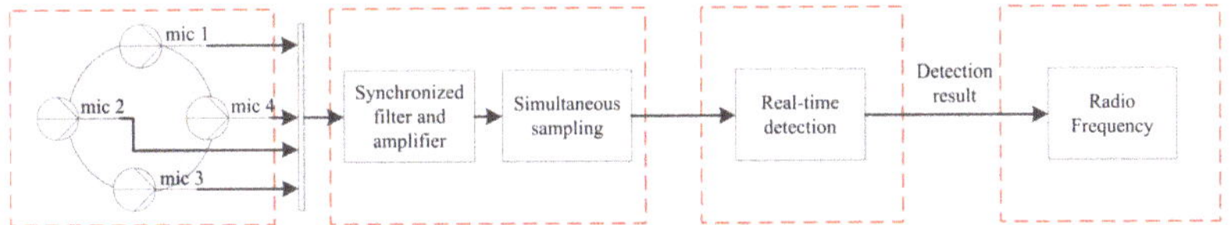

Figure 1. The diagram of the system hardware architecture.

Four ADMP504 MEMS microphones which are produced by Analog Devices are placed uniformly on the main circuit board. The device for AD sampling is MAXIM MAX11043, a 4-channel 16-bit simultaneous ADC [32]. The DSP chip, ASDP21479 is used for real-time data processing and acquisition. The printed circuit board layout is shown in Figure 2. A more detailed description of the hardware structure implemented in the modules can be found in [31].

2.2. Dataset

The acoustic signals are collected with microphone arrays in the intelligent sensor system deployed in the field. The vehicle recorded includes a small wheeled vehicle, a large wheeled vehicle and a tracked vehicle. The sensors are deployed 30 m, 50 m, 80 m and 150 m away from the road for the small wheeled vehicle. For the tracked vehicle and the large wheeled vehicle, the sensors are deployed 200 m, 250 m and 300 m away from the road. The length of road is 700 m, 350 m on each side of the microphone arrays. The recording scene is illustrated in Figure 3.

All the recordings are collected at a sample rate of 8k and a bit rate of 16 bits. For each experiment, the start time and end time of the vehicle are recorded. Therefore, the acoustic signals can be truncated by the start time and the end time. The signals of duration from the start time and end time are labeled as 1 for vehicle, while the remaining parts of the signals are labeled as 0 for non-vehicle. There are overall 445 recordings in the dataset; 191 recordings are non-vehicle, the average duration of which are about 104 s. A total of 91 recordings are from the small wheeled vehicle, 101 recordings are from the large wheeled vehicle, and 62 recordings are from the tracked vehicle, and the average duration of them are 40 s, 70 s and 150 s, respectively. The dataset composition is shown in Table 1.

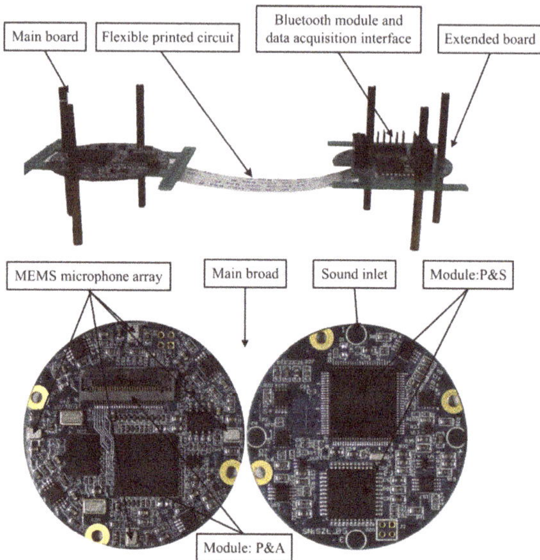

Figure 2. The system hardware circuits layout.

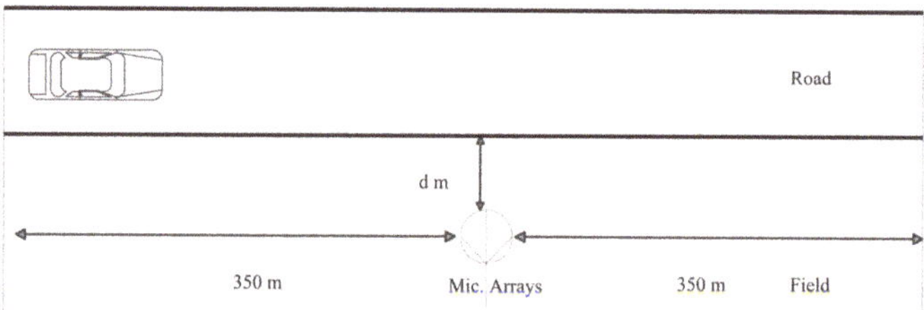

Figure 3. The recording scene.

Table 1. The dataset composition.

Vehicle Type	Avg Duration (s)	Distance (m)	Recording Num	Overall Num
small wheeled vehicle	40	30	25	91
		50	25	
		80	25	
		150	16	
large wheeled vehicle	70	200	45	101
		250	46	
		300	10	
tracked vehicle	150	200	21	62
		250	21	
		300	10	
non-vehicle	104	/	191	191

2.3. Feature Extraction

Mel-scale frequency cepstral coefficients (MFCC) features are extracted as the input features for the binary classifier. MFCC is widely used in acoustic tasks such as voice activity detection [33]. The diagram of MFCC extraction is illustrated in Figure 4. The steps of MFCC extractions are:

1. Pre-emphasis is used to compensate and amplify the high-frequency part from the acoustic signal [34]. This is calculated by:

$$s'(n) = s(n) - \alpha \cdot s(n-1) \tag{1}$$

  where $\alpha = 0.97$ in our case, $s(n)$ is the input acoustic signal, and $s'(n)$ is the output signal.

2. The signals are split into short parts by windowing. In our case, the window length of each frame is set to 200 milliseconds, the window step is 200 milliseconds, and no overlap is applied to each frame. A rectangular window is chosen for short time Fourier Transformation.

3. Mel filter banks are applied and a logarithm is taken to the extracted mel frequency features. The mel cepstral coefficients are calculated as follows for a given f in Hz:

$$Mel(f) = 2595 \cdot \log 10(1 + f/700) \tag{2}$$

4. Discrete cosine transformation is applied.
5. The zeroth cepstral coefficient is replaced with the log of the total frame energy.
6. Delta, a first order difference calculation and double-delta, a second order difference calculation, are finally calculated.

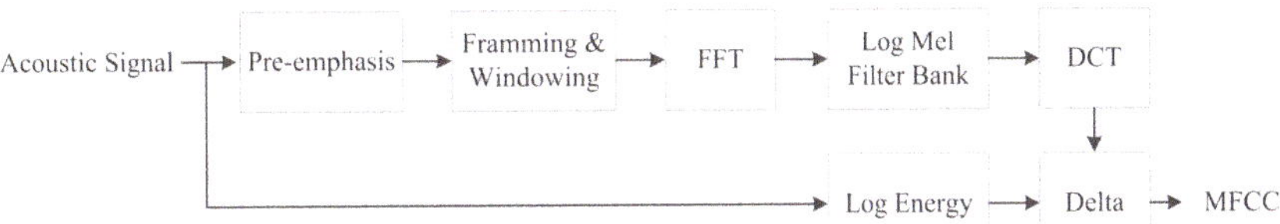

Figure 4. The diagram of MFCC extraction.

For each frame, 13 cepstral coefficients are extracted, and the output dimension of one frame is 39-dimensional after the delta step. Overall 100,000 samples are kept for the training set, the duration of which is about 5.6 h. A total of 20,000 samples are extracted for the validation set, and 20,000 samples are extracted for the test set. For the training set, validation set, and training set, half of the features are labeled as vehicle, and the others are labeled as non-vehicle.

2.4. Data Augmentation

Data augmentation is a strategy to increase the diversity of available data and make it possible to train models without collecting new data [35]. Our augmentation method is applied to the mel spectrogram domain. Frequency masks are applied to the mel spectrogram. Frequency masking is applied so that f consecutive mel frequency channels $[f_0, f_0 + f)$ are masked, where f is first chosen from a uniform distribution from 0 to the frequency mask parameter F, and f_0 is chosen from $[0, v - f)$; v is the number of mel frequency channels [26]. The mean value and standard deviation of the mel spectrogram of the training data are calculated. Then, the frequency masking coefficient X is generated with a Gaussian distribution of the same mean value and standard deviation of the original training set. The formulas can be written as:

$$Mel(f_m) = X, \; f_0 \leq f_m < f_0 + f \tag{3}$$

where $f \sim \mathcal{U}(0, F)$, $f_0 \sim \mathcal{U}(0, v - f)$, F is a frequency mask parameter, v is the number of mel frequency channels, $X \sim \mathcal{N}(\mu, \sigma^2)$, μ is the mean value, and σ is the standard deviation of the mel spectrogram in the training data.

We mainly apply the masking procedure on the frequency domain rather than the time domain because the environment noise such as wind noise has a large influence on some specific frequency bands, and we aim to increase the robustness against environment noise and expect the system to detect correctly even if a frequency band is masked or interrupted.

Figure 5 shows the original and masked log mel spectrogram of a recording. The upper figure is the original log mel spectrogram, and the lower is the masked log mel spectrogram. For the augmented data, the cepstral features ranging from 512 Hz to 1024 Hz are masked. After the log result of the mel spectrogram is calculated and discrete cosine transform is applied to the log-mel spectrogram, augmented MFCC features are calculated. Then, the augmented data are appended to the original training data. Finally 100,000 samples are augmented, and there are overall 200,000 samples in the training set.

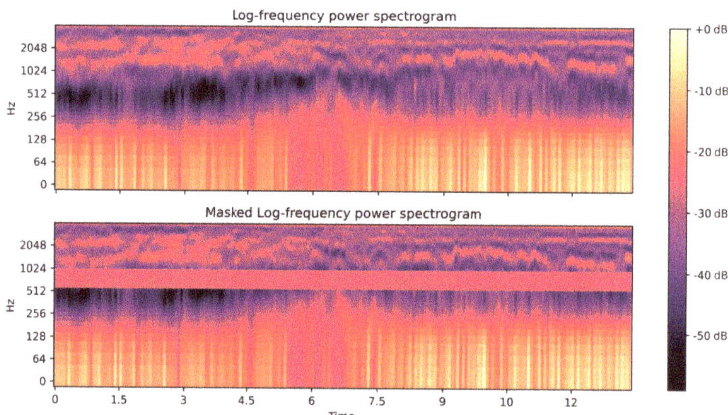

Figure 5. The original log-mel spectrogram of a vehicle recording and the masked log-mel spectrogram.

2.5. Depthwise Separable Convolution

Depthwise separable (DS) convolutions are a form of factorized convolutions which factorize a standard convolution into a depthwise convolution and a 1×1 convolution called a pointwise convolution [29]. The key insight is that different filter channels in regular convolutions are strongly coupled and may involve plenty of redundancy [36].

Depthwise convolution with one filter per input channel can be written as:

$$\hat{G}_{k,l,m} = \sum_{i,j} \hat{K}_{i,j,m} \times F_{k+i-1,l+j-1,m} \tag{4}$$

where $\hat{K}$ is the depthwise convolution kernel of size $D_K^p \times M$, and m_{th} filter in $\hat{K}$ is applied to m_{th} channel in a feature map F to produce the m_{th} channel of the filtered output feature map $\hat{G}$.

The standard convolutions have the computational cost of:

$$D_K^p \times M \times N \times D_F^p \tag{5}$$

where D_K is the kernel size, $p = 1$ for 1-dimensional convolution, $p = 2$ for 2-dimensional convolution, M is the number of input channels, N is the number of output channels, and D_F is the spatial width.

The depthwise separable convolutions have the cost of :

$$D_K^p \times M \times D_F^p + M \times N \times D_F^p \tag{6}$$

Therefore, after applying depthwise separable convolutions, we obtain the reduction in computation of:

$$1/N + 1/D_K^p \tag{7}$$

2.6. Two-Stage Detection Method

The older version of the algorithm in our system for vehicle detection is based on a two-stage detection method by log-sum detection and subspace-based target detection (SBTD) [32].

The first stage is to compare the log-sum energy of the high-frequency part of the acoustic signal and the low-frequency part of the acoustic signal [32]. If the log-sum energy of the high-frequency part is less than the low-frequency part, a result of non-vehicle is returned. Otherwise, the program will proceed to the next stage, subspace-based target detection (SBTD). The steps of the subspace-based target detection(SBTD) are:

1. Estimate the covariance matrix $\hat{R}$:

$$\hat{R} = \frac{1}{L} X X^H \tag{8}$$

   where X is the received signal, and H denotes the Hermitian transpose.
2. Obtain the eigenvalues λ of the covariance matrix $\hat{R}$ by eigenvalue decomposition.
3. Estimate the number of acoustic emissions K by the eigenvalues of the matrix $\hat{R}$, according to some signal number estimation criterion such as minimum description length (MDL) [37].
4. Estimate the total signal power:

$$\hat{P}_S = \frac{\sum\limits_{i=1}^{K} \lambda_i - K\lambda_{K+1}}{M} \tag{9}$$

   where K is the number of acoustic emissions, and M is the number of channels.
5. Estimate the noise power:

$$\hat{P}_N - \frac{\sum\limits_{i=K+1}^{K} \lambda_i + K\lambda_{K+1}}{M} \tag{10}$$

6. Compute the SNR by $SNR = 10 \log(\hat{P}_S / \hat{P}_N)$. If the estimated SNR is larger than the threshold T, we regard it as a target invasion, otherwise we consider it as non-target.

The result of the two-stage detection method is compared with the new proposed method in Section 3.

2.7. Experiment Setup

The two-stage detection method is set up as a baseline system. The optimal threshold for the SBTD stage of the two-stage detection method is decided by maximum likelihood criterion. The calculated optimal threshold is 9.9 dB.

For the proposed deep learning method, the dimension of the input matrix for training is $200,000 \times 39$, with $100,000 \times 39$ original features and $100,000 \times 39$ augmented features. For each feature, cepstral mean and variance normalization [38] is applied for feature normalization and avoiding gradient exploding.

To train a model, a cross-entropy loss function is chosen, and stochastic gradient descent is used as the optimizer [39]. The batch size is 128. Dropout layers are applied to the fully connected layer to avoid overfitting [40]. Each model is trained for 100 epochs. The learning rate is set to 0.01 constantly.

A fully connected neural network is built for comparison. The deep neural network has three hidden layers. A ReLU activation function and a random dropout of 0.2 for regularization are applied in each layer. The framework structure of the fully connected neural network is shown in Table 2.

Table 2. The fully-connected neural network structure.

Layer	Parameters
Fully Connected	39×64
Relu	-
Dropout	0.2
Fully Connected	64×32
Relu	-
Dropout	0.2
Fully Connected	32×8
Relu	-
Dropout	0.2
Fully Connected	8×2

The CNN architecture is comprised of three convolution layers with two max-pooling layers between the three convolution layers and two fully connected layers for the output. The input channel numbers for the first, second and third convolution layers are 1, 16 and 32 respectively; the output channels are 16, 32 and 16, and the kernel sizes are 3, 3 and 3. For each layer, the stride and padding sizes are all set to 1. The kernel sizes for max pooling are 2. The framework structure of the CNN is shown in Table 3.

Table 3. The CNN structure.

Layer	Parameters
Conv1d	$1 \times 16 \times 3$
Max Pooling	2
Conv1d	$16 \times 32 \times 3$
Max Pooling	2
Conv1d	$32 \times 16 \times 3$
Flatten	-
Fully Connected	144×16
Relu	-
Dropout	0.3
Fully Connected	16×2

A depthwise separable CNN architecture is trained for comparison with the same parameter settings as the original CNN structure. The convolution steps are replaced with depthwise separable convolution.

3. Results

3.1. Detection Accuracy

The frame-level accuracy and performance of the proposed method are evaluated on the test set of the vehicle recordings.

The training loss and validation loss of the DS CNN are shown in Figure 6. Figure 6A shows the training loss for each iteration, and Figure 6B shows the validation loss for each epoch. The decaying trends for the loss function of the training set and the validation set are consistent. The batch size is 128, and there are overall 100 epochs and 156,250 iterations. It can be seen that the loss function starts to converge at the 60th epoch, and therefore it is reasonable to choose the 100th epoch to stop training. Figure 7A shows the accuracy of the validation set for each epoch of the DS CNN. The confusion matrix of the DS CNN is illustrated in Figure 7B. The precision rate is 92.87%, the recall rate is 96.70%, and the false alarm rate is 7.42%.

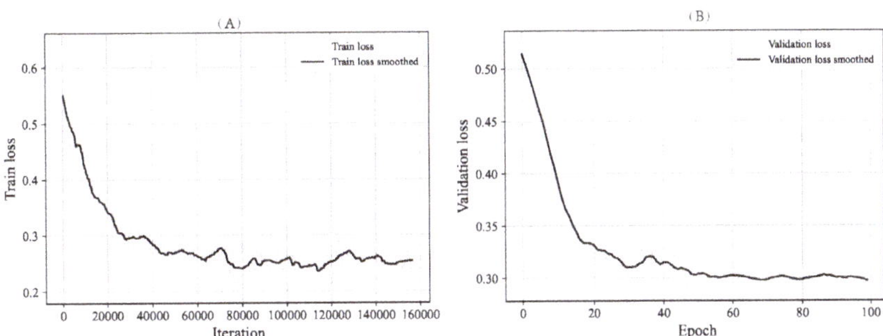

Figure 6. (**A**) The training loss vs. iteration; (**B**) the validation loss vs. epoch.

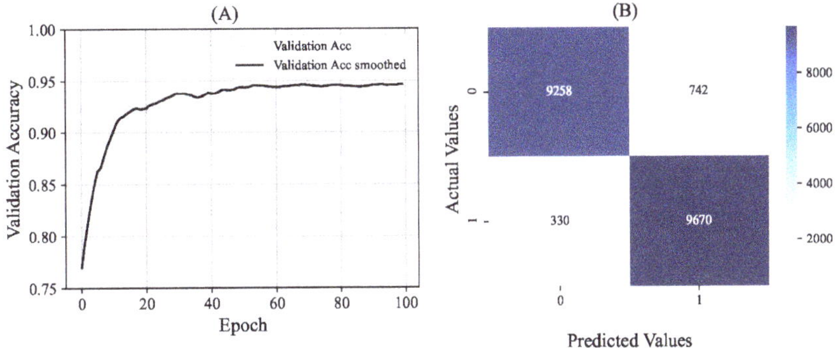

Figure 7. (**A**) The accuracy of the validation data vs. epoch; (**B**) the confusion matrix of the DS CNN.

The frame-level classification results of the proposed models are given in Table 4. The accuracy of our baseline system, two-state detection is 93.65%. The classification accuracy results for the DNN, the CNN and the depthwise separable CNN models, are 89.88%, 93.02% and 92.58%, respectively. The accuracy results with data augmentation for the DNN, the CNN and the depthwise separable CNN models, are 92.14%, 95.11% and 94.64%, respectively. It can be seen that the classification accuracy is improved with augmentation.

Table 4. The overall detection accuracy of each model.

Framework	Classification Accuracy (%)
Two-stage Detection	93.65
DNN	89.88
CNN	93.02
DS CNN	92.58
DNN (Spec Augmentation)	92.14
CNN (Spec Augmentation)	95.11
DS CNN (Spec Augmentation)	94.64

To test the models' ability to detect different types of vehicle, a test was conducted on different types of vehicles separately, and the result is shown in Table 5. The numbers in the brackets of the first column are the numbers of vehicle recordings of different types. All the accuracy results are in frame-level. The traditional subspace-based target detection method has high accuracy towards the large wheeled vehicle and the tracked vehicle because the two types of vehicles make louder sounds when starting, leading to a higher SNR, and the threshold is optimized for these cases. However, the traditional method does not have a good performance for the small wheeled vehicle, as it makes a lower sound, especially when the sensors are placed far from the moving target, causing a low SNR. The DS CNN structure outperform the traditional method on both recall rate and false alarm rate.

Table 5. The ability to detect different types of vehicle of the models.

Method	Two-Stage	DNN	CNN	DS CNN	DNN	CNN	DS CNN
Remark	(9.9 dB)	without SpecAug			with SpecAug		
SWV(91)	74.09	85.93	87.55	86.11	86.94	90.01	89.45
LWV(101)	96.45	90.41	93.88	93.63	93.01	96.06	95.58
TV(62)	96.81	91.01	94.48	94.28	93.49	96.36	95.93
Recall rate(254)	96.61	89.62	94.31	93.61	92.91	96.98	96.70
False alarm rate	9.31	9.86	8.27	8.45	8.63	6.76	7.42

3.2. Complexity Calculation

In the original CNN structure, there are three layers of CNN networks. According to Equations (5) and (6), the computation cost, C , for the first convolution layer is:

$$C = D_K^p \times M \times N \times D_F^p = 3^1 \times 1 \times 16 \times 1^1 = 48 \tag{11}$$

The computation cost for the first depthwise separable convolution layer is:

$$C = D_K^p \times M \times D_F^p + M \times N \times D_F^p = 3^1 \times 1 \times 1^1 + 1 \times 16 \times 1^1 = 19 \tag{12}$$

According to Equation (7), the computation ratio, R, is:

$$R = 1/N + 1/D_k^p = 1/16 + 1/3^1 = 19/48 = 39.58\% \tag{13}$$

The computation costs including the remaining two layers are shown in Table 6. It can be seen that overall cost is reduced 61.96%in the convolution steps.

Table 6. The computation cost of each convolution layer.

Convolution Layer	Original Cost	DS Cost	Reduction Rate(%)
1	48	19	60.42
2	1536	560	63.54
3	1536	608	60.42
All	3120	1187	61.96

According to Table 7, the overall parameter number is 5538 for the original CNN network, and the parameter number is 3654 after applying the depthwise separable CNN. The number of parameters reduced by 34.02% with only a small reduction of accuracy of 0.47%.

Table 7. The number of parameters of each model.

Model	Number of Parameters
DNN	4922
CNN	5538
DS CNN	3654

4. Discussion

The final model migrated to the chips of the sensors is the depthwise separable CNN. The model is lightweight and can be run efficiently on the chips of the sensors. For each frame, the average processing time is about 20 ms; thus, the real-time rate for each frame is 10%. The remaining computational resources can be utilized for other functions such as direction of arrival. The other reason for choosing a depthwise separable convolution network is to prolong the battery life. The intelligent sensor system has to be placed outdoors in the field over weeks. Therefore, the power consumption has to be limited. There is a trade-off between accuracy and model size, and finally the decrease in the accuracy is totally acceptable.

Figure 8 shows the signal and actual detection result of a sample. Figure 8A is the original time-domain signal of a large wheeled vehicle sample. Figure 8B shows some detection errors exist near the border region between the silence part and vehicle moving stage. Figure 8C shows the detection result after applying a smoothing function. Figure 8D represents the ground truth. The recorded moving time of the vehicle is from the 16th second to the 89th second. It can be seen that most classification errors occur at the border region between the silence part and the vehicle moving stage. This can be solved subsequently using a moving window to smooth. The detection algorithm is processed once every 200 milliseconds for each frame, and the detection result is transmitted every 1 s through the transmission module. Therefore, the following strategy is taken for smoothing: the final detection result follows the majority results of the five frames over a second.

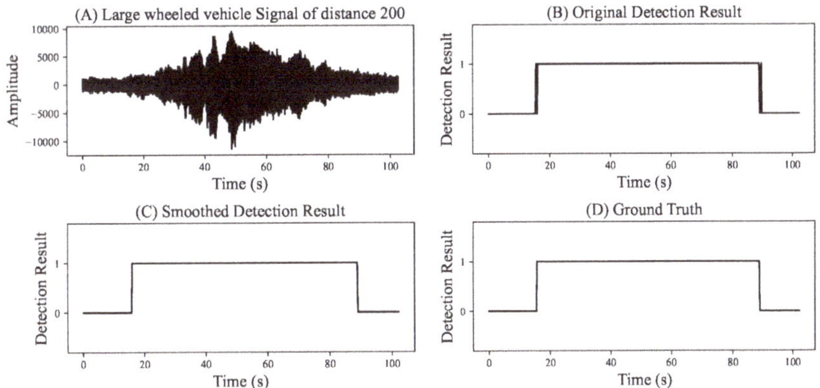

Figure 8. Example of the original signal of a recording and its detection results: (**A**) the original signal of a recording; (**B**) the detection result of the recording; (**C**) the smoothed result; (**D**) the detection ground truth.

Other classification errors occur when strong environment noise such as wind noise exists, and the distance between sensors and the vehicle is too long. In such cases, the signal-to-noise ratio becomes low, especially for a small wheeled vehicle, and the classification accuracy becomes affected. In the future, we intend to solve this problem by exploring signal processing methods including filtering and signal enhancement.

5. Conclusions

This paper proposes a CNN architecture with spectrogram augmentation for vehicle detection. A fully connected network and convolution neural networks are compared, and the CNN structure outperforms the other one. The depthwise separable CNN structure reduces the computational cost. Spectrogram augmentation also shows a huge improvement in the overall model performance. Experiments show that the DS CNN increases the recall rate of detection and reduces the false alarm rate simultaneously compared with the older two-stage method. The accuracy, recall rate and false alarm rate are 94.64%, 96.70% and 7.42%. Finally, the trained model is migrated to the chips of our intelligent sensor systems. The lightweight CNN model can be run efficiently on the these systems. Experiments show the structure has a robust and efficient performance on the sensors. In the future, we intend to discover some practical signal processing methods including filtering and a deep-learning-based signal denoising method to make the system more robust to wind noise and enhance the SNR.

Author Contributions: Conceptualization, C.W. and H.L. (Huawei Liu); data curation, C.W. and Y.S.; investigation, C.W. and H.L. (Haolong Liu); methodology, C.W. and B.L.; supervision, C.W., J.L. and X.Y.; writing—original draft, C.W. and B.L.; writing—review and editing, C.W. and X.Y. All authors have read and agreed to the published version of the manuscript.

Funding: This study was funded by Science and Technology on Micro-system Laboratory, Shanghai Institute of Microsystem and Information Technology, Chinese Academy of Sciences.

Data Availability Statement: Data are not publicly available due to privacy and confidentiality agreement. Not applicable.

Acknowledgments: The research is supported by Science and Technology on Micro-system Laboratory, Shanghai Institute of Microsystem and Information Technology, Chinese Academy of Sciences. I would like to thank Yaozhe Song, Haolong Liu, Huawei Liu, Jianpo Liu, Baoqing Li, Xiaobing Yuan and all the other group members in Science and Technology on Micro-system Laboratory, Shanghai Institute of Microsystem and Information Technology, Chinese Academy of Sciences.

Conflicts of Interest: The authors have no competing interests to declare that are relevant to the content of this article.

Abbreviations

The following abbreviations are used in this manuscript:

MFCC	Mel Frequency Cepstral Coefficients
DNN	Deep Neural Network
DS	Depthwise Separable
CNN	Convolution Neural Network
SBTD	Subspace-Based Target Detection
SNR	Signal-to-Noise Ratio

References

1. Dawton, B.; Ishida, S.; Arakawa, Y. C-AVDI: Compressive measurement-based acoustic vehicle detection and identification. *IEEE Access* **2021**, *9*, 159457–159474. [CrossRef]
2. Dawton, B.; Ishida, S.; Hori, Y.; Uchino, M.; Arakawa, Y.; Tagashira, S.; Fukuda, A. Initial evaluation of vehicle type identification using roadside stereo microphones. In Proceedings of the IEEE Sensors Applications Symposium (SAS), Kuala Lumpur, Malaysia, 9–11 March 2020; pp. 1–6.
3. Dawton, B.; Ishida, S.; Hori, Y.; Uchino, M.; Arakawa, Y. Proposal for a compressive measurement-based acoustic vehicle detection and identification system. In Proceedings of the IEEE 92nd Vehicular Technology Conference (VTC2020-Fall), Virtual, 18 November–16 December 2020; pp. 1–6.
4. Fang, J.; Meng, H.; Zhang, H.; Wang, X. A low-cost vehicle detection and classification system based on unmodulated continuous-wave radar. In Proceedings of the IEEE Intelligent Transportation Systems Conference, Bellevue, DC, USA, 30 September–3 October 2007; pp. 715–720.
5. Wang, X. Vehicle image detection method using deep learning in UAV video. *Comput. Intell. Neurosci.* **2022**, *2022*. [CrossRef]
6. Kumari, S.; Agrawal, D. A Review on Video Based Vehicle Detection and Tracking using Image Processing. *Int. J. Res. Publ. Rev.* **2022**, *2582*, 7421.
7. Allegro, G.; Fascista, A.; Coluccia, A. Acoustic Dual-function communication and echo-location in inaudible band. *Sensors* **2022**, *22*, 1284. [CrossRef] [PubMed]
8. Gencoglu, O.; Virtanen, T.; Huttunen, H. Recognition of acoustic events using deep neural networks. In Proceedings of the 22nd European signal processing conference (EUSIPCO), Lisbon, Portugal, 1–5 September 2014; pp. 506–510.
9. Bae, S.H.; Choi, I.K.; Kim, N.S. Acoustic scene classification using parallel combination of LSTM and CNN. In Proceedings of the Detection and Classification of Acoustic Scenes and Events 2016, Budapest, Hungary, 3 September 2016; pp. 11–15.
10. Fu, R.; He, J.; Liu, G.; Li, W.; Mao, J.; He, M.; Lin, Y. Fast seismic landslide detection based on improved mask R-CNN. *Remote Sens.* **2022**, *14*, 3928. [CrossRef]
11. Li, H.; Lu, J.; Tian, G.; Yang, H.; Zhao, J.; Li, N. Crop classification based on GDSSM-CNN using multi-temporal RADARSAT-2 SAR with limited labeled data. *Remote Sens.* **2022**, *14*, 3889. [CrossRef]
12. Li, S.; Fu, X.; Dong, J. Improved ship detection algorithm based on YOLOX for SAR outline enhancement image. *Remote Sens.* **2022**, *14*, 4070. [CrossRef]
13. Adapa, S. Urban sound tagging using convolutional neural networks. *arXiv* **2019**, arXiv:1909.12699.
14. Sharma, G.; Umapathy, K.; Krishnan, S. Trends in audio signal feature extraction methods. *Appl. Acoust.* **2020**, *158*, 107020. [CrossRef]

15. Vikaskumar, G.; Waldekar, S.; Paul, D.; Saha, G. Acoustic scene classification using block based MFCC features. In Proceedings of the IEEE AASP Challenge on Detection and Classification of Acoustic Scenes and Events (DCASE), Budapest, Hungary, 3 September 2016.

16. Ma, Y.; Liu, M.; Zhang, Y.; Zhang, B.; Xu, K.; Zou, B.; Huang, Z. Imbalanced underwater acoustic target recognition with trigonometric loss and attention mechanism convolutional network. *Remote Sens.* **2022**, *14*, 4103. [CrossRef]

17. Chaudhary, M.; Prakash, V.; Kumari, N. Identification vehicle movement detection in forest area using MFCC and KNN. In Proceedings of the 2018 International Conference on System Modeling & Advancement in Research Trends (SMART), Moradabad, India, 23–24 November 2018; pp. 158–164.

18. Pons, J.; Serra, X. Randomly weighted cnns for (music) audio classification. In Proceedings of the ICASSP 2019–2019 IEEE International Conference on Acoustics, Speech and Signal Processing (ICASSP), Brighton, UK, 12–17 May 2019; pp. 336–340.

19. Stowell, D.; Plumbley, M.D. Automatic large-scale classification of bird sounds is strongly improved by unsupervised feature learning. *PeerJ* **2014**, *2*, e488. [CrossRef] [PubMed]

20. Kinnunen, T.; Chernenko, E.; Tuononen, M.; Fränti, P.; Li, H. Voice activity detection using MFCC features and support vector machine. In Proceedings of the Int. Conf. on Speech and Computer (SPECOM07), Moscow, Russia, 4–10 August 2007; Volume 2, pp. 556–561.

21. Thomas, S.; Ganapathy, S.; Saon, G.; Soltau, H. Analyzing convolutional neural networks for speech activity detection in mismatched acoustic conditions. In Proceedings of the 2014 IEEE International Conference on Acoustics, Speech and Signal Processing (ICASSP), Florence, Italy, 4–9 May 2014; pp. 2519–2523.

22. Tokozume, Y.; Ushiku, Y.; Harada, T. Learning from between-class examples for deep sound recognition. *arXiv* **2017**, arXiv:1711.10282.

23. Salamon, J.; Bello, J.P. Deep convolutional neural networks and data augmentation for environmental sound classification. *IEEE Signal Process. Lett.* **2017**, *24*, 279–283. [CrossRef]

24. Ko, T.; Peddinti, V.; Povey, D.; Khudanpur, S. Audio augmentation for speech recognition. In Proceedings of the Sixteenth Annual Conference of the International Speech Communication Association, Dresden, Germany, 6–10 September b2015.

25. Piczak, K.J. Environmental sound classification with convolutional neural networks. In Proceedings of the IEEE 25th international workshop on machine learning for signal processing (MLSP), Boston, MA, USA, 17–20 September 2015; pp. 1–6.

26. Park, D.S.; Chan, W.; Zhang, Y.; Chiu, C.C.; Zoph, B.; Cubuk, E.D.; Le, Q.V. Specaugment: A simple data augmentation method for automatic speech recognition. *arXiv* **2019**, arXiv:1904.08779.

27. Krizhevsky, A.; Sutskever, I.; Hinton, G.E. Imagenet classification with deep convolutional neural networks. *Adv. Neural Inf. Process. Syst.* **2012**, *25*. [CrossRef]

28. Denil, M.; Shakibi, B.; Dinh, L.; Ranzato, M.; De Freitas, N. Predicting parameters in deep learning. *Adv. Neural Inf. Process. Syst.* **2013**, *26*.

29. Howard, A.G.; Zhu, M.; Chen, B.; Kalenichenko, D.; Wang, W.; Weyand, T.; Andreetto, M.; Adam, H. Mobilenets: Efficient convolutional neural networks for mobile vision applications. *arXiv* **2017**, arXiv:1704.04861.

30. Huang, J.; Zhang, X.; Guo, F.; Zhou, Q.; Liu, H.; Li, B. Design of an acoustic target classification system based on small-aperture microphone array. *IEEE Trans. Instrum. Meas.* **2014**, *64*, 2035–2043. [CrossRef]

31. Zhang, X.; Huang, J.; Song, E.; Liu, H.; Li, B.; Yuan, X. Design of small MEMS microphone array systems for direction finding of outdoors moving vehicles. *Sensors* **2014**, *14*, 4384–4398. [CrossRef]

32. Guo, F.; Huang, J.; Zhang, X.; Cheng, Y.; Liu, H.; Li, B. A two-stage detection method for moving targets in the wild based on microphone array. *IEEE Sensors J.* **2015**, *15*, 5795–5803. [CrossRef]

33. Zhang, X.L.; Wu, J. Deep belief networks based voice activity detection. *IEEE Trans. Audio, Speech Lang. Process.* **2012**, *21*, 697–710. [CrossRef]

34. Picone, J.W. Signal modeling techniques in speech recognition. *Proc. IEEE* **1993**, *81*, 1215–1247. [CrossRef]

35. Bahmei, B.; Birmingham, E.; Arzanpour, S. CNN-RNN and data augmentation using deep convolutional generative adversarial network for environmental sound classification. *IEEE Signal Process. Lett.* **2022**, *29*, 682–686. [CrossRef]

36. Guo, J.; Li, Y.; Lin, W.; Chen, Y.; Li, J. Network decoupling: From regular to depthwise separable convolutions. *arXiv* **2018**, arXiv:1808.05517.

37. Zhao, L.; Krishnaiah, P.R.; Bai, Z. On detection of the number of signals in presence of white noise. *J. Multivar. Anal.* **1986**, *20*, 1–25. [CrossRef]

38. Strand, O.M.; Egeberg, A. Cepstral mean and variance normalization in the model domain. In Proceedings of the COST278 and ISCA Tutorial and Research Workshop (ITRW) on Robustness Issues in Conversational Interaction, Norwich, UK, 30–31 August 2004.

39. Bottou, L. Large-scale machine learning with stochastic gradient descent. In Proceedings of the COMPSTAT'2010, Paris, France, 22–27 August 2010; pp. 177–186.

40. Srivastava, N.; Hinton, G.; Krizhevsky, A.; Sutskever, I.; Salakhutdinov, R. Dropout: A simple way to prevent neural networks from overfitting. *J. Mach. Learn. Res.* **2014**, *15*, 1929–1958.

Article

Retrieval of Live Fuel Moisture Content Based on Multi-Source Remote Sensing Data and Ensemble Deep Learning Model

Jiangjian Xie [1,2], Tao Qi [1,2], Wanjun Hu [1,2], Huaguo Huang [2,3], Beibei Chen [3] and Junguo Zhang [1,2,*]

[1] School of Technology, Beijing Forestry University, Beijing 100083, China
[2] Research Center for Biodiversity Intelligent Monitoring, Beijing Forestry University, Beijing 100083, China
[3] The College of Forestry, Beijing Forestry University, Beijing 100083, China
* Correspondence: zhangjunguo@bjfu.edu.cn

Abstract: Live fuel moisture content (LFMC) is an important index used to evaluate the wildfire risk and fire spread rate. In order to further improve the retrieval accuracy, two ensemble models combining deep learning models were proposed. One is a stacking ensemble model based on LSTM, TCN and LSTM-TCN models, and the other is an Adaboost ensemble model based on the LSTM-TCN model. Measured LFMC data, MODIS, Landsat-8, Sentinel-1 remote sensing data and auxiliary data such as canopy height and land cover of the forest-fire-prone areas in the Western United States, were selected for our study, and the retrieval results of different models with different groups of remote sensing data were compared. The results show that using multi-source data can integrate the advantages of different types of remote sensing data, resulting in higher accuracy of LFMC retrieval than that of single-source remote sensing data. The ensemble models can better extract the nonlinear relationship between LFMC and remote sensing data, and the stacking ensemble model with all the MODIS, Landsat-8 and Sentinel-1 remote sensing data achieved the best LFMC retrieval results, with R^2 = 0.85, RMSE = 18.88 and ubRMSE = 17.99. The proposed stacking ensemble model is more suitable for LFMC retrieval than the existing method.

Keywords: live fuel moisture content; deep learning; ensemble learning; multi-source remote sensing

Citation: Xie, J.; Qi, T.; Hu, W.; Huang, H.; Chen, B.; Zhang, J. Retrieval of Live Fuel Moisture Content Based on Multi-Source Remote Sensing Data and Ensemble Deep Learning Model. *Remote Sens.* **2022**, *14*, 4378. https://doi.org/10.3390/rs14174378

Academic Editor: Gwanggil Jeon

Received: 22 August 2022
Accepted: 1 September 2022
Published: 3 September 2022

Publisher's Note: MDPI stays neutral with regard to jurisdictional claims in published maps and institutional affiliations.

1. Introduction

Live fuel moisture content (LFMC) is the ratio of vegetation water content to its dry weight [1]. The research shows that there is a clear correlation between the probability of fire and LFMC [2,3], which is an important index affecting the occurrence probability and propagation rate of forest wildfire. To put it another way, accurate and dynamic retrieval of LFMC is extremely valuable to realize the fire risk assessment and spatial modeling of fire behavior [4]. Remote sensing satellite can provide large-scale, multi-band and near-real-time image data, which makes remote sensing technology one of the main methods to estimate LFMC on a large scale [5]. The method of estimating LFMC based on optical remote sensing data is the most widely studied [6–8]. MODIS optical remote sensing data are commonly used in the early stage. Myoung et al. [9] developed an empirical model function of LFMC using an aqua-enhanced vegetation index based on MODIS satellite data for wildfire risk assessment in Southern California. Carmine et al. [10] developed a new spectral index, the perpendicular moisture index (PMI), which is sensitive to LFMC based on MODIS satellite data. The experimental results show that PMI had a linear relationship with LFMC, and the highest R^2 was 0.87. Landsat-8 can provide higher spatial resolution than MODIS, which has been introduced to estimate LFMC in recent years. Considering the complexity of upper tree canopy and lower grass canopy, Quan et al. [11] predicted the forest FMC of a two-layer canopy structure in Southwest China by coupling a radiative transfer model and a Landsat-8 product. Mbulisi et al. [5] used Landsat-5 and Landsat-8 data to quantitatively retrieve vegetation canopy FMC in six study areas based on on PROSAIL and PROGeoSAIL radiative transfer models. These methods based on optical remote sensing depend on the absorption characteristics of leaf water at near-infrared (NIR) or short-wave infrared (SWIR) wavelengths [12]. Optical and infrared reflectance are highly sensitive to vegetation characteristics such as canopy structure [13] and leaf area index [14,15], and so these models are often only applicable for very specific sites, and the generalization ability of different regions is limited [16,17].

The wavelength of microwave remote sensing is longer than that of optical remote sensing by four orders of magnitude. Microwaves can penetrate the clouds and enter the vegetation canopy, which enables microwave remote sensing to acquire the dynamic changes in vegetation moisture better than optical remote sensing [18–20]. In recent years, the prediction ability of active microwave remote sensing technology represented by synthetic aperture radar (SAR) for fire-related variables has been verified [21,22]. Wang et al. [20] coupled the soil backscatter linear model with the vegetation backscatter water cloud model, achieving forest FMC retrieval based on Sentinel-1 SAR data and a better performance than that obtained using Landsat-8 data and empirical methods.

Different remote sensing data have different sensitivities to vegetation water and biomass, and the effect of single-source remote sensing data retrieval of LFMC is limited. Using multi-source remote sensing data to estimate LFMC can avoid the limitations of single-source remote sensing data and provide more comprehensive data for extracting the parameters required for LFMC retrieval [23]. Deep learning can approach the complex nonlinear relationship between various biological, geophysical parameters and remote sensing data through multi-layer learning [24,25], which provides a data-driven alternative for large-scale LFMC retrieval. Rao et al. [19] performed LFMC retrieval based on a long short-term memory (LSTM) network with fused data, i.e., Landsat-8 data, SAR data, terrain, slope and other auxiliary variables. The retrieval of fused data achieved $R^2 = 0.63$, RMSE = 25%, which is better than that of single-source remote sensing data ($R^2 = 0.44$, RMSE = 31.8%). Zhu et al. [26] proposed the LFMC retrieval architecture TempCNN-LFMC based on temporal convolutional networks (TCNs). With MODIS, altitude, slope and other auxiliary data as the input fused data, the retrieval achieved $R^2 = 0.64$, RMSE = 22.74%. The above research shows that the fused data are helpful in improving the performance of LFMC retrieval.

A single model cannot completely extract the features of remote sensing variables in LFMC. To improve the accuracy of LFMC retrieval, it is worthwhile to combine multiple models to extract the features of multi-source remote sensing in time and space dimensions at the same time [25]. Therefore, based on deep learning and ensemble learning methods, this study discusses the LFMC retrieval performance using multi-source remote sensing data. The contents of this study include the following aspects:

(1) We explore the advantages of LFMC retrieval utilizing multi-source remote sensing data obtained from combing MODIS, Landsat-8, Sentinel-1 and auxiliary data such as canopy height and land cover as data sources, which can provide more comprehensive data and avoid the limitations of single-source remote sensing data.

(2) We propose a LFMC retrieval model integrating the LSTM and TCN, which exploits the long-time memory capability of LSTM and the superior feature extraction capability of TCN, and finally performs better than LSTM and the TCN alone.

(3) Based on LSTM, TCN and TCN-LSTM models, two ensemble models (the stacking and Adaboost ensemble models) are designed, and the advantages of stacking ensemble model are confirmed by comparative experiments.

2. Data and Methods

2.1. Study Area

The Western United States was selected as the study area (shown in Figure 1), where wildfires occurred frequently. This area covers more than 3.7 million square kilometers, containing different climates and terrains. The vegetation types are abundant, including broadleaf deciduous forests, needleleaf evergreen forests, shrublands, grasslands and sparse vegetation areas, which made it an ideal area for studying LFMC prediction methods. Considering the integrity and generality of the data, the selected study period was from 1 January 2015 to 31 December 2018.

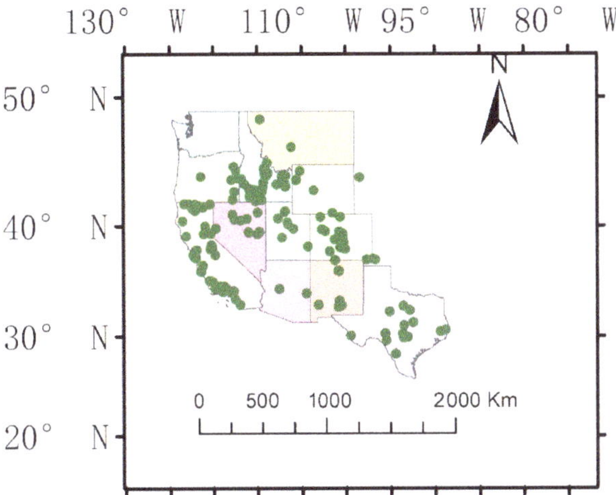

Figure 1. Geographical location of study area and sample point distribution.

2.2. Research Data

2.2.1. LFMC Data

The National Fuel Moisture Database (NFMD) [27] is a web-based query system. There have been over 200,000 actual measurements of fuel moisture data since 1977. The database regularly updates monitored fuel moisture data, covering 976 samples mainly located in the Western United States, each covering an area of 5 acres. The measurements were taken in the mid-afternoon and on dry days with no dew or precipitation. In this paper, 133 representative samples were selected, and the specific location is marked by circular points in Figure 1. During the study period, the value of LFMC varied from 16% to 320%, which covers the common water state of live fuels.

2.2.2. MODIS Data

The MODIS data came from MODIS Terra and Aqua joint observation of the MCD43A4 product [28]. The product was the nadir bidirectional reflectance distribution function (BRDF)-adjusted reflectance (NBAR) data, the spatial resolution of which is 500 m. BRDF was fitted using 16-day Terra and Aqua MODIS data and applied to the original reflectance to obtain NBAR. In this study, Band1–Band7 of NBAR reflectivity data were selected as the model input.

In addition, snow cover will lead to abnormal reflectivity. Thus, the snow pixels need to be deleted. The MODIS snow product (MOD10A2-V6) [29] was used to determine whether there was snow. MOD10A2 is a snow cover product synthesized every eight days from the first day of each year. In MOD10A2, if a pixel is classified as snow on any day of the eight days, the pixel is identified as snow.

2.2.3. Landsat Data

Landsat data came from the 16-day surface reflectance data of Landsat-8 [30], which are Level 1T products with a spatial resolution of 30 m. There is a strong correlation between the normalized difference water index (NDWI) and LFMC [31]. Considering that water mainly absorbs the energy of near-infrared (NIR) and short-wave infrared (SWIR) spectral regions, the original band reflectances of red, green, blue, near-infrared and short-wave infrared channels were selected to directly reflect the change in water [32]. The normalized difference vegetation index (NDVI) is a simple, effective and empirical measurement of surface vegetation, and it is also a key factor affecting the prediction of LFMC [33]. The near-infrared vegetation index (NIRV) is an indicator of vegetation biomass level because it is related to carbon assimilation of photosynthesis, so it may help to separate the effects of biomass and LFMC on Sentinel-1 backscattering [34]. To sum up, three vegetation indexes, NDWI, NDVI and NIRV, and the original band reflections of red, green, blue, near-infrared and short-wave infrared channels were selected from Landsat-8 data.

2.2.4. Sentinel-1 Data

Sentinel-1 is a 5.4 GHz C-band synthetic aperture radar (SAR) with a 12-day revisit cycle in the Western United States. The Sentinel-1 data used in this study were derived from the ground-range detector (GRD) data of Sentinel-1, and the data were collected in the wide-strip mode of interferometric measurement with vertical–vertical (VV) and vertical–horizontal (VH) polarization on land [35]. Since the microwave signal has a longer wavelength, is less sensitive to atmospheric conditions, is not susceptible to cloud pollution and can detect deeper vegetation canopy, the microwave remote sensing data can provide more continuous global observation [36]. At the same time, the absorption and scattering of the microwave signal by the surface (including vegetation and soil) is mainly determined by the microwave backscatter σ [37], and microwave backscatter is mainly affected by the moisture content, so the microwave signal is sensitive to vegetation water content [38]. Therefore, in this paper, σ_{VV}, σ_{VH} and $\sigma_{VV} - \sigma_{VH}$ were used as the microwave input for the model.

2.2.5. Auxiliary Data

Seven kinds of static auxiliary data were chosen to help the model learn the radiative transfer process between time-varying input and LFMC. The specific data can be divided into the following three categories:

The first category is soil data, including silt, sand and clay content, which was used to control the sensitivity of microwave backscattering to soil moisture, so that the retrieval model could separate vegetation-related information from microwave backscattering. The soil data come from the North American soil map of Liu et al. [39].

Vegetation canopy water content has a certain sensitivity in the near-infrared and short-wave infrared bands [5], and the sensitivity of different vegetations to remote sensing data is also different. The canopy height measured by the Global Laser Altimeter System lidar [40] and the land cover information of 300 m spatial resolution obtained by GLOBCOVER [41] were selected as the second auxiliary data.

The third category is terrain data; considering that the local incidence angle will affect the parameterization of backward scattering on vegetation water [37], it was necessary to use the elevation and slope of the National Elevation Dataset [42] to help the model calibrate the local terrain.

Table 1 summarizes all the inputs used in the model.

Table 1. Input variables of LFMC retrieval model.

MODIS	Landsat-8	Sentinel-1	Auxiliary Variables
Band1	red	σ_{VV}	Silt content
Band2	green	σ_{VH}	Sand content
Band3	blue	$\sigma_{VV} - \sigma_{VH}$	Clay content
Band4	NIR		Canopy height (m)
Band5	SWIR		Land cover
Band6	NDWI		Altitude (m)
Band7	NDVI		Slope (°)
	NIR$_V$		

2.3. Data Process

Given the presence of numerous vegetation species at certain sample points and the lack of information on the abundance of these species, directly averaging the LFMCs of different species will result in significant inaccuracies. We adopted the same strategy as [19], excluding sampling points with multiple species unless the LFMCs of multiple species were similar during the research period (Pearson r between any two species $\geq$ 0.5). Thus, 2934 samples from 133 sampling points were included in total (shown in Figure 1).

Because the spatial and temporal resolutions of MODIS, Landsat-8 and Sentinel-1 are not the same, spatial and temporal consistency processing was needed. The remote sensing variable data of the sample points were extracted by the Google Earth Engine (GEE) according to the latitude and longitude coordinates. According to the latitude and longitude coordinates, the spatial synchronization of ground data and remote sensing data could be realized. The remote sensing data were unified to the resolution of 250 m using bilinear interpolation.

The sampling period of live fuel samples in each location was roughly one month, so the time series input was linearly interpolated to the end of each month to ensure that the data had the same time phase. The maximum changes in MODIS, Landsat-8 and Sentinel-1 data were only 2.3%, 6.7% and 3.0%, respectively. It can be considered that the interpolation operation had little effect on the input data.

2.4. Dataset

In this work, three-fold cross-validation was used to test the model. To ensure that the performance of the model was tested on samples that were completely different from the training sample points by separating data by sample points, the data were first stratified randomly sampled into training and test sets by a ratio of 2:1 to ensure that the distribution of land cover types in the training and test sets remained the same. This implies that the training set was made up of data from two-thirds of the locations (89 sample points), while the test set was made up of data from the remaining one-third (44 sample points). In addition, the training set was divided into three folds, two for training and one for validation. Finally, the results presented in the paper were calculated based on the estimated value of the test set.

2.5. LFMC Retrieval Models

2.5.1. TCN-LSTM Model

LSTM can effectively deal with the dynamic dependence of complex long-term time series. The TCN has simple structure and strong feature extraction ability. Combined with the ability of TCN feature extraction and LSTM long-time series memory, the TCN-LSTM network is designed to predict the LFMC. The structure of the TCN-LSTM network is shown in Figure 2.

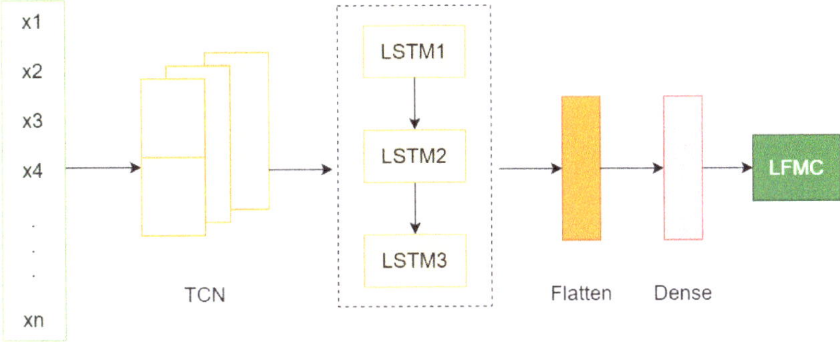

Figure 2. Structure of TCN-LSTM model.

The retrieval process is as follows:

(1)   Firstly, the LFMC data and selected input variables (x1, . . . , xn) are fed into the TCN. The features of remote sensing variables and LFMC are extracted through the causal convolution layer contained in the TCN.
(2)   Then, multiple LSTM layers combined with the dropout mechanism are used for prediction, which can prevent over fitting.
(3)   Through the flatten layer, the output matrix is compressed into one dimension to facilitate the connection of the later dense layer.
(4)   The nonlinear relationship is mapped to the output space through the dense layer to achieve the LFMC prediction results.

2.5.2. Stacking Ensemble Model

In order to further improve the performance of LFMC retrieval, a two-layer stacking ensemble model integrating LSTM, TCN and TCN-LSTM was further proposed. The model structure is shown in Figure 3.

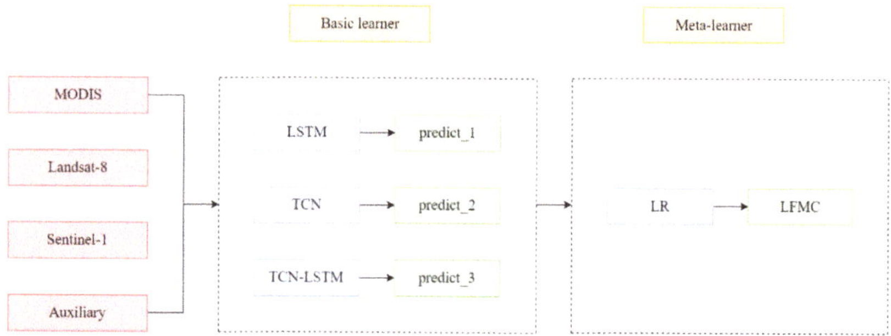

Figure 3. Structure of stacking model.

The first layer extracts the features from the original split dataset through three basic learners. The basic learners of the stacking model should be "accurate but different", that is, the prediction accuracy of each basic learner is required to be high, and the types of basic learners should also be diverse. So, LSTM, the TCN and TCN-LSTM were introduced as the base learners. In order to avoid over fitting, a simple linear regression (LR) was selected as the meta-learner of the second layer.

2.5.3. Adaboost Ensemble Model

Unlike stacking ensemble, Adaboost ensemble trains several weak learners based on different training subsets randomly selected from the original training dataset. Adaboost ensemble is based on homogeneous integration, which is composed of the same type of basic learners. In this work, the TCN-LSTM model was selected as the weak learner to construct the Adaboost ensemble model. Figure 4 shows the structure of the Adaboost ensemble model.

In each training process, the initial weights are assigned to the samples at first, and the weights are updated after each iteration. The samples with a high error rate obtain higher weights, which makes the algorithm focus on the samples that are more difficult to learn. The sample weight is adjusted to Dn, and passed to the next weak learner Gn for better prediction. Therefore, the features extracted by G1 are transmitted to G2, and then the features estimated with high error can be corrected in the transmission process, which is helpful to improve the prediction accuracy. At last, the weighted average method is utilized to obtain the strong learner HM, the output of which is the final prediction result. Considering the computational efficiency, the number of TCN-LSTM, that is, the number of iterations t, was set to 3.

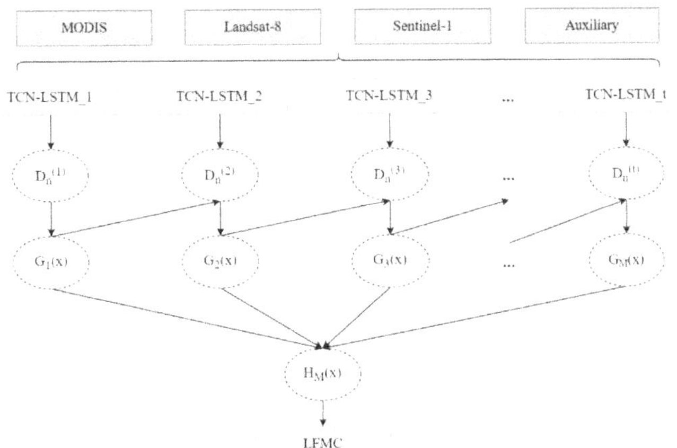

Figure 4. Structure of Adaboost ensemble model.

2.5.4. Model Settings

There are three basic models, LSTM, the TCN and TCN-LSTM. Table 2 lists the architectures of these models. In the ensemble models, the same architectures were used. All the proposed models estimated LFMC for each month using input variables of three previous months. Although predicting one month averaged LFMC value can be error-prone due to the variations in LFMC, we were constrained by the temporal resolution of the remote sensing data.

Table 2. The architectures of the used basic prediction models.

LSTM		TCN		TCN-LSTM	
Layer	**Output Shape**	**Layer**	**Output Shape**	**Layer**	**Output Shape**
LSTM	(32,4,10)	Conv1D	(32,365,64)	Conv1D	(32,4,32)
LSTM	(32,4,10)	AvgPool	(32,182,64)	Conv1D	(32,4,32)
LSTM	(32,10)	Conv1D	(32,182,64)	MaxPool	(32,2,32)
Dense	(32,1)	AvgPool	(32,60,64)	Flatten	(32,64)
		Conv1D	(32,60,64)	RepeatVector	(32,1941,64)
		MaxPool	(32,15,64)	LSTM	(32,1941,10)
		Flatten	(32,960)	LSTM	(32,1941,10)
		Dense	(32,256)	LSTM	(32,10)
		Dense	(32,1)	Dense	(32,1)

3. Experiments and Results

3.1. Experimental Setup

The hardware environment of the experiments was: CPU: Intel (R) Core (TM) i7-8565U, Memory: 8 GB. The software environment was: Windows 10 64 operating system, deep learning framework Tensorflow2.3.0 and python 3.7. Adam optimizer was used, and the parameters were the default values. The batch size was 32, the learning rate was 0.01, and the epoch was 300. In order to avoid over fitting, early stopping based on the loss of the validation set was used [40], and patience was 30.

3.2. Evaluating Indicator

Bias, determination coefficient R^2, root mean square error (RMSE) and unbiased root mean square error (ubRMSE) between estimated and measured LFMCs were chosen to quantitatively evaluate the performance of the models. When R^2 was closer to 1 and the

RMSE value was lower, the model accuracy was higher and the model was more accurate. The calculations of *RMSE* and *ubRMSE* are shown in Formulas (1) and (2):

$$RMSE = \sqrt{\frac{1}{N} \sum_{i}^{N} \left(LFMC_{i,m} - LFMC_{i,e} \right)^2} \tag{1}$$

$$ubRMSE = \sqrt{\frac{1}{N} \sum_{i}^{N} \left(LFMC_{i,m} - LFMC_{i,e} - \left(\overline{LFMC_m} - \overline{LFMC_e} \right) \right)^2} \tag{2}$$

where N is the number of measurements; $LFMC_{i,m}$ and $LFMC_{i,e}$ are the ith measured and estimated LFMC, respectively; $\overline{LFMC_m}$ and $\overline{LFMC_e}$ are the averages of measured and estimated LFMC, respectively.

3.3. Comparison of Different Deep Learning Models

The performances of three different models, LSTM, the TCN and TCN-LSTM, with different remote sensing data were compared. The retrieval results are shown in Figure 5.

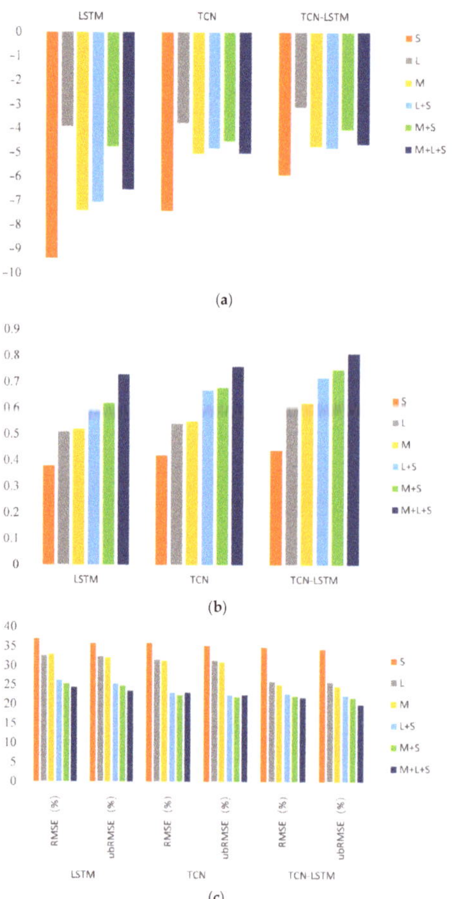

Figure 5. Evaluation indicators of LFMC retrieval results based on LSTM, TCN and TCN-LSTM. (a) Bias; (b) R^2; (c) RMSE and ubRMSE. S represents Sentinel-1 data, L represents Landsat8 data, M represents MODIS data.

By comparing and analyzing the results in Figure 5 and Table 3, it can be concluded that:

Table 3. Evaluation results of LFMC retrieval results based on LSTM, TCN and TCN-LSTM.

Data	Model	Bias (%)	R^2	RMSE (%)	ubRMSE (%)
S	LSTM	−9.39	0.38	37.07	35.86
	TCN	−7.42	0.42	35.97	35.2
	TCN-LSTM	−5.93	0.44	34.81	34.3
L	LSTM	−3.93	0.51	32.67	32.43
	TCN	−3.8	0.54	31.58	31.35
	TCN-LSTM	**−3.12**	0.60	25.83	25.64
M	LSTM	−7.39	0.52	33.01	32.17
	TCN	−5.04	0.55	31.39	30.98
	TCN-LSTM	−4.74	0.62	25.11	24.66
L+S	LSTM	−7.06	0.60	26.32	25.35
	TCN	−4.83	0.67	23.05	22.54
	TCN-LSTM	−4.81	0.72	22.78	22.31
M+S	LSTM	−4.76	0.62	25.33	24.88
	TCN	−4.53	0.68	22.43	21.97
	TCN-LSTM	−4.05	0.75	22.21	21.71
M+L+S	LSTM	−6.53	0.73	24.39	23.5
	TCN	−5.03	0.76	23.03	22.48
	TCN-LSTM	−4.66	**0.81**	**21.73**	**19.93**

(1) The bias of all the three models was negative, indicating that all the models underestimated LFMC as a whole. The TCN-LSTM model had the lowest bias among all the models on the same dataset. The bias of Sentinel-1 was the largest, and that of Landsat-8 was the lowest. Although microwave remote sensing (Sentinel-1) is more penetrating due to its high sensitivity to surface moisture, it is difficult to distinguish between vegetation and bare soil backscatter only using microwave remote sensing data, which leads to higher bias. The multi-source remote sensing data fuse the microwave remote sensing and optical remote sensing together, which can be essentially seen as the integration of the microwave backscattering characteristics and optical characteristic. Therefore, the retrieval performances of multi-source remote sensing data were higher than those of the single-source remote sensing data.

(2) The R^2, RMSE and ubRMSE of the TCN-LSTM model were also better than those of the LSTM and TCN models. The retrieval accuracy of the TCN-LSTM model with all three kinds of remote sensing data was the highest at $R^2 = 0.81$, RMSE = 21.73 and ubRMSE = 19.93, which means that TCN-LSTM can incorporate the advantages of LSTM and the TCN and effectively extract the features of multi-source remote sensing.

A comparison with the retrieval results of references is shown in Table 4; the TCN-LSTM model with multi-source remote sensing achieved the best results for LFMC retrieval. Compared with the best results of the existing method [20], R^2 and RMSE were improved by 26.56% and 4.44%, respectively.

Table 4. Comparison of different LFMC retrieval methods.

Method	R^2	RMSE (%)
LSTM (Landsat+SAR) [19]	0.63	25
TempCNN-LFMC (MODIS+Auxiliary data) [20]	0.64	22.74
TCN-LSTM model	**0.81**	**21.73**

3.4. Comparison of Different Ensemble Learning Models

Table 5 shows the performance comparison of different ensemble learning models with different remote sensing data. It can be seen that the performances of two ensemble learning models with all three kinds of remote sensing data were better than with other data. The retrieval results for LFMC based on the stacking ensemble model with MODIS, Landsat-8 and Sentinel-1 are the best. This is mainly due to the integration of the advantages of three different models. While the Adaboost ensemble model only uses one kind of basic learner, its performance was poorer than that of the stacking ensemble model.

Table 5. Performances of different ensemble learning models.

Data	Stacking				Adaboost			
	Bias (%)	R^2	RMSE (%)	ubRMSE (%)	Bias (%)	R^2	RMSE (%)	ubRMSE (%)
S	−5.75	0.53	31.87	31.35	−4.59	0.53	31.62	31.29
L	−3.35	0.7	23.26	23.21	−1.65	0.65	23.6	23.54
M	−4.55	0.74	23.82	23.39	−4.16	0.68	22.53	22.14
L+S	−1.55	0.81	19.96	19.95	−2.61	0.76	22	21.31
M+S	−1.43	0.81	19.86	19.81	−2.7	0.8	20.5	20.32
M+L+S	**−0.542**	**0.85**	**18.88**	**17.99**	**−0.563**	**0.83**	**19.7**	**18.8**

Together with Table 3, it was found that with different groups of remote sensing data, the trend in the performances of the single model and the ensemble model was almost the same. The more data used, the better the performance. Additionally, based on multi-source data (M+L+S), compared with the TCN-LSTM model, the R^2, RMSE and ubRMSE of the stacking ensemble model realized an improvement of 4.9%, 15.1% and 10.8%, respectively.

4. Discussion

4.1. Explanation of Estimated LFMC Value

In general, when using Landsat-8, Sentinel-1 and MODIS for LFMC retrieval, the estimated value is higher than the measured value, and the linear fitting is good when the LFMC is low. With the increase in LFMC, the estimated value is lower than the actual value, the points are discrete, and the overall correlation is high [11]. Figure 6 shows the LFMC retrieval results and measured values based on two ensemble learning models combined with MODIS, Landsat-8 and Sentinel-1. We can see that two proposed ensemble learning models underestimated high LFMC values (>120%), and there was a systematic bias for phenological periods with high LFMC values. This can be partly explained by the limited sensitivity of the optical sensing data to wet vegetation and the tendency of the proposed method to globally optimize the solution at the cost of underestimation at high values. Similar underestimations have been observed in other studies using physical or data-driven methods [45]. However, such underestimation is not significant when considering the cause of the fire hazard or behavior [46]. Experience has shown that when LFMC is high (>120%), the probability of fire occurrence is comparatively low, or fire movement through this area is limited, so this has less of an impact on fire managers, who might use this model to assess LFMC.

The proposed models also overestimated low values (<30%), which may have been due to the presence of dead combustibles, such as grass fuel [47] and leaf litter. Nevertheless, the magnitude of the positive bias was very small (as shown in Figure 6). Moreover, when the LFMC value is lower than 60%, the likelihood of fire occurrence increasing dramatically [48]. So, if the LFMC value is less than 30%, fire managers will be more aggressive with the estimated results. The impact introduced by the minor error on fire managers who may use the model is limited to the extent that this is overestimated in the range (<30%).

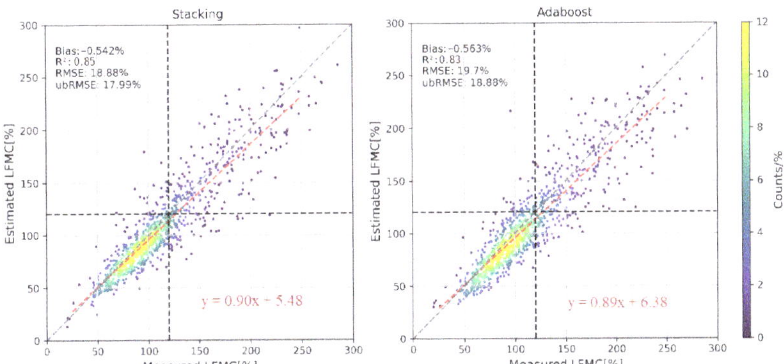

Figure 6. LFMC retrieval results based on MODIS, Landsat-8 and Sentinel-1. The gray dotted line and the black dotted line represent the 1:1 line and the 120% fire risk based on level of moisture, respectively, and the red dotted line is the fitting line of the model retrieval results.

4.2. Advantages of Multi-Source Remote Sensing Data and Ensemble Learning

Due to the different shortcomings of different remote sensing data, as expected, the LFMC retrieval results with all the Landsat-8, Sentinel-1 and MODIS remote sensing data are much higher than those of other data when using the same model, which can be attributed to the fact that multi-source remote sensing data can reduce the uncertainty of single-source data and provide more valuable features derived from the complementarity of different data.

Furthermore, the ensemble learning method comprises several basic learners together to obtain better performance. The Adaboost ensemble model is a sequential ensemble technique, in which the final prediction is based on the weighted average results of three weak learners (TCN-LSTM) trained on different training subsets, while the stacking ensemble model combines three parallel basic learners (LSTM, TCN and TCN-LSTM) in the first layer to extract abundant features, and then concatenates straightforward logistic regression as the second learner. Three different basic learners combined with sequential concatenation operation produced better features and an improved retrieval performance over the Adaboost ensemble model.

To summarize, the combination of multi-source data fusion and ensemble learning can significantly improve retrieval performance and provide considerable potential for accurate LFMC estimation.

4.3. Limitations of the Proposed Method with Processed Data

As we all know, the estimation of LFMC using remote sensing data (such as optical and microwave data) has the same issue since remote sensing data are dependent not only on LFMC but also on other bio- and geophysical characteristics [26]. Previous studies tried to find the empirical relationships or physical models between LFMC and other factors. Despite the satisfactory results of these methods, carefully handcrafted input variables chosen based on our understanding of radiative transfer processes must be selected; additionally, corresponding field data are needed, making these models challenging to generalize and operationalize on large-scale sites.

In our work, we introduced deep learning to capture the complicated nonlinear relationship among the LFMC and the remote sensing data, hoping to avoid the selection of carefully handcrafted input variables and the collection of corresponding field data, making it easier to realize large-scale LFMC estimation. The results demonstrate that this method performs admirably in large-scale sites (133 sampling points) with diverse vegetations, while during the data processing, considering the time resolution of remote sensing data and the frequency of measured LFMC, we interpolated the data to the end of each month, which means the time resolution was one month, resulting in the misrepresentation of daily or weekly fluctuations in LFMC. However, this limitation would be solved by gathering data with a smaller resolution.

Furthermore, we are all aware that the kind of vegetation has a direct impact on various remote sensing data [19,26]. Here, we simply delegated the task of classifying vegetation types implicitly to the deep learning model. Figure 7 presents the RMSEs of LFMC retrieval results of different vegetations, demonstrating that the RMSEs of four single-vegetations are often lower than that of mixed vegetations. In particular, the worst predictions were made for mixed shrub–grassland. This suggests that our previous strategy of selecting sample points makes sense, and that using the selected samples to train the model is beneficial in improving the accuracy of the predictions. Nevertheless, a fully data-driven model would very likely result in mistakes if detailed vegetation distribution data were not included. In practical application, a feasible option is to collect more data and then create more advanced models.

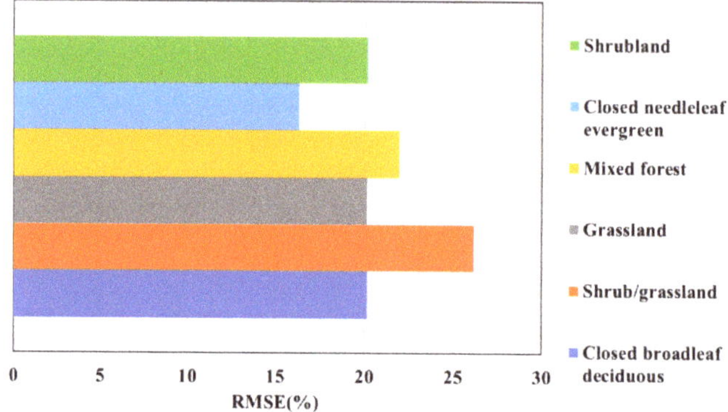

Figure 7. RMSEs of LFMC retrieval results for different vegetations.

Finally, while it is widely acknowledged that deep learning is a data-driven nonlinear model with high automated learning and generalization capabilities that have the potential to be applied to other regions, the efficacy of its application in other locations requires more data for validation.

5. Conclusions

In this study, a TCN-LSTM model was firstly designed to improve the effect of feature extraction, and further, two ensemble models were proposed based on the TCN-LSTM model to achieve more accurate retrieval of LFMC. Considering the different shortcomings of separate Landsat-8, Sentinel-1 and MODIS remote sensing data, all the three data were utilized together to obtain higher performance. The results of the experiments on the LFMC data from the Western United States show that the stacking ensemble model with all three remote sensing data achieved the best performance. The proposed stacking ensemble model was trained on historical data, which can automatically extract the nonlinear correlation between remote sensing data and LFMC. This enabled the proposed model's good generalization ability. Our model is data-driven, which means it has the potential to realize significant accuracy in LFMC estimation for other locations with appropriate training data. Meanwhile, our results reveal that our proposed models had a mixture of predictions with low and high amounts of bias. We believe this is because different vegetations are not explicitly considered. We will study and improve our model on more available data in the future.

Author Contributions: T.Q. analyzed the data and wrote the manuscript. W.H. checked and revised the manuscript. J.X., J.Z., H.H. and B.C. provided guidance for experiments and revised the manuscript. All authors have read and agreed to the published version of the manuscript.

Funding: This work was jointly supported by the National Key Research and Development Program of China (No. 2020YFC1511601), the Beijing Municipal Natural Science Foundation (No. 6214040), the Fundamental Research Funds for the Central Universities (No. 2021ZY70) and China Scholarship Council (202106515010).

Data Availability Statement: Not applicable.

Acknowledgments: We would like to express our gratitude to the National Key Research and Development Program of China (no. 2020YFC1511601), the Beijing Municipal Natural Science Foundation (no. 6214040), the Fundamental Research Funds for the Central Universities (no. 2021ZY70) and China Scholarship Council (202106515010).

Conflicts of Interest: The authors declare no conflict of interest.

References

1. Cunill Camprubí, À.; González-Moreno, P.; Resco de Dios, V. Live fuel moisture content mapping in the Mediterranean Basin using random forests and combining MODIS spectral and thermal data. *Remote Sens.* **2022**, *14*, 3162. [CrossRef]
2. Chuvieco, E.; Gonzalez, I.; Verdu, F.; Aguado, I.; Yebra, M. Prediction of fire occurrence from live fuel moisture content measurements in a Mediterranean ecosystem. *Int. J. Wildland Fire* **2009**, *18*, 430–441. [CrossRef]
3. Nolan, R.H.; Boer, M.M.; de Dios, V.R.; Caccamo, G.; Bradstock, R.A. Large-scale, dynamic transformations in fuel moisture drive wildfire activity across southeastern Australia. *Geophys. Res. Lett.* **2016**, *43*, 4229–4238. [CrossRef]
4. Chuvieco, E.; Aguado, I.; Jurdao, S.; Pettinari, M.L.; Yebra, M.; Salas, J.; Hantson, S.; de la Riva, J.; Ibarra, P.; Rodrigues, M.; et al. Integrating geospatial information into fire risk assessment. *Int. J. Wildland Fire* **2014**, *23*, 606–619. [CrossRef]
5. Sibanda, M.; Onisimo, M.; Dube, T.; Mabhaudhi, T. Quantitative assessment of grassland foliar moisture parameters as an inference on rangeland condition in the mesic rangelands of southern Africa. *Int. J. Remote Sens.* **2021**, *42*, 1474–1491. [CrossRef]
6. Yebra, M.; Dennison, P.E.; Chuvieco, E.; Riaño, D.; Zylstra, P.; Hunt, E.R., Jr.; Danson, F.M.; Qi, Y.; Jurdao, S. A global review of remote sensing of live fuel moisture content for fire danger assessment: Moving towards operational products. *Remote Sens. Environ.* **2013**, *136*, 455–468. [CrossRef]
7. Garcia, M.; Chuvieco, E.; Nieto, H.; Aguado, I. Combining AVHRR and meteorological data for estimating live fuel moisture content. *Remote Sens. Environ.* **2008**, *112*, 3618–3627. [CrossRef]
8. García, M.; Riaño, D.; Yebra, M.; Salas, J.; Cardil, A.; Monedero, S.; Ramirez, J.; Martín, M.P.; Vilar, L.; Gajardo, J.; et al. A live fuel moisture content product from Landsat TM satellite time series for implementation in fire behavior models. *Remote Sens.* **2020**, *12*, 1714. [CrossRef]
9. Myoung, B.; Kim, S.H.; Nghiem, S.V.; Jia, S.; Whitney, K.; Kafatos, M.C. Estimating live fuel moisture from MODIS satellite data for wildfire danger assessment in Southern California USA. *Remote Sens.* **2018**, *10*, 87. [CrossRef]
10. Maffei, C.; Menenti, M. A MODIS-based perpendicular moisture index to retrieve leaf moisture content of forest canopies. *Int. J. Remote Sens.* **2014**, *35*, 1829–1845. [CrossRef]
11. Quan, X.; He, B.; Yebra, M.; Yin, C.; Liao, Z.; Li, X. Retrieval of forest fuel moisture content using a coupled radiative transfer model. *Environ. Model. Softw.* **2017**, *95*, 290–302. [CrossRef]
12. Yebra, M.; Van Dijk, A.; Leuning, R.; Huete, A.; Guerschman, J.P. Evaluation of optical remote sensing to estimate actual evapotranspiration and canopy conductance. *Remote Sens. Environ.* **2013**, *129*, 250–261. [CrossRef]
13. Song, C.H. Optical remote sensing of forest leaf area index and biomass. *Prog. Phys. Geogr. -Earth Environ.* **2013**, *37*, 98–113. [CrossRef]
14. Quan, X.; He, B.; Li, X.; Liao, Z. Retrieval of Grassland live fuel moisture content by parameterizing radiative transfer model with interval estimated LAI. *IEEE J. Sel. Top. Appl. Earth Obs. Remote Sens.* **2016**, *9*, 910–920. [CrossRef]
15. Xiao, Z.; Song, J.; Yang, H.; Sun, R.; Li, J. A 250 m resolution global leaf area index product derived from MODIS surface reflectance data. *Int. J. Remote Sens.* **2022**, *43*, 1409–1429. [CrossRef]
16. Al-Moustafa, T.; Armitage, R.P.; Danson, F.M. Mapping fuel moisture content in upland vegetation using airborne hyperspectral imagery. *Remote Sens. Environ.* **2012**, *127*, 74–83. [CrossRef]
17. Houborg, R.; Anderson, M.; Daughtry, C. Utility of an image-based canopy reflectance modeling tool for remote estimation of LAI and leaf chlorophyll content at the field scale. *Remote Sens. Environ.* **2009**, *113*, 259–274. [CrossRef]
18. Fan, L.; Wigneron, J.P.; Xiao, Q.; Al-Yaari, A.; Wen, J.; Martin-StPaul, N.; Dupuy, J.-L.; Pimont, F.; Al Bitar, A.; Fernandez-Moran, R.; et al. Evaluation of microwave remote sensing for monitoring live fuel moisture content in the Mediterranean region. *Remote Sens. Environ.* **2018**, *205*, 210–223. [CrossRef]
19. Rao, K.; Williams, A.P.; Flefil, J.F.; Konings, A.G. SAR-enhanced mapping of live fuel moisture content. *Remote Sens. Environ.* **2020**, *245*, 111797. [CrossRef]

20. Wang, L.; Quan, X.W.; He, B.B.; Yebra, M.; Xing, M.; Liu, X. Assessment of the dual polarimetric Sentinel-1A data for forest fuel moisture content estimation. *Remote Sens.* **2019**, *11*, 1568. [CrossRef]
21. Bai, X.; He, B.; Li, X.; Zeng, J.; Wang, X.; Wang, Z.; Zeng, Y.; Su, Z. First assessment of Sentinel-1A data for surface soil moisture estimations using a coupled water cloud model and advanced integral equation model over the Tibetan Plateau. *Remote Sens.* **2017**, *9*, 714. [CrossRef]
22. Bai, X.; He, B. Potential of Dubois model for soil moisture retrieval in prairie areas using SAR and optical data. *Int. J. Remote Sens.* **2015**, *36*, 5737–5753. [CrossRef]
23. Jiao, W.; Wang, L.; McCabe, M.F. Multi-sensor remote sensing for drought characterization: Current status, opportunities and a roadmap for the future. *Remote Sens. Environ.* **2021**, *256*, 112313. [CrossRef]
24. Reichstein, M.; Camps-Valls, G.; Stevens, B.; Jung, M.; Denzler, J.; Carvalhais, N.; Prabhat. Deep learning and process understanding for data-driven Earth system science. *Nature* **2019**, *566*, 195–204. [CrossRef]
25. Yuan, Q.Q.; Shen, H.F.; Li, T.W.; Li, Z.; Li, S.; Jiang, Y.; Xu, H.; Tan, W.; Yang, Q.; Wang, J.; et al. Deep learning in environmental remote sensing: Achievements and challenges. *Remote Sens. Environ.* **2020**, *241*, 111716. [CrossRef]
26. Zhu, L.; Webb, G.I.; Yebra, M.; Scortechini, G.; Miller, L.; Petitjean, F. Live fuel moisture content estimation from MODIS: A deep learning approach. *ISPRS J. Photogramm. Remote Sens.* **2021**, *179*, 81–91. [CrossRef]
27. United States Forest Sevices. *National Fuel Moisture Database*; United States Forest Services: Washington, DC, USA, 2010.
28. Nietupski, T.C.; Kennedy, R.E.; Temesgen, H.; Kerns, B.K. Spatiotemporal image fusion in Google Earth Engine for annual estimates of land surface phenology in a heterogenous landscape. *Int. J. Appl. Earth Obs. Geoinf.* **2021**, *99*, 102323. [CrossRef]
29. Faiz, M.A.; Liu, D.; Tahir, A.A.; Li, H.; Fu, Q.; Adnan, M.; Zhang, L.; Naz, F. Comprehensive evaluation of 0.25° precipitation datasets combined with MOD10A2 snow cover data in the ice-dominated river basins of Pakistan. *Atmos. Res.* **2020**, *231*, 104653. [CrossRef]
30. Vermote, E.; Justice, C.; Claverie, M.; Franch, B. Preliminary analysis of the performance of the Landsat 8/OLI land surface reflectance product. *Remote Sens. Environ.* **2016**, *185*, 46–56. [CrossRef]
31. Roberts, D.A.; Dennison, P.E.; Peterson, S.; Sweeney, S.; Rechel, J. Evaluation of airborne visible/infrared imaging spectrometer (AVIRIS) and moderate resolution imaging spectrometer (MODIS) measures of live fuel moisture and fuel condition in a shrubland ecosystem in southern California. *J. Geophys. Res. -Biogeosci.* **2006**, *111*, 1–16. [CrossRef]
32. Colombo, R.; Merom, M.; Marchesi, A.; Busetto, L.; Rossini, M.; Giardino, C.; Panigada, C. Estimation of leaf and canopy water content in poplar plantations by means of hyperspectral indices and inverse modeling. *Remote Sens. Environ.* **2008**, *112*, 1820–1834. [CrossRef]
33. Chuvieco, E.; Aguado, I.; Dimitrakopoulos, A.P. Conversion of fuel moisture content values to ignition potential for integrated fire danger assessment. *Can. J. For. Res.* **2004**, *34*, 2284–2293. [CrossRef]
34. Badgley, G.; Field, C.B.; Berry, J.A. Canopy near-infrared reflectance and terrestrial photosynthesis. *Sci. Adv.* **2017**, *3*, e1602244. [CrossRef]
35. Torres, R.; Snoeij, P.; Geudtner, D.; Bibby, D.; Davidson, M.; Attema, E.; Potin, P.; Rommen, B.; Floury, N.; Brown, M.; et al. GMES Sentinel-1 mission. *Remote Sens. Environ.* **2012**, *120*, 9–24. [CrossRef]
36. Keane, R.E. Fuel concepts. In *Wildland Fuel Fundamentals and Applications*; Springer: Cham, Switzerland, 2015; pp. 175–184
37. Brocca, L.; Crow, W.T.; Ciabatta, L.; Massari, C.; de Rosnay, P.; Enenkel, M.; Hahn, S.; Amarnath, G.; Camici, S.; Tarpanelli, A.; et al. A review of the applications of ASCAT soil moisture products. *IEEE J. Sel. Top. Appl. Earth Obs. Remote Sens.* **2017**, *10*, 2285–2306. [CrossRef]
38. Datta, S.; Das, P.; Dutta, D.; Giri, R.K. Estimation of surface moisture content using Sentinel-1 C-band SAR data through machine learning models. *J. Indian Soc. Remote Sens.* **2020**, *49*, 887–896. [CrossRef]
39. Liu, S.; Wei, Y.; Post, W.M.; Cook, R.B.; Schaefer, K.; Thornton, M.M. *NACP MsTMIP: Unified North American Soil Map*; ORNL Distributed Active Archive Center: Oak Ridge, TN, USA; p. 2014.
40. Simard, M.; Pinto, N.; Fisher, J.B.; Baccini, A. Mapping forest canopy height globally with spaceborne lidar. *J. Geophys. Res. -Biogeosci.* **2011**, *116*, G04021. [CrossRef]
41. Arino, O.; Ramos, J.; Kalogirou, V.; Bontemps, S.; Defourny, P.; Van Bogaert, E. *Global Land Cover Map for 2009 (GlobCover 2009)*; European Space Agency (ESA): Paris, France; Université catholique de Louvain (UCL), PANGAEA: Ottignies-Louvain-la-Neuve, Belgium. [CrossRef]
42. USGS-NED. *National Elevation Dataset*; US Geological Survey: Lafayette, LA, USA, 2004.
43. Pelletier, C.; Webb, G.I.; Petitjean, F. Temporal convolutional neural network for the classification of satellite image time series. *Remote Sens.* **2019**, *11*, 523. [CrossRef]
44. Jia, S.; Kim, S.H.; Nghiem, S.V.; Kafatos, M. Estimating live fuel moisture using SMAP L-band radiometer soil moisture for Southern California, USA. *Remote Sens.* **2019**, *11*, 1575. [CrossRef]
45. Yebra, M.; Quan, X.; Riano, D.; Larraondo, P.R.; van Dijk, A.I.J.M.; Cary, G.J. A fuel moisture content and flammability monitoring methodology for continental Australia based on optical remote sensing. *Remote Sens. Environ. Interdiscip. J.* **2018**, *212*, 260–272. [CrossRef]
46. Pimont, F.; Ruffault, J.; Martin-StPaul, N.K.; Dupuy, J.-L. Why is the effect of live fuel moisture content on fire rate of spread underestimated in field experiments in shrublands? *Int. J. Wildland Fire* **2019**, *28*, 127–137. [CrossRef]

47. Yebra, M.; Scortechini, G.; Badi, A.; Beget, M.E.; Boer, M.M.; Bradstock, R.; Chuvieco, E.; Danson, F.M.; Dennison, P.; de Dios, V.R.; et al. Globe-LFMC, a global plant water status database for vegetation ecophysiology and wildfire applications. *Sci. Data* **2019**, *6*, e155. [CrossRef] [PubMed]
48. Jurdao, S.; Chuvieco, E.; Arevalillo, J.M. Modelling fire ignition probability from satellite estimates of live fuel moisture content. *Fire Ecol.* **2012**, *8*, 77–97. [CrossRef]

Article

An Empirical Study on Retinex Methods for Low-Light Image Enhancement

Muhammad Tahir Rasheed [1,†], **Guiyu Guo** [1,†], **Daming Shi** [1,*], **Hufsa Khan** [1] and **Xiaochun Cheng** [2]

1   College of Computer Science and Software Engineering, Shenzhen University, Shenzhen 518060, China
2   Computer Science Department, Middlesex University, Hendon, London NW4 4BT, UK
*   Correspondence: dshi@szu.edu.cn
†   These authors contributed equally to this work.

Abstract: A key part of interpreting, visualizing, and monitoring the surface conditions of remote-sensing images is enhancing the quality of low-light images. It aims to produce higher contrast, noise-suppressed, and better quality images from the low-light version. Recently, Retinex theory-based enhancement methods have gained a lot of attention because of their robustness. In this study, Retinex-based low-light enhancement methods are compared to other state-of-the-art low-light enhancement methods to determine their generalization ability and computational costs. Different commonly used test datasets covering different content and lighting conditions are used to compare the robustness of Retinex-based methods and other low-light enhancement techniques. Different evaluation metrics are used to compare the results, and an average ranking system is suggested to rank the enhancement methods.

Keywords: low-light image enhancement; retinex theory; deep learning; remote-sensing

1. Introduction

Citation: Rasheed, M.T.; Guo, G.; Shi, D.; Khan, H.; Cheng, X. An Empirical Study on Retinex Methods for Low-Light Image Enhancement. *Remote Sens.* **2022**, *11*, 1608. https://doi.org/10.3390/rs14184608

Academic Editor: Gwanggil Jeon

Received: 7 August 2022
Accepted: 11 September 2022
Published: 15 September 2022

Low-light enhancement methodologies try to recover buried details, remove the noise, restore the color details, and increase the dynamic range and contrast of the low-light images. Low light has inescapable effects on remote monitoring equipment and computer vision tasks. Low signal-to-noise ratio (SNR) causes severe noise in low-light imaging and makes it difficult to extract features for interpreting remote-sensing via computer vision tasks, whereas the performance of computer vision tasks entirely depends on accurate feature extraction [1]. Remote-sensing image enhancement has a wide range of applications in object detection [2,3], object tracking [4–7], video surveillance [8,9], military applications, daily life [10–14], atmospheric sciences [15], driver assistance systems [16], and agriculture. Earth is continuously being monitored by analyzing the images taken by satellites. Analyzing remotely taken images to help in fire detection, flood prediction, and understanding other environmental issues. Low-light enhancement of these images is playing a vital role in understanding these images in a better way. Even the accuracy of other remote sensing algorithms, such as classification and object detection, depends entirely on the image's quality. In the literature, different methodologies exist for enhancing such degraded low-light images. Retinex theory-based enhancement methods are widely accepted among these enhancement methodologies due to their robustness. The main purpose of this study is to compare the Retinex-based methods with other non-Retinex-based enhancement methods experimentally. For comparison, we have categorized all the enhancement methods into two major groups (i.e., Retinex-based and non-Retinex-based methods). The Retinex group includes classical and deep learning-based Retinex enhancement methods. Meanwhile, the non Retinex group includes histogram equalization, gamma correction, fusion, and deep learning-based enhancement methods.

According to Retinex theory [17], an image can be decomposed into reflectance and illumination component. The reflectance component is considered an intrinsic component of the image and remains consistent under any lighting condition, whereas the illumination component represents the different lighting conditions. Later on, different Retinex theory based methods, such as single-scale retinex (SSR), [18] multiscale retinex with color restoration (MSRCR) [19], simultaneous reflectance and illumination estimation (SRIE) [20], and low-light illumination map estimation (LIME) [21] were developed for low-light enhancement. These methods produce promising results but may require fine-tuning of parameters and may fail to decompose the image correctly into reflectance and illumination parts. Wei et al. is the first one to introduce a real low/normal-light LOw-Light (LOL) dataset and Retinex theory-based deep network (Retinex-Net) in [22]. Retinex-Net comprises Decom-Net for decomposing the image into reflectance and illumination parts and an Enhance-Net for illumination adjustment. Later on, different Retinex theory-based deep learning methods were developed for low-light image enhancement algorithm [22–25].

Non-Retinex method such as histogram equalization is one of the simplest methods for enhancing low-light images. It flattens the distribution of pixel values throughout the image to improve contrast. In addition, using entire histogram information may over brighten some regions of the image, deteriorate its visual quality and introduce some artifacts in it. Different histogram-based methods such as local histogram equalization [26] and dynamic histogram equalization [27] were introduced to address these issues. However, these methods require higher computation power, the quality of the output depends on the fine-tuning of parameters, and in case of severe noise, it may produce artifacts. On the other hand, gamma correction based methods [28–30] apply the pixel-wise nonlinear operation to enhance the image. The main drawback of these methods is that each pixel is considered an individual entity, and their relationship with neighbor pixels is entirely ignored. Due to this, the output may be inconsistent with real scenes. Lore et al. [31] is the first to propose a learning-based enhancement network named LLNet using a synthetic dataset. Later on, different low-light training datasets (e.g., LOL [22], SID [32], SICE [33], VV (https://sites.google.com/site/vonikakis/datasets (accessed on 7 July 2021)), TM-DIED (https://sites.google.com/site/vonikakis/datasets (accessed on 7 July 2021)), and LLVIP [34]) were developed in order to assist the development of learning-based architectures [35–38].

Wang et al. [39] present a technical evaluation of different methods for low-light imaging. Most of the methods reviewed are classical, and comparing evaluations on five images is quite unfair. Later on, Qi et al. , in [40], provide an overview of low-light enhancement techniques, whereas the quantitative analysis of a few methods only on the synthetic dataset (without noise) is provided. Noise is the most critical part of low-light enhancement and a single synthetic low-light dataset cannot compare performance. In [41], Li et al. propose a low-light image and video dataset to examine the generalization of existing deep learning-based image and video enhancement methods. In sum, low-light enhancement has a wide range of applications and is one of the most important image processing fields. To the best of our knowledge, no such study paper is present in the literature mentioned above that extensively provides the technical evaluation of low-light enhancement methods.

The main purpose of this research is to fairly compare the performance of Retinex-based enhancement methods with non-Retinex enhancement methods on a wide range of test datasets covering different contents and lighting conditions. For a fair comparison, the experimental evaluation criteria are defined first, and then all the methods are compared based on the criteria. In addition, an average ranking system is suggested to rank the enhancement methods based on their robustness. Computational complexity analysis of methods is also carried out on four different image sizes for real-time application. This experimental comparison and suggested ranking system of enhancement methods help the research community to understand their shortcomings and to design more robust models in the future.

The main contribution of this research can be summarized as follows:

- A comprehensive literature review is presented for Retinex-based and non-Retinex methods.
- A detailed experimental analysis is provided for a variety of Retinex-based and non-Retinex methods on a variety of publicly available test datasets using well-known image quality assessment metrics. Experimental results provide a holistic view of this field and provide readers with an understanding of the advantages and disadvantages of existing methodologies. In addition, the inconsistency of commonly used evaluation metrics is pointed out.
- An analysis of the computational effectiveness of enhancement methods is also conducted on images of different sizes. As a result of this computation cost, we can determine which enhancement methods are more suitable for real-time applications.
- Publicly available low-light test datasets were ranked based on experimental analysis. In developing more robust enhancement methods, the reader will benefit from this ranking of benchmarking test datasets.

The rest of the paper is organized as follows. Section 2 presents the relevant background knowledge of non-Retinex-based, and Retinex-based classical and advanced low-light enhancement methodologies. Section 3 presents the objectives of overall paper. In Section 4, experimental setup is defined, a detailed discussion of the qualitative, quantitative, and computational analysis of the classical and advanced low-light enhancement methodologies are provided. Section 5, reports the challenges and the future trends. Finally, the conclusion is drawn in Section 6.

2. Fundamentals

A thorough review of the literature related to Retinex-based and non-Retinex-based classical and advanced learning-based low-light enhancement methods is presented in this section. The following subsections contain literature on each of the categories mentioned above.

2.1. Retinex-Based Methods

Classical Retinex-based methods: The Retinex theory was developed by Land after he studied the human retina-and-cortex system in detail [17]. According to the presented theory, an image can be decomposed into two parts: reflectance and illumination. Reflectance is considered an intrinsic property and remains the same regardless of the lighting condition. Illumination is determined by the intensity of light. The following representation can be used to explain it:

$$S(x,y) = R(x,y) \circ I(x,y), \tag{1}$$

where S, R and I represent the source image, reflectance and illumination, respectively and the operator $\circ$ denotes the element-wise multiplication between R and I. As time progressed, different implementations of Retinex theory were proposed in the literature. Path-based implementation of the Reinex [42–47] uses different geometry to calculate the relative brightness of adjacent pixels to obtain the reflection component. Marini and Rizzi proposed a biologically inspired implementation of Retinex for dynamic adjustment and color constancy in their article [45]. In [44], the authors examine the different path-wise approaches in detail and propose a mathematical formula to analyze these approaches. It is worth noting that the number of paths has a significant impact on the accuracy of the results. As a result, these path-wise implementations of Retinex theory suffer from a high degree of dependency on the path and sampling noise, as well as a high cost of computation when fine-tuning parameters.

The new method, random spray Retinex (RSR), was developed by Provenzi after replacing the paths with 2-D pixels sprays in [48]. When paths are replaced with 2-D random points distributed across the image, it is possible to determine the locality of color perception. Even though this approach is faster, the spray radius, radial density function, number of sprays, and pixels per spray must be adjusted. Jobson et al. , in [18], used a single-scale Retinex (SSR) to implement Retinex for color constancy, and lightness and color rendition of grayscale images. It is not possible for the SSR to provide both dynamic range compression (small scale) and tonal rendition (large scale) simultaneously. However, it can only perform one of these tasks. Later, the authors of SSR extended their idea to multiscale retinex with color restoration (MSRCR) [19]. As a result of MSRCR, dynamic range compression, color consistency, and tonal rendition can be provided. SSR and MSRCR both improve lighting and scene restoration for digital images, but halo artifacts are visible near edges [49]. The majority of Retinex-based algorithms ignore the illumination component and only extract the reflection component as an enhanced result, but this results in unnaturalness. Enhancing an image is not just about enhancing details but also about maintaining its natural appearance. To solve the unnatural appearance, Wang et al. [50] make three contributions: (1) lightness-order-error metrics are proposed to measure objective quality, (2) bright-pass filters decompose images into reflectance and illumination, and (3) bi-log transformations to map illumination while maintaining the balance between details and naturalness.

Zosso et al. reviewed Retinex-based methods and classified them into five broad categories in [51]. Additionally, a two-step non-local unifying framework is proposed to enhance the results and address the Retinex problem. In the first step, a quasi gradient filter is obtained which satisfies gradient-sparsity and gradient-fidelity prior constraints. As a second step, additional constraints are applied to the calculated quasi-gradient filter in order to make it fit the reflectance data. Guo et al. devised a method named low-light illumination map estimation (LIME) [21] to estimate the illumination of each pixel first; then, apply a structure to that illumination map and use it as the final illumination map. A variational based framework (VF) was introduced for Retinex for the first time by Kimmel et al. [52]. In accordance with previous methods, the objection function is based on the assumption that the illumination field is smooth. On the other hand, this model lacks information regarding reflectance. Later on, different variational approaches to Retinex theory are presented [53–55]. In [56], a total variational model (TVM) for Retinex is proposed, assuming spatial smoothness of illumination and piecewise continuity of reflection. In order to minimize TVM, a split Bregman iteration is used. VF and TVM differ primarily in that TVM also takes into account reflection.

Fu et al. proposed a linear domain probabilistic method for simultaneous illumination and reflectance estimation (PM-SIRE) [49]. By using an alternating direction multiplier method, maximum a posteriori (MAP) is employed to estimate illumination and reflectance effectively. Later, Fu et al. presented a weighted variational model for simultaneous illumination and reflectance estimation (WV-SIRE) [20]. A WV-SIRE model is capable of preserving more details about the estimated reflectance as well as suppressing noise more effectively than a log-transformed model. The PM-SIRE and WV-SIRE both assume that illumination changes smoothly over time, which may lead to incorrect illumination estimation. Based on the luminous source, different surfaces are illuminated in different directions.

A fusion-based method for enhancing weakly illuminated images is proposed in [57]. This fusion method decomposes a weakly illuminated image into a reflectance map and an illumination map. By using sigmoid and adaptive histogram equalization functions, the illumination map is further decomposed into luminance-improved and contrast-enhanced versions, and two weights are designed for each. Finally, an enhanced image is obtained by combining the luminance-improved and contrast-improved versions with their corresponding weights in a multi-scale manner. For the purpose of preserving intrinsic and extrinsic priors, Cai et al. proposed a joint intrinsic-extrinsic prior (JieP) model [58]. In JieP, shape prior is used to preserve structure information, texture prior is used to estimate illumination with fine details, and illumination prior is used to capture luminous information. Ying et al. [59] simulate the camera response model (CRM) by investigating the relationship between two different exposure images and use the illumination estimation to estimate the exposure ratio map. Later, the CRM and exposure ratio map are used to produce the enhanced image. According to the CRM algorithm, some dark parts of the body, such as the hair, are misinterpreted as dark backgrounds, and they are over-enhanced as well.

Advanced Retinex-based methods: The robustness of Retinex theory makes it applicable to deep learning methods as well. Wei et al. were the first to combine the idea of Retinex theory with deep learning by proposing the Retinex-Net network. Retinex-Net consists of a Decom-Net for decomposing the image into reflectance and illumination parts and an Enhance-Net for adjusting illumination. Furthermore, they introduce a real low/normal-light Low-Light (LOL) dataset [22]. As a further development of the Retinex theory, Zhang et al. proposed the kindling the darkness (KinD) network in [36]. There are three components of KinD: layer decomposition, reflectance restoration, and illumination adjustment. As a result of layer decomposition, the input image is divided into reflectance and illumination elements, the reflectance part is improved by reflecting restoration and the illumination part is smoothed piece-by-piece by illumination adjustment. By combining the outputs of the reflectance and illumination modules, the final result is achieved. Artifacts, overexposure, and uneven lighting are common problems with KinD outputs. For mitigating these effects, Zhange et al. proposed an improved version of KinD in [60]. This improved version of KinD implements a multi-scale illumination attention module, known as KinD++. KinD++ has improved the quality of output images, but it has a lower computational efficiency than KinD. In [61], a Retinex-based real-low to real-normal network (R2RNet) was proposed. R2RNet consists of a decomposition network, a denoise network, and a relight network, each of which is trained separately using decomposition loss, denoise loss, and relight loss, respectively. As a result of decomposition, illumination and reflectance maps are produced. The denoise-net uses the illumination map as a constraint to reduce the noise in the reflectance map, and the relight-net utilizes the denoised illumination map and reflectance map in order to produce an enhanced output. It is noteworthy that three separately trained networks are utilized to solve the low-light enhancement problem, which is not an optimal strategy. Decomposing an image into illumination and reflectance is a computationally inefficient process. Retinex-based transfer functions were introduced by Lu and Zhange in [23] to solve this decomposition problem. As opposed to decomposing the image, the network learn the transfer function to obtain the enhanced image. Liu et al. [62] introduces reference free Retinex-inspired unrolling with architecture search (RUAS) to reduce computational burden and construct lightweight yet effective enhancement. First, RAUS exploits the intrinsic underexposed structure of low-light images; then, it unrolls the optimization process to establish a holistic propagation model. Wang et al. [63] presents paired seeing dynamic scene in the dark (SDSD) datasets. A self-supervised end-to-end framework based on Retinex is also proposed in order to simultaneously reduce noise and enhance illumination. This framework consists of modules for progressive alignment, self-supervised noise estimation, and illumination map prediction. With progressive alignment, temporal information is utilized to produce blur-free frames, self-supervised noise estimation estimates noise from aligned feature maps of the progressive module, and illumination estimation estimates illumination maps consistent with frame content.

Retinex theory is also used in semi-supervised and zero-shot learning-based techniques for enhancing low light visibility. In Zhang et al. [24], a self-supervised maximum entropy Retinex (ME-Retinex) model is presented. In the ME-Retinex model, a network for enhancing image contrast is coupled with a network for re-enhancing and denoising. Zhao et al. [64] proposed a zero-reference framework named RetinexDIP that draws inspiration from the concept of a deep image prior (DIP). The Retinex decomposition is carried out in a generative manner in RetinexDIP. From random noises as input, RetinexDIP generates both reflectance and illumination maps simultaneously, and enhances the illumination map resulting from this process. The proposed model generalizes well to various scenes, but producing an illumination map requires hundreds of iterations. This iterative learning approach consumes a lot of time to produce optimized results. The robust retinex decomposition network (RRDNet) is a three-branch zero-shot network that is proposed in RRDNet [25] to decompose low-light input images into illumination, reflectance, and noise. RRDNet weights are updated by a zero-shot scheme using a novel non-reference loss function. In the proposed loss function, there are three components: the first part reconstructs the image, the second part enhances the texture of the dark region, and the third part suppresses noise in the dark regions. Qu et al. , in [65], segmented an image into sub-images, applied deep reinforcement learning to learn the local exposure for each sub-image and finally adversarial learning is applied to approximate the global aesthetic function. It is also proposed to learn discriminators asynchronously and reuse them as value functions.

2.2. Non-Retinex Methods

Histogram equalization (HE) [66] is one of the earlier methods used for enhancing the dynamic range of low-light images. It is a well-known method due to its simplicity. When the entire image histogram is balanced, the visual quality of the image is deteriorated, false contours are introduced, and annoying artifacts are introduced into the image [67]. As a result, some uniform regions become saturated with very bright and very dark intensities [68]. Gamma correction [69] is a non-linear classical technique that is used for image enhancement. It increases the dark portion of the image while suppressing the bright portion. During gamma correction, each pixel is treated as an individual. It is possible that some regions of the image will be under- or over-enhanced due to a single transformation function used for each pixel.

In later years, deep learning has been applied to my field of study. Lore et al. [31] were the first one to use a stacked sparse based autoencoder approach called LLNet for joint enhancement and noise reduction. There is evidence that deeper networks perform better than non-deeper networks; however, deeper networks suffer from gradient vanishing problems. To use a deeper network and solve the gradient vanishing problem, Tao et al. in LLCNN [70] proposed a special module to utilize multiscale feature maps for low-light enhancement. A multi-branch low-light enhancement network (MBLLEN) is designed by Lv et al. in [71] to extract features of different levels, enhance these multi-level features, and fuse them in order to produce an enhanced image. Additionally, Lv et al. also propose a novel loss function that takes into account the structure information, context information, and regional differences of the image. Wang et al. , in [72], propose the global illumination-aware and detail-preserving network (GLADNet). In the first step, GLADNet uses an encoder-decoder network to estimate the global illumination and then reconstructs the details lost during the rescaling process. The major disadvantage of LLNet, LLCN, MBBLEN and GLADNet is that they were trained on synthetically darkened and noise-added datasets. Chen et al. [32] used a Unet based pipeline for enhancing and denoising extremely low-light images using the RAW training see-in-the-dark (SID) dataset. This Unet-based pipeline is designed specifically for images in RAW format. Practically, the most common image format is sRGB. The majority of previous methods have used pixel-wise reconstruction losses and failed to provide effective regularization of the local structure of the image, which in turn undermines the network's performance. The pixel-to-pixel deterministic mapping results in improperly exposed regions, introduces artifacts, and fails to describe the visual distance between the reference and the enhanced image. A flow-based low-light enhancement method (LLFlow) has been proposed by Wang et al. [38] to address this pixel-to-pixel mapping issue. It is possible to map multi-modal image manifolds into latent distributions using the normalizing flow. Effectively enhanced manifolds can be constructed using the latent distribution.

Getting low-light and normal-light images paired can be difficult, expensive, or impractical. An unpaired low-light enhancement method called EnlightenGAN is proposed by Jiang et al. [73] to eliminate the need for paired training datasets. A global-local discriminator structure and an easy-to-use attention U-net generator are proposed in EnlightenGAN. By designing the attention U-net only to enhance the dark regions more, the image is neither overexposed nor underexposed. A dual global-local discriminator strategy contributes to the balance between local and global enhancement of low-light images. Xiong et al. [74] considered low-light enhancements as two subtasks: illumination enhancement and noise reduction. A two-stage framework referred to as decoupled networks is proposed for handling each task. In decoupled networks, there are two encoder-decoder architectures, the first architecture enhances illumination, and the second architecture suppresses noise by taking the original input along with the enhanced output from stage one. To facilitate unsupervised learning, an adaptive content loss and pseudo triples are proposed. Xia et al. [75] used two images of the scene taken in quick succession (with and without a flash) to generate a noise-free and accurate display of ambient colors. Using a neural network, an image taken without flash is analyzed for color and mood, while an image taken with a flash is analyzed for surface texture and details. One of the major disadvantages of this method is that paired images with and without flash are not generally available.

The camera sensors on mobile phones perform poorly in low-light conditions. An improved face verification method using a semisupervised decomposition and reconstruction network is proposed in [76] to improve accuracy for low-light images of faces. Yang et al. [77] proposes a deep semi-supervised recursive band network (DRBN) to address the decreased visibility, intensive noise, and biased color of low-light images. DRBN learns in two stages, the first stage involves learning the linear band representation by comparing low- and normal-light images, and the second stage involves recomposing the linear band representation from the first stage to fit the visual properties of high-quality images through adversarial learning. Further improvement of the DRBN is impeded by the separation of supervised and unsupervised modules. Qiao et al. [78] further improved DRBN performance by introducing a joint training based semi-supervised algorithm. Wu et al. [79] proposed the lightweight two stream method to overcome the limitations of the training data due to sample bias and the hurdle of the large number of parameters in real-time deployment. Additionally, a self-supervised loss function is proposed to overcome the sample bias of the training data.

Guo et al. [80] proposes zero-reference deep curve estimation (Zero-DCE) rather than performing image-to-image mapping. In order to preserve the contrast of the neighboring pixels, Zero-DCE creates high-order curves from low-light images and then adjusts low-light images pixel-by-pixel using these high-order curves. It is superior to existing GAN-based methods since it does not require paired or unpaired data for its training. Enhanced images are produced with four non-reference loss functions: spatial consistency loss, exposure control loss, color constancy loss, and illumination smoothness loss. The re-design and reformulation of the network structure were subsequently carried out by Li et al. , who introduced Zero-DCE++, which is an accelerated and lighter version of Zero-DCE.

3. Objectives of Experimental Study

This research study aims to address the following questions:

1. It has been noted that although there have been a large number of algorithms developed for low-light enhancement, Retinex theory-based models are gaining more attention due to their robustness. Retinex theory is even used in deep learning-based models. Specifically, this paper attempts to compare the performance of Retinex theory-based classical and deep learning low-light enhancement models with other state-of-the-art models.

2. Several low-light enhancement methods perform well on some test datasets but fail in real-world scenarios. An extensive range of real-world images should be used to test the robustness of the low-light enhancement models. As a means of assessing the robustness of enhancement methods in real-world scenarios, various test datasets spanning a wide range of lighting conditions and contents need to be selected, and the performance of Retinue-based models needs to be compared with that of other enhancement techniques on these test datasets.

3. The trend of real-time cellphone night photography is increasing day by day. Therefore, analyzing the computational costs associated with low-light enhancement methods is necessary. Comparison of not only the parameters of these methods but also the processing time for the images of four different sizes (i.e., $400 \times 600 \times 3$, $640 \times 960 \times 3$,

2304 × 1728 × 3 and 2848 × 4256 × 3) is required. A computational analysis of different sizes of images will enable the researchers to determine whether the computational cost increases linearly or exponentially as the image size increases.

4. The quality of low-light enhancement methods needs to be evaluated using a variety of image quality assessment (IQA) methods. Every metric aims to identify the particular quality of the predicted image. The LOE measures the naturalness of the image, whereas the information entropy measures the information contained in the image. What is the most effective method of comparing the robustness of low-light enhancement methods when comparing results based on these evaluation metrics?

4. Quantitative and Qualitative Analysis

This subsections of this section present the experimental setup for farily comparing the methods, qualitative, quantitative comparison, and computational cost analysis of enhancement methods. In addition, it also discusses the evaluation metrics and test datasets.

4.1. Experimental Criteria for Enhancement Methods Comparison

To conduct a fair comparison to analyze the enhancement methods generalization, we have selected the nine different publicly available test datasets widely used in the literature for comparing the performance of enhancement methods [64,73,81]. The selected datasets include LIME [21], LOL [22], DICM [82], VV (https://sites.google.com/site/vonikakis/datasets (accessed on 7 July 2021)), MEF [83], NPE [50], LSRW [61], SLL [84] and ExDark [85]. The main purpose of selecting these different nine test datasets is to cover diversified scenes, camera devices, lighting conditions (i.e., weak lighting, under exposure, twilight, dark), and contents. In summary, each test dataset covers a different aspect of low-lighting, scene or content. Therefore, these test datasets are useful to compare the performance of enhancement methods from different aspects.

The four most commonly used no-reference metrics for the quantitative evaluation of low-light enhancement methods are used. These metrics include entropy [86], BRISQUE [87], NIQE [88], and LOE [50]. The entropy measures the information content of an image. a higher value of entropy indicates richer details and a higher contrast level of an image. Blind/referenceless image spatial quality evaluator (BRISQUE) is another commonly used model to quantify the quality of low-light enhancement methods. It does not compute the distortion specific feature, but instead it uses the scene statistics to quantify the loss of naturalness in an image due to the presence of distortion. BRISQIE uses a space vector machine (SVM) regressor to predict the quality of the image. Natural image quality evaluator (NIQE) quantifies the quality of the distorted image by measuring the distance of natural scene statistic (NSS) feature model and the multivariate Gaussian (MVG) feature model of distorted image. Lightness order error (LOE) is designed to measure the order of lightness. The order of lightness represents the direction of the light source and helps to quantify the naturalness preservation. LOE can be defined as follows:

$$LOE = \frac{1}{m * n} \sum_{x=1}^{m} \sum_{y=1}^{n} (U(Q(i,j), Q(x,y)) \oplus U(Q_r(i,j), Q_r(x,y))), \tag{2}$$

where $U(x,y)$ is a unit step function. It returns 1 if $x > y$ and returns 0 otherwise. m, n represents height and width of the image, respectively. Moreover, $Q(i,j)$ and $Q_r(i,j)$ are maximum values among the three color channels at location (i,j) for the original image and enhanced image, respectively.

In this study, the performance of 17 Retinex-based methods and 17 non-Retinex will be compared. We consider the publicly available codes and recommended settings of these methods to have a fair comparison. The higher value of entropy indicates better quality and for the other three methods (i.e., LOE, NIQE, and BRISQUE) lower values of entropy indicate the better image quality. To show a better understanding the comparison, average ranking has been suggested to enhancement methods based on these IQA methods. For example, the enhancement methods that got the highest average score of entropy on all test datasets are given rank 1 and vice versa. Similarly, the enhancement methods show the lowest average score according to LOE or NIQE or BRISQUE are assigned rank 1 and the highest average score is assigned the highest rank. Rank 1 indicates the best performance and the rank with higher value indicates the worst performance.

In addition, we compare the computational complexity of classical methods on images of four different sizes. The classical codes computational complexity is computed on CPU, whereas those of deep learning-based methods on NVIDIA Titan Xp GPU.

4.2. Qualitative Evaluation of Enhancement Methods

In this section, we provide a detailed description of the qualitative evaluation of enhancement methods. The comparative visual results of the top ten classical and advanced methods on six publicly available test datasets are shown in Figures 1 and 2, respectively. These figures' first to sixth columns indicate the enhancement results of different methods on LIME, LARW, DICM, ExDark, LOL, and SLL datasets, respectively. For simplicity, deep learning and classical methods are discussed one by one. It is encouraged to zoom in to compare the results.

Zero-shot learning-based methods (i.e., ZeroDCE and RetinexDIP) produce darker and noisy images compared to other methods. The results of GLADNet, TBEFN, and LLFlow are more realistic, sharper, less noisy, and have accurate color rendition. The output images of MBLLEN are over-smoothed and darker but less dark than ZeroDCE. GLADNet, TBEFN, LLFlow, MBLLEN, and KinD are trained on paired data. The supervised learning-based models achieved the appropriate restoration of color and textures, noise suppression, and better generalization. However, no method has produced good results on all the datasets. For example, GLADNet results on DICM are too noisy and produce artifacts on the ExDark image. Similarly, strange artifacts on DICM images are produced by TBEFN. LLFLow produces greenish color around the edges of LSRM image. As it can be seen, StableLLVE has a lighter washed-out effect and smoothed edges on all the results. KinD results look realistic, but some parts of the image look too dark, such as the background chairs in the LOL image. SS-Net produces a good result on the VV test image but produces poor results on DICM and ExDark. Moreover, the strange pattern, missing color information, and other details can be observed easily on the ExDark image. The results of Retinex-based methods (i.e., TBEFN, KinD, SS-Net, RetinexDIP) look more natural and real.

The classical methods (i.e., CVC, DHE, BIMEF, IAGC, and AGCWD) shown in the Figure 2 belong to the non-Retinex category, and PM-SIRE, WV-SRIE, JieP, EFF, and NPE belongs to Retinex theory. If we closely observe their visual results, one thing that is common among majority of these methods is noise. Except for BIMEF and EFF, most results can easily observe noise. The average brightness of BIMEF is too low and does not enhance the overall image. On the other hand, EFF produces higher brightness results, but the image's details are not too sharp. CVC and IAGC do not accurately render the color information, making their results look black and white. Although some classical methods' results quality is good, their results are still darker than deep learning-based methods. The results produced by BIMEF, IAGE, and CVC are darker as compared to other classical methods. Over-enhancement, severe noise and loss of color information can be seen in the results of DHE. The results of CVC are not only darker but also lost color information. AGCWD produces low contrast and less bright images, and some parts of the image are too dark (for example background buildings in the LIME test image and the background wall in the LOL test image). Gamma correction-based methods (i.e., AGCWD and IAGC) enhance some parts of the image while darker parts become darker. Strange artifacts around the fire can be easily seen in the IAGC result on the ExDark image.

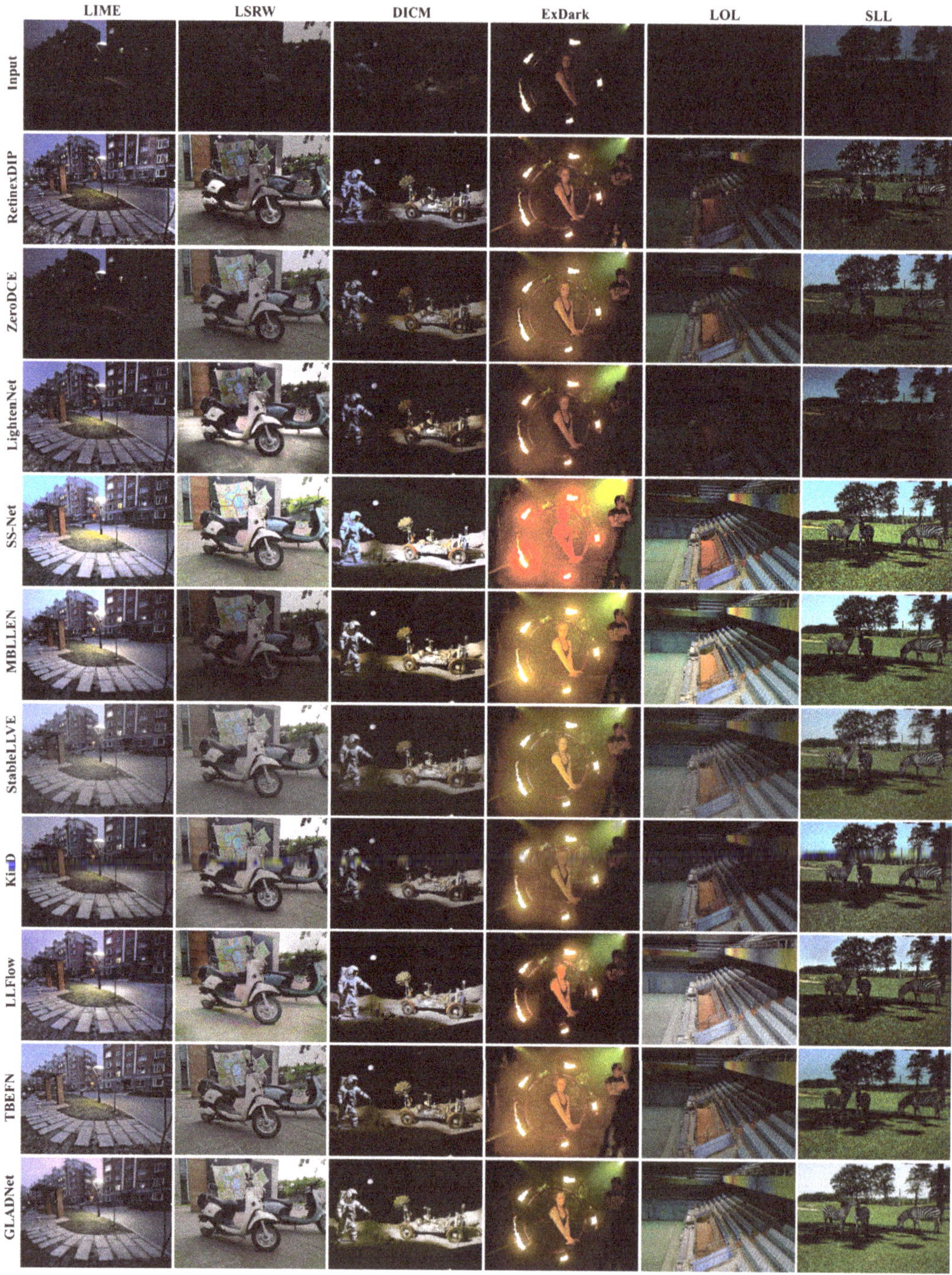

Figure 1. A visual representation of results from top ten deep learning methods on six datasets. The rows are showing the results produced by different algorithms, whereas the columns are showing datasets.

Figure 2. A visual representation of results from top ten classical methods on six datasets. The rows are showing the results produced by different algorithms, whereas the columns are showing datasets.

The results of Retinex-based methods (NPE and WV-SIRE) enhance the image's brightness, contrast, and sharpness, but fail to suppress the noise. The major issue with the majority of traditional methods is noise suppression. Histogram-based methods work to balance the histogram of the image to increase the brightness and contrast, but there is no such mechanics to remove the noise. Meanwhile, gamma correction-based methods treat each pixel individually and fail to exploit their relationship with neighbor pixels, which results in different artifacts and noise. In contrast, Retinex theory-based methods create different algorithms for successfully decomposing low-light images into reflectance and illumination components. In the case of severe noise, decomposing the image becomes difficult. The noise is not considered a major factor in any of these approaches. Therefore, noise dominates the visual results of these methods. When Figures 1 and 2 are compared, it is evident that deep learning-based methods produce brighter, sharper, cleaner, and higher contrast results. There is still some noise in some results, but compared to traditional methods, it is very low. Contrary to this, traditional visual results have many shortcomings. For example, some results have a lower average brightness, a lesser contrast level, a lesser degree of sharpness, failure to remove noise, and serious color shifts. Some of them enhance the image and the noise associated with it.

4.3. Quantitative Comparison of Enhancement Methods

Four non-reference evaluation metrics were used for the quantitative comparison. There are two reasons for using no-reference-based IQA metrics: (1) the majority of widely used test datasets are no-reference, and (2) unsupervised methods are emerging. Metrics adopted for evaluation include NIQE [88], BRISQUE [87], LOE [50], and Entropy. Low NIQE, BRISQIE, and LOE values indicate better image quality. In contrast, higher values of entropy indicate richer information. Tables 1–4 provide quantitative results for these metrics. Red indicates the best scores obtained on each dataset, while blue and green indicate the second and third best scores. The LOE indicates that non-Retinex methods perform better, whereas the other three metrics show that performance is uniform across both categories (i.e., Retinex and non-Retinex). Each method is evaluated by four metrics. There is no winner on all four metrics. To determine which method generalizes well, the enhancement methods score on all test data is averaged. The last column of the aforementioned tables represents the average score of enhancement methods on all test datasets. Based on averaged score, ranking number is assigned to each method and we summarize these rankings in Figure 3. Ranking 1 goes to the method with the best average score, and ranking 31 to the method with the worst average score. Different metrics rank enhancement methods differently. For instance, AGCWD ranked first according to LOE metric, whereas the same method is ranked as fifth, eighteenth, and twenty-ninth according to BRISQIE, NIQE, and entropy, respectively. Instead of analyzing the enhancement methods based on different metrics, we have taken the average of the ranking assigned based on the mentioned metrics and discussed the results of this average ranking.

Table 1. Quantitative comparison of enhancement algorithms on nine test datasets using LOE metric. A lower value of the LOE metric indicates better performance. The first, second, and third best scores are highlighted with red, blue, and green colors, respectively.

Methods	Datasets	LIME	LOL	DICM	VV	MEF	NPE	LSRW	SLL	ExDark	Average
Non-Retinex Methods	HE [89]	290.280	423.910	283.980	280.750	406.930	184.590	122.84	753.990	408.76	358.222
	DHE [27]	7.663	22.227	75.608	21.013	7.852	23.974	13.930	10.177	138.049	35.610
	BPDHE [68]	6.960	125.046	14.936	4.110	5.480	7.643	5.985	382.146	134.774	76.342
	CVC [90]	99.386	286.840	135.324	91.217	97.464	131.478	124.946	324.260	189.896	164.534
	CLAHE [91]	183.094	397.432	386.183	209.867	224.280	379.588	242.572	504.013	252.236	308.807
	AGCWD [29]	10.075	0.1325	57.482	14.777	6.046	31.432	1.463	6.132	137.990	31.932
	IAGC [92]	63.028	170.190	53.502	55.943	66.710	41.488	77.123	278.054	165.790	113.600
	BIMEF [93]	136.898	141.159	239.271	102.891	155.616	225.588	117.777	480.848	237.563	212.589
	MBLLEN [71]	122.188	302.577	176.580	79.013	131.243	123.871	168.128	484.809	190.384	207.076
	GLADNet [72]	123.603	349.720	285.239	145.034	199.632	203.488	204.887	518.189	262.524	254.702
	DLN [81]	132.594	264.065	404.673	325.572	189.831	-	176.527	528.411	212.723	-
	Zero-DCE [80]	135.032	209.426	340.803	145.435	164.262	312.392	219.127	539.673	315.084	280.775
	Exposure Correction [94]	242.461	438.420	362.552	220.876	275.476	314.833	288.659	588.132	307.881	349.604
	StableLLVE [95]	134.130	267.686	476.374	192.262	198.069	394.811	179.101	344.573	248.400	287.660
	LightenNet [96]	681.834	387.204	772.380	328.510	896.201	714.390	930.978	924.638	636.000	698.788
	White-box [97]	90.876	125.682	195.516	124.115	96.704	120.687	84.279	370.972	135.606	156.695
	LLFlow [38]	365.530	367.153	563.765	300.058	430.534	538.078	685.344	764.261	445.274	511.808
Retinex-based Methods	LIME [21]	559.618	404.114	818.660	460.440	618.480	870.215	434.485	1103.98	575.987	649.553
	NPE [50]	300.505	317.399	264.604	352.294	344.953	257.010	216.597	690.829	345.754	323.818
	JieP [58]	249.137	314.798	287.305	137.026	292.798	305.435	216.597	189.09	193.194	142.148
	PM-SIRE [49]	113.631	73.558	152.779	113.031	166.640	104.945	143.945	236.846	220.823	158.856
	WV-SRIE [20]	106.308	83.806	162.224	69.480	210.261	155.683	131.724	1211.11	676.415	1115.43
	MSRCR [19]	842.029	1450.95	1185.11	1280.68	973.893	1252.07	893.216	1211.11	352.672	314.419
	CRM [59]	271.652	21.818	450.102	174.751	285.250	534.275	119.712	619.537	352.672	314.419
	EFF [98]	136.898	141.159	239.271	102.891	155.616	255.588	117.777	480.848	237.563	207.512
	pmea [99]	491.663	725.647	477.792	318.569	679.002	610.183	418.046	1005.66	529.189	595.511
	RetinexNet [22]	472.189	770.105	636.160	391.745	708.250	838.310	591.278	950.895	548.905	679.456
	KinD [36]	214.893	434.595	261.771	134.844	275.474	241.221	379.899	479.139	308.869	303.412
	RetinexDIP [64]	767.042	1084.35	852.782	396.417	926.948	1099.39	572.429	1283.77	633.489	856.197
	RRDNet [25]	72.917	21.438	261.429	168.601	100.735	-	136.011	380.747	1.100	-
	KinD++ [60]	573.877	720.025	493.882	258.744	629.841	-	727.695	555.363	484.989	-
	IBA [100]	14.657	0.1616	445.574	169.714	12.823	364.810	137.727	21.758	284.333	179.613
	Self-supervised Network [24]	241.639	322.628	737.847	282.273	311.342	581.691	261.280	467.892	333.842	412.349
	TBEFN [23]	289.754	464.947	617.100	271.871	419.666	527.675	386.583	859.878	389.558	492.160
	Average	178.196	342.070	387.311	227.201	313.656	378.930	286.698	548.053	320.401	-

Table 2. Quantitative comparison of enhancement algorithms on nine test datasets using NIQE metric. A lower value of the NIQE metric indicates better performance. The first, second, and third best scores are highlighted with red, blue, and green colors, respectively.

	Methods \ Datasets	LIME	LOL	DICM	VV	MEF	NPE	LSRW	SLL	ExDark	Average
Non-Retinex Methods	Input	4.357	6.748	4.274	3.524	4.263	3.717	5.391	5.358	5.128	4.800
	HE [89]	3.884	8.413	3.850	2.662	3.870	3.535	3.963	6.438	4.752	4.685
	DHE [27]	3.914	8.987	3.780	2.648	3.518	3.510	3.626	6.292	4.518	4.610
	BPDHE [68]	3.827	NaN	3.786	2.857	3.902	3.531	3.935	NaN	4.727	-
	CVC [90]	4.029	8.014	3.823	2.692	3.636	3.498	4.127	5.828	4.662	4.535
	CLAHE [91]	3.907	7.268	3.792	2.784	3.606	3.461	4.581	5.756	4.734	4.490
	AGCWD [29]	4.032	7.528	3.868	2.970	3.629	3.544	3.733	5.660	4.582	4.434
	IAGC [92]	3.951	7.418	4.015	3.012	3.652	3.598	3.963	5.740	4.557	4.494
	BIMEF [93]	3.859	7.515	3.845	2.807	3.329	3.540	3.879	5.747	4.514	4.397
	MBLLEN [71]	4.513	4.357	4.230	4.179	4.739	3.948	4.722	3.979	4.478	4.329
	GLADNet [72]	4.128	6.475	3.681	2.790	3.360	3.522	3.397	5.066	3.767	4.009
	DLN [81]	4.341	4.883	3.789	3.228	4.022	-	4.419	4.376	4.415	-
	Zero-DCE [80]	3.769	7.767	3.567	3.216	3.283	3.582	3.720	5.998	3.917	4.381
	Exposure Correction [94]	4.215	7.886	3.588	3.078	4.456	3.414	3.820	4.942	4.357	4.443
	StableLLVE [95]	4.234	4.372	4.061	3.420	3.924	3.486	4.367	4.185	4.053	3.984
	LightenNet [96]	3.731	7.323	3.539	2.995	3.350	3.407	3.583	5.453	4.025	4.209
	White-box [97]	4.598	7.819	4.630	3.558	4.622	4.004	4.314	7.138	5.534	5.202
	LLFlow [38]	3.956	5.445	3.765	3.026	3.441	3.498	3.564	4.722	4.094	3.944
Retinex-based Methods	LIME [21]	4.109	8.129	3.860	2.494	3.576	3.658	3.655	6.372	4.588	4.542
	NPE [50]	3.578	8.158	3.736	2.471	3.337	3.426	3.576	5.771	4.220	4.337
	JieP [58]	3.719	6.872	3.678	2.765	3.390	3.522	4.015	5.622	4.215	4.260
	PM-SIRE [49]	4.050	7.506	3.978	3.010	3.450	3.531	3.984	5.435	4.383	4.410
	WV-SRIE [20]	3.786	7.286	3.898	2.849	3.474	3.450	3.826	5.453	4.241	4.310
	MSRCR [19]	3.939	8.006	3.948	2.814	3.688	3.780	3.872	5.574	4.904	4.573
	CRM [59]	3.854	7.686	3.801	2.617	3.264	3.562	3.721	6.008	4.525	4.391
	EFF [98]	3.859	7.515	3.845	2.807	3.329	3.540	3.879	5.747	4.514	4.390
	pmea [99]	3.843	8.281	3.836	2.573	3.431	3.598	3.694	6.237	4.296	4.493
	RetinexNet [22]	4.597	8.879	4.415	2.695	4.410	4.464	4.150	7.573	4.551	5.142
	KinD [36]	4.763	4.709	4.150	3.026	3.876	3.557	3.543	4.450	4.340	3.956
	RetinexDIP [64]	3.735	7.096	3.705	2.496	3.245	3.638	4.081	5.8828	4.234	4.297
	RRDNet [25]	3.936	7.436	3.637	2.814	3.508	-	4.126	5.524	4.010	-
	KinD++ [60]	4.385	4.616	3.804	2.660	3.730	-	3.354	5.090	4.018	
	IBA [100]	4.062	7.884	3.723	3.310	3.536	3.630	3.728	5.837	4.273	4.490
	Self-supervised Network [24]	4.819	3.753	4.717	3.548	4.351	4.602	4.061	5.400	4.048	4.310
	TBEFN [23]	3.954	3.436	3.503	2.884	3.227	3.292	3.478	4.648	3.621	3.511
	Average	3.935	6.728	3.889	2.956	3.698	3.626	3.933	5.409	4.403	-

Table 3. Quantitative comparison of enhancement algorithms on nine test datasets using entropy [86] metric. A higher value of the entropy metric indicates better performance. The first, second, and third best scores are highlighted with red, blue, and green colors, respectively.

Methods	Datasets	LIME	LOL	DICM	VV	MEF	NPE	LSRW	SLL	ExDark	Average
Non-Retinex Methods	Input	6.148	4.915	6.686	6.715	6.075	7.017	5.415	5.616	5.744	6.023
	HE [89]	7.342	7.184	7.221	7.383	7.118	7.756	6.874	6.662	6.708	7.113
	DHE [27]	7.097	6.749	7.141	7.225	6.913	7.512	6.531	6.741	6.613	6.930
	BPDHE [68]	6.610	5.932	6.968	6.977	6.420	7.348	6.260	5.191	6.188	6.413
	CVC [90]	6.875	6.409	7.055	7.216	6.755	7.402	6.318	6.549	6.465	6.772
	CLAHE [91]	6.764	5.679	7.088	7.056	6.583	7.408	6.033	6.591	6.302	6.595
	AGCWD [29]	6.792	6.415	6.925	7.021	6.648	7.398	6.394	6.278	6.248	6.666
	IAGC [92]	6.991	6.247	7.015	7.193	6.878	7.351	6.318	6.698	6.554	6.782
	BIMEF [93]	7.006	6.145	7.029	7.243	6.898	7.311	6.516	6.452	6.464	6.760
	MBLLEN [71]	7.164	7.303	7.255	7.333	7.081	7.386	7.236	7.197	7.132	7.240
	GLADNet [72]	7.502	7.356	7.404	7.447	7.408	7.452	7.393	7.581	7.250	7.412
	DLN [81]	7.121	7.277	7.250	7.535	7.255	-	7.202	7.576	7.129	-
	Zero-DCE [80]	7.166	6.531	7.224	7.572	7.093	7.402	7.035	6.545	6.932	7.042
	Exposure Correction [94]	7.112	7.244	7.256	6.962	6.955	7.531	7.039	7.247	6.907	7.142
	StableLLVE [95]	7.227	6.625	7.010	7.385	7.241	7.042	6.846	7.439	7.129	7.090
	LightenNet [96]	7.234	6.119	7.263	7.411	7.308	7.398	7.599	6.130	6.688	6.990
	White-box [97]	5.984	5.925	6.051	5.475	5.391	7.380	6.352	5.460	5.275	5.914
	LLFlow [38]	7.468	7.462	7.425	7.565	7.366	7.564	7.343	7.304	7.125	7.394
Retinex-based Methods	LIME [21]	7.315	7.129	6.946	7.395	7.139	7.332	7.279	6.418	6.582	7.031
	NPE [50]	7.368	6.971	7.208	7.550	7.405	7.446	7.318	6.418	6.772	7.139
	JieP [58]	7.087	6.443	7.218	7.457	7.104	7.427	6.794	6.473	6.631	6.943
	PM-SIRE [49]	7.006	6.322	7.084	7.309	6.894	7.404	6.696	6.325	6.441	6.812
	WV-SRIE [20]	6.999	6.348	7.088	7.401	6.942	7.386	6.663	6.190	6.463	6.812
	MSRCR [19]	6.563	6.841	6.677	6.957	6.455	6.762	6.895	5.936	6.319	6.605
	CRM [59]	6.487	4.971	6.640	6.559	6.203	7.026	5.494	6.068	5.921	6.115
	EFF [98]	7.006	6.145	7.029	7.243	6.898	7.311	6.516	6.452	6.464	6.760
	pmea [99]	7.284	6.824	7.220	7.479	7.273	7.449	7.074	6.638	6.725	7.088
	RetinexNet [22]	7.489	7.233	7.413	7.575	7.448	7.463	7.243	7.385	7.273	7.379
	KinD [36]	7.388	7.017	7.211	7.498	7.328	7.435	7.209	7.408	6.905	7.251
	RetinexDIP [64]	6.974	5.375	7.214	7.557	6.661	7.381	6.352	6.213	6.668	6.678
	RRDNet [25]	6.646	5.457	7.142	7.275	6.453	-	6.775	6.077	6.426	-
	KinD++ [60]	7.486	7.065	7.332	7.627	7.463	-	7.316	7.452	7.034	-
	IBA [100]	5.905	4.913	6.826	7.255	5.749	7.035	7.146	5.465	6.971	6.420
	Self-supervised Network [24]	7.497	7.404	6.675	7.298	7.469	6.997	7.397	7.484	7.296	7.253
	TBEFN [23]	7.436	6.875	7.328	7.507	7.383	7.366	7.047	7.519	7.313	7.292
	Average	7.000	6.481	7.072	7.247	6.904	7.340	6.798	6.605	6.659	-

Table 4. Quantitative comparison of enhancement algorithms on nine test datasets using BRISQUE metric. A lower value of the BRISQUE metric indicates better performance. The first, second, and third best scores are highlighted with red, blue, and green colors, respectively.

	Methods / Datasets	LIME	LOL	DICM	VV	MEF	NPE	LSRW	SLL	ExDark	Average
Non-Retinex Methods	Input	25.142	21.929	28.115	29.380	29.066	26.673	32.726	25.304	34.015	28.401
	HE [89]	21.411	39.559	25.359	18.937	25.313	25.444	28.219	40.015	29.034	28.985
	DHE [27]	22.336	37.866	25.993	24.380	21.466	27.008	26.477	38.248	28.951	28.719
	BPDHE [68]	21.728	NaN	25.0972	25.183	22.345	26.425	25.129	NaN	27.417	-
	CVC [90]	22.589	27.101	24.620	21.766	19.285	25.693	26.808	29.007	26.979	25.126
	CLAHE [91]	23.274	29.463	24.248	23.480	22.701	25.368	29.570	31.579	28.543	26.825
	AGCWD [29]	21.964	28.421	24.725	23.961	19.420	26.4117	23.367	29.740	26.161	25.276
	IAGC [92]	24.314	24.058	27.026	26.617	21.843	26.044	23.854	32.813	27.429	26.211
	BIMEF [93]	23.135	27.651	26.811	22.542	20.220	25.504	24.077	34.982	27.910	26.174
	MBLLEN [71]	30.386	23.078	31.603	35.076	32.389	29.423	30.328	22.103	29.012	29.127
	GLADNet [72]	22.286	26.073	26.253	24.068	22.908	24.969	22.802	33.754	24.765	25.657
	DLN [81]	27.715	28.985	26.914	29.782	28.378	-	33.597	26.798	31.187	-
	Zero-DCE [80]	23.334	30.305	30.653	30.786	25.484	30.159	25.827	36.572	26.761	29.568
	Exposure Correction [94]	27.483	28.357	29.847	31.694	29.597	26.768	26.391	28.632	32.520	29.204
	StableLLVE [95]	28.885	32.194	28.150	28.295	28.475	25.662	30.563	25.850	27.749	28.367
	LightenNet [96]	19.523	28.062	28.791	23.502	21.469	27.667	25.144	28.055	25.924	26.077
	White-box [97]	28.807	31.721	33.212	35.733	33.599	26.671	25.081	39.450	37.429	32.862
	LLFlow [38]	22.856	29.709	25.072	23.157	25.673	25.392	22.011	28.041	26.133	25.649
Retinex-based Methods	LIME [21]	23.572	33.973	27.137	25.394	25.158	28.576	27.658	35.829	28.704	28.986
	NPE [50]	22.506	33.858	25.493	24.654	22.320	24.986	27.195	33.861	28.452	27.539
	JieP [58]	22.193	27.087	23.633	22.941	21.214	25.498	23.421	30.207	25.309	24.914
	PM-SIRE [49]	24.659	27.694	27.597	24.287	24.321	27.342	25.345	30.014	26.676	26.635
	WV-SRIE [20]	24.181	27.611	27.698	24.434	22.088	25.760	24.700	28.281	26.750	25.894
	MSRCR [19]	19.384	30.345	25.799	19.282	19.091	24.189	25.789	30.300	25.415	24.957
	CRM [59]	23.477	29.599	26.601	22.368	20.716	25.726	24.396	37.723	28.733	26.939
	EFF [98]	23.135	27.651	26.811	22.542	20.220	25.504	24.077	34.982	27.910	26.174
	pmea [99]	21.390	32.913	25.832	24.972	21.756	26.358	25.358	38.132	28.321	27.874
	RetinexNet [22]	26.101	39.586	26.656	22.459	26.036	29.086	29.021	41.506	30.170	30.565
	KinD [36]	26.773	26.645	30.696	28.887	30.438	27.753	26.763	30.539	29.256	28.872
	RetinexDIP [64]	21.723	19.679	25.199	25.338	23.605	26.671	25.081	32.618	32.175	26.296
	RRDNet [25]	24.499	26.834	29.621	23.396	17.750	-	27.100	29.205	27.606	-
	KinD++ [60]	20.025	25.086	27.852	28.164	30.024	-	26.973	34.978	31.775	-
	IBA [100]	24.336	31.117	32.103	34.646	23.748	29.933	25.826	32.537	26.639	29.569
	Self supervised Network [41]	30.192	19.768	29.529	30.183	28.355	29.159	26.205	32.016	27.990	27.901
	TBEFN [23]	25.720	17.346	23.606	23.651	24.435	24.0355	22.929	30.676	25.064	23.968
	Average	23.009	27.752	27.267	25.841	24.312	26.621	26.280	31.267	28.425	-

The red line in Figure 3 represents the average ranking achieved by enhancement methods on all test datasets. The average ranking puts GLADNet, TBEFN, and LLFlow in first, second, and third, respectively. GLADNet generalizes well despite being trained on 5000 synthetic images using L_1 loss. Retinex-based methods TBEFN, WV-SIRE, JieP, and KinD also generalized well and received the 2nd, 4th, 5th, and 6th rankings, respectively. TBEFN [23] is trained on a mixture of 14,531 patches collected from SICE [33] and LOL [22] datasets using SSIM, total variation, and VGG loss. KinD is based on Retinex theory and trained on LOL. A self-supervised network and a zero-shot-based Retinex method (i.e., self-supervised network and RetinexDIP) ranked 12th and 21st, respectively. MBLLEN is also a supervised learning-based networks and ranked 10th. Meanwhile, MBLLEM is a multi-branch fusion network trained on the PASCAL VOC dataset [101]. Zero-shot learning-based methods such as Zero-DCE got 19th. Among all deep learning-based methods, the top six methods are supervised learning-based methods. It is also worth noticing that among the top ten methods, five are Retinex-based methods and 5 are non-Retinex methods.

4.4. Computational Complexity Analysis of Enhancement Methods

The computational complexity analysis of classical methods and deep learning-based methods is presented in Tables 5 and 6, respectively. The analysis is conducted on four different test datasets (i.e., LOL, LSRW, VV, and SID). Tables 5 and 6 report the average time taken and the resolution of a single image for each dataset. For each of these tables, red, blue, and green colors are used to indicate the best, second best, and third best performance, respectively. Results shown in Table 5 have been obtained using a CPU, while results shown in Table 6 have been obtained using an NVIDIA Titan Xp GPU. HE has the shortest runtime of all classical methods. HE just takes around 20.3ms to process an image of resolution $2848 \times 4256 \times 3$. The majority of HE-based methods, such as BPDHE, WAHE, LDR, CVC, and BiHE, are time-efficient, except for DHE. DHE continuously divides an image into several sub-histogram units in order to avoid leaving a dominant portion in newly created sub-histograms. Due to the continual dividing process, this method is the slowest of all the HE-based methods mentioned. Gamma correction-based methods also have good computational efficiency. IAGC takes relatively longer than other methods because it truncates an image's cumulative distribution function (CDF) and adaptively corrects each truncated CDF.

Furthermore, Retinex-based methods are more computationally expensive than HE and gamma correction-based methods. NPE, PM-SIRE, and WV-SRIE are among the Retinex-based methods that experience significant increases in computation costs with increasing image size. These methods are computationally inefficient due to their iterative approach to finding the optimal solution and use of Gaussian filtering operations. The efficiency of deep learning-based methods depends on the number of parameters that are used. Zero-DCE is the fastest deep learning-based method due to its simplest network architecture and fewer parameters. The majority of deep learning-based methods' average runtime is between 1.7 ms and 2.57 s. RRDNet iteratively minimizes the error to produce the final enhanced output. The number of iterations varies for different inputs. The iteratively solving the problem makes it the slowest among all the networks. RetinexDIP is another zero-shot learning-based method and performs 300 iterations on each input to produce the final output. The iteratively solving problem makes RetinexDIP and RRDNet the slowest methods. A scatter plot of methods' performance versus time taken on CPU and GPU is shown in Figures 4 and 5, respectively. We consider CPU methods with less than 1s processing time and GPU methods with less than 0.5s. Methods closer to the origin have a lower computational cost and better performance.

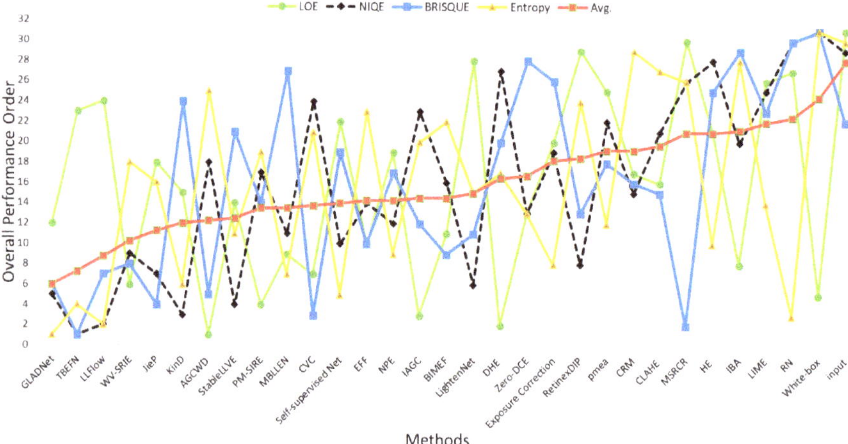

Figure 3. Different IQA metrics are used to rank the enhancement methods. Rank values range from 1 to 31. A rank value of 1 indicates the highest performance based on a particular IAQ method, and a rank value of 1 indicates the worst performance. The average rank is shown in red.

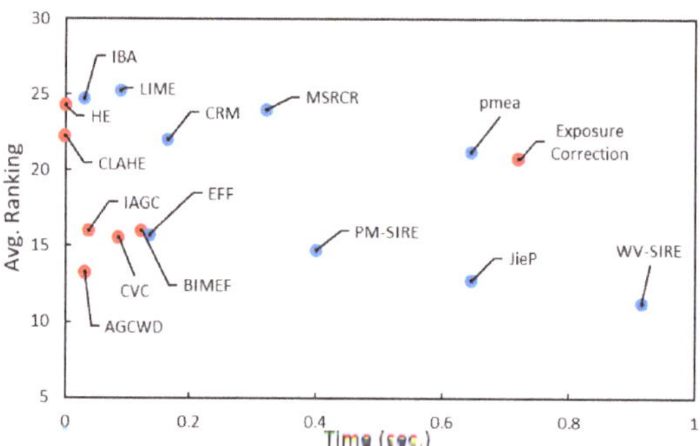

Figure 4. Avg. ranking versus Time is shown for each enhancement method. Only the methods take less than 1 s on CPU (Intel(R) Core(TM) i7-6700 CPU @ 3.40 GHz 3.41 GHz) with 16 GB RAM to process the image of size $400 \times 600 \times 3$ is shown in the figure. Red dots represent non-Retinex methods, while blue dots represent Retinex methods.

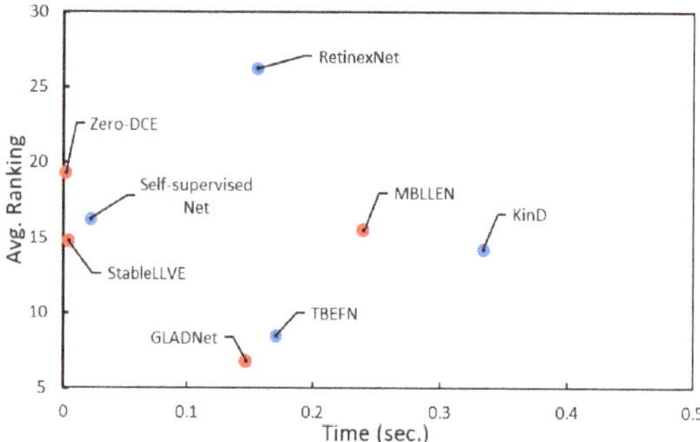

Figure 5. Avg. ranking versus Time is shown for each enhancement method. Only the methods take less than 0.5 s on GPU (NVIDIA Titan Xp GPU) to process the image of size $400 \times 600 \times 3$ is shown in the figure. Red dots represent non-Retinex methods, while blue dots represent Retinex methods.

Table 5. Computational time of classical methods in terms of seconds on CPU is reported. The red, blue, and green colors are used to indicate the best, second best, and third best performance, respectively.

	Methods \ Image Size	400 × 600 × 3	640 × 960 × 3	2304 × 1728 × 3	2848 × 4256 × 3	Avg.
Non-Retinex Methods	HE [89]	0.00079	0.0014	0.0071	0.0203	0.00742
	DHE [27]	23.590	59.625	409.628	1253.897	436.685
	BPDHE [68]	0.078	0.338	1.630	3.318	1.341
	CVC [90]	0.086	0.230	1.150	3.533	1.250
	CLAHE [91]	0.00033	0.00099	0.0058	0.0226	0.00743
	AGCWD [29]	0.031	0.053	0.344	1.079	0.377
	IAGC [92]	0.038	0.155	1.025	2.253	0.867
	BIMEF [93]	0.123	0.359	1.811	5.101	1.848
	Exposure Correction [94]	0.721	0.778	0.903	18.501	5.226
	LightenNet [96]	3.091	7.126	45.990	137.835	48.510
	LLFlow [38]	24.740	60.022	363.281	1403.92	462.991
Retinex-based Methods	LIME [21]	0.090	0.296	1.506	4.650	1.635
	NPE [50]	13.061	31.025	213.168	648.832	226.522
	JieP [58]	0.646	0.874	2.307	6.597	2.606
	PM-SIRE [49]	0.402	1.340	28.948	28.423	14.778
	WV-SRIE [20]	0.915	3.136	40.701	182.267	56.755
	MSRCR [19]	0.322	0.704	2.787	8.531	3.086
	CRM [59]	0.166	0.436	2.626	8.134	2.840
	EFF [98]	0.136	0.407	1.973	5.422	1.984
	pmea [99]	0.646	0.874	2.307	6.597	2.606
	IBA [100]	0.032	0.0829	0.512	1.385	0.503

Table 6. The computation time (seconds) and number of parameters (millions) for deep learning-based methods on GPUs (NVIDIA TITAN Xp) are reported. The red, blue, and green colors are used to indicate the best, second best, and third best performance, respectively.

	Methods \ Image Size	400 × 600 × 3	640 × 960 × 3	2304 × 1728 × 3	2848 × 4256 × 3	Avg.	Parameters.
Non-Retinex	StableLLVE [95]	0.0047	0.005	0.0076	0.097	0.020	4.310 M
	MBLLEN [71]	0.240	0.327	1.601	8.133	2.575	0.450 M
	GLADNet [72]	0.147	0.161	0.676	2.772	0.939	0.930 M
	White-box [97]	6.040	6.483	9.833	15.200	9.389	8.560 M
	DLN [81]	0.009	0.015	0.058	0.197	0.070	0.700 M
	Zero-DCE [80]	0.0025	0.0026	0.021	0.043	0.017	0.079 M
Retinex	RetinexNet [22]	0.155	0.162	0.591	1.289	0.549	0.440 M
	KinD [36]	0.334	0.604	3.539	5.213	2.423	0.255 M
	RetinexDIP [64]	33.924	37.015	63.443	112.545	61.732	0.707 M
	RRDNet [25]	59.479	128.217	893.0	3003.5	1021.1	0.128 M
	KinD++ [60]	0.337	0.857	5.408	19.746	6.587	8.275 M
	Self-supervised Net [24]	0.022	0.054	0.366	1.212	0.414	0.485 M
	TBEFN [23]	0.171	0.166	0.550	0.887	0.444	0.490 M

4.5. Difficulty Analysis of Test Datasets

Results of enhancement methods have also been used to rank the nine test datasets. The last row of Tables 1–4 shows the average score of different enhancement methods on the test datasets based on LOE, NIQE, Entropy, and BRISQUE, respectively. Figure 6 shows the difficulty rank for each dataset across IQA methods. A red line shows the average of all rankings in Figure 6. As determined by the average ranking score, VV is the easiest test dataset, while SLL is the most challenging. SLL is the synthetic dataset with severe noise added. There is too much noise to remove and produce better results. Meanwhile, VV has a lower noise level, making it the easiest dataset. LOL and ExDark are the second and third most difficult datasets. A test dataset's difficulty level is determined by its noise level. The higher the noise level, the harder it is to recover color details and other information.

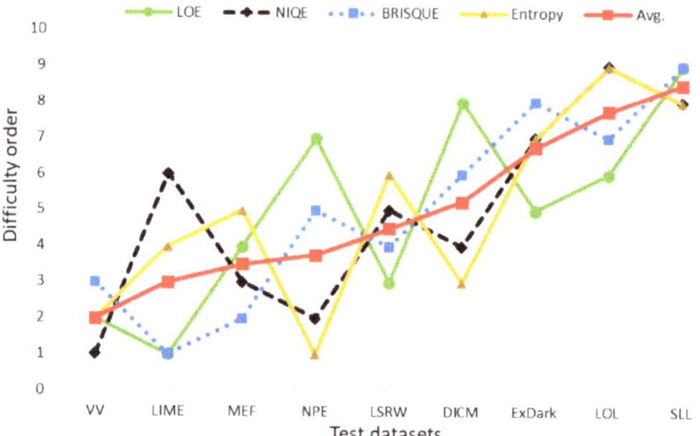

Figure 6. Each test dataset has been ranked based on its difficulty. Rank values range from 1 to 9. A lower rank indicates less difficulty, a higher rank indicates more difficulty.

4.6. Evaluation IQA methods

To analyze the objective quality of different enhancement methods, we have used LOE, NIQE, Entropy, and BRISQUE as described in Tables 1–4, respectively. We can easily identify the differences in the evaluations of these metrics if we compare their values among themselves. The best result was produced by the BPDHE enhancement method, according to LOE and NIQE, whereas BRISQUE evaluated MSRCR and Entropy evaluated GLADnet as best methods. Each metric measures a different aspect of the predicted image, which makes their results different. To easily depict and analyze the overall performance of enhancement methods, we have suggested the average rank from 1 to 31 (depending on how many methods are compared) to compare their performance. The best performance is ranked 1, and the worst performance is ranked 31. These rankings can be seen in Figure 3, where the *x*-axis represents the enhancement methods, and the *y*-axis represents the ranking. Green, dotted black, blue, and yellow lines in Figure 3 show the rankings of enhancement methods based on LOE, NIQE, BRISQUE, and Entropy metrics, respectively. Moreover, the red line in Figure 3 represents the average of all the rankings mentioned earlier (given based on different metrics). The best method can be chosen based on this average ranking system.

5. Discussion

In this section, we summarize the results obtained and the findings of the overall paper.

i   The enhancement methods are evaluated using four evaluation metrics. No method has emerged as the clear winner on all four metrics (LOE, entropy, NIQE, BRISQUE). This is due to the fact that each evaluation method measures a different aspect of enhancement methods (e.g., LOE measures naturalness, entropy measures information content, and NIQE measures distortion). A suggested average ranking system is found to be the most reliable method of comparing the overall performance of enactment methods.

ii   In the average ranking system, it has been observed that the three most successful enhancement methods (GLADNet, TBEFN, LLFlow) are based on supervised learning. Among the top ten methods, five are based on Retinex. In comparison to classical, advanced self-supervised, and zero-short learning methods, supervised learning is more effective. Denosing is the most challenging part in enhancement. Noise can be observed in the visual results of outperforming methods.

iii There is no Retinex-based method among the top three fastest methods. As a result of the image decomposition, these methods are more time consuming. As the size of the image increases, the computational time of classical Retinex-based methods increases dramatically. Zero-DCE is the fastest learning-based method, taking approximately 0.017 s to process an image of size $2848 \times 4256 \times 3$. However, it ranks 20th in terms of performance. GLADNet, on the other hand, is ranked first, but it takes approximately 2.772 s to process an image of the same size.

iv The average ranking of all enhancement methods is observed in a broader sense. The results indicate that five methods in the top ten are based on Retinex theory (i.e., TBEFN, WV-SRIE, JieP, KinD, and PM-SIRE). The remaining five fall into different categories (i.e., HE, gamma correction, deep learning). When it comes to real-world scenarios, Retinex theory algorithms are more robust. In contrast, decomposing the image into illumination and reflectance makes them more computationally intensive and, therefore, slower. Computational complexity is the bottleneck for their development in real-world scenarios.

6. Conclusions

In this study, we present an experimental comparison of Retinex-based methods with other non-Retinex methods on nine diversified datasets. According to this study, five out of the top 10 methods are based on Retinex. Researchers are aiming to develop methods that can be generalized and produce enhanced, denoised, color rendered results in real time. Based on the comparisons, ZeroDCE is considered to be the fastest method for processing high-resolution images within 17 milliseconds. However, ZeroDCE ranked 19th and its results were darker and noisy. In contrast, Retinex-based methods have a greater degree of robustness and generalization. The decomposition of the image is a time-consuming process and is a bottleneck in the processing time of Retinex-based methods. Based on the overall ranking, supervised learning methods (e.g., GLADNet, TBEFN, LLFLow) perform better than all other methods. Training images for GLADNet and patches for TBEFN are 5000 images and 14,531 patches, respectively. Both GLADNet and TBEFN are able to generalize well due to their large training data, as well as their Unet architecture which makes them more efficient as compared to other heavy network designs. Moreover, this research evaluated the results of enhancement methods on four different metrics and suggested a method for ranking enhancement methods according to their performance. This research study may help the research community develop more robust and lightweight models for real-time photography and video shooting.

Author Contributions: Conceptualization, M.T.R.; methodology, M.T.R.; software, M.T.R.; validation, M.T.R. and G.G.; formal analysis, M.T.R.; investigation, G.G.; resources, D.S.; data curation, H.K.; writing—original draft preparation, G.G.; writing—review and editing, H.K.; visualization, H.K.; supervision, D.S.; project administration, X.C.; funding acquisition, D.S. All authors have read and agreed to the published version of the manuscript.

Funding: This work is supported by Ministry of Science and Technology China (MOST) Major Program on New Generation of Artificial Intelligence 2030 No. 2018AAA0102200. It is also supported by Natural Science Foundation China (NSFC) Major Project No. 61827814 and Shenzhen Science and Technology Innovation Commission (SZSTI) project No. JCYJ20190808153619413.

Data Availability Statement: Not applicable.

Conflicts of Interest: The authors declare no conflict of interest.

References

1. Wang, J.; Wang, W.; Wang, R.; Gao, W. CSPS: An adaptive pooling method for image classification. *IEEE Trans. Multimed.* **2016**, *18*, 1000–1010. [CrossRef]
2. Zhao, Q.; Sheng, T.; Wang, Y.; Tang, Z.; Chen, Y.; Cai, L.; Ling, H. M2det: A single-shot object detector based on multi-level feature pyramid network. In Proceedings of the AAAI Conference on Artificial Intelligence, Atlanta, GA, USA, 8–12 October 2019; Volume 33, pp. 9259–9266.
3. Rezatofighi, H.; Tsoi, N.; Gwak, J.; Sadeghian, A.; Reid, I.; Savarese, S. Generalized intersection over union: A metric and a loss for bounding box regression. In Proceedings of the IEEE/CVF Conference on Computer Vision and Pattern Recognition, Long Beach, CA, USA, 15–20 June 2019; pp. 658–666.
4. Bertinetto, L.; Valmadre, J.; Henriques, J.F.; Vedaldi, A.; Torr, P.H. Fully-convolutional siamese networks for object tracking. In Proceedings of the European Conference on Computer Vision, Munich, Germany, 8–14 September 2016; Springer: Berlin/Heidelberg, Germany, 2016; pp. 850–865.
5. He, A.; Luo, C.; Tian, X.; Zeng, W. A twofold siamese network for real-time object tracking. In Proceedings of the IEEE Conference on Computer Vision and Pattern Recognition, Salt Lake City, UT, USA, 18–22 June 2018; pp. 4834–4843.
6. Luo, W.; Sun, P.; Zhong, F.; Liu, W.; Zhang, T.; Wang, Y. End-to-end active object tracking via reinforcement learning. In Proceedings of the International Conference on Machine Learning, PMLR, Stockholm, Sweden, 10–15 July 2018; pp. 3286–3295.
7. Ristani, E.; Tomasi, C. Features for multi-target multi-camera tracking and re-identification. In Proceedings of the IEEE Conference on Computer Vision and Pattern Recognition, Salt Lake City, UT, USA, 18–23 June 2018; pp. 6036–6046.
8. Saini, M.; Wang, X.; Atrey, P.K.; Kankanhalli, M. Adaptive workload equalization in multi-camera surveillance systems. *IEEE Trans. Multimed.* **2012**, *14*, 555–562. [CrossRef]
9. Feng, W.; Ji, D.; Wang, Y.; Chang, S.; Ren, H.; Gan, W. Challenges on large scale surveillance video analysis. In Proceedings of the IEEE Conference on Computer Vision and Pattern Recognition Workshops, Salt Lake City, UT, USA, 18–23 June 2018; pp. 69–76.
10. Ko, S.; Yu, S.; Kang, W.; Park, C.; Lee, S.; Paik, J. Artifact-free low-light video enhancement using temporal similarity and guide map. *IEEE Trans. Ind. Electron.* **2017**, *64*, 6392–6401. [CrossRef]
11. Rasheed, M.T.; Shi, D. LSR: Lightening super-resolution deep network for low-light image enhancement. *Neurocomputing* **2022**, *505*, 263–275. [CrossRef]
12. Khan, H.; Wang, X.; Liu, H. Handling missing data through deep convolutional neural network. *Inf. Sci.* **2022**, *595*, 278–293. [CrossRef]
13. Khan, H.; Wang, X.; Liu, H. Missing value imputation through shorter interval selection driven by Fuzzy C-Means clustering. *Comput. Electr. Eng.* **2021**, *93*, 107230. [CrossRef]
14. Khan, H.; Liu, H.; Liu, C. Missing label imputation through inception-based semi-supervised ensemble learning. *Adv. Comput. Intell.* **2022**, *2*, 1–11. [CrossRef]
15. Ellrod, G.P. Advances in the detection and analysis of fog at night using GOES multispectral infrared imagery. *Weather. Forecast.* **1995**, *10*, 606–619. [CrossRef]
16. Negru, M.; Nedevschi, S.; Peter, R.I. Exponential contrast restoration in fog conditions for driving assistance. *IEEE Trans. Intell. Transp. Syst.* **2015**, *16*, 2257–2268. [CrossRef]
17. Land, E.H. The retinex theory of color vision. *Sci. Am.* **1977**, *237*, 108–129. [CrossRef]
18. Jobson, D.J.; Rahman, Z.U.; Woodell, G.A. Properties and performance of a center/surround retinex. *IEEE Trans. Image Process.* **1997**, *6*, 451–462. [CrossRef] [PubMed]
19. Jobson, D.J.; Rahman, Z.U.; Woodell, G.A. A multiscale retinex for bridging the gap between color images and the human observation of scenes. *IEEE Trans. Image Process.* **1997**, *6*, 965–976. [CrossRef] [PubMed]
20. Fu, X.; Zeng, D.; Huang, Y.; Zhang, X.P.; Ding, X. A weighted variational model for simultaneous reflectance and illumination estimation. In Proceedings of the IEEE Conference on Computer Vision and Pattern Recognition, Las Vegas, NV, USA, 26 June–1 July 2016; pp. 2782–2790.
21. Guo, X.; Li, Y.; Ling, H. LIME: Low-light image enhancement via illumination map estimation. *IEEE Trans. Image Process.* **2016**, *26*, 982–993. [CrossRef] [PubMed]
22. Wei, C.; Wang, W.; Yang, W.; Liu, J. Deep retinex decomposition for low-light enhancement. *arXiv* **2018**, arXiv:1808.04560.
23. Lu, K.; Zhang, L. TBEFN: A two-branch exposure-fusion network for low-light image enhancement. *IEEE Trans. Multimed.* **2020**, *23*, 4093–4105. [CrossRef]
24. Zhang, Y.; Di, X.; Zhang, B.; Li, Q.; Yan, S.; Wang, C. Self-supervised Low Light Image Enhancement and Denoising. *arXiv* **2021**, arXiv:2103.00832.
25. Zhu, A.; Zhang, L.; Shen, Y.; Ma, Y.; Zhao, S.; Zhou, Y. Zero-shot restoration of underexposed images via robust retinex decomposition. In Proceedings of the 2020 IEEE International Conference on Multimedia and Expo (ICME), London, UK, 6–10 July 2020; pp. 1–6.
26. Gonzalez, R.C. *Digital Image Processing*, 2nd ed.; Addison-Wesley: Boston, MA, USA, 1992.
27. Abdullah-Al-Wadud, M.; Kabir, M.H.; Dewan, M.A.A.; Chae, O. A dynamic histogram equalization for image contrast enhancement. *IEEE Trans. Consum. Electron.* **2007**, *53*, 593–600. [CrossRef]
28. Rahman, S.; Rahman, M.M.; Abdullah-Al-Wadud, M.; Al-Quaderi, G.D.; Shoyaib, M. An adaptive gamma correction for image enhancement. *EURASIP J. Image Video Process* **2016**, *35*, 2016. [CrossRef]

29. Huang, S.C.; Cheng, F.C.; Chiu, Y.S. Efficient contrast enhancement using adaptive gamma correction with weighting distribution. *IEEE Trans. Image Process.* **2012**, *22*, 1032–1041. [CrossRef]

30. Wang, Z.G.; Liang, Z.H.; Liu, C.L. A real-time image processor with combining dynamic contrast ratio enhancement and inverse gamma correction for PDP. *Displays* **2009**, *30*, 133–139. [CrossRef]

31. Lore, K.G.; Akintayo, A.; Sarkar, S. LLNet: A deep autoencoder approach to natural low-light image enhancement. *Pattern Recognit.* **2017**, *61*, 650–662. [CrossRef]

32. Chen, C.; Chen, Q.; Xu, J.; Koltun, V. Learning to see in the dark. In Proceedings of the IEEE Conference on Computer Vision and Pattern Recognition, Salt Lake City, UT, USA, 18–23 June 2018; pp. 3291–3300.

33. Cai, J.; Gu, S.; Zhang, L. Learning a deep single image contrast enhancer from multi-exposure images. *IEEE Trans. Image Process.* **2018**, *27*, 2049–2062. [CrossRef] [PubMed]

34. Jia, X.; Zhu, C.; Li, M.; Tang, W.; Zhou, W. LLVIP: A Visible-infrared Paired Dataset for Low-light Vision. In Proceedings of the IEEE/CVF International Conference on Computer Vision, Montreal, QC, Canada, 10–17 October 2021; pp. 3496–3504.

35. Park, J.; Lee, J.Y.; Yoo, D.; Kweon, I.S. Distort-and-recover: Color enhancement using deep reinforcement learning. In Proceedings of the IEEE Conference on Computer Vision and Pattern Recognition, Salt Lake City, UT, USA, 18–23 June 2018; pp. 5928–5936.

36. Zhang, Y.; Zhang, J.; Guo, X. Kindling the darkness: A practical low-light image enhancer. In Proceedings of the 27th ACM International Conference on Multimedia, Nice, France, 21–25 October 2019; pp. 1632–1640.

37. Zheng, C.; Shi, D.; Shi, W. Adaptive Unfolding Total Variation Network for Low-Light Image Enhancement. In Proceedings of the IEEE/CVF International Conference on Computer Vision, Montreal, QC, Canada, 10–17 October 2021; pp. 4439–4448.

38. Wang, Y.; Wan, R.; Yang, W.; Li, H.; Chau, L.P.; Kot, A.C. Low-Light Image Enhancement with Normalizing Flow. *arXiv* **2021**, arXiv:2109.05923.

39. Wang, W.; Wu, X.; Yuan, X.; Gao, Z. An experiment-based review of low-light image enhancement methods. *IEEE Access* **2020**, *8*, 87884–87917. [CrossRef]

40. Qi, Y.; Yang, Z.; Sun, W.; Lou, M.; Lian, J.; Zhao, W.; Deng, X.; Ma, Y. A Comprehensive Overview of Image Enhancement Techniques. *Arch. Comput. Methods Eng.* **2021**, *29*, 583–607. [CrossRef]

41. Li, C.; Guo, C.; Han, L.H.; Jiang, J.; Cheng, M.M.; Gu, J.; Loy, C.C. Low-light image and video enhancement using deep learning: A survey. *IEEE Trans. Pattern Anal. Mach. Intell.* **2021**, *1*. doi: 10.1109/TPAMI.2021.3126387. [CrossRef]

42. Land, E.H.; McCann, J.J. Lightness and retinex theory. *Josa* **1971**, *61*, 1–11. [CrossRef]

43. Land, E.H. Recent advances in retinex theory and some implications for cortical computations: Color vision and the natural image. *Proc. Natl. Acad. Sci. USA* **1983**, *80*, 5163. [CrossRef]

44. Provenzi, E.; De Carli, L.; Rizzi, A.; Marini, D. Mathematical definition and analysis of the Retinex algorithm. *JOSA A* **2005**, *22*, 2613–2621. [CrossRef]

45. Marini, D.; Rizzi, A. A computational approach to color adaptation effects. *Image Vis. Comput.* **2000**, *18*, 1005–1014. [CrossRef]

46. Land, E.H. An alternative technique for the computation of the designator in the retinex theory of color vision. *Proc. Natl. Acad. Sci. USA* **1986**, *83*, 3078–3080. [CrossRef] [PubMed]

47. Cooper, T.J.; Baqai, F.A. Analysis and extensions of the Frankle-McCann Retinex algorithm. *J. Electron. Imaging* **2004**, *13*, 85–92. [CrossRef]

48. Provenzi, E.; Fierro, M.; Rizzi, A.; De Carli, L.; Gadia, D.; Marini, D. Random spray Retinex: A new Retinex implementation to investigate the local properties of the model. *IEEE Trans. Image Process.* **2006**, *16*, 162–171. [CrossRef] [PubMed]

49. Fu, X.; Liao, Y.; Zeng, D.; Huang, Y.; Zhang, X.P.; Ding, X. A probabilistic method for image enhancement with simultaneous illumination and reflectance estimation. *IEEE Trans. Image Process.* **2015**, *24*, 4965–4977. [CrossRef] [PubMed]

50. Wang, S.; Zheng, J.; Hu, H.M.; Li, B. Naturalness preserved enhancement algorithm for non-uniform illumination images. *IEEE Trans. Image Process.* **2013**, *22*, 3538–3548. [CrossRef] [PubMed]

51. Zosso, D.; Tran, G.; Osher, S.J. Non-Local Retinex—A Unifying Framework and Beyond. *SIAM J. Imaging Sci.* **2015**, *8*, 787–826. [CrossRef]

52. Kimmel, R.; Elad, M.; Shaked, D.; Keshet, R.; Sobel, I. A variational framework for retinex. *Int. J. Comput. Vis.* **2003**, *52*, 7–23. [CrossRef]

53. Ma, W.; Osher, S. A TV Bregman iterative model of Retinex theory. *Inverse Probl. Imaging* **2012**, *6*, 697. [CrossRef]

54. Ma, W.; Morel, J.M.; Osher, S.; Chien, A. An L 1-based variational model for Retinex theory and its application to medical images. In Proceedings of the CVPR, Colorado Springs, CO, USA, 20–25 June 2011; pp. 153–160.

55. Fu, X.; Zeng, D.; Huang, Y.; Ding, X.; Zhang, X.P. A variational framework for single low light image enhancement using bright channel prior. In Proceedings of the 2013 IEEE Global Conference on Signal and Information Processing, Austin, TX, USA, 3–5 December 2013; pp. 1085–1088.

56. Ng, M.K.; Wang, W. A total variation model for Retinex. *SIAM J. Imaging Sci.* **2011**, *4*, 345–365. [CrossRef]

57. Fu, X.; Zeng, D.; Huang, Y.; Liao, Y.; Ding, X.; Paisley, J. A fusion-based enhancing method for weakly illuminated images. *Signal Process.* **2016**, *129*, 82–96. [CrossRef]

58. Cai, B.; Xu, X.; Guo, K.; Jia, K.; Hu, B.; Tao, D. A joint intrinsic-extrinsic prior model for retinex. In Proceedings of the IEEE International Conference on Computer Vision, Venice, Italy 22–29 October 2017; pp. 4000–4009.

59. Ying, Z.; Li, G.; Ren, Y.; Wang, R.; Wang, W. A new low-light image enhancement algorithm using camera response model. In Proceedings of the IEEE International Conference on Computer Vision Workshops, Venice, Italy, 22–29 October 2017; pp. 3015–3022.

60.   Zhang, Y.; Guo, X.; Ma, J.; Liu, W.; Zhang, J. Beyond brightening low-light images. *Int. J. Comput. Vis.* **2021**, *129*, 1013–1037. [CrossRef]

61.   Hai, J.; Xuan, Z.; Yang, R.; Hao, Y.; Zou, F.; Lin, F.; Han, S. R2RNet: Low-light Image Enhancement via Real-low to Real-normal Network. *arXiv* **2021**, arXiv:2106.14501.

62.   Liu, R.; Ma, L.; Zhang, J.; Fan, X.; Luo, Z. Retinex-inspired unrolling with cooperative prior architecture search for low-light image enhancement. In Proceedings of the IEEE/CVF Conference on Computer Vision and Pattern Recognition, Nashville, TN, USA, 20–25 June 2021; pp. 10561–10570.

63.   Wang, R.; Xu, X.; Fu, C.W.; Lu, J.; Yu, B.; Jia, J. Seeing Dynamic Scene in the Dark: A High-Quality Video Dataset With Mechatronic Alignment. In Proceedings of the IEEE/CVF International Conference on Computer Vision, Montreal, QC, Canada, 10–17 October 2021; pp. 9700–9709.

64.   Zhao, Z.; Xiong, B.; Wang, L.; Ou, Q.; Yu, L.; Kuang, F. RetinexDIP: A Unified Deep Framework for Low-light Image Enhancement. *IEEE Trans. Circuits Syst. Video Technol.* **2021**, *32*, 1076–1088. [CrossRef]

65.   Yu, R.; Liu, W.; Zhang, Y.; Qu, Z.; Zhao, D.; Zhang, B. Deepexposure: Learning to expose photos with asynchronously reinforced adversarial learning. In Proceedings of the 32nd International Conference on Neural Information Processing Systems, Montreal, QC, Canada, 3–8 December 2018; pp. 2153–2163.

66.   Cheng, H.; Shi, X. A simple and effective histogram equalization approach to image enhancement. *Digit. Signal Process.* **2004**, *14*, 158–170. [CrossRef]

67.   Kim, Y.T. Contrast enhancement using brightness preserving bi-histogram equalization. *IEEE Trans. Consum. Electron.* **1997**, *43*, 1–8.

68.   Ibrahim, H.; Kong, N.S.P. Brightness preserving dynamic histogram equalization for image contrast enhancement. *IEEE Trans. Consum. Electron.* **2007**, *53*, 1752–1758. [CrossRef]

69.   Guan, X.; Jian, S.; Hongda, P.; Zhiguo, Z.; Haibin, G. An image enhancement method based on gamma correction. In Proceedings of the 2009 Second International Symposium on Computational Intelligence and Design, Changsha, China, 12–14 December 2009; Volume 1, pp. 60–63.

70.   Tao, L.; Zhu, C.; Xiang, G.; Li, Y.; Jia, H.; Xie, X. LLCNN: A convolutional neural network for low-light image enhancement. In Proceedings of the 2017 IEEE Visual Communications and Image Processing (VCIP), St. Petersburg, FL, USA, 10–13 December 2017; pp. 1–4.

71.   Lv, F.; Lu, F.; Wu, J.; Lim, C. MBLLEN: Low-Light Image/Video Enhancement Using CNNs. In Proceedings of the BMVC, Newcastle, UK, 3–6 September 2018; p. 220.

72.   Wang, W.; Wei, C.; Yang, W.; Liu, J. GLADNet: Low-light enhancement network with global awareness. In Proceedings of the 2018 13th IEEE International Conference on Automatic Face & Gesture Recognition (FG 2018), Jodhpur, India, 15–18 December 2018; pp. 751–755.

73.   Jiang, Y.; Gong, X.; Liu, D.; Cheng, Y.; Fang, C.; Shen, X.; Yang, J.; Zhou, P.; Wang, Z. Enlightengan: Deep light enhancement without paired supervision. *IEEE Trans. Image Process.* **2021**, *30*, 2340–2349. [CrossRef]

74.   Xiong, W.; Liu, D.; Shen, X.; Fang, C.; Luo, J. Unsupervised real-world low-light image enhancement with decoupled networks. *arXiv* **2020**, arXiv:2005.02818.

75.   Xia, Z.; Gharbi, M.; Perazzi, F.; Sunkavalli, K.; Chakrabarti, A. Deep Denoising of Flash and No-Flash Pairs for Photography in Low-Light Environments. In Proceedings of the IEEE/CVF Conference on Computer Vision and Pattern Recognition, Nashville, TN, USA, 20–25 June 2021; pp. 2063–2072.

76.   Le, H.A.; Kakadiaris, I.A. SeLENet: A semi-supervised low light face enhancement method for mobile face unlock. In Proceedings of the 2019 International Conference on Biometrics (ICB), Crete, Greece, 4–7 June 2019; pp. 1–8.

77.   Yang, W.; Wang, S.; Fang, Y.; Wang, Y.; Liu, J. From fidelity to perceptual quality: A semi-supervised approach for low-light image enhancement. In Proceedings of the IEEE/CVF Conference on Computer Vision and Pattern Recognition, Seattle, WA, USA, 13–19 June 2020; pp. 3063–3072.

78.   Qiao, Z.; Xu, W.; Sun, L.; Qiu, S.; Guo, H. Deep Semi-Supervised Learning for Low-Light Image Enhancement. In Proceedings of the 2021 14th International Congress on Image and Signal Processing, BioMedical Engineering and Informatics (CISP-BMEI), Online, 23–25 October 2021; pp. 1–6.

79.   Wu, W.; Wang, W.; Jiang, K.; Xu, X.; Hu, R. Self-Supervised Learning on A Lightweight Low-Light Image Enhancement Model with Curve Refinement. In Proceedings of the ICASSP 2022–2022 IEEE International Conference on Acoustics, Speech and Signal Processing (ICASSP), Singapore, 22–27 May 2022; pp. 1890–1894.

80.   Guo, C.G.; Li, C.; Guo, J.; Loy, C.C.; Hou, J.; Kwong, S.; Cong, R. Zero-reference deep curve estimation for low-light image enhancement. In Proceedings of the IEEE Conference on Computer Vision and Pattern Recognition (CVPR), Seattle, WA, USA, 13–19 June 2020; pp. 1780–1789.

81.   Wang, L.W.; Liu, Z.S.; Siu, W.C.; Lun, D.P. Lightening network for low-light image enhancement. *IEEE Trans. Image Process.* **2020**, *29*, 7984–7996. [CrossRef]

82.   Lee, C.; Lee, C.; Kim, C.S. Contrast enhancement based on layered difference representation of 2D histograms. *IEEE Trans. Image Process.* **2013**, *22*, 5372–5384. [CrossRef]

83.   Ma, K.; Zeng, K.; Wang, Z. Perceptual quality assessment for multi-exposure image fusion. *IEEE Trans. Image Process.* **2015**, *24*, 3345–3356. [CrossRef] [PubMed]

84. Lv, F.; Li, Y.; Lu, F. Attention guided low-light image enhancement with a large scale low-light simulation dataset. *arXiv* **2019**, arXiv:1908.00682.

85. Loh, Y.P.; Chan, C.S. Getting to know low-light images with the exclusively dark dataset. *Comput. Vis. Image Underst.* **2019**, *178*, 30–42. [CrossRef]

86. Gonzalez, R.C. *Digital Image Processing*; Pearson Education India: Noida, India, 2009.

87. Mittal, A.; Moorthy, A.K.; Bovik, A.C. No-reference image quality assessment in the spatial domain. *IEEE Trans. Image Process.* **2012**, *21*, 4695–4708. [CrossRef] [PubMed]

88. Mittal, A.; Soundararajan, R.; Bovik, A.C. Making a "completely blind" image quality analyzer. *IEEE Signal Process. Lett.* **2012**, *20*, 209–212. [CrossRef]

89. Papasaika-Hanusch, H. *Digital image PROCESSING Using Matlab*; Institute of Geodesy and Photogrammetry, ETH Zurich: Zurich, Switzerland, 1967; Volume 63.

90. Celik, T.; Tjahjadi, T. Contextual and variational contrast enhancement. *IEEE Trans. Image Process.* **2011**, *20*, 3431–3441. [CrossRef]

91. Pizer, S.M. Contrast-limited adaptive histogram equalization: Speed and effectiveness stephen m. pizer, r. eugene johnston, james p. ericksen, bonnie c. yankaskas, keith e. muller medical image display research group. In Proceedings of the First Conference on Visualization in Biomedical Computing, Atlanta, GA, USA, 22–25 May 1990; Volume 337.

92. Cao, G.; Huang, L.; Tian, H.; Huang, X.; Wang, Y.; Zhi, R. Contrast enhancement of brightness-distorted images by improved adaptive gamma correction. *Comput. Electr. Eng.* **2018**, *66*, 569–582. [CrossRef]

93. Ying, Z.; Li, G.; Gao, W. A bio-inspired multi-exposure fusion framework for low-light image enhancement. *arXiv* **2017**, arXiv:1711.00591.

94. Afifi, M.; Derpanis, K.G.; Ommer, B.; Brown, M.S. Learning Multi-Scale Photo Exposure Correction. In Proceedings of the IEEE/CVF Conference on Computer Vision and Pattern Recognition, Nashville, TN, USA, 20–25 June 2021; pp. 9157–9167.

95. Zhang, F.; Li, Y.; You, S.; Fu, Y. Learning Temporal Consistency for Low Light Video Enhancement From Single Images. In Proceedings of the IEEE/CVF Conference on Computer Vision and Pattern Recognition (CVPR), Nashville, TN, USA, 20–25 June 2021; pp. 4967–4976.

96. Li, C.; Guo, J.; Porikli, F.; Pang, Y. LightenNet: A convolutional neural network for weakly illuminated image enhancement. *Pattern Recognit. Lett.* **2018**, *104*, 15–22. [CrossRef]

97. Hu, Y.; He, H.; Xu, C.; Wang, B.; Lin, S. Exposure: A white-box photo post-processing framework. *ACM Trans. Graph. (TOG)* **2018**, *37*, 1–17. [CrossRef]

98. Ying, Z.; Li, G.; Ren, Y.; Wang, R.; Wang, W. A new image contrast enhancement algorithm using exposure fusion framework. In Proceedings of the International Conference on Computer Analysis of Images and Patterns, Ystad, Sweden, 22–24 August 2017; Springer: Berlin/Heidelberg, Germany, 2017; pp. 36–46.

99. Pu, T.; Wang, S. Perceptually motivated enhancement method for non-uniformly illuminated images. *IET Comput. Vis.* **2018**, *12*, 424–433. [CrossRef]

100. Al-Ameen, Z. Nighttime image enhancement using a new illumination boost algorithm. *IET Image Process.* **2019**, *13*, 1314–1320. [CrossRef]

101. Everingham, M.; Van Gool, L.; Williams, C.K.; Winn, J.; Zisserman, A. The pascal visual object classes (voc) challenge. *Int. J. Comput. Vis.* **2010**, *88*, 303–338. [CrossRef]

 *remote sensing*

Article

Towards Robust Semantic Segmentation of Land Covers in Foggy Conditions

Weipeng Shi [1], Wenhu Qin [1,*] and Allshine Chen [2]

[1] School of Instrument Science and Engineering, Southeast University, Nanjing 210096, China
[2] Health Sciences Center, University of Oklahoma, Oklahoma City, OK 73106, USA
* Correspondence: qinwenhu@seu.edu.cn

Abstract: When conducting land cover classification, it is inevitable to encounter foggy conditions, which degrades the performance by a large margin. Robustness may be reduced by a number of factors, such as aerial images of low quality and ineffective fusion of multimodal representations. Hence, it is crucial to establish a reliable framework that can robustly understand remote sensing image scenes. Based on multimodal fusion and attention mechanisms, we leverage HRNet to extract underlying features, followed by the Spectral and Spatial Representation Learning Module to extract spectral-spatial representations. A Multimodal Representation Fusion Module is proposed to bridge the gap between heterogeneous modalities which can be fused in a complementary manner. A comprehensive evaluation study of the fog-corrupted Potsdam and Vaihingen test sets demonstrates that the proposed method achieves a mean $F1_{score}$ exceeding 73%, indicating a promising performance compared to State-Of-The-Art methods in terms of robustness.

Keywords: semantic segmentation; attention mechanism; robust deep learning; remote sensing; data fusion

Citation: Shi, W.; Qin, W.; Chen, A. Towards Robust Semantic Segmentation of Land Covers in Foggy Conditions. *Remote Sens.* **2022**, *14*, 4551. https://doi.org/10.3390/rs14184551

Academic Editor: Gwanggil Jeon

Received: 5 August 2022
Accepted: 8 September 2022
Published: 12 September 2022

Publisher's Note: MDPI stays neutral with regard to jurisdictional claims in published maps and institutional affiliations.

1. Introduction

Computer vision has emerged as a powerful and labor-saving tool for automatic scene parsing of remote sensing images (RSIs). Land cover classification (LCC) of aerial imagery, which is also known as semantic labeling or segmentation, assigns a class label to each pixel in RSIs. As an integral part of computer vision, semantic segmentation plays a pivotal role in remote sensing for rapid and accurate detection. A considerable amount of literature has been published concerning applications like landslide extraction [1], road extraction [2], collapsed building detection [3], and so on. Semantic segmentation models can be summarized into three categories [4], namely FCNets [5] which yield a coarse feature map directly from the low-resolution representation, UNets [6] which perform high-resolution recovery from the downsampled representation as well as HRNets [4] which retain the high-resolution representation during all procedures. Despite plenty of research exploring semantic segmentation based on natural scene images (NSIs), there is still a lack of scientific literature specifically focusing on robust remote sensing in foggy conditions.

RSIs with fog are distinguished from NSIs by a number of challenging characteristics that may result in decreased classification robustness. It is believed that there exist two major challenges when it comes to LCC with fog. First, model robustness is susceptible to fog corrupted RSIs, which refer to a series of issues [7], including intra-class heterogeneity, inter-class homogeneity, geometric size diversity, and so on. In terms of intra-class heterogeneity, the models tend to classify objects with distinctive appearances as disparate species; they may belong to the identical yet [8]. For instance, various materials and structures may lead to different appearances and textures. Whereas, fog-covered objects affiliated with diverse species frequently exhibit close characteristics when they are made up of the same material. The concrete building and the impervious surface in Figure 1 appear similar. Models assume they belong to the same class by mistake accidentally, which is inter-class homogeneity. Additionally, objects under fog exhibit geometric size diversity and only a robust model can capture multi-scale attributes in RSIs. Overall, RSIs with low quality can exert negative impacts on classification robustness. Second, when dealing with dense fog, a single sensor is not always effective. It is difficult for a single optical camera to classify objects robustly when they are partially obscured. Figure 1 illustrates the failing cases, such as the shadow of buildings, fog coverage, and cars parked under the trees. The optical input is rich in semantic details, while Digital Surface Model (DSM) provides discriminative height information. It is imperative to excavate informative cues from multimodal inputs.

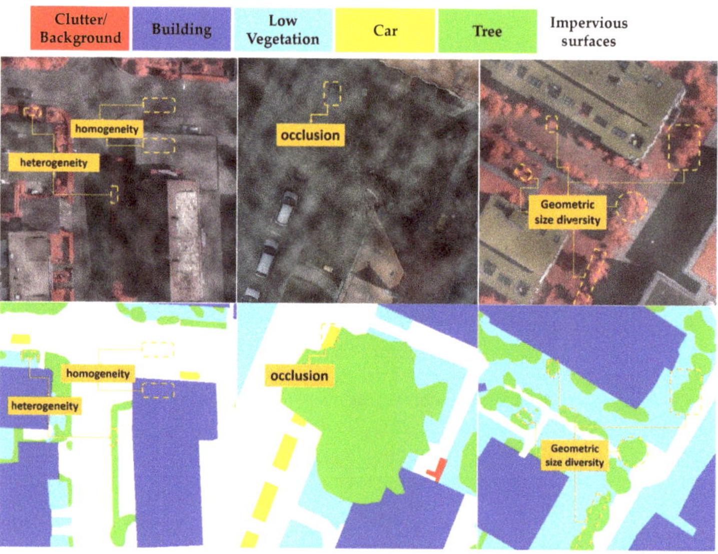

Figure 1. Challenges associated with robust LCC in the areas covered with fog. The first row is the fog corrupted image and the second row is the corresponding ground truth (GT).

To address the listed challenges of LCC in foggy conditions, we propose a framework with superior robustness based on attention mechanisms and multimodal fusion. We adopt HRNet as the backbone. Through compiling representations from all the high-to-low resolution streams in parallel, HRNet is robust to intra-class heterogeneity and geometric size diversity. The proposed Spectral and Spatial Representation Learning (SSRL) module probes into the relationship between spectral channels and spatial locations to improve robustness to intra-class heterogeneity. Thus, the output representation is gifted with semantic information and spatial accuracy. The introduced Multimodal Representation Fusion Module (MRFM) investigates the fusion of multimodal remote sensing data to learn the boundary connectivity and contour closure in RSIs to cope well with object occlusion and inter-class homogeneity issues. In summary, the main contributions are as follows:

- Based on multimodal fusion and attention mechanisms, we propose a robust end-to-end model that can fuse the optical and DSM input for LCC.
- Adopting HRNet as the backbone, we propose and incorporate SSRL and MRFM into the framework. To enhance the semantic information, a lightweight SSRL is inserted to capture the long-range dependencies and explore the interactions between various spectral channels. MRFM is employed for the effective fusion of multimodal remote sensing data. All the components cooperate and contribute to the classification robustness.
- We conduct an ablation study to evaluate the effectiveness of the proposed framework, including functions of different modules and modal inputs.

- We compare our model with SOTA methods to demonstrate the robustness against natural noise.

2. Related Work

There is a large volume of published studies describing how to conduct LCC. Most publications concentrate on accuracy instead of robustness, which is also vital in daily application. Thus far, several studies investigating robustness are predominantly associated with NSIs, such as scenes in ImageNet, Cityscape, BDD100k, and so on. Nonetheless, there is still a lack of relevant research focusing on LCC robustness. RSIs captured in foggy conditions are characterized by low quality, which poses challenges to robustness. This paper aims to design a robust model which can improve the classification performance in harsh environments. Our work refers to LCC, model robustness, and multimodal fusion. This section discusses the related work from these three perspectives.

2.1. Land Cover Classification

LCC actually refers to semantic segmentation of land covers using computer vision. There has been a great deal of research into semantic segmentation focusing on classification accuracy. Conventional segmentation algorithms are normally put forward on the premise of basic image attributes, e.g., grey-scale mutations are utilized to detect edges. Comparable grey scale values are partitioned into several regions according to the predefined criteria. However, it is extremely complex to detect boundaries when there exist substantial grey-scale changes. Considerable evidence has accumulated to show that deep learning-based models are more suitable for the semantic segmentation of NSIs. These models can usually be divided into three groups [4], namely, FCNet [5] type, UNet [6,9] type, and HRNet [4] type. FCNets learn the representation from high to low resolution in series to extract coarse feature maps. This group includes models like Deeplab [10], DenseASPP [11], PSPNet [12], and so on. UNets learn the encoded low-resolution representation and then recover to the high-resolution representation. Analogous models are DeepLabV3+ [13], SegNet [14], and so on. Moreover, ref. [9] inserts a cascaded dilated convolution in UNet to capture objects of diverse shapes, which is an effective approach to enhance robustness to multi-scale issues. Different from [9], we alleviate the influence of diverse shapes by adopting HRNet as the backbone because it retains a high-resolution representation throughout the process [15].

2.2. Model Robustness

Although neural networks are highly accurate for classification, they are not always as robust as human beings while actually applied [16]. Ref. [16] suggests that building multimodal and multitasking systems based on multi-sensor fusion is indispensable for robust decisions. Noises are categorized into three main groups [17], mainly adversarial noises, systematic noises, and natural noises. By comparing the robustness of three types of models, ref. [17] shows that CNN is more robust under natural noise and systematic noise, while Transformer is more robust against adversarial noise. Adversarial noise is the result of ambiguous decision-making at boundaries due to the limited training dataset and the inability to cover the whole sampling space. A small perturbation will always lead to completely distinct results. Ref. [18] constructed ImageNet-P and ImageNet-C datasets on top of ImageNet to facilitate researchers to evaluate and test the corruption and perturbation robustness. Based on this work, ref. [19] investigates the robustness of semantic segmentation. Researchers find that the Atrous Spatial Pyramid Pooling module significantly improves robustness, while the generalization performance depends heavily on the corruption degrees. Furthermore, there are some recent works on the adversarial noise robustness of Visual Transformers (ViTs). Ref. [20] found that shallow features in ViTs enable it to possess a better generalization than CNN, thus, better coping with adversarial noise. Meanwhile, the ensemble operation of CNN and ViTs can also improve the model robustness [21].

2.3. Attention Mechanism

A large and growing body of literature has investigated the role of attention mechanism [22] in deep visual models. Ref. [23] proposes an efficient channel module to explore the cross-channel interactions without dimension reduction. Ref. [24] designs a global context module to model the long-range dependencies with significantly less computation. There is a consensus among researchers that multi-head attention in ViTs[25] has acquired SOTA due to the uniform representation. Swin transformer [26] is capable of modeling input of different scales flexibly and the complexity is linear with input sizes. SETR [27] presents a multi-level feature fusion module to classify each pixel at a fine-grained level. Segformer [28] removes the complex position encoding binding a lightweight multi-layer perceptron to output feature maps of various sizes. Volo [29] introduces a novel outlook attention to grasp both coarse and fine-grained representations.

2.4. Multimodal Fusion

Data collected by a single sensor are often flawed and multi-sensor fusion is fundamental for accurate and robust decisions. Ref. [30] put forward a top-down pyramid fusion architecture for multimodal fusion. It is lightweight and can extract complementary features from multi-sources. Ref. [31] proposes a new and lightweight depth fusion transformer network for LCC, with different backbones extracting features from various inputs. Ref. [32] explores the pros and cons of early fusion along with late fusion to show that they all can utilize the complementarity of multimodal inputs. To calibrate features of the current modality from spatial and channel dimensions, ref. [33] has developed a Cross-Modal Feature Rectification Module for feature extraction. A multimodal fusion module was proposed in [15] to explore complementary features of heterogeneous inputs. In contrast, our method exploits the discriminate representation of each input from the perspective of the channel and spatial location before multimodal fusion, which greatly enhances the robustness of low-quality RSIs.

Overall, most studies remain narrow in focusing only on NSIs instead of RSIs. RSIs are characteristic of challenges of high resolution, multi-scales, class imbalance, occlusion, and so on. The attention mechanism is capable of capturing long-range dependencies. We extend ideas from deep learning and RGB-D semantic segmentation as well as an attention mechanism to establish a practical framework that can robustly classify land covers in foggy conditions.

3. Core of the Framework

3.1. Overview

The overall framework is illustrated in Figure 2. HRNetV2-W48 is adopted as the backbone for feature extraction. UperHead from [34] serves as the decoder. Each batch combines an Optical image ($\mathbf{X_{RGB}} \in \mathbb{R}^{C \times H \times W}$) with the corresponding DSM ($\mathbf{X_{DSM}} \in \mathbb{R}^{1 \times H \times W}$), which contains the height information of land covers. H and W denote the height and width of RSIs, respectively. There is a considerable amount of noise in low-quality RSIs. We intend to design SSRL in such a manner that it would extract useful and discriminate representation efficiently without incurring excessive computational costs. Conventional methods simply aggregate two modalities without obtaining complementary features effectively. By contrast, MRFM, which is dedicated to multimodal fusion, exploits the complementarity between heterogeneous data. The backbone and decoder can be replaced by the other models. We will dig into the detailed design of HRNet, SSRL, and MRFM in Sections 3.2–3.4, respectively.

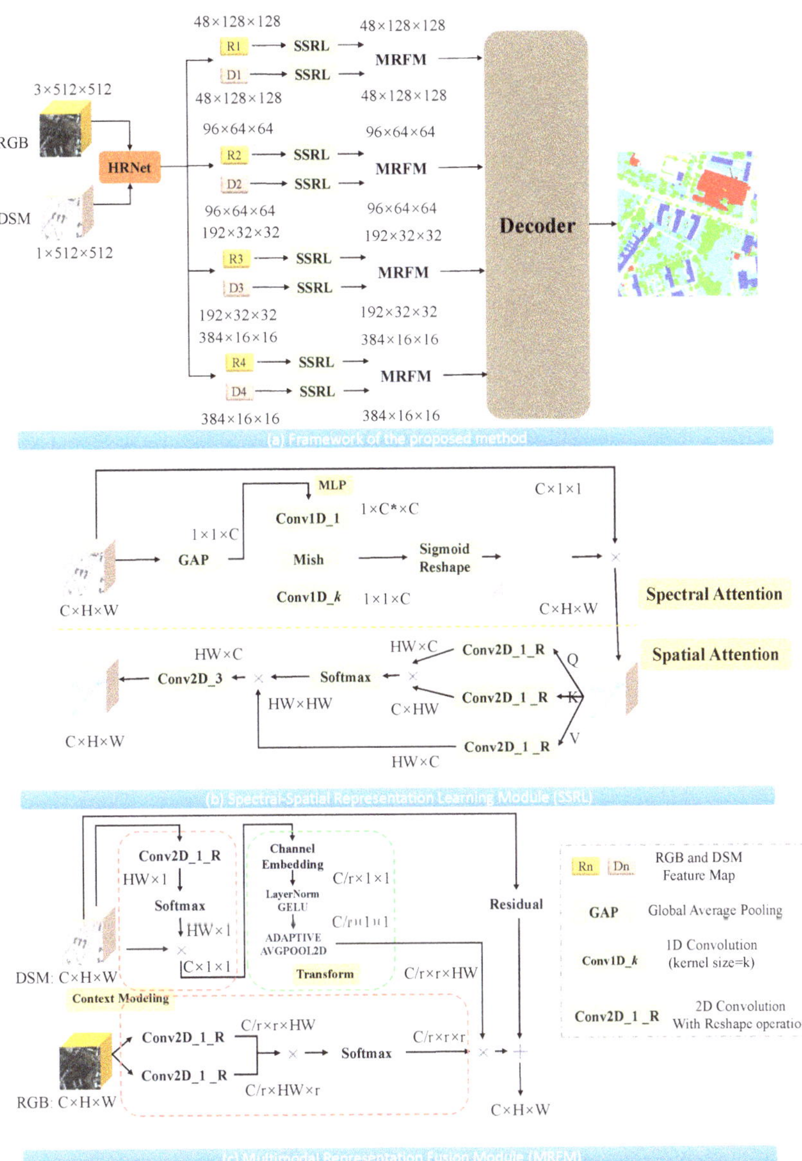

Figure 2. (**a**) Illustration of the proposed framework for LCC in foggy conditions. The input consists of a pair of the optical image and DSM. UperHead is selected as the decoder. (**b**) Detailed design of SSRL, which is composed of Spectral Attention and Spatial Attention modules. C and C^* are different channel numbers. (**c**) The framework of MRFM is to explore the complementarity of heterogeneous inputs.

3.2. Backbone

Low-quality RSIs collected under fog suffer from the issues of intra-class heterogeneity and diverse geometric shapes, which impairs classification robustness. LCC is actually a dense pixel prediction task that requires a strong backbone with powerful modeling capability. Semantic segmentation networks are often constructed based on encoder-decoder architecture, where ResNet [35] is usually applied as the encoder. It is characterized by the accurate prediction of the spatial location at the low-level stage but is limited by a small receptive field that lacks consistent semantic information. This could lead to blurry classification. At the high-level stage, the network possesses a larger receptive field to make fine semantic predictions, but is deficient in the global representation. Consequently, conventional CNNs are susceptible to the above issues which are attributed to a loss of spatial details with the degradation of resolution.

HRNet is composed of four parallel branches pertaining to different resolutions. They constantly exchange information across multiple scales. High-resolution representations contain more spatial details, while the low-resolution representations are more capable of fine-grained classification. By maintaining a high-resolution representation, HRNet can generate spatially accurate feature maps that contain semantic information abundantly. Accordingly, HRNet-W48 is selected as our backbone for robust semantic segmentation. It can learn discriminative and distinct representations efficiently. Furthermore, HRNet is capable of integrating both local and global features with multiple scales, thereby increasing robustness in the presence of fog corruptions.

The hierarchical structure of HRNet is illustrated in Figure 3, which consists of 4 multi-resolution branches, each with resolutions of $1/4, 1/8, 1/16, 1/32$. Each branch can be partitioned into 4 stages, and the output channel number of each branch is $C, 2C, 4C, 8C$. Between each stage, there are blocks of multi-resolution fusion which consist of a 3×3 stride convolution integrating with a 1×1 upsampling layer, represented by the crossed lines. The fusion module serves as a mechanism for transferring feature information between branches of different resolutions. HRNet can be applied to semantic segmentation by accessing a 1×1 convolution for mixing and merging the representations from four branch outputs to align channel numbers. Detailed convolution parameters are shown in Table 1 where parameters for each stage are in the form $[a \times a, nC] \times b \times c$. '[]' represents the residual connection unit. Parameters a,b represent kernel size and duplication times of the residual unit separately. c means to repeat entire modularized part c times. Four basic blocks like $\begin{bmatrix} 3 \times 3, & nC \\ 3 \times 3, & nC \end{bmatrix}$ accompanying with fusion modules constitute each branch of HRNet.

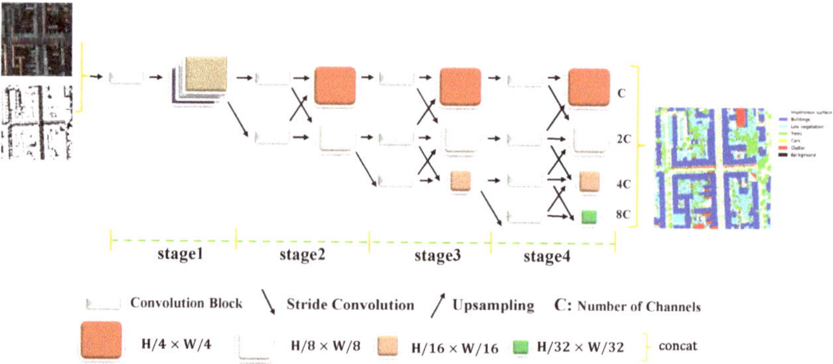

Figure 3. Overview of the backbone HRNet. HRNet maintains high-resolution representations and exchanges information throughout branches by means of 1×1 and 3×3 convolution. It can cope well with intra-class heterogeneity and multi-scale issues in RSIs with fog.

Table 1. Detailed HRNET specifications about every stage and channels.

Downsp. Rate	Stage1	Stage2	Stage3	Stage4
4×	$\begin{bmatrix} 1 \times 1, 64 \\ 3 \times 3, 64 \\ 1 \times 1, 256 \end{bmatrix} \times 4 \times 1$	$\begin{bmatrix} 3 \times 3, C \\ 3 \times 3, C \end{bmatrix} \times 4 \times 1$	$\begin{bmatrix} 3 \times 3, C \\ 3 \times 3, C \end{bmatrix} \times 4 \times 1$	$\begin{bmatrix} 3 \times 3, C \\ 3 \times 3, C \end{bmatrix} \times 4 \times 3$
8×	$\begin{bmatrix} 3 \times 3, 2C \\ 3 \times 3, 2C \end{bmatrix} \times 4 \times 1$	$\begin{bmatrix} 3 \times 3, 2C \\ 3 \times 3, 2C \end{bmatrix} \times 4 \times 4$	$\begin{bmatrix} 3 \times 3, 2C \\ 3 \times 3, 2C \end{bmatrix} \times 4 \times 3$	
16 ×		$\begin{bmatrix} 3 \times 3, 4C \\ 3 \times 3, 4C \end{bmatrix} \times 4 \times 4$	$\begin{bmatrix} 3 \times 3, 4C \\ 3 \times 3, 4C \end{bmatrix} \times 4 \times 3$	
32 ×				$\begin{bmatrix} 3 \times 3, 8C \\ 3 \times 3, 8C \end{bmatrix} \times 4 \times 3$

3.3. Spectral and Spatial Representation Learning

Generally, RSIs captured in foggy conditions are of low quality, with a significant amount of noise existing in heterogeneous inputs. Merely incorporating encoded representations into the decoder will bring excess redundancy to the network, which reduces the classification robustness. There have been some proposals to enhance semantic features through an attention mechanism. The non-local operation proposed by [22] can obtain the attention map corresponding to a specific query tensor for modeling the global context. After rigorous experiments in [24], researchers argue that the gap between attention activation maps corresponding to different query locations is narrow, illustrating the non-necessity of query weights. In addition, ViT [25] can also enhance the semantic information through multi-head attention. It is featured with numerous parameters, high complexity as well as overfitting. This computationally intensive approach, however, ignores the correlation between various spectral channels. Additionally, the channel attention mechanism [23,36] has also been proposed to explore the interaction between different channels. Nevertheless, simply regarding 2D images as 1D disrupts the dependencies between different positions, which reduces the robustness in capturing long-range relationships. Inspired by CBAM [37], we propose SSRL which is composed of Spectral Attention and Spatial Attention. Spectral attention is to explore interdependencies between different spectral channels, thereby improving semantic representations. Spatial attention is to capture long-range dependencies. Thus, SSRL can generate a global context and acquire correlations between various pixels and spectral channels to improve the robustness of intra-class heterogeneity.

Spectral Attention is illustrated in the upper half of Figure 2b. It is constructed for acquiring the spectral-level representation weight $\Re_{spe}$ ($\Re_{spe} \in \mathbb{R}^{C \times 1 \times 1}$). SSRL firstly transforms the dimension of $\mathbf{X}_{RGB}$ or $\mathbf{X}_{DSM}$ ($\mathbf{X}_{RGB} \in \mathbb{R}^{C \times H \times W}$ or $\mathbf{X}_{DSM} \in \mathbb{R}^{C \times H \times W}$) to $1 \times 1 \times C$ (C is the channel number) using a global average pooling layer (GAP). The compressed tensor is fed into MLP ($\mathbf{W}_{mlp}$) to compute the interaction between k adjacent channels. The MLP consists of two one-dimensional convolution layers with kernel size 1 and k, denoted as $Conv1D\_1(\mathbf{W}_{1D\_1})$ and $Conv1D\_k(\mathbf{W}_{1D\_k})$, respectively. $Conv1D\_1$ is applied for the dimension reduction, converting the channel number from C to C^*. The same learnable weight values are shared among channels, where the efficiency is significantly improved because only k values are noted. Detailed formulas about Spectral Attention are as the following:

$$\Re_{spe} = \sigma \left(\mathbf{W}_{mlp} \frac{1}{WH} \sum_{i=1, j=1}^{W, H} \mathcal{X}_{ij} \right) \tag{1}$$

$$\mathbf{W}_{mlp}(x) = \mathbf{W}_{1D\_k}[F_{Mish}(\mathbf{W}_{1D\_1}(x))] \tag{2}$$

$$F_{Mish}(y) = x \cdot \tanh(\ln(1 + e^y)) \tag{3}$$

In Equation (1), σ is the sigmoid activation function. Local cross-channel attention interaction coverage is adaptively and dynamically adjusted according to the input channel numbers. There is a nonlinear mapping between the total number of channels C and the kernel size k of $Conv1D$. k increases with the number of channels. **odd** means to select the nearest odd number. $\mathbf{F}_{spe}$ means the feature map acquired through the Spectral Attention. The specific correspondence is as follows:

$$k = \psi(C) = \left| \frac{\log 2(C)}{2} + \frac{1}{2} \cdot \frac{1}{1 + e^{-C}} + \frac{1}{2} \right|_{odd} \tag{4}$$

$$\mathbf{F}_{spe} = \mathfrak{R}_{spe} \otimes X \tag{5}$$

Spatial Attention is illustrated in the lower panel of Figure 2b. This part is constructed for acquiring the spatial-level weight $\mathfrak{R}_{spe}$ ($\mathfrak{R}_{spe} \in \mathbb{R}^{HW \times HW}$). Through the linear transformation of $\mathfrak{R}_{spa}$, we can get three weight matrixes W_q, W_k, W_v. By calculating the interaction between positions (softmax operation, Equation (6)), the long range dependencies $\mathfrak{R}_{spa}$ can be captured. Finally, the feature map $\mathbf{F}_{spa}$ after Spectral Attention and Spatial Attention is obtained through matrix multiplication.

$$\mathfrak{R}_{spa} = \frac{\exp\left(\langle W_q \mathbf{F}_{spe}, W_k \mathbf{F}_{spe} \rangle\right)}{\sum\limits_{m} \exp\left(\langle W_q \mathbf{F}_{spe}, W_k \mathbf{F}_{spe} \rangle\right)} \tag{6}$$

$$\mathbf{F}_{spa} = W_c \left[\mathfrak{R}_{spa} \otimes (W_v \mathbf{F}_{spe}) \right] \tag{7}$$

3.4. Multimodal Representation Fusion Module

RSIs are collected by satellites or drones far from the ground, which will inevitably cause pixel loss owing to the atmosphere and clouds. Meanwhile, the terrestrial environment is a three-dimensional space and there exist complicated interactions between land covers. The quality of RSIs with fog is low as some of the land covers will be obscured by fog, resulting in a single optical sensor failing. This phenomenon reduced the classification robustness by a large margin. DSM built with lidar will not suffer from this. We fuse multiple inputs by designing an effective MRFM to explore the respective characteristics of each modality. Different from the early and late fusion strategy in [32], we exploit semantic representation of different modalities through the cross attention mechanism. This can improve robustness to inter-class homogeneity and object occlusion issues.

The structure of MRFM is illustrated in Figure 2c. Red and green dashed boxes in Figure 2c function as context modeling and transform respectively. We extract the coarse representation $\mathfrak{R}_{DSM}$ from DSM since it contains the height information of each land cover. Firstly, utilizing 1×1 convolution (W_v) with softmax in Equation (8) to obtain the global semantic key weight from DSM in the batch, this step is to obtain the coarse correlation feature maps between different locations. N_p signifies the number of all pixels. j is for pixel indexing. Then, the computational cost is reduced by bottleneck. r is the reduction coefficient, which is set to 16 by default. Layer Normalization and GELU activation functions are integrated which enable a faster convergence as well as a stable training process. This step plays a role of transform in exploring channel-wise features while Channel Embedding (W_{CE}) can enhance the nonlinearity as well as reduce the dimension. Finally, the linked adaptive average pooling layer (AAP) is utilized for the late fusion. γ, β in F_{LN} are trainable vectors, which is for affine transformation and ϵ is for numerical stability. E and Var mean expectation and standard deviation separately. We extract the fine-grained feature map from RGB input as a result of the abundant semantic features in this modality. Like non-local operation [22], we obtain the RGB feature weight $\mathfrak{R}_{RGB}$ through linear transformation and softmax. To exploit the complementary representation, we fuse both modalities using matrix multiplication accompanied by the residual connection. We likewise incorporate the residual structure into MRFM to make it easier for information to flow between layers, including providing feature reuse during forwarding propagation and mitigating the gradient vanishing phenomenon during backward propagation. Adding the original DSM input, dependencies between different positions obtained from DSM are fused with the exhaustive global information obtained from the optical so that each coarse position in DSM has an element-wise corresponding response generated.

$$\alpha(j) = \frac{e^{W_v X_{Dj}}}{\sum_{m=1}^{N_p} e^{W_v X_{Dm}}} \tag{8}$$

$$F_{GELU}(x) = 0.5x\left(1 + \tanh\left[\sqrt{2/\pi}\left(x + 0.044715x^3\right)\right]\right) \tag{9}$$

$$F_{LN}(x) = \frac{x - E[x]}{\sqrt{Var[x] + \epsilon}} * \gamma + \delta \tag{10}$$

$$\Re_{DSM} = AAP\langle F_{LN}\{F_{Gelu}[W_{CE}\sum_{j=1}^{N_p}\alpha(j)X_{Dj}]\}\rangle \tag{11}$$

$$\Re_{RGB} = \frac{\exp\left(\langle W_q\mathbf{X_{RGB}}, W_k\mathbf{X_{RGB}}\rangle\right)}{\sum_m \exp\left(\langle W_q\mathbf{X_{RGB}}, W_k\mathbf{X_{RGB}}\rangle\right)} \tag{12}$$

$$\mathbf{F}_{MRFM} = \Re_{RGB} \otimes \Re_{DSM} + X_{DSM} \tag{13}$$

3.5. Loss Function

From the analysis in Section 4.1, we can observe that the class imbalance issue exists in the ISPRS dataset. The proportion of impervious surface is sixteen times higher than the vehicle which is the tail class. Many deep models are heavily biased towards the dominant class during the training process and fail to classify the tail classes instead. Drawing on an extensive range of sources, we propose a unified loss function for our framework following a series of studies like [38,39]. Three elements constitute the unified loss. It is generalized for RSIs with long-tail class imbalance. FC, FT, and CE in Equation (14) stand for the modified focal loss [40], focal Tversky loss [41], and cross-entropy loss [42], respectively. *f* refers to the final loss originating from [38], whose input is the prediction result of MRFM. *aux* is the auxiliary loss employed to supervise the coarse object area estimation of SSRL output. x stands for the input data and y_{gt} is the ground truth. y_{pred} is the prediction output values. *back* denotes the background class. *gt, pred,* and *coar* are ground truth, prediction, and coarse feature map, respectively.

$$\mathcal{L}_{unified} = \alpha.\mathcal{L}_{\widehat{FC}}^f + (1-\alpha).\mathcal{L}_{\widehat{FT}}^f + \lambda \sum_{x \in \{rgb,d\}} \mathcal{L}_{CE}^{aux}\left(y_{pred}^{coar}, y_{gt}\right) \tag{14}$$

$$\mathcal{L}_{\widehat{FC}}^f = \frac{1}{N_c+1}[\sum_{j=1}^{N_c}\delta\mathcal{L}_{CE} + (1-\delta)(1-y_{pred}^{back})^{\gamma_1}\mathcal{L}_{CE}] \tag{15}$$

$$\mathcal{L}_{\widehat{FT}}^f = \frac{1}{N_c+1}[\sum_{j=1}^{N_c}(1 - DSC(x^j))^{1-\gamma_2} + (1 - DSC(x^{back}))] \tag{16}$$

$$DSC(x) = \frac{TP + \varepsilon}{TP + \delta FN + (1-\delta)FP + \varepsilon} \tag{17}$$

$$\mathcal{L}_{CE} = -\sum_{j=1}^{N_c}[y_{gt}log(y_{pred}) + (1-y_{gt})log(1-y_{pred})] \tag{18}$$

α (e.g., 0.5) is designed to balance the relative weights of the final loss while λ is the weight for auxiliary bootstrap loss. N_c is the number of classes. δ is the threshold parameter (e.g., 0.7) related to the proportion of positive and negative samples. γ controls the degree of down-weighting of easy samples while enhancing the rare. $\gamma_1|\gamma_2$ are 2 and 0.75 by default. ε is the small number for numerical stability in DSC, which acts similarly to the Tversky index to control the optimizing for output imbalance.

4. Experiment

4.1. Dataset Overview

ISPRS provided true orthophotos (TOP) of Potsdam and Vaihingen for model training and validation, whose resolutions are 6000×6000 and average 2500×2500, respectively. Ground sampling distance (GSD) is 5 cm and 9 cm. We need to eliminate the Image Index 7–10 in Potsdam dataset due to the error in GT. Each TOP has six classes: car, low vegetation, tree, building, impervious surface, and cluttered background. We split both datasets into the training, test, and validation sets according to the ratios of 0.8, 0.1, and 0.1. DSM, which is acquired through lidars, contains the height information for land covers in 32-bit floating type. In addition to DSM, we use RGB and IRRG optical bands of Potsdam and Vaihingen for training and inference. nDSM is the normalized DSM which supplies the height of each pixel accompanied by ground elevation subtracted.

In order to intuitively observe the composition of different classes, we pull out all pixels from GT for an exploratory data analysis. Class proportions are shown in Figure 4. In both datasets, the ratios of each class are similar, with the proportion of cars being the smallest, less than 2%. The volume of buildings along with impervious surfaces, which are difficult to distinguish, is between 26% and 28%. Potsdam has a proportion of low vegetation that is 8.92% higher than the number of trees, whereas the difference between the two classes in Vaihingen is only 2.35%.

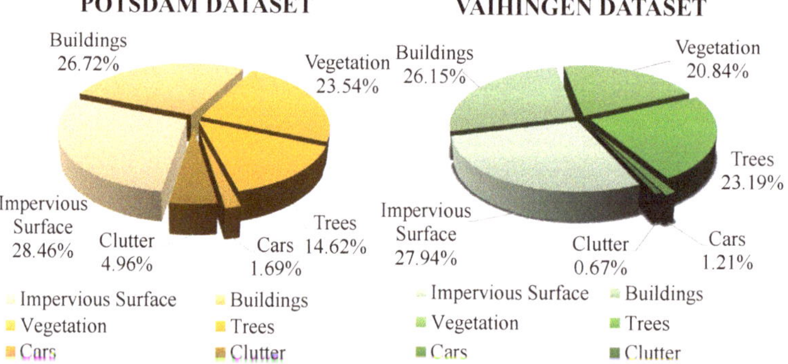

Figure 4. Illustrations about the composition of each class in ISPRS datasets.

4.2. Implementation Details

ISPRS datasets consist of high-resolution RSIs. Limited by GPU memory, we crop the optical image into 512×512 for model training and inference. In view of reducing overfitting, we augment the training set. Each pair of the image and the corresponding GT are rotated in arbitrary directions. Basic attributes such as contrast, brightness, saturation, and so on are randomly set for the augmentation. Reflection padding is conducted at the edges after cropping, this adjustment is particularly effective for urban complexes like buildings. Details are inevitably lost when cropping randomly, hence the symmetry of the buildings is well preserved by adopting reflection padding. Figure 5 illustrates samples for two datasets.

The hardware and software environment is listed in Table A1. We choose HRNet-W48 as the backbone, whose four branches yield feature maps ($R1, D1 \sim R4, D4$) with 1/4, 1/8, 1/16, 1/32 of the original size. UPerHead [34] is selected as the decoder. The learning rate is set to 0.00006 with the AdamW optimizer. The weight decay is 2×10^{-2} and the power of poly optimization strategy is 1. A mixed precision scheme is employed to reduce memory usage. The model is trained for 40k iterations loaded with weight pretrained on the ImageNet.

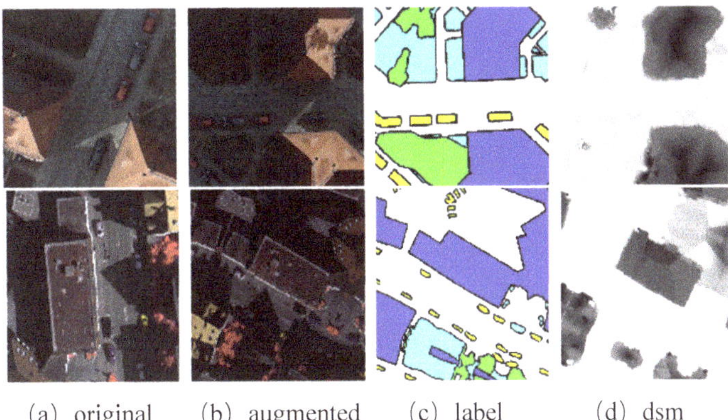

(a) original   (b) augmented   (c) label   (d) dsm

Figure 5. Illustrations about samples of Potsdam and Vaihingen dataset. *Augmented* means the image after the augmentation operation.

4.3. Test Set Transformation

The original dataset was captured in normal weather conditions. It is necessary to augment the test set for robustness evaluation. Inspired by the generation of corrupted ImageNet data sets in [18], we also render the corrupted RSIs test set with different degrees of fog and average the classification results for judging the robustness performance. We model the fog corruption through a diamond square algorithm which is to create a weighted heat map blended with the clean image. Thus, we can acquire corrupted test sets of Potsdam and Vaihingen, which are employed to measure the classification robustness in clean and foggy conditions. Corrupted samples are displayed in Figure 6, with fog generated corresponding to five levels of severity. This facilitates the robustness evaluation in various foggy conditions. Evaluation values are averaged over all five severity levels. As can be observed from Figure 6, the fog-covered region is indistinguishable from the actual scenario.

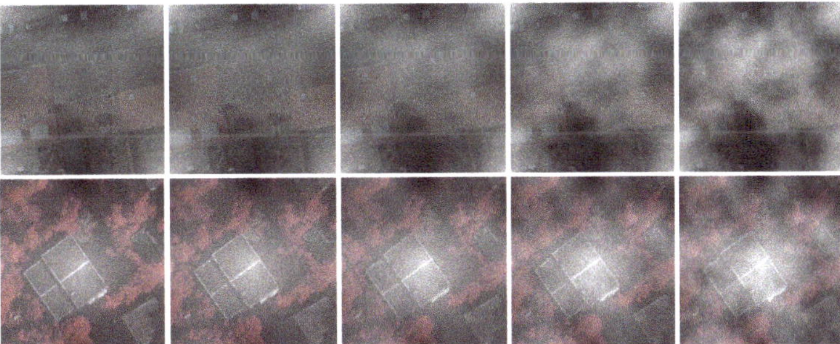

Figure 6. Illustrations of five severity levels of fog rendered the ISPRS dataset. First row: Potsdam. Second row: Vaihingen

4.4. Metrics

Metrics like overall accuracy (OA), $F1_{score}$ ($F1_{score}$) are selected to evaluate the classification accuracy. Following the robustness evaluation in [18,19], we take Corruption degradation (CD) and relative corruption degradation (rCD) into consideration for measuring LCC robustness. Specifically, metrics for accuracy evaluation are defined in Equation (19) to Equation (21). TP, FP, TN, FN represent true positive, false positive, true negative, and false negative classifications, respectively. Higher $F1_{score}$ and OA indicate a better classification accuracy.

$$Precision = \frac{TP}{TP + FP} \quad Recall = \frac{TP}{TP + FN} \tag{19}$$

$$OA = \frac{TP + TN}{TP + FP + TN + FN} \tag{20}$$

$$F1_{score} = 2 \times \frac{Precision \times Recall}{Precision + Recall} \tag{21}$$

From Equation (22) to Equation (24), ref stands for the baseline which is regarded as a reference model in the ablation and comparison experiments. D refers to the degree of degradation. S signifies obtaining the mean value across different corruption degrees. f is the selected model. *Clean* and F represents clean and corrupted datasets. $\widetilde{D}^{F}_{s,f}$ means that we acquire the average degradation value using model f on RSIs across different degrees of fog corruption.

$$D = 1 - F1_{score} \tag{22}$$

$$CD^{f}_{F} = \frac{\widetilde{D^{f}_{s,F}}}{D^{ref}_{s,F}} \times 100\% \tag{23}$$

$$rCD^{f}_{F} = \frac{\widetilde{D^{f}_{s,F}} - D^{f}_{clean}}{D^{ref}_{s,F} - D^{ref}_{clean}} \times 100\% \tag{24}$$

CD is a measure of absolute robustness. CD greater than 100% indicates a decline in robustness compared to the reference. The part over 100% represents the degradation in performance. To evaluate the relative robustness, rCD takes the performance on the clean dataset into account. Based on the reference, it is a proportional measure of the degradation in robustness relative to the clean data. When $rCD < 100\%$, it indicates that the performance degradation in foggy conditions is less than that of the corresponding reference value compared with the clean. When CD or $rCD > 100\%$, it means that model is not as robust as the reference. Robustness is better when both values are lower.

5. Result

5.1. Impact of Different Modules

To verify the effectiveness of each component, we conduct an ablation study on Vaihingen by removing or replacing the original part. From the qualitative visualization in Figure 7, it can be noted that when SSRL is added alone, the model can enhance capturing the global context and semantic features for classifying land cover edges. However, owing to the lack of height information, it is tough to grasp the correlation in the vertical space precisely (e.g., car in the box in Figure 7d). When MRFM is incorporated, the model can obtain the complementary information of multiple modalities, yet boundary labeling is coarse due to the absence of semantic representation details. The integration of both increases the model robustness and allows it to classify various land covers in foggy conditions more accurately.

In Tables 2–7, ✓ indicates the element that we incorporate on top of the baseline. Values in the *fog* column refer to the average value across five severity levels of fog corruption. We first illustrate the effectiveness of each constituent. To demonstrate the robustness and accuracy of HRNet-W48, ResNet50 is selected for comparison in view of the comparable size. H and R in Tables 2–4 correspond to HRNet and ResNet respectively. When only SSRL is integrated, we directly transfer the Optical and DSM feature maps following convolutional layers into the decoder. When only MRFM is available, we exclude SSRL and transfer both modalities into MRFM. U and C in *loss* column represent the Unified loss and Cross-entropy loss. *ImpSurf** and *LowVeg** signify the impervious surface and low vegetation.

Table 2. Ablation study results with different components integrated based on the clean and fog corrupted variants of the Vaihingen test set. The $F1_{score}$ is averaged across five severity levels. The best results are marked in bold.

Method				Per-Class $F1_{score}$ (%)										Mean F1 (%)		OA (%)	
				Imp Surf*		Building		Low Veg*		Tree		Car					
Baseline	SSRL	MRFM	Loss	Clean	Fog	Clean	Fog	Clean	Fog	Clean	Fog	Clean	Fog	Clean	Fog	Clean	Fog
H	✓		U	84.93	72.18	88.71	76.71	81.70	68.32	82.51	70.93	83.46	63.82	84.26	70.39	85.92	73.12
H		✓	U	88.13	75.10	90.97	80.21	82.66	70.38	84.36	71.32	85.79	65.92	86.38	72.58	86.91	75.16
R	✓	✓	U	85.94	75.86	88.85	77.04	82.96	65.71	83.83	69.01	84.09	63.65	85.13	70.25	86.63	72.53
H	✓	✓	C	91.74	80.70	92.05	80.08	81.86	71.47	86.97	73.74	84.21	64.08	87.36	74.01	88.55	76.97
H	✓	✓	U	**92.98**	**80.85**	**93.19**	**81.54**	**83.39**	**73.50**	**87.91**	**75.28**	**92.12**	**72.67**	**89.92**	**76.77**	**90.49**	**79.15**

Table 3. CD in ablation study with different components integrated based on clean and fog corrupted variants of the Vaihingen test set. Our framework is regarded as the reference and values above 100% denote a decrease in robustness compared with the reference. The highest CD is bold.

Method				CD for Per-Class $F1_{score}$ (%)										Mean CD (%)		CD for OA (%)	
				Imp Surf*		Building		Low Veg*		Tree		Car					
Baseline	SSRL	MRFM	Loss	Clean	Fog	Clean	Fog	Clean	Fog	Clean	Fog	Clean	Fog	Clean	Fog	Clean	Fog
H	✓		U	**214.50**	**145.23**	**165.73**	**126.15**	**110.16**	119.53	**144.69**	117.61	**209.78**	132.41	**156.06**	127.44	**148.08**	128.94
H		✓	U	169.01	130.01	132.58	107.19	104.39	111.78	129.40	116.04	180.21	124.74	135.05	118.01	137.69	119.14
R	✓	✓	U	200.23	126.05	163.72	124.36	102.58	**129.40**	133.78	**125.37**	201.80	**133.03**	147.44	**128.04**	140.65	**131.75**
H	✓	✓	C	117.66	100.77	116.64	107.90	109.22	107.66	107.81	106.24	200.27	131.45	125.30	111.86	120.36	110.48
H	✓	✓	U	100.00	100.00	100.00	100.00	100.00	100.00	100.00	100.00	100.00	100.00	100.00	100.00	100.00	100.00

Table 4. The relative CD in ablation study with different components integrated based on clean and fog corrupted variants of the Vaihingen test set. Lower rCD suggests an improvement of robustness in the presence of fog corruption. The highest rCD is bold.

| Method | | | | rCD for Per-Class $F1_{score}$ (%) | | | | | Mean rCD (%) | rCD for OA (%) |
Baseline	SSRL	MRFM	Loss	Imp Surf*	Building	Low Veg*	Tree	Car		
H	✓		U	105.12	**103.01**	135.26	91.70	101.03	105.49	112.89
H		✓	U	**107.42**	92.35	124.19	103.25	102.24	104.94	103.59
R	✓	✓	U	83.10	101.36	**174.43**	**117.31**	**105.13**	**113.17**	**124.28**
H	✓	✓	C	90.99	102.78	105.05	104.74	103.54	101.55	102.19
H	✓	✓	U	100.00	100.00	100.00	100.00	100.00	100.00	100.00

Table 5. Quantitative evaluation for clean and fog corrupted variants of the Vaihingen test set about using different input data. Every $F1_{score}$ is averaged over all five severity levels. The best results are marked in bold.

| Method | | Per-Class $F1_{score}$ (%) | | | | | | | | | | Mean F1 (%) | | OA (%) | |
| Optical | DSM | Imp Surf* | | Building | | Low Veg* | | Tree | | Car | | | | | |
		Clean	Fog	Clean	Fog	Clean	Fog	Clean	Fog	Clean	Fog	Clean	Fog	Clean	Fog
✓		90.70	77.05	91.58	83.49	82.32	70.91	85.32	73.56	90.63	65.87	88.11	74.17	88.69	73.25
	✓	85.50	69.28	88.66	79.17	76.21	59.70	83.59	65.12	83.89	55.85	83.57	65.82	80.18	65.21
✓	✓	**92.98**	**80.85**	**93.19**	**81.54**	**83.39**	**73.50**	**87.91**	**75.28**	**92.12**	**72.67**	**89.92**	**76.77**	**90.49**	**79.15**

Table 6. Quantitative evaluation of CD on clean and fog corrupted variants of the Vaihingen test set about using different input data. Our framework is regarded as the reference and values above 100% denote a decrease in robustness compared with the reference. The highest CD is bold.

| Method | | CD for Per-Class $F1_{score}$ (%) | | | | | | | | | | Mean CD (%) | | CD for OA (%) | |
| Optical | DSM | Imp Surf* | | Building | | Low Veg* | | Tree | | Car | | | | | |
		Clean	Fog	Clean	Fog	Clean	Fog	Clean	Fog	Clean	Fog	Clean	Fog	Clean	Fog
✓		132.42	119.84	123.59	89.42	106.41	109.76	121.47	106.96	118.83	124.91	117.91	111.16	118.93	128.32
	✓	**206.37**	**160.39**	**166.46**	**112.84**	**143.21**	**152.07**	**135.73**	**141.11**	**204.35**	**161.56**	**162.92**	**147.11**	**208.43**	**166.88**
✓	✓	100.00	100.00	100.00	100.00	100.00	100.00	100.00	100.00	100.00	100.00	100.00	100.00	100.00	100.00

Table 7. Quantitative evaluation of *rCD* on clean and fog corrupted variants of the Vaihingen test set in ablation study about using different input data evaluated. Our framework is regarded as the reference and lower *rCD* indicates an improvement of robustness in the presence of fog corruption. The highest *rCD* is bold.

Method		rCD for Per-Class $F1_{score}$ (%)					Mean rCD (%)	rCD for OA (%)
Optical	DSM	Imp Surf*	Building	Low Veg*	Tree	Car		
✓		112.56	69.44	115.40	93.08	127.37	105.98	136.18
	✓	**133.76**	**81.49**	**166.94**	**146.25**	**144.21**	**134.98**	**132.04**
✓	✓	100.00	100.00	100.00	100.00	100.00	100.00	100.00

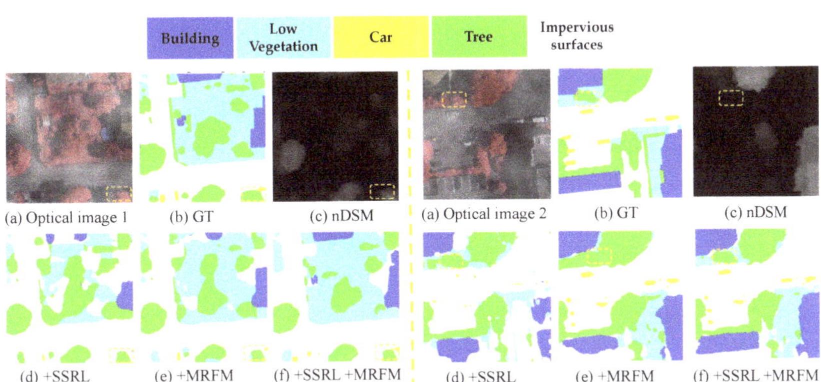

Figure 7. Illustrations about some ablation results of each component in the framework. In this case, optical images are corrupted by fog, which belongs to severity level 2.

Combining Tables 2–4 with Figure 8, we observe that the incorporation of SSRL alone could improve the accuracy, but for impervious surface (CD = 145.23%, rCD = 105.12%) and low vegetation (CD = 119.53%, rCD = 135.26%), values of both exceed 100% and robustness is still inferior to the others. In the absence of effective multimodal fusion, SSRL is more biased towards the regular shape and small objects in Vaihingen. Moreover, the addition of MRFM is conducive to the improvement of robustness. Misclassification result of tail-end distributed cars ($F1_{score}$ = 64.08%, CD = 131.45%) manifests if just cross-entropy loss function is utilized. When compared with the Unified loss, $F1_{score}$ is reduced by 8.59%, while CD is increased by 31.45%, indicating that UFL improves the robustness of imbalanced distributed objects ($F1_{score}$ = 72.67%).

5.2. Impact of Multimodal Fusion

We also perform an ablation study with different inputs to investigate the improvement of robustness under multimodal fusion. ✓ represents the input modality. In the case of single modal input, the original multimodality is replaced by the identical modal input. From Tables 5–7, we can conclude that utilizing a single modality alone is less effective than multimodal fusion. When using DSM alone, the model performs poorly because DSM contains fewer semantic features compared with the optical. There is an 8.35% and 35.95% performance loss compared to the corresponding result of the optical input. Fusing multimodalities improves accuracy and robustness in foggy environments. Compared to the single optical modal input, the accuracy is 5.9% higher and *rCD* is 36.18% lower. From Figure 9, we can observe that our model is capable of classifying edges of the cars robustly in dense fog.

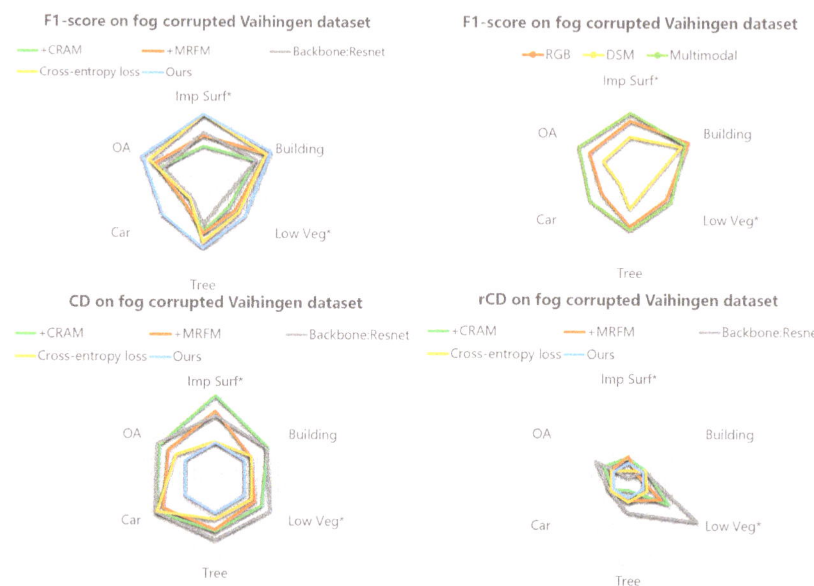

Figure 8. Radar plot visualization of the ablation study results. The first in top two is the classification result when different modules are integrated into the backbone. The second is about different inputs. Based on the radar plots regarding *CD* and *rCD*, the smaller envelope range is indicative of a model that is more robust on a fog-corrupted test set.

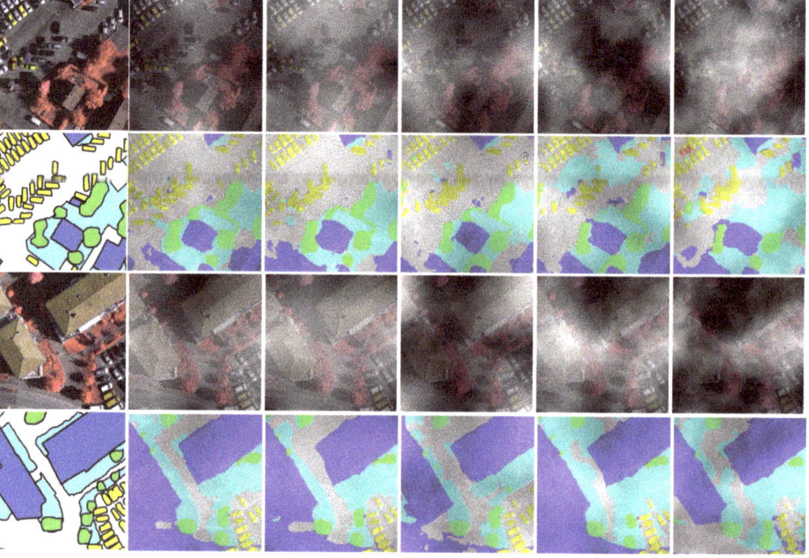

Figure 9. Comparing the LCC results across different corruption levels. 1st column displays the clean image and ground truth. Others are images with fog of various severity levels, accompanied by the corresponding semantic labeling result. From left to right, the fog intensity increases gradually.

6. Discussion

To further elucidate the model robustness, we select some existing SOTA methods to conduct a comparison experiment on the Potsdam dataset. Models in the experiment can be grouped into CNNs and Transformers. Specifically, CNN-based models contain FCN [5], UNet [6], PSPNet [12], DeepLabV3+[13], CCNet [43], OCRNet [44], TRM [15], and Transformer-based models include SETR [27], Segmenter [45], and Segformer [28]. To ensure that models in the experiment have comparable parameters, we adopt ResNet101 [35] as the backbone for FCN, UNet, PSPNet, DeepLabV3+, and CCNet. TRM, OCRNet, and our framework are built on top of HRNet-W48 [4]. Encoder backbones selected for SETR, Segmenter, and Segformer are DeiT-B [46], DeiT-B, and MiT-B5 [28], respectively. Thus, the parameter of each model is around 70–90M in size. To utilize multimodal data, we stack the optical and DSM inputs in the channel dimension for all models excluding ours.

Results obtained from Figures 10 and 11 and Table 8 show that Transformers perform better compared to CNNs on the clean test set with comparable sizes. However, the robustness of most ViTs is significantly reduced on the fog corrupted test set, with the exception of Segformer. There is a coarse classification of cars and edges in the box regions. Multimodal fusion allows our model to precisely learn the hierarchical features of inter-modal and the relationship between neighboring objects and the global. In this way, edges and interiors can be accurately classified. Compared to our previously proposed algorithm TRM, the accuracy and robustness LCC have been improved as a result of SSRL and MRFM, which enhance the ability to capture semantic information in low-quality images with a more effective data fusion approach. Specifically speaking, the $F1_{score}$ on the corrupted dataset improved by 1.3%, while there is a reduction of over 3% on both CD and rCD.

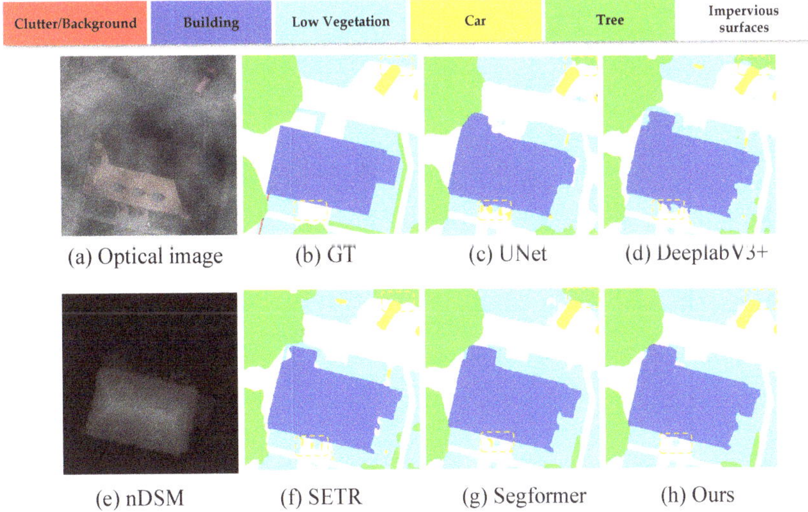

Figure 10. Qualitative comparisons between different methods applied to semantic segmentation of RSIs. The optical image is corrupted by the third severity level of fog.

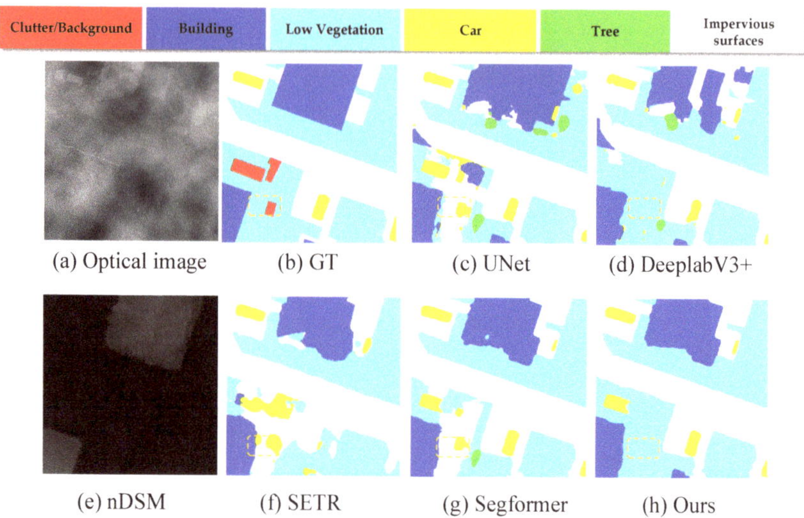

Figure 11. Qualitative comparisons between different methods applied to semantic segmentation of RSIs. The optical image is corrupted by the fourth severity level of fog.

It can be concluded from Tables 8–10 that performance on the clean and fog corrupted test set varies significantly with regard to different models, whereas CD and rCD results are generally stable. ViTs perform better than CNNs on the clean test set (Segformer is 0.31% better on the Mean $F1_{score}$ and 0.71% better on OA), which can be attributed to the capability of capturing global information. However, in terms of robustness, CNNs represented by OCRNet are stronger than the best-performing ViTs (OCRNet's CD decreased by 0.36% and rCD decreased by 2.23% in the mean $F1_{score}$ compared to Segformer), which is attributable to the fact that ViTs require more sophisticated training strategies with data augmentation. We are able to achieve a balance between accuracy and robustness compared to several SOTAs, as shown in Figure 12, where our model encircles a relatively smaller area (CD decreases by 3.96% and rCD decreases by 2.87% on OA compared to OCRNet). The balanced classification result of each class also reflects the robustness. This illustrates the effectiveness of the proposed framework in generalizing and handling with class imbalance.

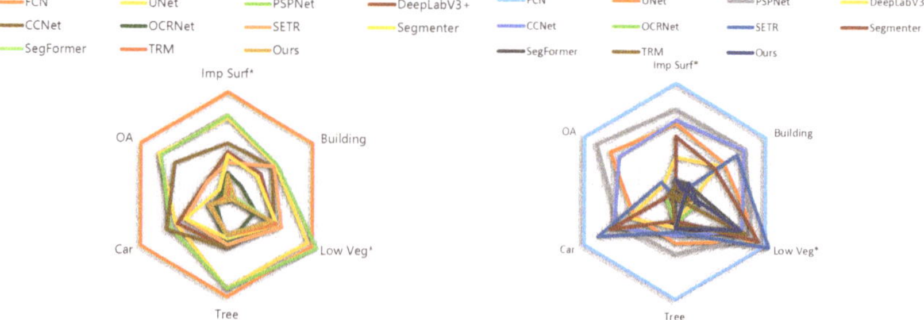

Figure 12. Radar plot for the robust performance of several SOTAs on the Potsdam test set. The envelope area of a robust model should be small and balanced. Although ViTs can boost the performance, CNNs manifest stronger robustness compared with ViTs.

Table 8. Quantitative comparison with SOTA methods on the clean and fog corrupted variants of the Potsdam test set. Each $F1_{score}$ is averaged over all severity levels. The best results are marked in bold.

| Method | Per-Class $F1_{score}$ (%) | | | | | | | | | | Mean F1 (%) | | OA (%) | |
| | Imp Surf* | | Building | | Low Veg* | | Tree | | Car | | | | | |
	Clean	Fog	Clean	Fog	Clean	Fog	Clean	Fog	Clean	Fog	Clean	Fog	Clean	Fog
FCN [5]	85.15	63.22	86.08	67.12	79.02	63.15	81.72	63.58	78.24	55.61	82.04	62.54	85.86	58.33
UNet [6]	85.61	66.93	88.17	72.79	79.31	63.99	79.16	64.77	80.61	62.89	82.57	66.27	86.25	61.85
PSPNet [12]	86.34	66.52	89.79	72.55	77.34	62.46	79.88	64.73	80.95	61.74	82.86	65.60	88.34	62.31
DeepLabV3+[13]	87.73	71.83	90.06	75.08	80.27	69.59	83.89	70.56	81.23	63.12	84.64	70.04	88.76	70.86
CCNet [43]	89.49	70.55	90.37	73.24	83.91	69.71	84.47	70.59	80.63	60.98	85.77	69.01	88.27	64.91
OCRNet [44]	88.77	74.86	90.69	78.19	84.02	**72.95**	84.11	71.02	85.33	**70.79**	86.58	73.56	89.31	74.92
SETR [27]	86.09	73.38	89.53	73.04	84.57	68.44	84.61	70.98	**85.93**	64.71	86.15	70.11	89.26	70.99
Segmenter [45]	89.94	72.30	90.22	75.61	**84.58**	69.21	84.77	71.90	84.85	64.83	86.87	70.77	89.88	73.42
SegFormer [28]	**89.53**	75.52	92.40	80.02	84.07	69.99	85.38	72.13	83.05	69.48	**86.89**	73.43	90.01	74.91
TRM [15]	89.11	**75.57**	91.45	80.61	83.83	69.29	85.56	72.37	83.58	69.62	86.71	73.49	89.92	75.27
Ours	89.31	**75.16**	**92.49**	**80.70**	84.23	70.15	**85.83**	**72.45**	82.08	69.83	86.79	**73.66**	**90.17**	**76.57**

Table 9. Quantitative evaluation of CD on clean and fog corrupted variants of Potsdam test set about different SOTA methods. FCN is regarded as the reference and values lower than 100% represent an improvement in the robust performance compared with the reference. The minimum in each *Fog* column is bold.

| Method | CD for Per-Class $F1_{score}$ (%) | | | | | | | | | | CD for Mean F1 (%) | | CD for OA (%) | |
| | Imp Surf* | | Building | | Low Veg* | | Tree | | Car | | | | | |
	Clean	Fog	Clean	Fog	Clean	Fog	Clean	Fog	Clean	Fog	Clean	Fog	Clean	Fog
FCN [5]	100.00	100.00	100.00	100.00	100.00	100.00	100.00	100.00	100.00	100.00	100.00	100.00	100.00	100.00
UNet [6]	96.90	89.91	84.99	82.76	98.62	97.72	114.00	96.73	89.11	83.60	97.05	90.02	97.24	91.55
PSPNet [12]	91.99	91.03	73.35	83.49	108.01	101.87	110.07	96.84	87.55	86.19	95.44	91.82	82.46	90.45
DeepLabV3+[13]	82.63	76.59	71.41	75.79	94.04	82.52	88.13	80.83	86.26	83.08	85.56	79.98	79.49	69.93
CCNet [43]	70.77	80.07	69.18	81.39	76.69	82.20	84.96	80.75	89.02	87.90	79.22	82.71	82.96	84.21
OCRNet [44]	75.62	68.35	66.88	66.33	76.17	73.41	86.93	79.57	67.42	65.80	74.71	70.57	75.60	60.19
SETR [27]	93.67	72.38	75.22	82.00	73.55	85.64	84.19	79.68	64.66	79.50	77.15	79.78	75.95	69.62
Segmenter [45]	67.74	75.31	70.26	74.18	73.50	83.55	83.32	77.16	69.62	79.23	73.10	78.02	71.57	63.79
SegFormer [28]	70.51	66.56	54.60	60.77	75.93	81.44	79.98	76.52	77.90	68.75	73.03	70.93	70.65	60.21
TRM [15]	73.33	**66.42**	61.42	58.97	77.07	83.34	78.99	75.86	75.46	68.44	74.03	70.76	71.29	59.35
Ours	71.99	67.54	53.95	**58.70**	75.17	**81.00**	77.52	**75.65**	82.35	**67.97**	73.57	**70.31**	69.52	**56.23**

Table 10. Quantitative evaluation of *rCD* on clean and fog corrupted variants of Potsdam test set about different SOTA methods. FCN is regarded as the reference and values lower than 100% represent an improvement in the robust performance compared to the reference. The minimum in each column is in bold.

Method	*rCD* for Per-Class $F1_{score}$ (%)					Mean *rCD* (%)	*rCD* for OA (%)
	Imp Surf*	Building	Low Veg*	Tree	Car		
FCN [5]	100.00	100.00	100.00	100.00	100.00	100.00	100.00
UNet [6]	85.18	81.12	96.53	79.33	78.30	83.55	88.63
PSPNet [12]	90.38	90.93	93.76	83.52	84.89	88.49	94.55
DeepLabV3+ [13]	72.50	79.01	**67.30**	73.48	80.03	74.85	65.02
CCNet [43]	86.37	90.35	89.48	76.52	86.83	85.92	84.85
OCRNet [44]	63.43	65.93	69.75	72.16	64.25	**66.76**	52.27
SETR [27]	57.96	86.97	101.64	75.14	93.77	82.21	66.36
Segmenter [45]	80.44	77.06	96.85	**70.95**	88.47	82.55	59.79
SegFormer [28]	63.89	65.30	88.72	73.04	59.96	68.99	54.85
TRM [15]	**61.74**	**57.17**	91.62	72.71	61.69	67.74	53.21
Ours	64.52	62.18	88.72	73.76	**54.13**	67.31	**49.40**

7. Conclusions

This study set out to design a robust model for LCC. The framework utilizes multimodal fusion and attention mechanisms to achieve a robust segmentation of RSIs in foggy conditions. We transfer heterogeneous data into HRNet, which serves as the backbone to maintain the high-resolution representation. Incorporating MRFM into the framework can exploit cross-modal complementary fusion. SSRL is deployed for exploring the correlations between different channels and positions. Unified loss helps to mitigate class imbalance issues. Multiple experiment analyses reveal that the proposed model has superior robustness on the fog-corrupted Potsdam and Vaihingen test sets. In addition, this study has also confirmed that in terms of robustness, ViTs are often inferior to CNNs in the presence of natural noises. Overall, this study highlights the importance of multimodal fusion and attention mechanisms for enhancing segmentation robustness. Our future research plans to investigate this topic further by combining ViTs with more fundamental attributes of RSIs.

Author Contributions: Conceptualization, W.S.; methodology, W.S.; validation, W.S.; writing—review and editing, W.S. and A.C.; project administration, W.Q. All authors have read and agreed to the published version of the manuscript.

Funding: This work was supported by Key R&D Program of Jiangsu Province under Grant BE2019311, Jiangsu Modern Agricultural Industry Key Technology Innovation Project under Grant CX(20)2013 and National Key Research and Development Program under Grant 2020YFB160070301.

Data Availability Statement: No new data were created or analyzed in this study.

Acknowledgments: A publicly available GitHub repository fastai was adapted for the experiment.

Conflicts of Interest: The authors declare no conflict of interest.

Abbreviations

The following abbreviations are used in this manuscript:

RSIs	Remote Sensing Images
NSIs	Natural Scene Images
LCC	Land Cover Classification
SSRL	Spectral and Spatial Representation Learning
MRFM	Multimodal Representation Fusion Module
DSM	Digital Surface Model
SOTA	State Of The Art
CNN	Convolution Neural Network
ViT	Visual Transformer
TOP	True Ortho Photos
GSD	Ground Sampling Distance

OA	Overall Accuracy
CD	Corruption Degradation
rCD	Relative Corruption Degradation
GT	Ground Truth
ISPRS	International Society for Photogrammetry and Remote Sensing

Appendix A

Table A1. Experiment Environment.

Software	Software Version	Hardware	Hardware Version
CUDA	10.2	CPU	i7-5930K CPU @ 3.50 GHz
cuDNN	7.6	GPU	2 × Titan XP(12G)
Pytorch	1.7	RAM	64 GB
Fast.ai	2.2.2	HARD DISK	Toshiba SSD 2T
Wandb	0.1.20	SYSTEM	Ubuntu 18.0.4

References

1. He, H.; Li, C.; Yang, R.; Zeng, H.; Li, L.; Zhu, Y. Multisource Data Fusion and Adversarial Nets for Landslide Extraction from UAV-Photogrammetry-Derived Data. *Remote Sens.* **2022**, *14*, 3059. [CrossRef]
2. Shao, S.; Xiao, L.; Lin, L.; Ren, C.; Tian, J. Road Extraction Convolutional Neural Network with Embedded Attention Mechanism for Remote Sensing Imagery. *Remote Sens.* **2022**, *14*, 2061. [CrossRef]
3. Ding, J.; Zhang, J.; Zhan, Z.; Tang, X.; Wang, X. A Precision Efficient Method for Collapsed Building Detection in Post-Earthquake UAV Images Based on the Improved NMS Algorithm and Faster R-CNN. *Remote Sens.* **2022**, *14*, 663. [CrossRef]
4. Wang, J.; Sun, K.; Cheng, T.; Jiang, B.; Deng, C.; Zhao, Y.; Liu, D.; Mu, Y.; Tan, M.; Wang, X.; et al. Deep High-Resolution Representation Learning for Visual Recognition. *arXiv* **2020**, arXiv:1908.07919.
5. Shelhamer, E.; Long, J.; Darrell, T. Fully Convolutional Networks for Semantic Segmentation. *IEEE Trans. Pattern Anal. Mach. Intell.* **2017**, *39*, 640–651. [CrossRef]
6. Ronneberger, O.; Fischer, P.; Brox, T. U-Net: Convolutional Networks for Biomedical Image Segmentation. *arXiv* **2015**, arXiv:1505.04597.
7. Xu, Q.; Yuan, X.; jun Ouyang, C.; Zeng, Y. Attention-Based Pyramid Network for Segmentation and Classification of High-Resolution and Hyperspectral Remote Sensing Images. *Remote Sens.* **2020**, *12*, 3501. [CrossRef]
8. Zhang, G.; Lei, T.; Cui, Y.; Jiang, P. A dual-path and lightweight convolutional neural network for high-resolution aerial image segmentation. *ISPRS Int. J. Geo-Inf.* **2019**, *8*, 582. [CrossRef]
9. Li, X.; Jiang, Y.; Peng, H.; Yin, S. An aerial image segmentation approach based on enhanced multi-scale convolutional neural network. In Proceedings of the 2019 IEEE International Conference on Industrial Cyber Physical Systems (ICPS), Taipei, Taiwan, 6–9 May 2019; pp. 47–52.
10. Chen, L.C.; Papandreou, G.; Kokkinos, I.; Murphy, K.; Yuille, A.L. DeepLab: Semantic Image Segmentation with Deep Convolutional Nets, Atrous Convolution, and Fully Connected CRFs. *arXiv* **2017**, arXiv:1606.00915.
11. Yang, M.; Yu, K.; Zhang, C.; Li, Z.; Yang, K. DenseASPP for Semantic Segmentation in Street Scenes. In Proceedings of the 2018 IEEE/CVF Conference on Computer Vision and Pattern Recognition, Salt Lake City, UT, USA, 18–23 June 2018; IEEE: Salt Lake City, UT, USA, 2018; pp. 3684–3692. [CrossRef]
12. Zhao, H.; Shi, J.; Qi, X.; Wang, X.; Jia, J. Pyramid Scene Parsing Network. *arXiv* **2017**, arXiv:1612.01105.
13. Chen, L.C.; Zhu, Y.; Papandreou, G.; Schroff, F.; Adam, H. Encoder-Decoder with Atrous Separable Convolution for Semantic Image Segmentation. *arXiv* **2018**, arXiv:1802.02611.
14. Badrinarayanan, V.; Kendall, A.; Cipolla, R. SegNet: A Deep Convolutional Encoder-Decoder Architecture for Image Segmentation. *arXiv* **2016**, arXiv:1511.00561.
15. Shi, W.; Qin, W.; Yun, Z.; Chen, A.; Huang, K.Y.; Zhao, T. Land Cover Classification in Foggy Conditions: Toward Robust Models. *IEEE Geosci. Remote Sens. Lett.* **2022**, *19*, 1–5. [CrossRef]
16. Zhang, X.Y.; Liu, C.L.; Suen, C.Y. Towards Robust Pattern Recognition: A Review. *Proc. IEEE* **2020**, *108*, 894–922. [CrossRef]
17. Tang, S.; Gong, R.; Wang, Y.; Liu, A.; Wang, J.; Chen, X.; Yu, F.; Liu, X.; Song, D.; Yuille, A.; et al. RobustART: Benchmarking Robustness on Architecture Design and Training Techniques. *arXiv* **2021**, arXiv:2109.05211.
18. Hendrycks, D.; Dietterich, T. Benchmarking Neural Network Robustness to Common Corruptions and Perturbations. *arXiv* **2019**, arXiv:1903.12261.
19. Kamann, C.; Rother, C. Benchmarking the Robustness of Semantic Segmentation Models. In Proceedings of the 2020 IEEE/CVF Conference on Computer Vision and Pattern Recognition (CVPR), Seattle, WA, USA, 13–19 June 2020; pp. 8825–8835. [CrossRef]
20. Shao, R.; Shi, Z.; Yi, J.; Chen, P.Y.; Hsieh, C.J. On the Adversarial Robustness of Visual Transformers. *arXiv* **2021**, arXiv:2103.15670.

21. Mahmood, K.; Mahmood, R.; van Dijk, M. On the Robustness of Vision Transformers to Adversarial Examples. *arXiv* **2021**, arXiv:2104.02610.

22. Wang, X.; Girshick, R.; Gupta, A.; He, K. Non-Local Neural Networks. *arXiv* **2018**, arXiv:1711.07971.

23. Wang, Q.; Wu, B.; Zhu, P.; Li, P.; Zuo, W.; Hu, Q. ECA-Net: Efficient Channel Attention for Deep Convolutional Neural Networks. *arXiv* **2020**, arXiv:1910.03151.

24. Cao, Y.; Xu, J.; Lin, S.; Wei, F.; Hu, H. GCNet: Non-local Networks Meet Squeeze-Excitation Networks and Beyond. *arXiv* **2019**, arXiv:1904.11492.

25. Dosovitskiy, A.; Beyer, L.; Kolesnikov, A.; Weissenborn, D.; Zhai, X.; Unterthiner, T.; Dehghani, M.; Minderer, M.; Heigold, G.; Gelly, S.; et al. An Image Is Worth 16x16 Words: Transformers for Image Recognition at Scale. *arXiv* **2021**, arXiv:2010.11929.

26. Liu, Z.; Lin, Y.; Cao, Y.; Hu, H.; Wei, Y.; Zhang, Z.; Lin, S.; Guo, B. Swin Transformer: Hierarchical Vision Transformer Using Shifted Windows. *arXiv* **2021**, arXiv:2103.14030.

27. Zheng, S.; Lu, J.; Zhao, H.; Zhu, X.; Luo, Z.; Wang, Y.; Fu, Y.; Feng, J.; Xiang, T.; Torr, P.H.S.; et al. Rethinking Semantic Segmentation from a Sequence-to-Sequence Perspective with Transformers. *arXiv* **2021**, arXiv:2012.15840.

28. Xie, E.; Wang, W.; Yu, Z.; Anandkumar, A.; Alvarez, J.M.; Luo, P. SegFormer: Simple and Efficient Design for Semantic Segmentation with Transformers. *arXiv* **2021**, arXiv:2105.15203.

29. Yuan, L.; Hou, Q.; Jiang, Z.; Feng, J.; Yan, S. VOLO: Vision Outlooker for Visual Recognition. *arXiv* **2021**, arXiv:2106.13112.

30. Gu, Y.; Hao, J.; Chen, B.; Deng, H. Top-Down Pyramid Fusion Network for High-Resolution Remote Sensing Semantic Segmentation. *Remote Sens.* **2021**, *13*, 4159. [CrossRef]

31. Yan, L.; Huang, J.; Xie, H.; Wei, P.; Gao, Z. Efficient Depth Fusion Transformer for Aerial Image Semantic Segmentation. *Remote Sens.* **2022**, *14*, 1294. [CrossRef]

32. Audebert, N.; Le Saux, B.; Lefèvre, S. Beyond RGB: Very High Resolution Urban Remote Sensing with Multimodal Deep Networks. *ISPRS J. Photogramm. Remote Sens.* **2018**, *140*, 20–32. [CrossRef]

33. Liu, H.; Zhang, J.; Yang, K.; Hu, X.; Stiefelhagen, R. CMX: Cross-Modal Fusion for RGB-X Semantic Segmentation with Transformers. *arXiv* **2022**, arXiv:2203.04838.

34. Xiao, T.; Liu, Y.; Zhou, B.; Jiang, Y.; Sun, J. Unified Perceptual Parsing for Scene Understanding. *arXiv* **2018**, arXiv:1807.10221.

35. He, K.; Zhang, X.; Ren, S.; Sun, J. Deep Residual Learning for Image Recognition. *arXiv* **2015**, arXiv:1512.03385.

36. Hu, J.; Shen, L.; Albanie, S.; Sun, G.; Wu, E. Squeeze-and-Excitation Networks. *arXiv* **2019**, arXiv:1709.01507.

37. Woo, S.; Park, J.; Lee, J.Y.; Kweon, I.S. CBAM: Convolutional Block Attention Module. *arXiv* **2018**, arXiv:1807.06521.

38. Yeung, M.; Sala, E.; Schönlieb, C.B.; Rundo, L. Unified Focal Loss: Generalising Dice and Cross Entropy-Based Losses to Handle Class Imbalanced Medical Image Segmentation. *Comput. Med Imaging Graph.* **2022**, *95*, 102026. [CrossRef] [PubMed]

39. Ma, J.; Chen, J.; Ng, M.; Huang, R.; Li, Y.; Li, C.; Yang, X.; Martel, A.L. Loss Odyssey in Medical Image Segmentation. *Med. Image Anal.* **2021**, *71*, 102035. [CrossRef]

40. Lin, T.Y.; Goyal, P.; Girshick, R.B.; He, K.; Dollár, P. Focal Loss for Dense Object Detection. *IEEE Trans. Pattern Anal. Mach. Intell.* **2020**, *42*, 318–327. [CrossRef] [PubMed]

41. Abraham, N.; Khan, N.M. A Novel Focal Tversky Loss Function With Improved Attention U-Net for Lesion Segmentation. In Proceedings of the 2019 IEEE 16th International Symposium on Biomedical Imaging (ISBI 2019), Venice, Italy, 8–11 April 2019; pp. 683–687.

42. Zhang, Z.; Sabuncu, M.R. Generalized Cross Entropy Loss for Training Deep Neural Networks with Noisy Labels. In Proceedings of the NeurIPS, Montreal, QC, Canada, 3–8 December 2018.

43. Huang, Z.; Wang, X.; Wei, Y.; Huang, L.; Shi, H.; Liu, W.; Huang, T.S. CCNet: Criss-Cross Attention for Semantic Segmentation. *arXiv* **2020**, arXiv:1811.11721.

44. Yuan, Y.; Chen, X.; Chen, X.; Wang, J. Segmentation Transformer: Object-Contextual Representations for Semantic Segmentation. *arXiv* **2021**, arXiv:1909.11065.

45. Strudel, R.; Garcia, R.; Laptev, I.; Schmid, C. Segmenter: Transformer for Semantic Segmentation. *arXiv* **2021**, arXiv:2105.05633.

46. Touvron, H.; Cord, M.; Douze, M.; Massa, F.; Sablayrolles, A.; Jégou, H. Training Data-Efficient Image Transformers & Distillation through Attention. *arXiv* **2021**, arXiv:2012.12877.

remote sensing

Article

Blind Restoration of Atmospheric Turbulence-Degraded Images Based on Curriculum Learning

Jie Shu, Chunzhi Xie * and Zhisheng Gao

School of Computer and Software Engineering, Xihua University, Chengdu 610039, China
* Correspondence: xcz_xihua@mail.xhu.edu.cn

Abstract: Atmospheric turbulence-degraded images in typical practical application scenarios are always disturbed by severe additive noise. Severe additive noise corrupts the prior assumptions of most baseline deconvolution methods. Existing methods either ignore the additive noise term during optimization or perform denoising and deblurring completely independently. However, their performances are not high because they do not conform to the prior that multiple degradation factors are tightly coupled. This paper proposes a Noise Suppression-based Restoration Network (NSRN) for turbulence-degraded images, in which the noise suppression module is designed to learn low-rank subspaces from turbulence-degraded images, the attention-based asymmetric U-NET module is designed for blurred-image deconvolution, and the Fine Deep Back-Projection (FDBP) module is used for multi-level feature fusion to reconstruct a sharp image. Furthermore, an improved curriculum learning strategy is proposed, which trains the network gradually to achieve superior performance through a local-to-global, easy-to-difficult learning method. Based on NSRN, we achieve state-of-the-art performance with PSNR of 30.1 dB and SSIM of 0.9 on the simulated dataset and better visual results on the real images.

Keywords: noise suppression deblurring; curriculum learning; image reconstruction; turbulence degradation

Citation: Shu, J.; Xie, C.; Gao, Z. Blind Restoration of Atmospheric Turbulence-Degraded Images Based on Curriculum Learning. *Remote Sens.* **2022**, *14*, 4797. https://doi.org/10.3390/rs14194797

Academic Editor: Dusan Gleich

Received: 2 August 2022
Accepted: 20 September 2022
Published: 26 September 2022

Publisher's Note: MDPI stays neutral with regard to jurisdictional claims in published maps and institutional affiliations.

1. Introduction

Under long-range imaging conditions such as ground-based space-target imaging and long-range air-to-air and air-to-ground military reconnaissance imaging, the captured images are always affected by atmospheric turbulence [1]. Restoration of these degraded images into sharp images requires efficient post-processing. It is generally believed that due to the long distance and the uncontrollable imaging environment, atmospheric turbulence degradation is a coupled degradation process with multiple factors [2]. Imaging is not only affected by turbulence blur caused by atmospheric turbulence [2–7], but also by motion blur caused by the relative motion of the camera [8,9] and defocus blur caused by lens aberration [10] during exposure. Moreover, the images are also disturbed by severe additive noise [2]. Therefore, the core problem of the restoration of images degraded by atmospheric turbulence is image deblurring in the case of noise interference.

Image deblurring, which is essentially the process of obtaining a potentially sharp image, has been addressed in several ways. Deblurring methods can be classified into blind deblurring [11,12] and non-blind deblurring [13,14] depending on whether the blur kernel is known. Non-blind deblurring requires prior knowledge of the blur kernel (point spread function) and blur parameters. However, in practical applications, the point spread function (PSF) cannot be obtained, and a single blurred image is usually the only input data obtainable. Therefore, in practical applications, blind deblurring is much more common than non-blind deblurring.

Traditional blind deblurring methods usually represent the blurring of the entire image as a single, unified model. The standard procedure for these methods is to estimate the blur kernel before non-blind deconvolution. Regularization priors [15,16] need to be introduced in this process due to the ill-posed nature of the problem. A popular approach is to add image priors such as sparse priors [17–20], Principal Component Analysis (PCA) [21], and gradient priors [22–24] in a MAximum Posterior framework (MAP). This method usually uses iterative alternating steps to complete the optimal solution of the equation. The first step estimates blur kernels, and the second step estimates potentially sharp images. Since the assumption based on traditional methods has deviated from the actual scene prior, these methods can only be applied to the restoration of single-mechanism-degraded and less degraded images (such as motion blur). In practical applications, images, especially turbulence-degraded images, are often affected by various degradation factors. Therefore, the above methods have difficulty achieving the expected effect.

It is difficult to design a regularization prior that is suitable for practical application scenarios and that can be optimally solved. Therefore, the use of deep neural networks to learn the intrinsic features of images from degraded images and to use these features to reconstruct sharp images has become a research hotspot in recent years, with gratifying results in practical application scenarios [2,25,26]. Such methods usually require designing an End-to-End (E2E) deep neural network model, which can be divided into two parts. The first part is an encoder for learning features from degradation, while the second part is a decoder for reconstructing sharp images [27,28]. Most existing neural network-based methods can only deal with a single mode of degradation, such as image de-moiré [29], denoising [30,31], JPEG artifact removal [32], deblurring [33–35], etc., or use only one model to complete the restoration of multiple single-mode degraded images [26]. However, degraded images in atmospheric turbulence environments are often affected by the coupling of various degradation factors, especially severe additive random noise, which greatly increases the sample space dimension of the input data. As the intensity of the noise increases, the performance of the above neural network-based methods decreases. Therefore, the impact of noise on the model has received more and more attention in the industry [36–38]. The denoiser prior [36,37] is an efficient solution to this problem and is split into two independent subtasks: denoising and deblurring.

We consider turbulence degradation to be a coupled degradation of multiple factors that are difficult to be decoupled individually [38]. Based on this idea, we propose a Noise Suppression-based Restoration Network (NSRN) for turbulence-degraded images that consists of a shallow feature extraction module, a noise reduction module, an asymmetric U-NET network, and a sub-network for image reconstruction. The noise suppression module is designed to learn low-rank subspaces from turbulence-degraded images. The attention-based Asymmetric U-NET (AU-NET) module is designed for blurred image deconvolution, and the FDBP is designed to fuse multi-level features for degraded-image reconstruction. The NSRN is based on the prior that additive noise and blur are tightly coupled and that the entire network is inseparable. To make the noise suppression module pay more attention to the removal of additive noise and to overcome the problem that the model is difficult to train in the case of heavy noise, a curriculum learning strategy (i.e., local-to-global and easy-to-difficult) is introduced into the NSRN. Therefore, the proposed method has the advantage of being robust to noise when used for blind deblurring of atmospheric turbulence-degraded images. The main contributions of this paper are as follows:

(1) For the tightly coupled priors of additive noise and blur, a noise suppression-based neural network model is designed for restoration of turbulence-degraded images. It achieves image deconvolution while suppressing additive noise to benefit the restoration of turbulence-degraded images.

(2) A local-to-global and easy-to-difficult curriculum learning strategy is proposed to ensure that the proposed neural network first focuses on noise suppression and then removes blur to achieve the reconstruction of turbulence-degraded images.

(3) A multi-scale fusion module and a non-local attention-based noise suppression module are designed and used in the NSRN so that the proposed network denoises

through multi-scale and multi-level non-local information fusion while preserving the image's intrinsic information.

(4)   The back-projection idea [39] is introduced and combined with the U-NET for the final refined reconstruction of the image.

The remainder of this paper is organized as follows. Research related to this paper is introduced in Section 2. In Section 3, the motivation and rationality of this method are analyzed from the physical meaning, and the detailed design process of NSRN is given. In Section 4, the construction protocol of the experimental data and the training method of the model are introduced, and a comparative experimental analysis of the model is carried out. Finally, Section 5 summarizes the conclusions of this study.

2. Related Work

Atmospheric turbulence-degraded images have severe noise and random blurring. The restoration of such degraded images is still a very difficult problem [40,41]. In this section, we introduce previous work related to the solving of this problem.

2.1. Model-Based Image Restoration

A model-based method regards image restoration as the inverse problem of image degradation and then designs the restoration and optimization objective function through the degradation model of the image. To obtain the objective function, these methods guide the maximum a posteriori probability through some assumed priors, such as incident light and reflectance regularizer [15], sparsity and gradients [16–18,22–24], group sparsity, and low-rank priors [42]. In particular, the method proposed in [43] simultaneously considers both internal and external non-local self-similarity priors to offer mutually complementary information. Plug-and-Play (PnP) regularization [44–46] has been a hot research topic in recent years. In PnP regularization, proximal mapping of the Alternating Direction Method of the Multiplier (ADMM) algorithm can be regarded as a single denoising step and used as an off-the-shelf denoiser [47] for image reconstruction [44]. In [45], a tuning-free PnP approximation algorithm is proposed that can automatically determine internal parameters such as penalty parameters, denoising strength, and termination time. PnP has achieved great empirical success; however, its theoretical convergence is not fully understood even for the simple linear denoiser [46].

2.2. End-to-End CNN-Based Methods

The powerful representation learning ability of a Convolutional Neural Network (CNN) can be exploited to learn intrinsic features in degraded images, and then the restored images can be reconstructed by these intrinsic features [2,25–28,30,48–50]. Gao et al. [2] developed a stacked encoder–decoder for single-frame image restoration and adopted a curriculum learning strategy to ensure the convergence of the network. Chen et al. [28,38] developed a noise suppression module to address the restoration of images disturbed by severe noise. In [30], residual learning was used to remove multiple types of noise and to obtain more detailed information. In [48], CNN was used for text-image deblurring for the first time. An encoder–decoder network with symmetric skip connections proposed for image restoration in [49]. Based on regional similarity, a region-based restoration algorithm named path-restore was proposed in [27]. An Attention-guided Denoising convolutional neural Network (ADNet) [31] is a model that can be used for the restoration of images degraded by multiple factors. MemNet [50] is an extended memory model that effectively utilizes multi-layer features for image restoration. Attention mechanisms have also been successfully applied to image restoration [25,26].

2.3. Plug-and-Play with Deep CNN Denoiser

Recent work reports the state-of-the-art performance of PnP-based algorithms using pre-trained deep neural networks as denoisers in many imaging applications. Zhang et al. trained a set of fast and efficient CNN denoisers and integrated them into a model-based optimization method to solve other inverse problems [37]. They further trained a highly flexible and efficient CNN denoiser and plugged it in as a module in an iterative algorithm based on semi-quadratic splitting to solve various image restoration problems [36]. In the Multiple Self-Similarity Network (MSSN) model [51], a recurrent neural network-based PnP denoising prior is designed, and self-similar matching is performed using a multi-head attention mechanism. A prior-based deep generative network was proposed in [52] for nonlinear blind image deconvolution. The Denoising Prior-driven Deep Neural Network (DPDNN) [53] is a denoising-based image restoration algorithm whose iterative process is expanded into a deep neural network consisting of multiple denoising modules interleaved with back-projection modules to ensure consistency of observations.

Traditional PnP-based algorithms have high computational and memory requirements and are not suitable for large-scale environments. Thus, an incremental variant of the widely used PnP-ADMM algorithm was proposed in [54]; it can be used in environments involving a large number of measurements. To ensure the convergence of the resulting iterative scheme obtained by PnP-based methods, an enhanced convergent PnP algorithm [55] has been proposed. Moreover, the rank-one network [56] is an efficient image restoration framework that combines traditional rank-one decomposition and neural networks. Although PnP ADMM has proven effective in many applications, it requires manual tuning of some parameters and a large number of iterations to converge [57]. Furthermore, PnP is a non-convex framework for which current theoretical analysis is insufficient even for the most basic problems such as convergence [58].

3. Proposed Method

3.1. Motivation

Most of the existing reconstruction algorithms for turbulence-degraded images are based on an ideal image degradation model for which the image is degraded by blur and additive noise, expressed as:

$$f(x,y) = g(x,y) * h(x,y) + n(x,y), \tag{1}$$

where $g(x,y)$ is the original image before degradation, $f(x,y)$ is the observed image, $*$ is the convolution, $h(x,y)$ is the PSF of atmospheric turbulence, and $n(x,y)$ is the noise function and is usually set to be Gaussian white noise. However, real-space target images are affected by various degradation factors such as turbulence blur, out of focus blur, and atmospheric noise. This multi-factor coupling degradation can be expressed as [2]:

$$f(x,y) = O(g(x,y) * h(x,y) * k(x,y) + \zeta(x,y)) + n(x,y), \tag{2}$$

where $\zeta(x,y)$ is the noise during the transmission of a given target image in space, $h(x,y)$ is the PSF of atmospheric turbulence, $k(x,y)$ is the PSF of the disturbance, $n(x,y)$ is the sensor system noise, and $O(\cdot)$ is adaptive optics correction. It can be seen that the space target image is affected by atmospheric turbulence blur and various noises. These factors are overlapping and coupled and cannot be simply expressed as a linear combination relationship. Therefore, the degradation of the coupling of multiple factors is the most important feature of the spatial target image, which makes restoration of the spatial target image more difficult.

The PnP method considers that the degradation factors of the image include noise-free degradation and additive noise [36]. The restored model is expressed as:

$$\hat{g} = \arg\min_{g} \frac{1}{2}\|f - \tau(g)\|^2 + \lambda R(z) + \frac{\mu}{2}\|z - g\|^2. \tag{3}$$

The solution of this equation can be decomposed into the following two alternate iterative steps by half-quadratic splitting [59]:

$$g_k = \arg\min_{g} \|f - \tau(g)\|^2 + \mu\|g - z_{k-1}\|^2$$
$$z_k = \arg\min_{z} \frac{\mu}{2}\|z - g_k\|^2 + \lambda R(z) \quad , \tag{4}$$

where $\tau(\cdot)$ represents a two-dimensional convolution, z is the auxiliary variable, and μ and λ are the penalty parameters, respectively. Thus, in Equation (4), the first term is deblurring, and the second term is additive noise removal. Therefore, the preconditions for this method to be effective are that the degradation process of the image conforms to Equation (1), and the noise level in each iteration is known. Directly training an E2E deep neural network is an easy solution to solve the image restoration problem of the degraded model described by Equation (2). However, for this type of method, our studies [2,28,38,60] and related studies [61] all show that E2E-based methods have great difficulty in model training, and the restored images are visually unnatural and prone to artifacts.

Our motivation is to solve the multi-factor-coupled degraded image restoration problem by combining these two ideas and exploiting their advantages. We tried training deep deblurring neural networks with multi-task regularization and achieved good restoration results, as reported in [62]. In this paper, we design a deep neural network with two modules of denoising and reconstruction to restore severely degraded images. Our method incorporates the task decomposition idea of PnP and reduces the difficulty of the problem by decomposing complex tasks into sub-tasks, which makes the proposed method both have the advantages of E2E and avoid the assumption that multiple degeneracy factors need to be linearly separable. Further, multi-factor weak decoupling is achieved through regularization constraints to better restore complex degraded images.

3.2. Proposed Network Model

Instead of trying to express the reconstruction of blur-degraded images as an analytical expression, we design a network model for turbulence-degraded image reconstruction based on the fact that the degradation of multi-factor coupling is inseparable, as shown in Figure 1. The main components of the proposed model include a Multi-Scale Denoising Block (MSDB), a Self-Attention Dense connection Block (SADB) for suppressing noise and preserving more detailed information, and an attention-based asymmetric U-NET module. In this way, the intrinsic features of the image can be extracted from the coupled degraded image by the model, and the image can be reconstructed using these intrinsic features. Further, two FDBPs are used to fuse these intrinsic features and reconstruct sharp images. The proposed restoration reconstruction model can be expressed as:

$$\hat{f} = F_2(F_1(R(S_M(f_p) \oplus S_A(f_p) \oplus f_p) + f_p) + f_p) \tag{5}$$

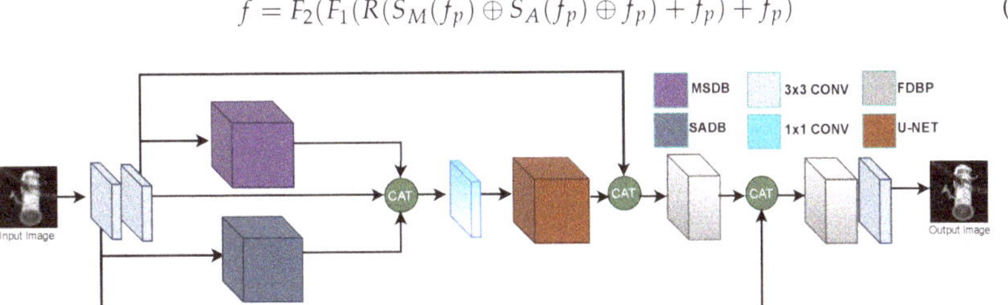

Figure 1. The proposed deep neural network model for the reconstruction of turbulence-degraded images.

Here, $\hat{f}$ is the reconstructed sharp image, f_p represents the result of the front-end preprocessing of the input-degraded image, $S_M(\cdot)$ represents MSDB, $S_A(\cdot)$ represents SADB, $R(\cdot)$ is for AU-NET, and $F(\cdot)$ is for FDBP. The proposed model first performs shallow feature extraction and denoising on the input image, and then the fused features are used as the input of U-Net. To ensure the reconstructed image has the same information distribution as the original one, this paper uses long skip connections to pass shallow features to the refined reconstruction layer. Thus, the entire model is still an E2E deep convolutional neural network. To make MSDB and SADB in the model mainly focus on removing image noise while the rest of the modules focus on image deblurring, a curriculum learning strategy from local-to-global learning is introduced. For details, see Section 3.3.

3.2.1. MSDB

The main task of this module is to achieve noise suppression by extracting multi-degree features from noisy images and reconstructing noise-free image features. As shown in Figure 2, the encoder of MSDB consists of two multi-scale convolutional layers, each of which consists of three-scale convolutions with kernels of 3×3, 5×5, and 7×7, respectively. The extracted multi-scale features are connected and then passed through a dimensionality reduction fusion layer with a convolution kernel of 1×1 to obtain the high-level features of the degraded image. The decoder of MSDB consists of four dilated convolutional layers, and each dilated convolution is followed by ReLU activation and batch normalization. Dilated convolution has shown good performance in image denoising [62] because it is more beneficial to use contextual information to reconstruct sharp images, and it can increase the receptive field while avoiding the loss of downsampling information. The scale factors of the four dilated convolutional layers of MSDB are 1, 2, 2, and 1, respectively.

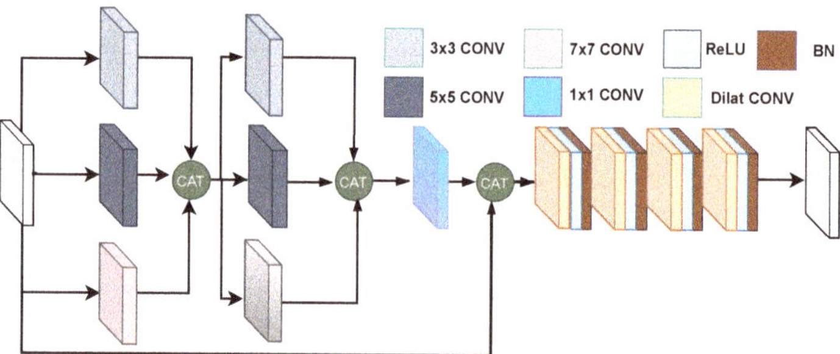

Figure 2. The MSDB in the NSRN model.

3.2.2. SADB

The idea of non-local was used in image denoising in BM3D [47] with remarkable success. To this day, the latest state-of-the-art methods still use non-local as a basic strategy [37,51]. The randomness of noise makes it easier to achieve noise removal by collaborative filtering of correlated regions. In our designed SADB, the self-attention mechanism [63,64] is introduced to realize non-locality. As shown in Figure 3, given an input tensor $X = (H, W, C)$, two 1×1 convolutions in parallel are used to change its shape to (HW, C) and (C, HW). Then, multiply these two matrices to get the (HW, HW) matrix and use the softmax activation to get the weighted (HW, HW) matrix. Then, multiply the feature (C, HW) matrix with the weighted (HW, HW) matrix to get the (C, HW) matrix. After changing its shape to (H, W, C), it is added to the initial feature map, and finally, the feature map with weight redistribution is obtained. Non-local attention can be expressed as:

$$\hat{x}_i = w\text{softmax}(< wx_i \cdot wx_j >)(wx_i) + x_i \tag{6}$$

where x is the input feature, $\hat{x}$ is the feature after non-local attention processing, and $<\cdot>$ represents the inner product; wx represents a one-dimensional linear embedding, implemented in this work by a convolution of 1×1.

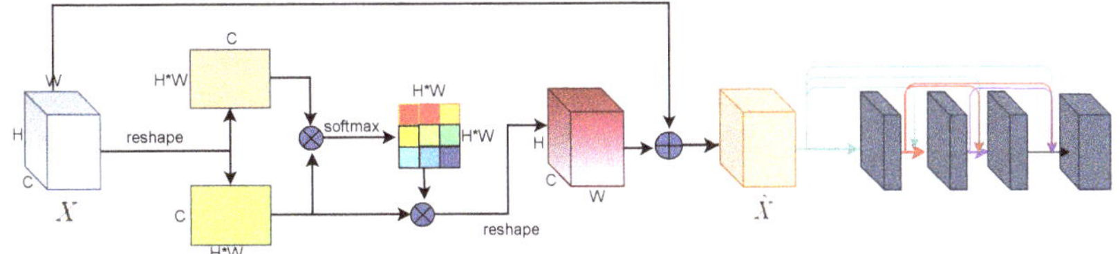

Figure 3. The SADB in the NSRN model ($\otimes$ denotes matrix multiplication, and $\oplus$ denotes element-wise addition).

Almost all denoising methods are based on the prior assumption that noise is high-frequency and sparse. Therefore, these algorithms tend to blur the image while removing noise. In the proposed SADB, the dense connection is adopted to solve this problem. SADB takes the weighted feature map as an input and passes it to each subsequent convolutional layer in turn, and dense transmission is also performed between the convolutional layers. This allows the feature map information to flow efficiently, which not only avoids the vanishing gradient but also reduces the depth of the network and allows the network to converge faster. The proposed SADB can better utilize the context information of each layer and retain more image details while removing noise.

3.2.3. AU-NET

Noise-suppressed feature maps are obtained after MSDB and SADB. To further extract effective features from degraded images and reconstruct a sharp image, an attention-based asymmetric U-Net is designed. It uses dilated convolution and batch normalization techniques in the first two layers of encoders to further suppress high-frequency noise in feature maps. Further, under the constraint of the loss function, the encoder has greater modeling ability, which means that its encoding efficiency is higher, and the encoded features are beneficial to the output of the decoder. Further, we use a channel attention mechanism to assign weights to the outputs of the encoder and decoder so that the features of the outputs are more beneficial to the subsequent reconstruction work.

To reduce the information loss caused by fixed downsampling and upsampling, a convolution with stride two is used for downsampling, and a transposed convolution is used for upsampling. Compared with the widely used pooling and interpolation, convolution not only achieves the same downsampling and upsampling effect but also makes the whole process learnable, especially when using the backpropagation algorithm to learn more accurate parameters. Furthermore, the corresponding encoders and decoders are connected by skipping to make the information flow better from shallow layers to deep layers, avoiding a vanishing gradient. Due to the use of noise-reduction processing in the encoding stage and the channel attention mechanism used at the end of encoding and decoding, the entire structure is no longer symmetric, so it is called an attention-based asymmetric U-Net, as shown in Figure 4.

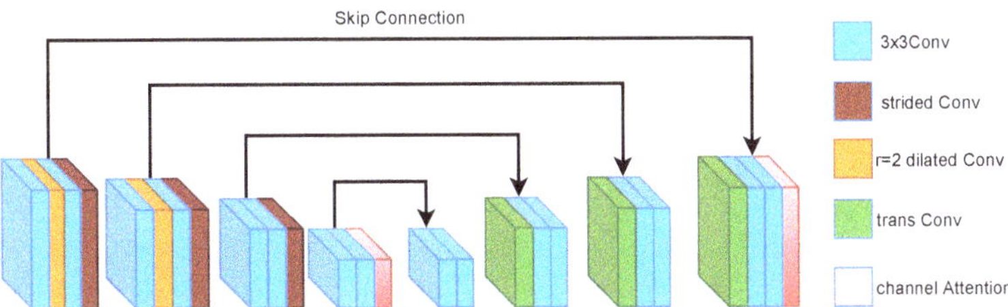

Figure 4. The attention-based asymmetric U-Net in the proposed model.

3.2.4. FDBP

The reconstruction of AU-Net is based on high-level features, and there is a potential risk of insufficient reconstruction of detailed texture information. To enhance the presentation ability of the network and restore clearer images, an FDBP is designed. Back-projection has been successfully applied in image super-resolution tasks [39], where it has been shown to have good reconstruction capabilities for texture details. Inspired by it, we design FDBP, which projects high-resolution features into low-resolution space through a downsampling unit, then projects low-resolution features into high-resolution space through an upsampling module, and finally guides network learning by the error between the old and new high-resolution features. The main operations in our designed FDBP are defined as:

$$up\ sample: x_l = (x_{l-1} * k_l)\uparrow_s, \tag{7}$$

$$down\ sample: x_l = (x_{l-1} * k_l)\downarrow_s, \tag{8}$$

$$residual: e_l = x_1 - x_{l-1}, \tag{9}$$

$$up\ residual\ sample: x_l = (x_{l-1} * k_l)\uparrow_s, \tag{10}$$

$$output: x_l = x_0 + x_l, \tag{11}$$

where x_0 represents the feature after convolution of the input, and $x * k$ is the convolution of 3×3. To enhance the flow of information and keep the reconstructed features consistent, we use two FDBP operations. The FDBP module we designed is shown in Figure 5 and can capture multi-scale context information well and downsample the feature map to a small space to save memory and speed up network training.

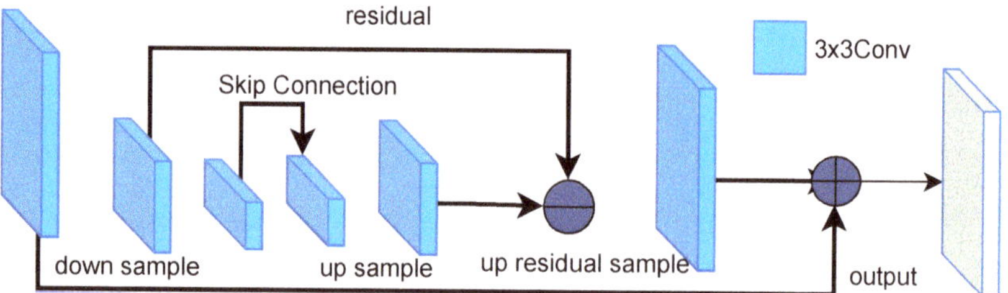

Figure 5. The FDBP for reconstruction in the model.

3.3. Curriculum Learning Strategy

Due to the randomness of various types of noise, the spatial dimension of the samples of multi-factor-coupled degraded images is very large, and its representation learning is very difficult. Therefore, a complex neural network needs to be designed to achieve its restoration. In such cases, due to the complexity of the problem and the scale of the parameters, the learning difficulty of the neural network is increased. Curriculum learning [65–67] is considered an effective way to address this problem. Aiming at the difficulty of multi-factor-coupled image restoration, a systematic curriculum learning strategy from local-to-global network and from easy-to-difficult data learning is designed.

3.3.1. Local-to-Global Network Learning

Multi-task decomposition is helpful to reduce the difficulty of the restoration of multi-factor-coupled images. Although the restoration of turbulence-degraded images is difficult to simply decompose into multiple independent tasks [60], we design the NSTR neural network based on the weak assumption that images are mainly affected by additive noise and turbulence blur. Since MSDB and SADB are primarily good at noise suppression, these two modules are trained separately. First, a new training set is constructed by adding Gaussian noise and Poisson noise to the blurred images, and the blurred images without noise are used as labels. Then, the output components are plugged into MSDB and SADB, respectively. Finally, MSDB and SADB are pre-trained to obtain weight parameters.

After completing the training of MSDB and SADB, their weights are transferred to the overall network model. This transfer learning strategy enables NSRN to have a certain ability to suppress noise from the beginning. To preserve the noise suppression ability of MSDB and SADB in the overall training of NSRN, the learning rate should be set to a small value. In our experiments, the overall learning rate is set to 0.1. By fine-tuning the learning rate, the proposed network not only maintains noise suppression effectively but also focuses a lot of attention on image deconvolution reconstruction. This kind of curriculum learning strategy of first local and then global features not only reduces the learning difficulty of the whole model but also avoids strict task decomposition.

3.3.2. Easy-to-Difficult Data Curriculum Learning

The main reason for the difficulty in restoring turbulence-blurred images is the high dynamics of turbulent flow, which results in a large spatial distribution of samples. We find it extremely difficult to train the network directly with severely turbulence-degraded images. Therefore, the easy-to-difficult learning strategy is used to train the NSRN network. By setting the value of the atmospheric coherence length r, data with different degrees of blur can be obtained. In this paper, three r values are used to obtain data with mild, moderate, and severe blur, respectively. First, initial network training is performed via the weight initialization method provided by He [68], and then the network is sequentially trained using datasets with varying degrees of blur from mild to severe. After the mild set converges, its weights are saved and used for weight initialization for training on the blurrier datasets. Through this easy-to-difficult training strategy, the proposed network can eventually learn more complex mappings and achieve better results.

NSRN uses the $L1$ loss function for training. The inputs to train the local modules MSDB and SADB are noisy blurred images, and the labels are blurred images without noise. The input to train the entire model is degraded images, and the labels are sharp images. The loss function in the network can be formulated as:

$$L(\Theta) = \frac{1}{N} \sum_{i=1}^{N} \|\hat{y} - y\|_1, \tag{12}$$

where $\|\cdot\|_1$ can restore better texture information. PyTorch was used to implement the proposed network model, and the whole network was trained using GTX 1080Ti under Ubuntu 16. The image block size used for training is 32×32, and the default setting for the batch size is 64. Since the input and output images of the network have the same resolution, any image resolution can be used in testing. To make the network converge faster, a learning-rate decay strategy is used; that is, the initial learning rate is set to 0.001 and decays to 0.5 times the previous learning rate every 50 epochs. Overall training used 250 epochs. A Mean Squared Error loss function (MSE) is used, and the Adam optimizer is used to constrain gradient descent. The learning algorithm of the proposed NSRN is shown in Algorithm 1. The experimental convergence curve of Algorithm 1 is shown in Figure 6. The restoration of mildly degraded images is less difficult, and the model converges well. As shown in Figure 6a, both training accuracy and validation accuracy converge to better positions. Both moderate degradation and severe degradation converge to low error levels due to the curriculum learning strategy (see Figure 6b,c). In moderate degradation, the validation curve indicates slight overfitting. In severe degradation, the validation curve indicates oscillation at the beginning and convergence after 125 epochs. The training time is 8.5 min/per epoch. When the test image size is 384×384 pixels, the inference time is 0.24 s/frame.

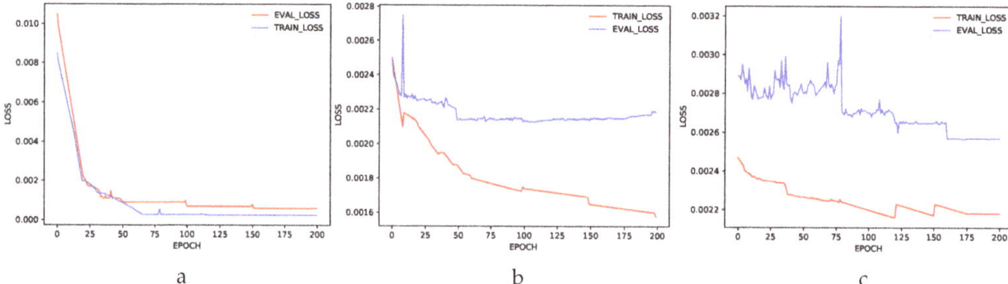

Figure 6. Convergence curves: (**a**) mildly degraded; (**b**) moderately degraded; and (**c**) severely degraded.

Algorithm 1 Systematic curriculum learning algorithm for NSRN

Require:
   B: number of MSDB and SADB training;
   $D = \{D_1, D_2, D_n\}$: NSRN training set;
   w_0:weight initialization.
Ensure:
   NSRN(w): parameters of NSRN.
  1: Begin:
   /* local-to-global learning */
  2: MSDB learning: $s_m(w_s) = S_m(f_p(B))$, where w_s is the parameter of MSDB
  3: SADB learning: $s_A(w_A) = S_A(f_p(B))$, where w_A is the parameter of SADB
   /* easy-to-difficult learning */
  4: Initialize MSDB in NSRN with w_s
  5: Initialize SADB in NSRN with w_A
  6: **for** each D_i **do**
  7:    NSRN learning: NSRN $(w_i) = F_2(F_1(R(S_M(f_p) \oplus S_A(f_p) \oplus f_p) + f_p) + f_p)$
  8: **end for**
  9: Initialize NSRN with w_n
  10: Train NSRN with all training data D
  11: Output: NSRN

---

4. Experiments and Discussions

4.1. Dataset

There are few public real-space target images, and ground-truth labels of degraded images are also difficult to obtain. Therefore, degraded image simulation is used to obtain training data to verify the effectiveness of the proposed method. The 3D models used to obtain images of simulated space objects are from STK (Satellite Tool Kit) [69], which provides various satellite models and turbulence degradation models. The reflected sunlight of space objects is refracted by atmospheric turbulence, which makes the images observed by ground-based telescopes blurred. This turbulence blur can be represented by the following model [28].

$$h(u,v) = e^{\left\{-3.44\left(\frac{\lambda f U}{r}\right)^{5/3}\right\}} \tag{13}$$

where $U = \sqrt{u^2 + v^2}$ is the frequency, (u,v) is the unit pulse, λ is the wavelength, f is the optical focal length, and r is the atmosphere coherence length. It can be seen that the larger the r, the stronger the atmospheric motion and the blurrier the image. Therefore, different degrees of turbulence blurred images can be generated by changing the size of r.

To obtain more diverse training data, clear satellite images with different attitude angles are obtained by rotating the 3D satellite model from STK. The acquired images are data-enhanced, including rotating 90, 180, and 270 degrees and flipping horizontally and vertically. Images are then blurred using the atmospheric turbulence long-exposure degradation function shown in Equation (13). By setting different r values in [0, 0.02], blurred image datasets with three levels contained in three subsets—mildly degraded ($r \in [0.005, 0.01)$), moderately degraded (r $\in$ [0.005, 0.015)), and severely degraded (r $\in$ [0.005, 0.02])—can be obtained. During atmospheric turbulence imaging, the turbulence blurring is also mixed with photon noise, dark noise, reset noise, and readout noise. These noises mainly obey Gaussian and Poisson distributions, so we add Gaussian noise and Poisson noise to the blurred image. The value range of the parameter of Gaussian noise is [35, 42], and the value range of the parameter of Poisson noise is [4, 7]. The real degradation model is expressed as:

$$f(x,y) = g(x,y) * h(x,y) + n(x,y) + p(x,y), \tag{14}$$

where f is the observed image, g is the original image, h is the PSF atmospheric turbulence, n represents Gaussian noise, and p represents Poisson noise. To ensure the generalization ability of the model and encourage the restoration model to learn the blur degradation mode and the corresponding restoration mode, we adopt the strategy of training on small images and verifying and testing on large images.

We cut the image at 20-pixel intervals to generate 32×32 image patches and then discarded samples in which more than 90% of the patches were black background area, resulting in 117,300 image patches for training the model. Some of the generated training samples are shown in Figure 7. A total of 56 large images that are not used to for the training set are used as the test set, and some test samples are shown in Figure 8. We also collected 17 real-world turbulence-degraded images from public sources as a test set, as shown in Figure 9. Detailed information about the dataset is show in Table 1. The spatial resolutions of the large images in the table are not uniform, and their ranges is [256×256, 1024×1024].

Figure 7. Some training data. From left to right: clear; mildly degraded; moderately degraded; and severely degraded.

Table 1. Composition details of dataset.

		Number of Large Images	Number of Image Patches
	mild	1358	117,300
Training set	moderate	1358	117,300
	severe	1358	117,300
	mild	100	/
Validation set	moderate	100	/
	severe	100	/
	mild	56	/
Simulated test set	moderate	56	/
	severe	56	/
Real test set	/	17	/

Figure 8. Some simulated data for testing. From left to right: clear; mildly degraded; moderately degraded; and severely degraded.

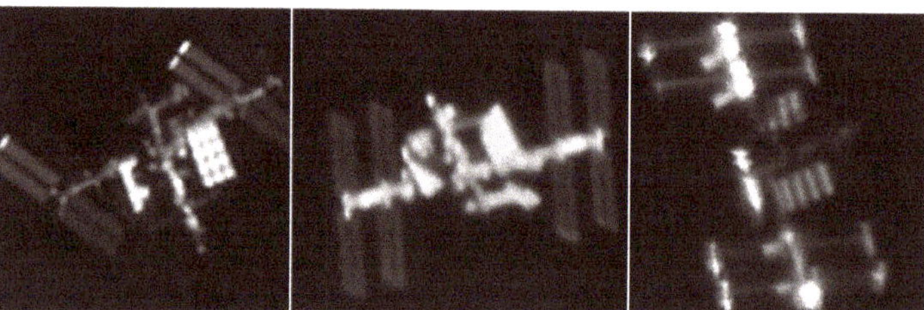

Figure 9. Some real-world turbulence-degraded data for testing.

4.2. Metrics for Evaluation and Methods for Comparison

The simulated images have labels, so the performance evaluation of the algorithm can be carried out by combining subjective methods and objective metrics. For objective metrics, peak signal-to-noise ratio (PSNR) and structural similarity (SSIM) are used to evaluate the restoration performance of each algorithm. For subjective metrics, the quality of the restored image is evaluated by human vision and the reference images. Moreover, for real images, due to the lack of reference images, only subjective evaluation and no-reference metrics can be used. In this paper, the no-reference evaluation metrics used are Brenner, Laplacian, SMD, Variance, Energy, Vollath, and Entropy. The calculation methods of these no-reference metrics can be found in [70].

Gao [2] conducted extensive analysis on traditional restoration methods for spatial images. The experimental results show that the traditional methods are not ideal for removing turbulence blur, so the proposed method is not compared using traditional methods. To better analyze and evaluate the performance of this method, some representative deep learning methods are selected for comparative experiments, namely Gao [2], Chen [38], Mao-30 [49], MemNet [50], CBDNet [48], ADNet [31], DPDNN [53], and DPIR [36]. For absolute fairness, for all comparison methods, we use the parameters given in the original text and train them with the training set of this paper.

4.3. Ablation Experiment

Our proposed model (Figure 1) uses an asymmetric U-NET as the backbone. To verify the effectiveness of the proposed model, an ablation experiment is performed. In this experiment, the backbone U-NET is named Model1, and Model1 to Model6 are formed by plugging MSDB, SADB, FDBP, and curriculum learning strategy (TNRS) into Model1, as shown in Table 2. When training models Model1–Model5, three training subsets with different blur degrees are directly merged as the final training set. Model6 is trained using the steps shown in Algorithm 1. The trained model is tested on three different degraded images; the results of the objective evaluation metric are shown in Table 2, and the partially restored images are shown in Figure 10.

Table 2. Performance of models with different components (The best results are shown in bold fonts).

		Model1	Model2	Model3	Model4	Model5	Model6
U-Net		√	√	√	√	√	√
MSDB			√		√	√	√
SADB				√	√	√	√
FDBP						√	√
TNRS							√
PSNR	mild	29.2092	29.8803	29.8666	30.0160	30.0587	**30.1817**
	moderate	27.9264	28.2895	28.0992	28.2989	28.3944	**28.6400**
	severe	25.9631	27.2224	27.1046	27.6352	27.8129	**28.0169**
SSIM	mild	0.8889	0.8923	0.8869	0.9001	0.8911	**0.9035**
	moderate	0.8430	0.8649	0.8757	0.8685	0.8701	**0.8732**
	severe	0.7052	0.8363	0.8218	0.8325	0.8341	**0.8545**

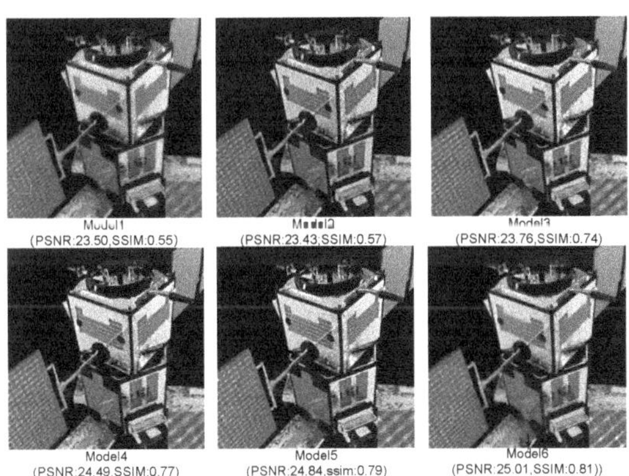

Figure 10. Restoration of severe turbulence blur using different modules (The red boxes represent the focus region).

It can be seen from Table 2 that: (1) Model1, which only contains the backbone U-NET, lacked sufficient representation power to learn intrinsic features from degraded images and reconstruct images well. (2) The PSNR of Model2 obtained by plugging MSDB into Model1 was significantly improved because MSDB enables U-NET to have better global and local information presentation capabilities. However, the PSNR of Model3 obtained by plugging SADB into Model1 decreased, but the image details are richer. (3) Model4 was obtained by plugging MSDB and SADB into Model1. Compared with Model1, Model2, and Model3, both the PSNR and the SSIM significantly improved in Model4. This is because Model4 has stronger noise suppression performance. (4) Model5, obtained by plugging FBPR into Model4, obtained more consistent results. (5) Model6 (NSRN) added the curriculum learning algorithm to Model5 to train the network. The performance of Model6 was further improved compared to Model5, which proves that the proposed model does have better generalization ability, and it is easier to capture the mapping relationship between sharp images and low-resolution images. Moreover, from the restored images of each model in Figure 10, the results of Model 6 have the best visual effect, and the edges and textures are clearer.

4.4. Experiments and Comparative Analysis of Simulated Images

(1) Model for mild degradation

We use the trained model for restoration experiments on test data with mild degradation, and the resulting averages of objective evaluation metrics are shown in Table 3. It can be seen that for PSNR, Mao, CBDNet, ADNet, DPDNN, DPIR, and the proposed method all achieve very good results. These methods all have more complex network models, so they have better presentation ability. For SSIM, DPDNN, DPIR, and our method have significantly better performance than the remaining methods, which shows that the method based on noise suppression has a better ability to restore textual details. Compared to the second-ranked method, our method improves PSNR by 0.16 and improves SSIM by 0.036. An example set of restored results is shown in Figure 11. It can be seen that for mildly degraded images, almost all methods achieve better visual effects.

Table 3. Average PSNR and SSIM of different state-of-the-art methods on mild degradation (The best results are shown in bold fonts).

Methods	PSNR	SSIM
Gao	27.5423	0.8337
Chen	28.0156	0.8431
Mao	29.3903	0.8387
MemNet	27.8413	0.8295
CBDNet	29.4395	0.8596
ADNet	29.7430	0.8828
DPDNN	30.0122	0.8999
DPIR	29.7316	0.8932
Ours	**30.1817**	**0.9035**

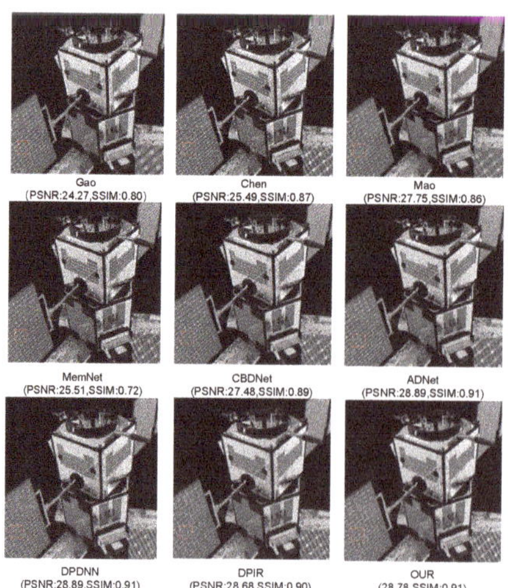

Figure 11. Restoration using different state-of-the-art methods on mild turbulence blur (The red boxes represent the focus region).

(2) Model for moderate degradation

The test results of all models on the moderately degraded dataset are shown in Table 4. It can be seen that for PSNR, DPIR, DPDNN, and Mao achieve competitive results. However, our method has the best performance and is nearly 0.3 higher than the second-ranked method, indicating that the proposed method does have a strong representation of learning ability by introducing modules such as FBPR. On SSIM, the best method is DPDNN, and our method is close to DPDNN. The restoration results of different methods on a typical moderately degraded image are shown in Figure 12. It can be seen that the visual effects of images restored by DPDNN, DPIR, Mao, and our method are similar. However, in contrast, DPDNN has sharper edges in some regions, and our method is more consistent.

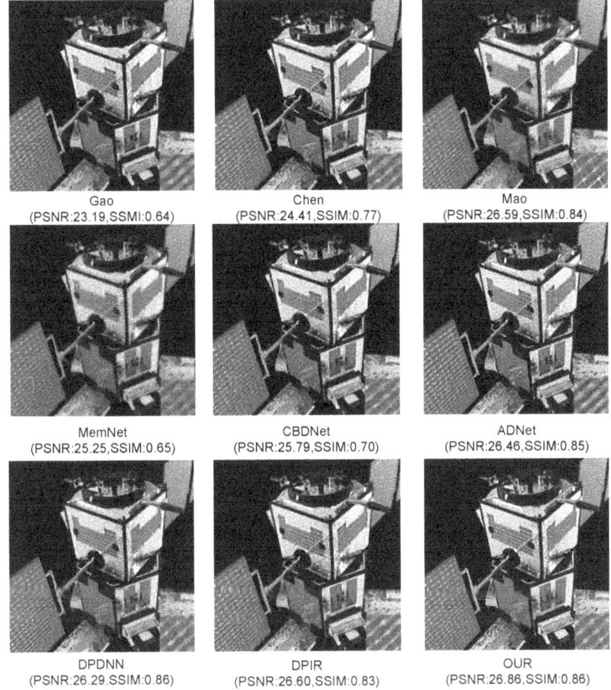

Figure 12. Restoration using different state-of-the-art methods on moderate turbulence blur (The red boxes represent the focus region).

Table 4. Average PSNR and SSIM of different state-of-the-art methods on moderate degradation (The best results are shown in bold fonts).

Methods	PSNR	SSIM
Gao	25.8558	0.7643
Chen	26.9923	0.8297
Mao	28.3321	0.8446
MemNet	26.4702	0.7480
CBDNet	27.7382	0.7817
ADNet	28.1007	0.8472
DPDNN	28.3600	**0.8766**
DPIR	28.3519	0.8284
Ours	**28.6400**	0.8732

(3) Model for severe degradation

The results of objective evaluation metrics of all restoration models on the severely degraded image test set are shown in Table 5. It can be seen that our method has obvious advantages in this dataset: the PSNR is higher than the second-ranked method by nearly 0.2, and the SSIM is higher than the second-ranked method by 0.007. Further, for PSNR, our method is the only one that exceeds 28. Our method is also the only method that shows the best performance in both metrics, which shows that for severely degraded images with severe noise and severe blur, the method that can specifically deal with the noise is more competitive. The restoration results of different methods on a typical severely degraded image are shown in Figure 13. From the visual effect, our method restores more texture details and has obvious advantages.

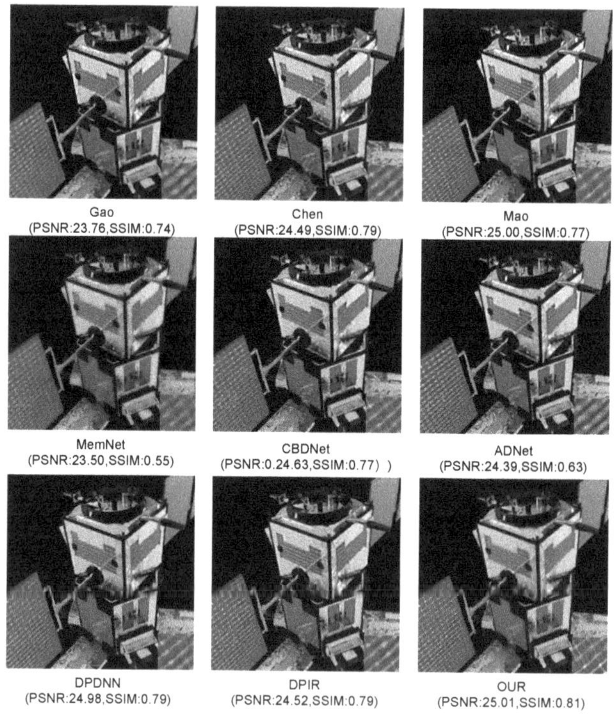

Figure 13. Restoration using different methods on severe turbulence blur (The red boxes represent the focus region).

Table 5. Average PSNR and SSIM of different state-of-the-art methods on severe degradation (The best results are shown in bold fonts).

Methods	PSNR	SSIM
Gao	26.7512	0.7934
Chen	27.1416	0.8250
Mao	27.1224	0.8190
MemNet	26.1868	0.7288
CBDNet	27.4253	0.8471
ADNet	27.1676	0.8346
DPDNN	27.8129	0.8431
DPIR	27.6249	0.8376
Ours	**28.0169**	**0.8545**

In general, the proposed method, DPDNN, and DPIR are the most competitive methods, while the Gao, Mao, and Chen models are too small to represent the huge sample space spanned by severely degraded images. This shows that a network that can restore heavily noisy and blurred severely degraded images not only needs sufficient representation ability but also some mechanism for learning features, such as attention. Moreover, as the model becomes more complex, the generalization ability and restoration ability of the network model can be improved by separately processing blur and noise.

To better compare the performance of each algorithm under different noise levels, an image is randomly selected from the test set and then mixed with different levels of noise for restoration experiments. As seen in Figure 14, DPIR and our method have similar performance on SSIM. DPDNN also has good performance when the noise intensity is greater than 35. Moreover, our method has the best PSNR at almost all noise levels.

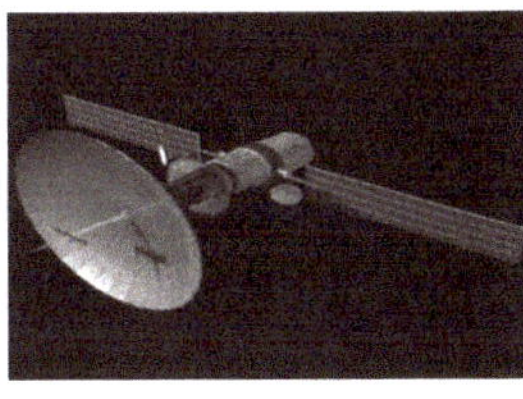

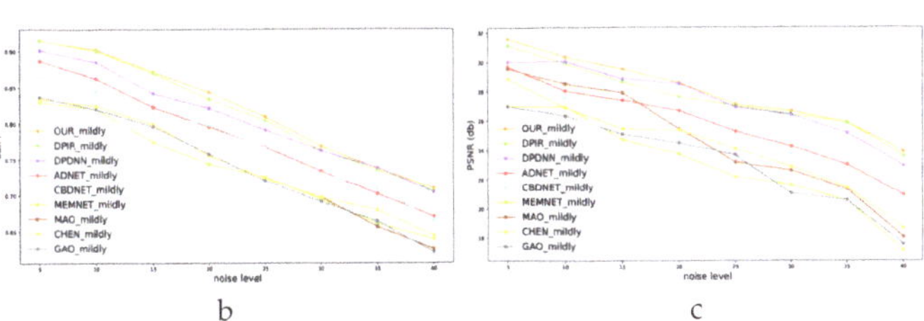

a b c

Figure 14. Results of different noise levels: (**a**) Test image; (**b**) SSIM; (**c**): PSNR.

4.5. Experiments and Comparative Analysis of Real Images

The results of the non-reference evaluation metrics of the restoration results obtained by all the compared methods on real data are shown in Table 6, and the restoration results on real data are shown in Figure 15. There was still a big difference between the simulated training data and the real image distribution, and all methods encountered cross-domain problems. However, under the same conditions, our method is the best in these numerical experiments and these evaluation metrics. Of course, the reliability of the no-reference evaluation and the consistency with human vision require further research [24]. The proposed method has a certain enhancement of texture and edges, so metrics such as upper edge and gradient have weak advantages over other methods. As shown in Figure 15, due to the weak network representation ability of the methods of Gao [2] and Mao [49], the restored image is still blurred. The rest of the methods can provide visually pleasing restoration. The visual effect restored by the method of Chen [38] is close to our method, indicating that our method has excellent performance for the restoration of severely degraded images. This is because it treated additive noise and blur degradation separately and designed special modules to denoise and perform blur deconvolution.

Table 6. Results of non-reference evaluation metrics on real test data (The best results are shown in bold fonts).

Method	Brenner (xe6)	Laplacian	SMD (xe4)	Variance (xe7)	Energy (xe6)	Vollath (xe7)	Entropy
ADNet	27.36	346.52	53.9847	17.477	19.42	17.05	2.58
CBDNet	23.07	310.00	49.80	17.42	16.85	17.06	2.51
Chen	27.62	419.92	56.31	17.57	19.92	17.13	**2.68**
Gao	24.45	231.94	52.34	17.41	16.53	17.05	2.61
Mao	16.71	220.832	43.61	16.83	12.48	16.58	2.32
MemNet	21.23	314.55	48.71	16.41	15.96	16.08	2.52
Zhang	19.26	242.84	46.14	17.85	13.90	17.55	2.49
DPDNN	15.65	183.75	42.31	16.31	11.29	16.07	2.57
Ours	**32.54**	**493.77**	**58.98**	**18.13**	**23.47**	**17.61**	2.41

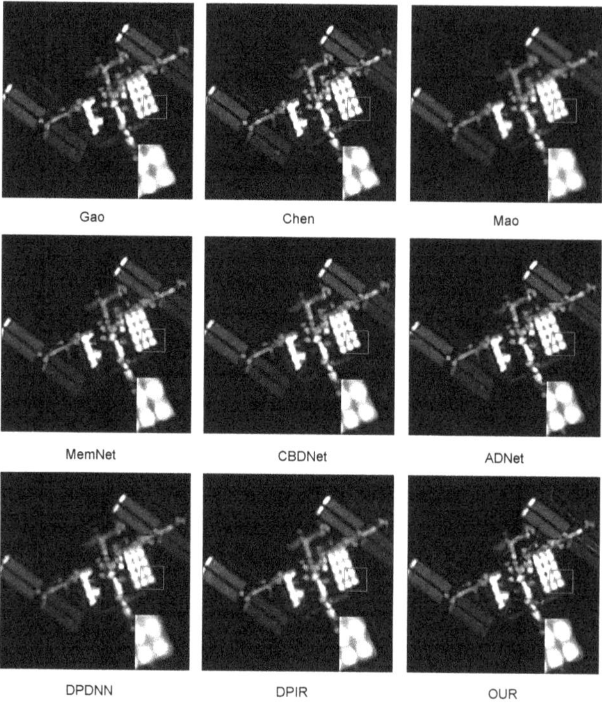

Figure 15. Restoration using different methods on real turbulence blur (The red boxes represent the focus region).

5. Conclusions

Atmospheric turbulence-blurred images are usually observed at long distances and contain severe noise. Therefore, the restoration of atmospheric turbulence-degraded images includes two tasks: deblurring and denoising. Although deblurring and denoising belong to the same underlying visual tasks, their internal principles are different. Denoising removes high-frequency noise in images, while deblurring using deconvolution to obtain high-frequency information from blurred images. Based on this knowledge, we design a deep neural network model for the restoration of atmospheric turbulence-degraded images based on curriculum learning. Noise suppression of degraded images is achieved by designing a dedicated denoiser without enforcing fully decoupled denoising and deblurring. The experimental results demonstrate the effectiveness of our method. However, the restoration of real turbulence-degraded images is still an open problem. The design of a GAN [71] model based on the ideas proposed in this paper to improve the restoration of real images will be the direction and focus of future research.

Author Contributions: Conceptualization, C.X.; methodology, C.X.; software, J.S.; validation, J.S.; formal analysis, Z.G.; investigation, J.S.; resources, J.S.; data curation, J.S.; writing—original draft preparation, J.S.; writing—review and editing, C.X.; visualization, C.X.; supervision, C.X.; project administration, C.X.; funding acquisition, C.X. All authors have read and agreed to the published version of the manuscript.

Funding: This work has been partially supported by the Sichuan Science and Technology Program (grant nos. 2021YFG0022, 2022YFG0095).

Institutional Review Board Statement: Not applicable.

Informed Consent Statement: Not applicable.

Data Availability Statement: Not applicable.

Acknowledgments: We thank anonymous reviewers and academic editors for their valuable comments.

Conflicts of Interest: The authors declare no conflict of interest.

References

1. Jefferies, S.M.; Hart, M. Deconvolution from wave front sensing using the frozen flow hypothesis. *Opt. Express* **2011**, *19*, 1975–1984. [CrossRef] [PubMed]
2. Gao, Z.; Shen, C.; Xie, C. Stacked convolutional auto-encoders for single space target image blind deconvolution. *Neurocomputing* **2018**, *313*, 295–305. [CrossRef]
3. Mourya, R.; Denis, L.; Becker, J.M.; Thiébaut, E. A blind deblurring and image decomposition approach for astronomical image restoration. In Proceedings of the 2015 23rd European Signal Processing Conference (EUSIPCO), Nice, France, 31 August–4 September 2015; IEEE: New York, NY, USA, 2015; pp. 1636–1640.
4. Yan, L.; Jin, M.; Fang, H.; Liu, H.; Zhang, T. Atmospheric-turbulence-degraded astronomical image restoration by minimizing second-order central moment. *IEEE Geosci. Remote Sens. Lett.* **2012**, *9*, 672–676.
5. Zhu, X.; Milanfar, P. Removing atmospheric turbulence via space-invariant deconvolution. *IEEE Trans. Pattern Anal. Mach. Intell.* **2012**, *35*, 157–170. [CrossRef]
6. Xie, Y.; Zhang, W.; Tao, D.; Hu, W.; Qu, Y.; Wang, H. Removing turbulence effect via hybrid total variation and deformation-guided kernel regression. *IEEE Trans. Image Process.* **2016**, *25*, 4943–4958. [CrossRef]
7. Gilles, J.; Dagobert, T.; De Franchis, C. Atmospheric Turbulence Restoration by Diffeomorphic Image Registration and Blind Deconvolution. In *Advanced Concepts for Intelligent Vision Systems*; Springer: Berlin/Heidelberg, Germany, 2008; pp. 400–409.
8. Jin, M.; Meishvili, G.; Favaro, P. Learning to extract a video sequence from a single motion-blurred image. In Proceedings of the IEEE Conference on Computer Vision and Pattern Recognition, Salt Lake City, UT, USA, 18–22 June 2018; pp. 6334–6342.
9. Xu, X.; Pan, J.; Zhang, Y.J.; Yang, M.H. Motion blur kernel estimation via deep learning. *IEEE Trans. Image Process.* **2017**, *27*, 194–205. [CrossRef]
10. Zhou, C.; Lin, S.; Nayar, S.K. Coded aperture pairs for depth from defocus and defocus deblurring. *Int. J. Comput. Vis.* **2011**, *93*, 53–72. [CrossRef]
11. Vasu, S.; Maligireddy, V.R.; Rajagopalan, A. Non-blind deblurring: Handling kernel uncertainty with cnns. In Proceedings of the IEEE Conference on Computer Vision and Pattern Recognition, Salt Lake City, UT, USA, 18–23 June 2018; pp. 3272–3281.
12. Zhang, J.; Pan, J.; Lai, W.S.; Lau, R.W.; Yang, M.H. Learning fully convolutional networks for iterative non-blind deconvolution. In Proceedings of the IEEE Conference on Computer Vision and Pattern Recognition, Honolulu, HI, USA, 21–26 July 2017; pp. 3817–3825.
13. Schuler, C.J.; Hirsch, M.; Harmeling, S.; Schölkopf, B. Learning to deblur. *IEEE Trans. Pattern Anal. Mach. Intell.* **2015**, *38*, 1439–1451. [CrossRef]
14. Zhang, Y.; Lau, Y.; Kuo, H.w.; Cheung, S.; Pasupathy, A.; Wright, J. On the global geometry of sphere-constrained sparse blind deconvolution. In Proceedings of the IEEE Conference on Computer Vision and Pattern Recognition, Honolulu, HI, USA, 21–26 July 2017; pp. 4894–4902.
15. Dai, C.; Lin, M.; Wu, X.; Zhang, D. Single hazy image restoration using robust atmospheric scattering model. *Signal Process.* **2020**, *166*, 107257. [CrossRef]
16. Hu, D.; Tan, J.; Zhang, L.; Ge, X.; Liu, J. Image deblurring via enhanced local maximum intensity prior. *Signal Process. Image Commun.* **2021**, *96*, 116311. [CrossRef]
17. Zhang, H.; Wipf, D.; Zhang, Y. Multi-image blind deblurring using a coupled adaptive sparse prior. In Proceedings of the IEEE Conference on Computer Vision and Pattern Recognition, Portland, OR, USA, 23–28 June 2013; pp. 1051–1058.
18. Xu, L.; Zheng, S.; Jia, J. Unnatural l0 sparse representation for natural image deblurring. In Proceedings of the IEEE Conference on Computer Vision and Pattern Recognition, Portland, OR, USA, 23–28 June 2013; pp. 1107–1114.
19. Rostami, M.; Michailovich, O.; Wang, Z. Image Deblurring Using Derivative Compressed Sensing for Optical Imaging Application. *IEEE Trans. Image Process.* **2012**, *21*, 3139–3149. [CrossRef] [PubMed]
20. He, R.; Wang, Z.; Fan, Y.; Fengg, D. Atmospheric turbulence mitigation based on turbulence extraction. In Proceedings of the 2016 IEEE International Conference on Acoustics, Speech and Signal Processing (ICASSP), Shanghai, China, 20–25 March 2016; pp. 1442–1446. [CrossRef]
21. Li, D.; Mersereau, R.M.; Simske, S. Atmospheric Turbulence-Degraded Image Restoration Using Principal Components Analysis. *IEEE Geosci. Remote Sens. Lett.* **2007**, *4*, 340–344. [CrossRef]
22. Krishnan, D.; Fergus, R. Fast image deconvolution using hyper-Laplacian priors. *Adv. Neural Inf. Process. Syst.* **2009**, *22*, 1033–1041.
23. Perrone, D.; Favaro, P. Total variation blind deconvolution: The devil is in the details. In Proceedings of the IEEE Conference on Computer Vision and Pattern Recognition, Columbus, OH, USA, 23–28 June 2014; pp. 2909–2916.
24. Pan, J.; Hu, Z.; Su, Z.; Yang, M.H. Deblurring text images via L0-regularized intensity and gradient prior. In Proceedings of the IEEE Conference on Computer Vision and Pattern Recognition, Columbus, OH, USA, 23–28 June 2014; pp. 2901–2908.

25. Mou, C.; Zhang, J. Graph Attention Neural Network for Image Restoration. In Proceedings of the 2021 IEEE International Conference on Multimedia and Expo (ICME), Taipei, Taiwan, 18–22 July 2022; IEEE: New York, NY, USA, 2021; pp. 1–6.
26. Anwar, S.; Barnes, N.; Petersson, L. Attention-Based Real Image Restoration. *IEEE Trans. Neural Netw. Learn. Syst.* **2021**, 1–11. [CrossRef] [PubMed]
27. Yu, K.; Wang, X.; Dong, C.; Tang, X.; Loy, C.C. Path-restore: Learning network path selection for image restoration. *IEEE Trans. Pattern Anal. Mach. Intell.* **2021**, *44*, 7078–7092. [CrossRef]
28. Chen, G.; Gao, Z.; Wang, Q.; Luo, Q. U-net like deep autoencoders for deblurring atmospheric turbulence. *J. Electron. Imaging* **2019**, *28*, 053024. [CrossRef]
29. Liu, B.; Shu, X.; Wu, X. Demoir\'eing of Camera-Captured Screen Images Using Deep Convolutional Neural Network. *arXiv Preprint* **2018**, arXiv:1804.03809.
30. Zhang, K.; Zuo, W.; Chen, Y.; Meng, D.; Zhang, L. Beyond a gaussian denoiser: Residual learning of deep cnn for image denoising. *IEEE Trans. Image Process.* **2017**, *26*, 3142–3155. [CrossRef] [PubMed]
31. Tian, C.; Xu, Y.; Li, Z.; Zuo, W.; Fei, L.; Liu, H. Attention-guided CNN for image denoising. *Neural Netw.* **2020**, *124*, 117–129. [CrossRef]
32. Retraint, F.; Zitzmann, C. Quality factor estimation of jpeg images using a statistical model. *Digit. Signal Process.* **2020**, *103*, 102759. [CrossRef]
33. Sim, H.; Kim, M. A deep motion deblurring network based on per-pixel adaptive kernels with residual down-up and up-down modules. In Proceedings of the IEEE/CVF Conference on Computer Vision and Pattern Recognition Workshops, Long Beach, CA, USA, 16–20 June 2019.
34. Zhang, H.; Dai, Y.; Li, H.; Koniusz, P. Deep stacked hierarchical multi-patch network for image deblurring. In Proceedings of the IEEE/CVF Conference on Computer Vision and Pattern Recognition, Long Beach, CA, USA, 16–20 June 2019; pp. 5978–5986.
35. Mao, Z.; Chimitt, N.; Chan, S.H. Accelerating Atmospheric Turbulence Simulation via Learned Phase-to-Space Transform. In Proceedings of the IEEE/CVF International Conference on Computer Vision, Montreal, QC, Canada, 10–17 October 2021; pp. 14759–14768.
36. Zhang, K.; Li, Y.; Zuo, W.; Zhang, L.; Van Gool, L.; Timofte, R. Plug-and-play image restoration with deep denoiser prior. *IEEE Trans. Pattern Anal. Mach. Intell.* **2021**, *44*, 6360–6376. [CrossRef]
37. Zhang, K.; Zuo, W.; Gu, S.; Zhang, L. Learning deep CNN denoiser prior for image restoration. In Proceedings of the IEEE Conference on Computer Vision and Pattern Recognition, Honolulu, HI, USA, 21–26 July 2017; pp. 3929–3938.
38. Chen, G.; Gao, Z.; Wang, Q.; Luo, Q. Blind de-convolution of images degraded by atmospheric turbulence. *Appl. Soft Comput.* **2020**, *89*, 106131. [CrossRef]
39. Haris, M.; Shakhnarovich, G.; Ukita, N. Deep back-projection networks for super-resolution. In Proceedings of the IEEE Conference on Computer Vision and Pattern Recognition, Salt Lake City, UT, USA, 18–23 June 2018; pp. 1664–1673.
40. Chatterjee, M.R.; Mohamed, A.; Almehmadi, F.S. Secure free-space communication, turbulence mitigation, and other applications using acousto-optic chaos. *Appl. Opt.* **2018**, *57*, C1–C13. [CrossRef] [PubMed]
41. Ramos, A.A.; de la Cruz Rodríguez, J.; Yabar, A.P. Real-time, multiframe, blind deconvolution of solar images. *Astron. Astrophys.* **2018**, *620*, A73. [CrossRef]
42. Zha, Z.; Wen, D.; Yuan, X.; Zhou, J.; Zhu, C. Image restoration via reconciliation of group sparsity and low-rank models. *IEEE Trans. Image Process.* **2021**, *30*, 5223–5238. [CrossRef]
43. Zha, Z.; Yuan, X.; Zhou, J.; Zhu, C.; Wen, B. Image restoration via simultaneous nonlocal self-similarity priors. *IEEE Trans. Image Process.* **2020**, *29*, 8561–8576. [CrossRef] [PubMed]
44. Venkatakrishnan, S.V.; Bouman, C.A.; Wohlberg, B. Plug-and-play priors for model based reconstruction. In Proceedings of the 2013 IEEE Global Conference on Signal and Information Processing, Austin, TX, USA, 3–5 December 2013; IEEE: New York, NY, USA, 2013; pp. 945–948.
45. Wei, K.; Aviles-Rivero, A.; Liang, J.; Fu, Y.; Schönlieb, C.B.; Huang, H. Tuning-free plug-and-play proximal algorithm for inverse imaging problems. In Proceedings of the International Conference on Machine Learning, PMLR, Virtual Event, 13–18 July 2020; pp. 10158–10169.
46. Nair, P.; Gavaskar, R.G.; Chaudhury, K.N. Fixed-point and objective convergence of plug-and-play algorithms. *IEEE Trans. Comput. Imaging* **2021**, *7*, 337–348. [CrossRef]
47. Dabov, K.; Foi, A.; Katkovnik, V.; Egiazarian, K. Image denoising by sparse 3-D transform-domain collaborative filtering. *IEEE Trans. Image Process.* **2007**, *16*, 2080–2095. [CrossRef]
48. Hradiš, M.; Kotera, J.; Zemčık, P.; Šroubek, F. Convolutional neural networks for direct text deblurring. In Proceedings of the BMVC, Swansea, UK, 7–10 September 2015; Volume 10.
49. Mao, X.; Shen, C.; Yang, Y.B. Image restoration using very deep convolutional encoder-decoder networks with symmetric skip connections. *Adv. Neural Inf. Process. Syst.* **2016**, *29*, 2810–2818.
50. Tai, Y.; Yang, J.; Liu, X.; Xu, C. Memnet: A persistent memory network for image restoration. In Proceedings of the IEEE International Conference on Computer Vision, Venice, Italy, 22–29 October 2017; pp. 4539–4547.
51. Song, G.; Sun, Y.; Liu, J.; Wang, Z.; Kamilov, U.S. A new recurrent plug-and-play prior based on the multiple self-similarity network. *IEEE Signal Process. Lett.* **2020**, *27*, 451–455. [CrossRef]

52. Asim, M.; Shamshad, F.; Ahmed, A. Blind image deconvolution using deep generative priors. *IEEE Trans. Comput. Imaging* **2020**, *6*, 1493–1506. [CrossRef]

53. Dong, W.; Wang, P.; Yin, W.; Shi, G.; Wu, F.; Lu, X. Denoising prior driven deep neural network for image restoration. *IEEE Trans. Pattern Anal. Mach. Intell.* **2018**, *41*, 2305–2318. [CrossRef]

54. Sun, Y.; Wu, Z.; Xu, X.; Wohlberg, B.; Kamilov, U.S. Scalable plug-and-play ADMM with convergence guarantees. *IEEE Trans. Comput. Imaging* **2021**, *7*, 849–863. [CrossRef]

55. Terris, M.; Repetti, A.; Pesquet, J.C.; Wiaux, Y. Enhanced convergent pnp algorithms for image restoration. In Proceedings of the 2021 IEEE International Conference on Image Processing (ICIP), Anchorage, AK, USA, 19–22 September 2021; IEEE: New York, NY, USA, 2021; pp. 1684–1688.

56. Gao, S.; Zhuang, X. Rank-One Network: An Effective Framework for Image Restoration. *IEEE Trans. Pattern Anal. Mach. Intell.* **2020**, *44*, 3224–3238. [CrossRef]

57. Jung, H.; Kim, Y.; Min, D.; Jang, H.; Ha, N.; Sohn, K. Learning Deeply Aggregated Alternating Minimization for General Inverse Problems. *IEEE Trans. Image Process.* **2020**, *29*, 8012–8027. [CrossRef]

58. Ryu, E.; Liu, J.; Wang, S.; Chen, X.; Wang, Z.; Yin, W. Plug-and-play methods provably converge with properly trained denoisers. In Proceedings of the International Conference on Machine Learning, PMLR, Long Beach, CA, USA, 9–15 June 2019; pp. 5546–5557.

59. Geman, D.; Yang, C. Nonlinear image recovery with half-quadratic regularization. *IEEE Trans. Image Process.* **1995**, *4*, 932–946. [CrossRef]

60. Chen, G.; Gao, Z.; Zhou, B.; Zuo, C. Optimization and regularization of complex task decomposition for blind removal of multi-factor degradation. *J. Vis. Commun. Image Represent.* **2022**, *82*, 103384. [CrossRef]

61. Wu, J.; Di, X. Integrating neural networks into the blind deblurring framework to compete with the end-to-end learning-based methods. *IEEE Trans. Image Process.* **2020**, *29*, 6841–6851. [CrossRef]

62. Anwar, S.; Barnes, N. Real image denoising with feature attention. In Proceedings of the IEEE/CVF International Conference on Computer Vision, Seoul, Korea, 27 October–2 November 2019; pp. 3155–3164.

63. Zhang, Y.; Li, K.; Li, K.; Zhong, B.; Fu, Y. Residual non-local attention networks for image restoration. *arXiv* **2019**, arXiv:1903.10082.

64. He, W.; Yao, Q.; Li, C.; Yokoya, N.; Zhao, Q.; Zhang, H.; Zhang, L. Non-local meets global: An integrated paradigm for hyperspectral image restoration. *IEEE Trans. Pattern Anal. Mach. Intell.* **2020**, *44*, 2089–2107. [CrossRef]

65. Graves, A.; Bellemare, M.G.; Menick, J.; Munos, R.; Kavukcuoglu, K. Automated curriculum learning for neural networks. In Proceedings of the International Conference on Machine Learning, PMLR, Sydney, Australia, 6–11 August 2017; pp. 1311–1320.

66. Jiang, L.; Zhou, Z.; Leung, T.; Li, L.J.; Fei-Fei, L. Mentornet: Learning data-driven curriculum for very deep neural networks on corrupted labels. In Proceedings of the International Conference on Machine Learning, PMLR, Stockholm, Sweden, 10–15 July 2018; pp. 2304–2313.

67. Yang, L.; Shen, Y.; Mao, Y.; Cai, L. Hybrid Curriculum Learning for Emotion Recognition in Conversation. *arXiv* **2021**, arXiv:2112.11718.

68. He, K.; Zhang, X.; Ren, S.; Sun, J. Delving deep into rectifiers: Surpassing human-level performance on imagenet classification. In Proceedings of the IEEE International Conference on Computer Vision, Santiago, Chile, 7–13 December 2015; pp. 1026–1034.

69. Caijuan, Z. STK and its application in satellite sys-tem simulation. *Radio Commun. Technol.* **2007**, *33*, 45–46.

70. Kuzmin, I.A.; Maksimovskaya, A.I.; Sviderskiy, E.Y.; Bayguzov, D.A.; Efremov, I.V. Defining of the Robust Criteria for Radar Image Focus Measure. In Proceedings of the 2019 IEEE Conference of Russian Young Researchers in Electrical and Electronic Engineering (EIConRus), Saint Petersburg/Moscow, Russia, 28–30 January 2019; IEEE: New York, NY, USA, 2019; pp. 2022–2026.

71. Zhu, J.Y.; Park, T.; Isola, P.; Efros, A.A. Unpaired image-to-image translation using cycle-consistent adversarial networks. In Proceedings of the IEEE International Conference on Computer Vision, Venice, Italy, 22–29 October 2017; pp. 2223–2232.

Article

Prediction of Sea Surface Temperature by Combining Interdimensional and Self-Attention with Neural Networks

Xing Guo [1], Jianghai He [2], Biao Wang [3] and Jiaji Wu [1,*]

1 School of Electronic Engineering, Xidian University, Xi'an 710071, China
2 School of Artificial Intelligence, Xidian University, Xi'an 710071, China
3 Science and Technology on Electromagnetic Scattering Laboratory, Shanghai 200438, China
* Correspondence: wujj@mail.xidian.edu.cn

Abstract: Sea surface temperature (SST) is one of the most important and widely used physical parameters for oceanography and meteorology. To obtain SST, in addition to direct measurement, remote sensing, and numerical models, a variety of data-driven models have been developed with a wealth of SST data being accumulated. As oceans are comprehensive and complex dynamic systems, the distribution and variation of SST are affected by various factors. To overcome this challenge and improve the prediction accuracy, a multi-variable long short-term memory (LSTM) model is proposed which takes wind speed and air pressure at sea level together with SST as inputs. Furthermore, two attention mechanisms are introduced to optimize the model. An interdimensional attention strategy, which is similar to the positional encoding matrix, is utilized to focus on important historical moments of multi-dimensional input; a self-attention strategy is adopted to smooth the data during the training process. Forty-three-year monthly mean SST and meteorological data from the fifth-generation ECMWF (European Centre for Medium-Range Weather Forecasts) reanalysis (ERA5) are collected to train and test the model for the sea areas around China. The performance of the model is evaluated in terms of different statistical parameters, namely the coefficient of determination, root mean squared error, mean absolute error and mean average percentage error, with a range of 0.9138–0.991, 0.3928–0.8789, 0.3213–0.6803, and 0.1067–0.2336, respectively. The prediction results indicate that it is superior to the LSTM-only model and models taking SST only as input, and confirm that our model is promising for oceanography and meteorology investigation.

Keywords: sea surface temperature; mutual information; LSTM; self-attention; interdimensional attention

Citation: Guo, X.; He, J.; Wang, B.; Wu, J. Prediction of Sea Surface Temperature by Combining Interdimensional and Self-Attention with Neural Networks. *Remote Sens.* 2022, 14, 4737. https://doi.org/10.3390/rs14194737

Academic Editor: Javier Marcello

Received: 31 July 2022
Accepted: 18 September 2022
Published: 22 September 2022

Publisher's Note: MDPI stays neutral with regard to jurisdictional claims in published maps and institutional affiliations.

1. Introduction

Sea surface temperature (SST) is the one of the most important and widely used parameters in the analysis of global climate change. It is also used as boundary conditions or assimilation information in the analysis of atmospheric circulation anomalies, atmospheric models, and sea–air coupled models [1]. In addition, SST constitutes important basic data for aquaculture industry environmental assurance [2].

Although observations of SST have a history of more than 200 years, it was not until 1853 when the Brussels International Conference on Nautical Meteorology decided to start the collection of global SST data and standardize the organization and analysis of SST data. In recent decades, SST observation has transitioned through bucket observation measurements, Engine Room Intake (ERI) observations, ship-sensing observations, and satellite remote-sensing observations [3]. The uneven spatial and temporal distribution of observations need to be solved to obtain long-term, accurate global SST information. For this purpose, the reanalysis takes advantage of data assimilation techniques to integrate SST data from various sources and types of observations with numerical forecast products [4]. A number of reanalysis products that provide accurate forecasts across broad spatial and temporal scales have been released. In recent years, there has been a large volume of published studies comparing the products from different aspects. In 2013, Baololqimuge summarized several commonly used SST observation methods and introduced four sets of SST data including the Hadley Centre Sea Ice and Sea Surface Temperature data set (HadISST), the International Comprehensive Ocean-Atmosphere Data Set (ICOADS), Extended Reconstructed Sea Surface Temperature (ERSST) and Optimum Interpolation Sea Surface Temperature Analysis (OISSTA) in detail [3]. In the same year, Jiang conducted a comparative statistical analysis of six different SST products [5]. In 2020, Wang compared the applicability of three sets of reanalysis data around China [6].

The traditional methods for predicting ocean elements are divided into three main categories: numerical models, artificial experience and statistical prediction [4]. Numerical models are obtained under parametric conditions for physical processes in the ocean and are more suited to large sea areas and long-term SST prediction [7]. Furthermore, the computational requirements for ocean simulations are immense and require high-performance computing (HPC) facilities [8]. Artificial empirical and statistical prediction methods are more affected by parameter settings and the degree of human cognition.

Over the years, with the development, launch, and application of a series of satellites for oceanology and meteorology, satellite data have been increasing in magnitude. Additionally, the improvements in ocean models together with increased computational capabilities led to a number of reanalysis data [8]. The accumulation of a large amount of data on marine environmental elements has laid the foundation for data-driven methods [9]. These models abandon the subjectivity of traditional machine learning that requires experts to design and create feature extractors through experiments; they automatically and objectively extract useful information from data. Thus, they bring new opportunities for intelligent analysis and mining of marine data.

SST prediction can essentially be regarded as a time-series regression problem. The traditional models for time-series prediction, such as autoregressive (AR), moving average (MA), the autoregressive integrated moving average model (ARIMA) and the regression model by machine learning, including support vector regression (SVR) [10], and multi-layer perceptron (MLP) have been widely used in SST prediction [11]. An atmospheric reflection Grey model was proposed to predict long-term SST [12]. More recent attention has focused on the deep learning models, which originated from the artificial neural network (ANN). In 2006, Hinton proposed the concept of deep learning, which promoted the implementation of a number of deep learning projects [13]. A recurrent neural network (RNN) is a deep model developed for modeling sequential data [14]. The RNN introduces hidden states to extract features from sequential data and convert them to outputs. Hochreiter proposed the long short-term memory (LSTM) model, which introduces the forgetting gate and the memory gate [15]. In 2017, Zhang first adopted the LSTM model to predict SST. In the same year, based on the convolutional LSTM (ConvLSTM) model [16], Xu proposed a sequence-to-sequence (Seq2Seq)-based regional sea level temperature anomaly prediction model [17]. In 2018, Yang introduced spatial information with the LSTM model to build a model for SST prediction, and applied it effectively in the SST data set of coastal China [18]. In 2019, Zhu applied the LSTM-RNN to SST prediction and constructed a model for SST time-series variation in the western Pacific sea area [19]. LSTM was applied to predict SST and high-water temperature occurrence [2]. The temporal convolutional network (TCN) was applied to obtain large-scale and long-term SST prediction [20]. LSTM was applied to short and midterm daily SST prediction for the Black Sea [21].

To date, a number of researchers have attempted to combine different models together to predict SST. A numerical model is combined with neural networks to predict site-specific SST [22]. In 2019, Xiao combined LSTM and AdaBoost for medium-and long-term SST prediction [23]. Later, to fully capture the information of SST across both space and time, the author combined the convolutional network with LSTM as the building block to predict SST [24]. He combined the seasonal-trend decomposition using loess (STL) and LSTM to predict SST [25]. Deep learning neural networks were combined with numerical estimators for daily, weekly, and monthly SST prediction [7]. To enhance the performance, a hybrid system which combines machine learning modes using residual forecasting was developed [26]. Jahanbakht designed an ensemble of two staked DNNs that used air temperature and SST to predict SST [27]. To forecast multi-step-ahead SST, a hybrid empirical model and gated recurrent unit was proposed [28]. Accuracy comparable to existing state of the art can be achieved by combining automated feature extraction and machine learning models [8]. Pedro evaluated the accuracy and efficiency of many deep learning models for time-series forecasting and show LSTM and CNN are the best alternatives [29].

LSTM has some advantages in sequence modeling owing to its long-time memory function; it is relatively simple to implement and solves the problem of gradient disappearance and gradient explosion that exists in the long sequence training process. However, it has disadvantages in parallel processing and always takes longer to train. A transformer, based on attention mechanisms, was proposed, which is parallelized and can significantly reduce the model's training time [30]. Li enhanced the locality and overcame the memory bottleneck on the transformer for a the time-series prediction problem [31]. Furthermore, an SST prediction model based on deep learning with an attention mechanism was proposed [32]. A transformer neural network based on self-attention was developed, which showed superior performance than other existing models, especially at larger time intervals [33]. The degrees of effect on the prediction result of the information at previous time steps differ; therefore, the addition of an attention mechanism can assign different levels of attention to the model enabling it to automatically handle the importance of different information [34].

Inspired by transformer's self-attention and positional encoding, the main contributions of this work can be summarized as follows:

1. The determining factors affecting SST distribution and variation, in other words, the input of the LSTM prediction model, is selected by the correlation analysis of mutual information.
2. To focus on important historical moments and important variables, a special matrix, that is similar to the position coding matrix, is obtained by multiplying the multi-dimensional data by a weight matrix W (where W is obtained by network training).
3. The input data are smoothed using a self-attention mechanism during the training process.

The remainder of this paper is organized as follows. Section 2 first presents the correlation analysis of SST and meteorological data based on mutual information, and then describes the proposed model combining LSTM with attention mechanism. The study area and data sets used, implementation detail, and experimental results are introduced in Section 3. Validation of the model and comparison of its performance with other models are presented in Section 4. Finally, Section 5 concludes this paper and outlines future plans.

2. Methodology

The ocean is a comprehensive and complex dynamic system, and many factors affect the distribution and variation of SST. In the process of multivariate time-series model building, when the dimensionality of the input variables increases to a certain degree, the accuracy of parameter estimation decreases, which significantly decreases the prediction accuracy of the model and generates a dimensional disaster. In addition, the number of learning samples required for training increases exponentially with the dimensionality, whereas in practice the samples available for training are often very limited. By contrast, the model input with an excessive number of irrelevant, redundant, or useless variables, tends to obscure the role of the important variables eventually leading to poor prediction results [35].

Therefore, to identify valid inputs for SST prediction models based on deep learning, it is necessary to analyze the correlation between SST and meteorological and marine factors that may affect SST distribution and variation. On the one hand, by analyzing the correlation between input variables and output variables, the relevant variables that contribute most to the model prediction can be identified. On the other hand, by analyzing whether there is some type of dependency between the input variables, redundant variables can be eliminated.

The present study involves the overall research plan shown in Figure 1. First, we used reanalysis data to construct a database of marine environmental elements, including SST, pressure, wind speed, solar irradiation, latitude, and longitude. Then, we perform quality analysis and corresponding preprocessing according to the analysis result. Because SST is affected by several factors simultaneously, to build a deep learning model, we determine the effective input of the model by analyzing the correlation among the influencing factors based on mutual information. Then, a hybrid model combining LSTM and attention mechanism is introduced. Subsequently, we evaluate the accuracy of the model for the surrounding sea areas of China.

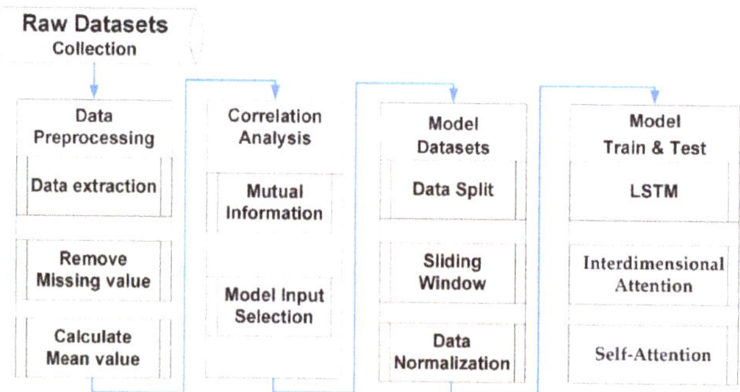

Figure 1. General flow chart of SST prediction based on a hybrid model combining LSTM and attention.

2.1. Correlation Analysis

In general, correlation is used to describe the closeness of the relationship between variables. Correlations include asymmetric causal and driving relationships, as well as symmetric correlations. Among the traditional statistical methods, the Pearson correlation coefficient, Spearman correlation coefficient, and Kendall correlation coefficient are commonly utilized [35]. The Pearson correlation coefficient is used to measure the degree of linear correlation between two variables and requires the corresponding variables to be bivariate normally distributed. The Spearman correlation coefficient is used to analyze a linear correlation using the rank order of two variables; it does not require the distribution of the original variables and is a nonparametric statistical method [36]. The Kendall correlation coefficient is an indicator used to reflect the correlation of categorical variables and is applicable to the case where both variables are ordered categorically.

Commonly applied methods of correlation analysis of multivariate data include Copula analysis, random forest, XGBoost, and mutual information analysis [37]. The definition of mutual information is derived from the concept of entropy in information theory, which is often also called information entropy or Shannon entropy. Entropy expresses the degree of uncertainty in the values of random variables in a numerical form, thus describing the magnitude of information content of variables.

Based on the definition of probability density of data, mutual information is a widely used method to describe the correlation of variables. This is because there is no special requirement for the distribution of data types, and it can be used for both linear and nonlinear correlation analysis [35].

The information entropy of discrete random variables is defined as

$$H(x) = -\sum_{i=1}^{N} p(x_i) \log(p(x_i))$$ (1)

where N is the number of samples and $p(x_i)$ is the frequency of x_i in the data sets.

The mutual information of variable X and variable Y is defined as

$$I(X,Y) = \iint p_{XY}(x,y) \log \frac{p_{XY}(x,y)}{p_X(x)p_Y(y)} dxdy \tag{2}$$

where $p_{XY}(x,y)$ is the joint probability density of X and Y, $p_X(x)$ and $p_Y(y)$ are the marginal probability density of X and Y, respectively.

According to the definition, when two variables X and Y are independent of each other or completely unrelated, their mutual information equals to 0, which implies that there is no jointly owned information between the two variables. When X and Y are highly dependent on each other, the mutual information will be large.

In practical problems, the joint probability density of the variables (X, Y) is usually not known, and the variables X and Y are generally discrete. Therefore, the histogram method is commonly used. It discretizes the values of continuous variables by dividing the bins in the range of variables, putting different values of variables into different bins, then counting their frequencies, and subsequently performing calculation using the formula of discrete information entropy. However, determining the range size of each bin is difficult, and it usually requires repeated calculations to obtain the optimal solution.

Another commonly used method is called k-nearest neighbor estimation, which was first proposed in 1987 [38]. In 2004, the mutual information calculation method for computing two continuous random variables was proposed [39].

$$I^{(1)}(X,Y) = \psi(k) - \langle \psi(n_x + 1) + \psi(n_y + 1) \rangle + \psi(N) \tag{3}$$

$$I^{(2)}(X,Y) = \psi(k) - 1/k - \langle \psi(n_x) + \psi(n_y) \rangle + \psi(N) \tag{4}$$

where $\langle \rangle$ is the mean value symbol and ψ is the Digamma function calculated by the following iterative formula

$$\begin{aligned} \psi(1) &= -0.5772516 \\ \psi(x+1) &= \psi(x) + 1/x \end{aligned} \tag{5}$$

The results obtained by the two calculation methods are similar in most cases. However, in general, the first method has smaller statistical errors and larger systematic errors, and the second method is more suitable for the calculation of high-dimensional mutual information quantity.

The calculation time of k nearest neighbor mutual information estimation mainly depends on the sample size, while it is less affected by the dimensionality of variable. Moreover, in general, the smaller the value of k is, the larger is the statistical error and the smaller is the systematic error. Usually, k is taken as 3.

2.2. Model Architecture

LSTM has been widely used in SST time-series prediction. However, the LSTM network requires a long training time because of the lack of parallelization ability. Further, the degree of effect at different time steps on the prediction result are different and varies dynamically with time. This cannot be handled by using LSTM exclusively. Inspired by the attention mechanism used in natural language processing, we added the attention structure into our model to enable it to automatically focus on important historical moments and important variables.

As Figure 2 shows, the model consists of five components. In addition to the necessary input and output module, a multivariate LSTM module is applied to capture the feature information in the time-series data. Integrating multi-dimensional information itself is difficult, because it is impossible to determine which dimension plays a more important role on the results. In addition, the importance of information tends to fluctuate with the time steps. Therefore, it is crucial to solve the problem reasonably linking multi-dimensional input data together to retain useful data and eliminate interfering data. The coefficient matrix W (green part) is determined in a way similar to positional encoding in the attention mechanism. In the blue part, whether the data are true is questioned. A self-attention approach is used to observe the difference between the true and predicted values of the adjacent data; based on this, the data in the current time step is fine-tuned. The weight of the data in the current time step is adjusted to make it closer to the true value in the next iteration.

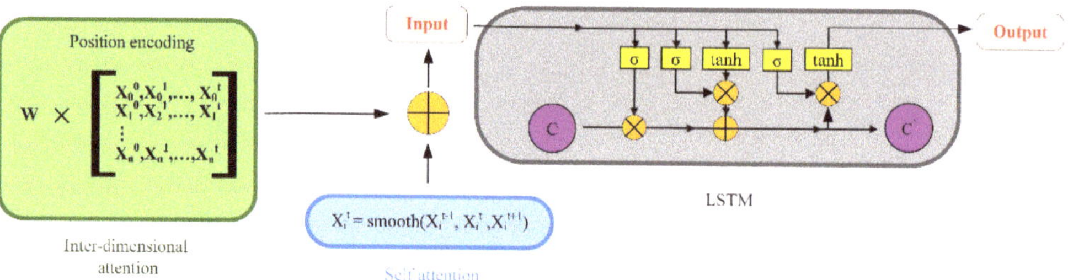

Figure 2. Structural diagram of the attention-based LSTM model.

2.2.1. Interdimensional Attention Strategy

One of the major advantages of a transformer network is that, for a single isolated data point, it not only mines that data point's information but also integrates multiple data together through positional encoding to mine the information between data. This approach is most suitable for text processing tasks, where a certain association between contexts exists and words are encoded in a uniform manner. However, single-dimensional time-series prediction tasks cannot be realized through this method. In case of SST prediction, owing to the scarcity of data, specific time-stamped information is often erased and only time-course data are retained as a set of information for any consecutive 12 months, rather than a fixed set of information for each month. The variation trend of the data differs for different starting months, even showing totally opposite trends, as illustrated in Figure 3. Moreover, determining the degree of correlation between the data from January 2010 and January 2011 is difficult. Therefore, positional encoding is not possible with only one-dimensional time-series data.

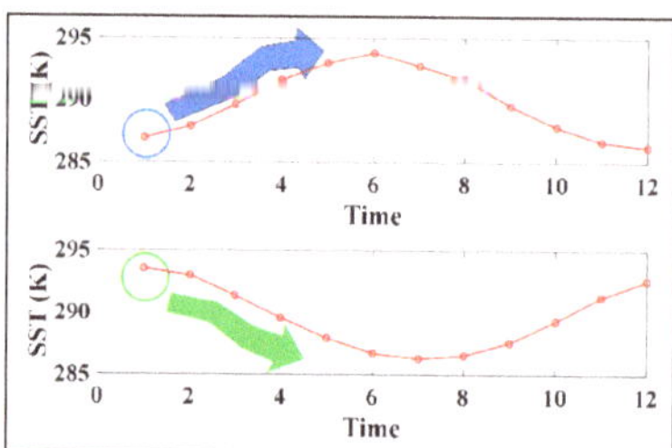

Figure 3. Variation trend of SST in a sliding window.

Note that SST at a specific location can be affected by various factors, including the wind speed, air pressure, and solar radiation, to varying degrees, and the general prediction algorithm, which often uses only the temperature data, has significant limitations. SST at a specific place has an implicit relationship with the meteorological factors with a high probability, which can be described by the following equation

$$T_{pred} = k \cdot T + v \cdot u_{10} + \ldots \tag{6}$$

where the coefficients k, v, and so on are unknown weight vectors, and their values cannot be directly determined on the basis of experience. The main problem is the possible contradiction and inconsistency of importance between parameters. Moreover, the parameters may vary with time steps. For example, the weight of temperature may be set to 0.8 and that of wind speed to 0.2 in January; however, in February, the temperature weight may become 0.7 and the wind speed weight may change to 0.3. Therefore, their values can only be obtained on the basis of the training of the neural network. For each time step, there is a matrix of corresponding coefficient matrix (k_i v_i, ...). Each coefficient vector is shown in the following equation.

$$k = [k_1, k_2, \cdots k_n] \tag{7}$$

where n is the length of time series. Together, these vector coefficients form the W matrix shown in Figure 2.

In this way, during the network training process, the W matrix gradually reveals the implicit connections between these different dimensional data. This combining of data of different dimensions produces an effect similar to the position encoding matrix in a transformer.

2.2.2. Self-Attention Smoothing Strategy

The adverse effects of inevitable systematic errors (e.g., temperature measurement errors, local temperature anomalies, weather anomalies at the time of temperature measurement, and human causes), can be reduced by requiring each data point to be self-conscious, as illustrated in Figure 4.

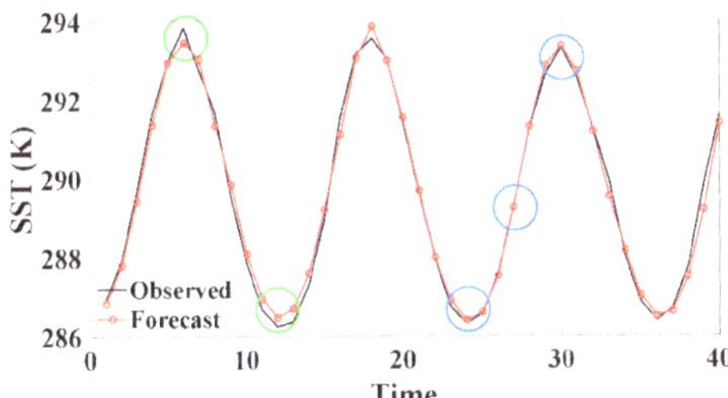

Figure 4. Self-attention smoothing strategy.

As shown in Figure 5, to determine whether it is smooth and fits the simulated curve, we need to calculate the relationship between the data at T_t, T_{t-1} and T_{t+1}. In specific implementations, the degree of fitting is judged by the difference between the predicted and actual values at the current time step is whether around the difference at the preceding and following time step or not. If it is smooth and fits the curve well enough, then the data are more reliable and the corresponding parameter k or v is increased accordingly (blue circle in Figure 4). If it is not smooth or does not fit the curve well enough (green circle in Figure 4), the data may be abnormal and the corresponding parameter k or v is reduced by a factor of 10%. A suitable value can be found after several training iterations.

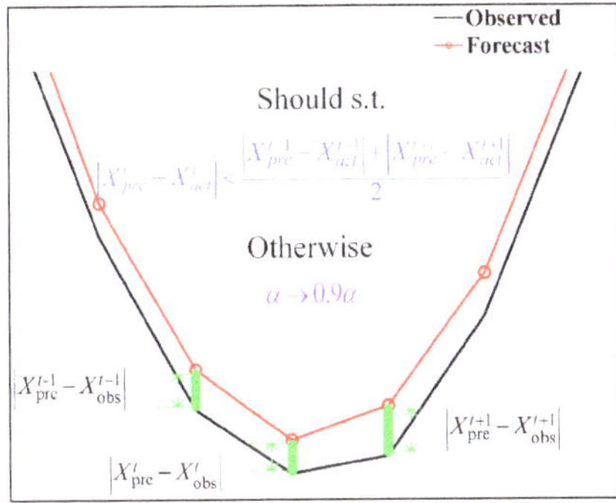

Figure 5. Specific execution process of self-attention smoothing.

2.3. Evaluation Metrics

The performance and reliability of the model are evaluated in terms of the statistics of the coefficient of determination (R^2), *RMSE*, mean absolute error (*MAE*), and mean average percentage error (*MAPE*). They are defined as Equations (8)–(11), respectively. Here, y_i represents the true SST values, $\hat{y}_i$ represents the predicted SST values, m is the length of the test data sets, and $\bar{y}_i$ is the mean value of the true SST.

$$R^2 = 1 - \frac{\sum\limits_{i=1}^{m}(\hat{y}_i - y_i)^2}{\sum\limits_{i=1}^{m}(\bar{y}_i - y_i)^2} \tag{8}$$

R^2 is in the range [0, 1]; 0 indicates that the model is poorly fitted, while 1 indicates that the model is error free. In general, the larger the R^2 is, the better the model.

$$RMSE = \sqrt{\frac{1}{m}\sum\limits_{i=1}^{m}(y_i - \hat{y}_i)^2} \tag{9}$$

$$MAE = \frac{1}{m}\sum\limits_{i=1}^{m}|\hat{y}_i - y_i| \tag{10}$$

$$MAPE = \frac{100\%}{m}\sum\limits_{i=1}^{m}\left|\frac{\hat{y}_i - y_i}{y_i}\right| \tag{11}$$

The *RMSE*, *MAE* and *MAPE* range in [0, +∞); 0 indicates that the predicted value exactly matches the true value, and the larger the error, the larger the value.

3. Model Implementation and Experiment Results

3.1. Study Area and Data Sets

The study area focuses on the sea areas around China (Figure 6); the specific locations and representative characteristics are shown in Table 1. The distribution and variation of SST depend on multiple meteorological elements. For example, solar radiation has a heating effect on the sea surface. Wind is the direct driver of the upper ocean circulation, which is an important factor to determine the flow of the upper layer and affects the distribution of SST [19].

This study considers the temporal and spatial resolution of the data and the completeness of the environmental variables fully, then the fifth-generation ECMWF (European Centre for Medium-Range Weather Forecasts) reanalysis (ERA5) is selected to construct a multi-physical field data set of marine environmental elements for the sea areas around China. ERA5 provides hourly, daily and monthly estimates for a large number of atmospheric, ocean-wave and land-surface quantities [40].

Table 1. Longitude and latitude range and characteristics of study areas.

ID	Ocean Region	Range		Average Depth (m)	Characteristics
		Longitude (E°)	Latitude (N°)		
1	Bohai Sea and North Yellow Sea	119~125	37~41	18	Nearly closed
2	South Yellow Sea	119~125	31~37	44	Semi-closed
3	East China Sea	121~125	29~31	370	Marginal sea
4		119~125	25~29		
5	Taiwan Strait	119~121	24~25	60	Narrow strait
6		117~120	22~24		
7	South China Sea	106~125	5~21	1212	Open sea area

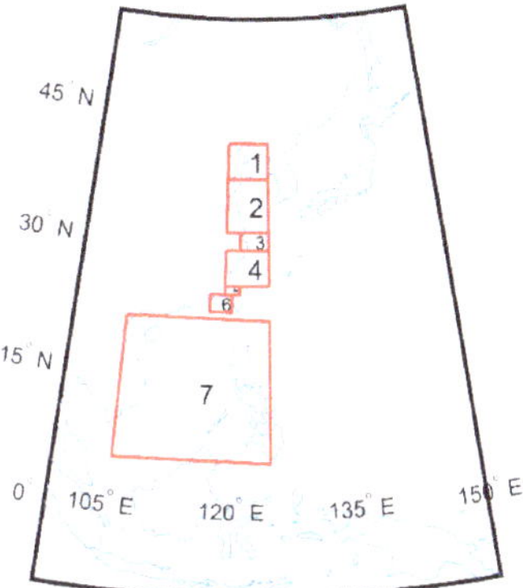

Figure 6. Location of the seven sea areas around China.

The temporal resolution of the data used in this study is monthly and the data sets cover the period 1979–2021. The spatial resolution in latitude and longitude is 0.25°. As illustrated in Table 2, in addition to SST data, data on meteorological factors including wind speed, sea surface pressure, and sea surface solar radiation have been collected.

Table 2. Variables affecting SST distribution and variation.

Parameters	Name	Unit
SST	Sea surface temperature	K
u_{10}	Eastward component of the 10 m wind	m/s
v_{10}	Northward component of the 10 m wind	m/s
msl	Mean sea level pressure	Pa
ssr	Surface net solar radiation	J/m^2
ssrc	Surface net solar radiation clear sky	J/m^2
str	Surface net thermal radiation	J/m^2
strc	Surface net thermal radiation clear sky	J/m^2
ssrd	Surface solar radiation downward	J/m^2
ssrdc	Surface solar radiation downward clear sky	J/m^2
strd	Surface solar radiation downwards	J/m^2
strdc	Surface thermal radiation downward clear sky	J/m^2

3.2. Implementation Detail

As shown in Table 2, the data sets consist of various parameters that have different units and ranges of values; thus, data normalization is necessary. The min-max normalization is utilized to scale the data between 0 and 1,

$$z_i = \frac{x_i - x_{min}}{x_{max} - x_{min}} \tag{12}$$

where x_i is the original data, x_{min} and x_{max} are the minimum and maximum of the original data, respectively.

Then, to transform the time series into input–output pairs required for model training, a sliding window with a fixed length is used as shown in Figure 7. In this study, the model receives an instance with a sliding window of length 12 as input and performs one-step predictions. The resulting samples are divided into training, validation, and test sets in a ratio of 6:1:2.

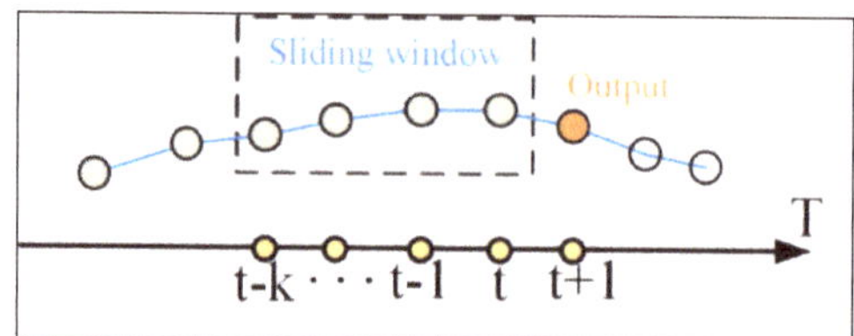

Figure 7. Sliding window procedure to obtain input–output pair of data sets.

As shown in Table 3, the training rate is improved by using a batch training method, with each batch containing 40 sample data sets. In addition, a random dropout layer is added after each layer of the LSTM network with a dropout rate setting as 0.1 to avoid overfitting. Next, the root mean squared error (*RMSE*) is chosen as the loss function for training, and the Adam algorithm is used to train the network. The number of maximum iterations (epoch) is set to 400.

Table 3. Key parameters of the model and training process.

Key Parameters	Model Methods or Values
Length of training data sets	300
Length of validation data sets	50
Length of testing data sets	100
Architecture of the model	Attention + LSTM + Dense
Input dimension	12×4
Output dimension	1
No. of neural of hidden layer	80
Optimizer	Adam
Epoch	400
Batch size	40
Dropout	0.1
Loss function	*RMSE*

All experiments are implemented using Keras 2.2.4 with TensorFlow 1.15.0 on a computer with an Intel i9-10900K CPU and an additional NVIDIA GeForce RTX 2080S GPU.

3.3. Experiment Results

3.3.1. SST Distribution and Variation

From Figures 8–10, it can be concluded that the latitudinal distribution of SST is obvious, i.e., the South China Sea has a lower latitude and a higher temperature all year round. The annual variation, except for the South China Sea, shows a pattern of synchronous change with the temperature, i.e., the highest in August and the lowest in January.

3.3.2. Correlation of SST with Other Meteorological Factors

In this study, mutual information is selected as a tool to analyze the correlations between different environmental factors and SST, which is further required for selecting the effective input variable for building a deep learning prediction model.

The mutual information of SST with each influencing factor was calculated using k-nearest neighbor-based mutual information, as shown in Figure 11. The figure indicates that, overall, the wind speed (u10, v10) and air pressure at sea level (msl) correlate more strongly with SST in different seas compared to radiation-related environmental variables.

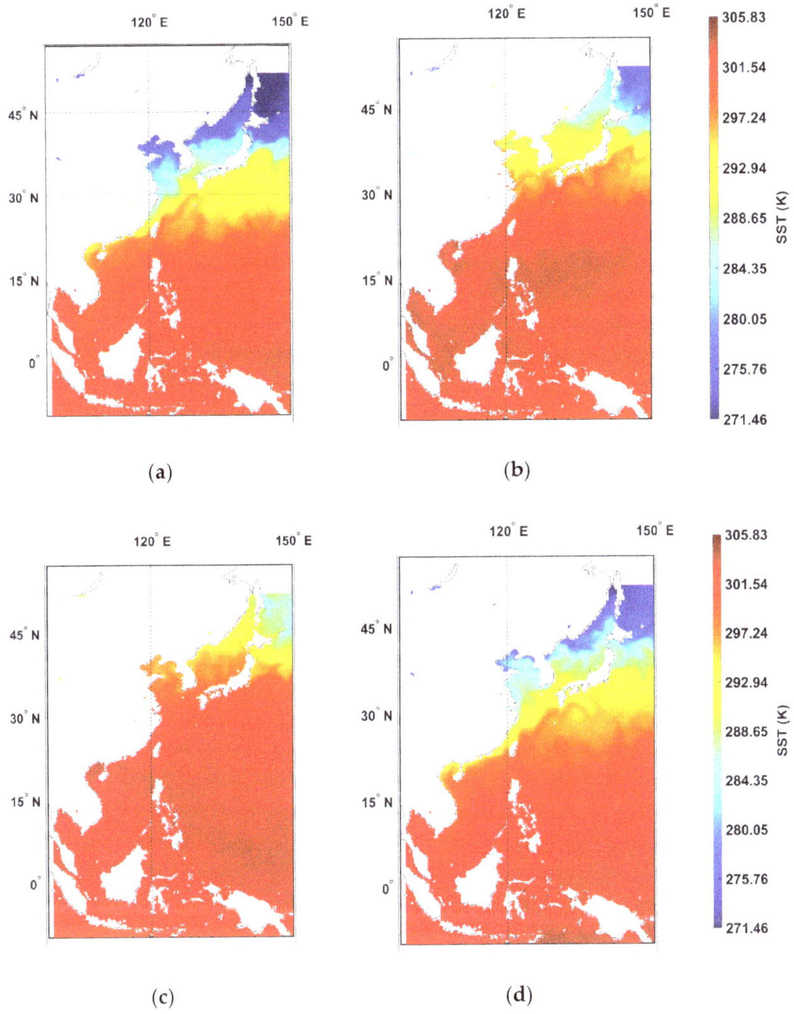

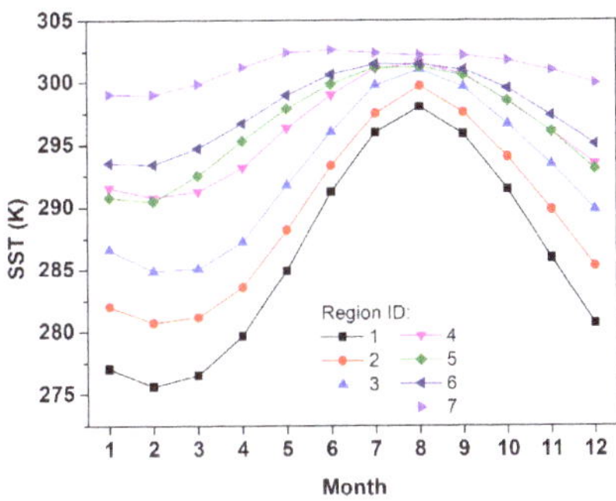

Figure 8. SST distribution of sea areas surrounding China in (**a**) March, (**b**) June, (**c**) September, and (**d**) December in 2021.

Figure 9. Variation of monthly mean SST for different sea areas around China during 1979–2021.

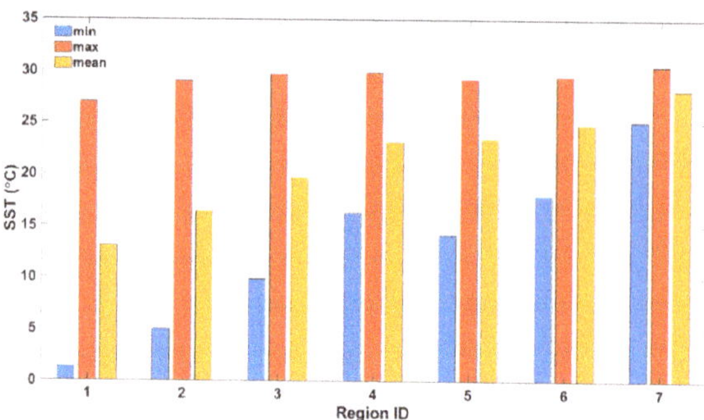

Figure 10. Statistics of monthly mean SST including minimum, maximum, and mean value for different parts of China's Surrounding Seas during 1979–2021.

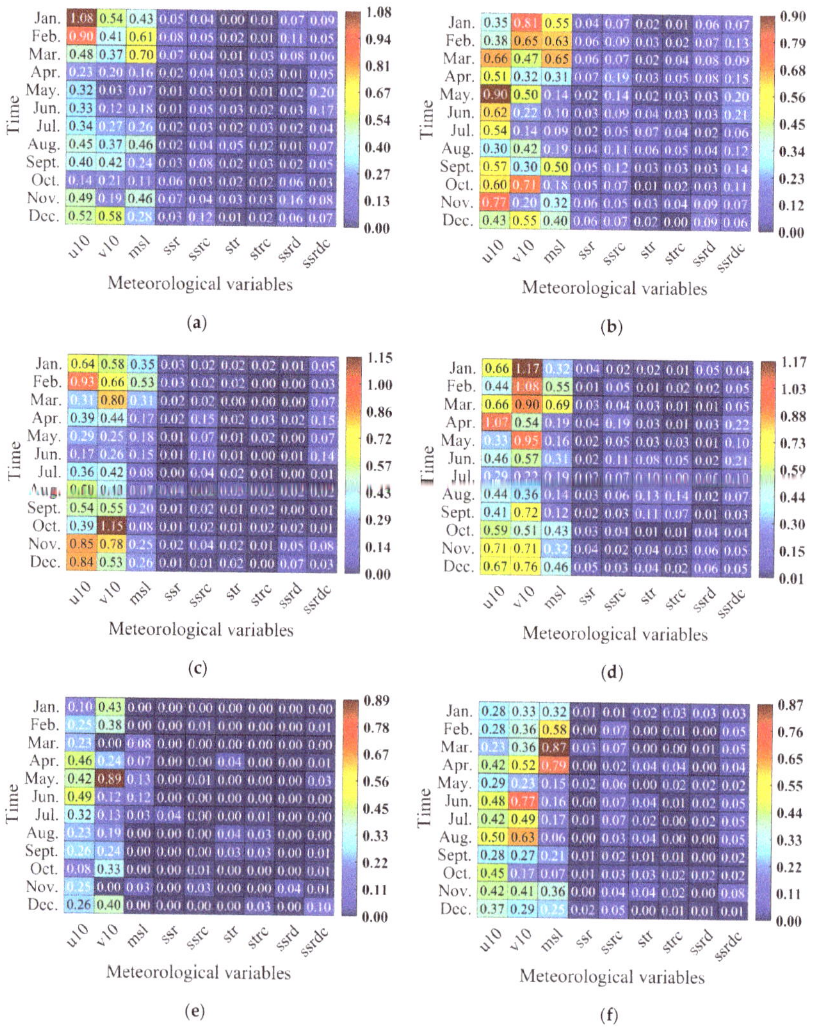

Figure 11. *Cont.*

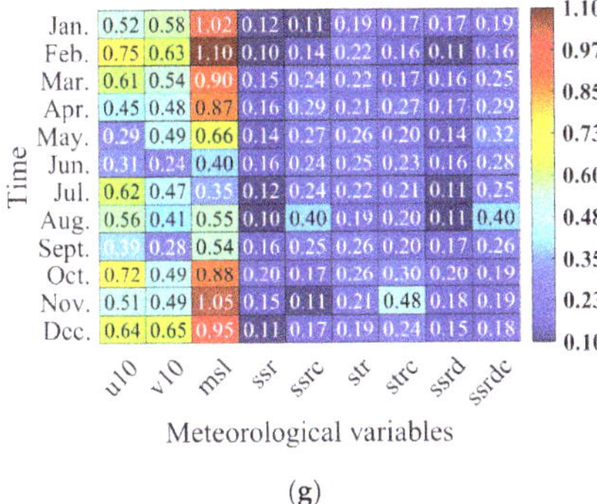

(**g**)

Figure 11. Heat map of mutual information for each meteorological variable and SST for the seas around China in different months (**a**) Bohai Sea and north part of Yellow Sea, (**b**) South part of Yellow Sea, (**c**) Part 1 (ID:3 in Table 1) of East China Sea, (**d**) Part 2 (ID:4 in Table 1) of East China Sea, (**e**) Part 1(ID:5 in Table 1) of Taiwan Strait, (**f**) Part 2(ID:5 in Table 1) of Taiwan Strait, and (**g**) South China Sea.

3.3.3. SST Prediction Results

The last 100 samples (about 8 years from 2014 to 2021) in the data sets are applied to test the model. The one-month ahead monthly mean SST prediction results of the sea areas around China are shown in Figure 12. The blue line represents the true values. Additionally, the red dot and the filled areas in the figure represent the average prediction results and the corresponding standard deviation for five runs. What should be noted is that the resolution of y axis is different for different regions.

Overall, the prediction results reveal a same trend between the true and predicted SST. However, for all regions, larger bias appears at the local extremums, because the model trained on the training data sets cannot capture the extremums of the test data sets. As SST of southern part of China, especially South China Sea, keeps high (approximately 300 K) all year round and fluctuates less, the model performs better.

To test the stability of the model, statistics including R^2, $RMSE$, MAE, and $MAPE$ for five runs are presented in Table 4 and Figure 13. From the perspective of $RMSE$, MAE and $MAPE$, the model performs better in the southern parts of the surrounding seas of China, especially the South China Sea, for which the SST varies less and maintains a high value all year round. The error of some isolated point probably leads to higher $RMSE$, MAE and $MAPE$. The fluctuation for region 5 (Taiwan Strait) is the smallest, which may indicate that, for narrow strait areas, we can trust the result more from arbitrary initialization conditions.

Table 4. Comparison of R^2, $RMSE$, MAE and $MAPE$ for seven study areas (average for five runs).

Region ID	R^2	$RMSE$	MAE	$MAPE$
1	0.9910	0.7551	0.6211	0.2170
2	0.9829	0.8789	0.6803	0.2336
3	0.9827	0.7547	0.5936	0.2029
4	0.9827	0.5120	0.4100	0.1382
5	0.9727	0.6515	0.5065	0.1711
6	0.9649	0.5666	0.4531	0.1521
7	0.9138	0.3928	0.3213	0.1067

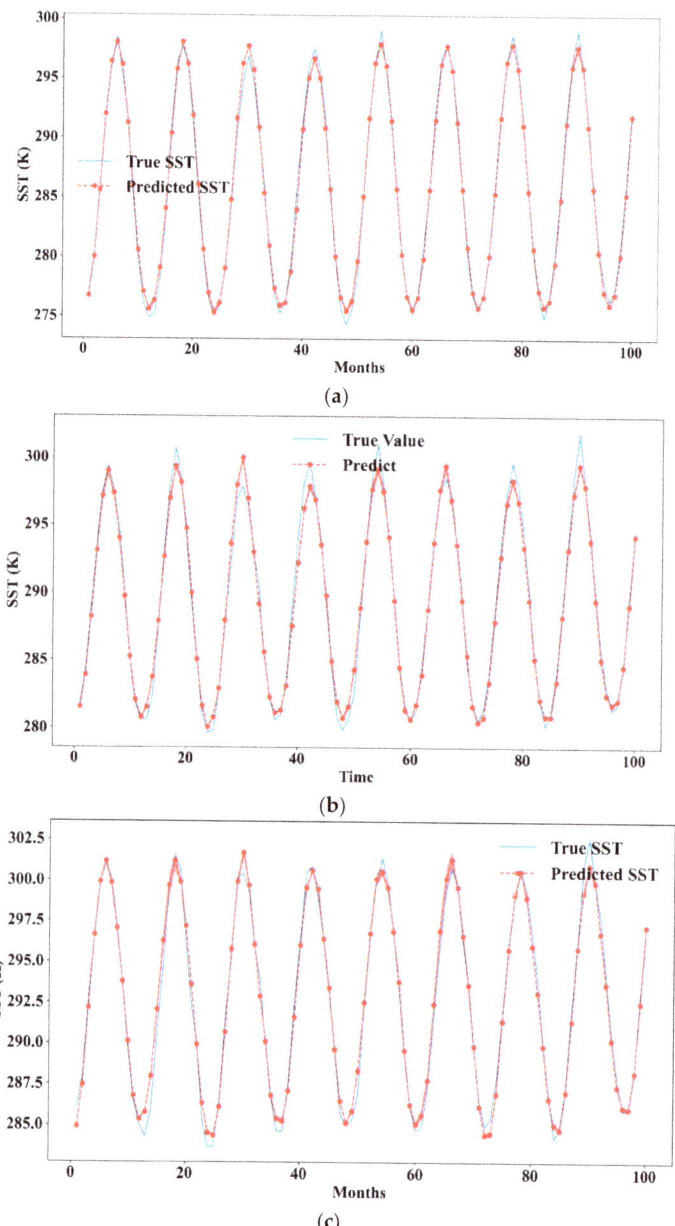

(a)

(b)

(c)

Figure 12. *Cont.*

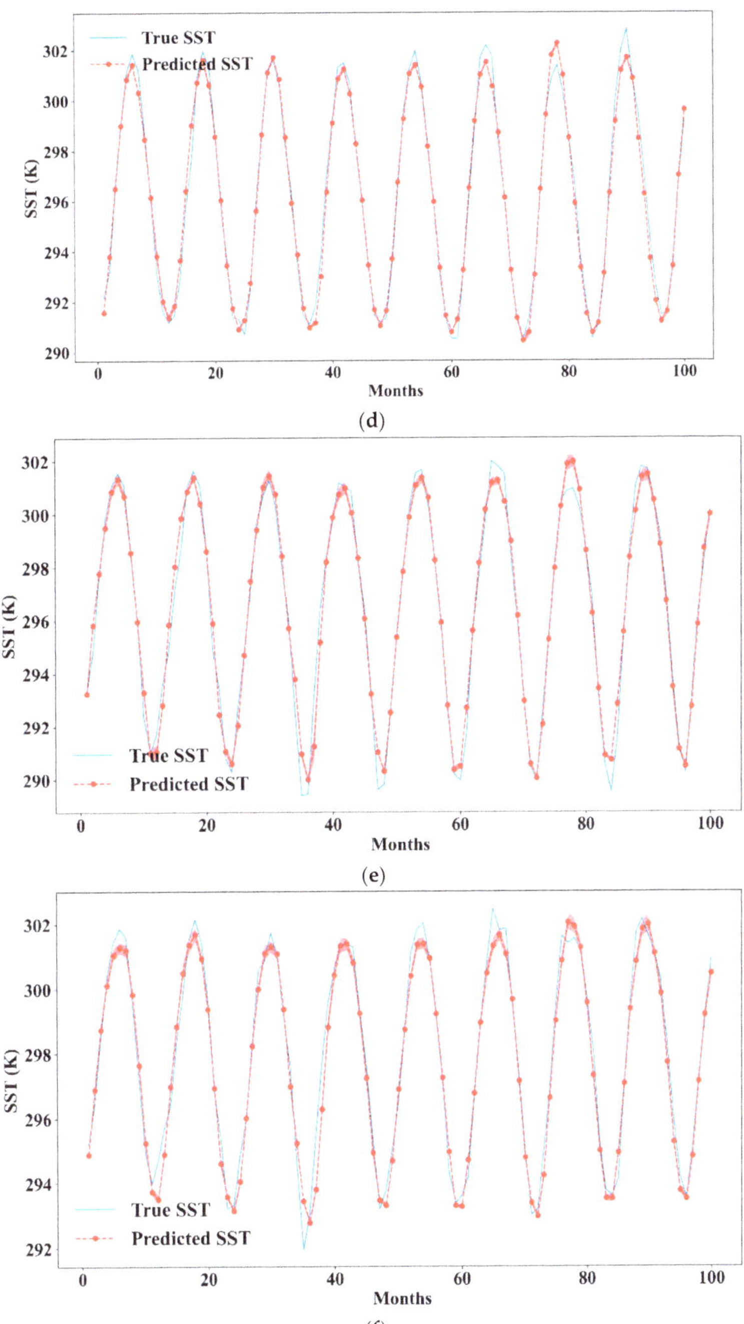

Figure 12. *Cont.*

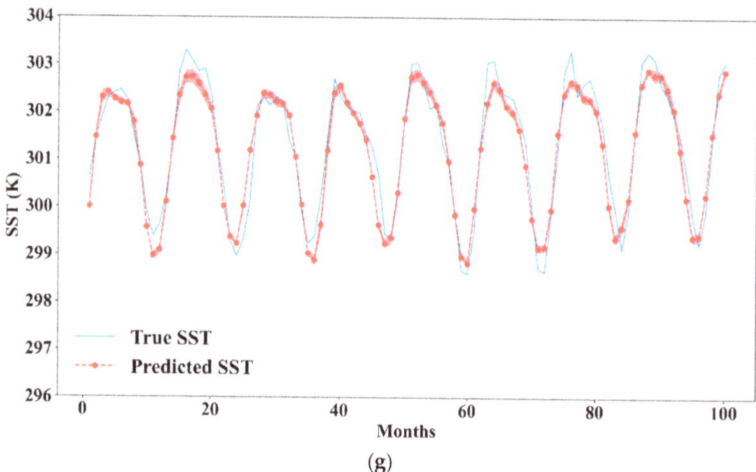

Figure 12. SST prediction results of the sea areas around China (**a**) Bohai Sea and north part of Yellow Sea, (**b**) South part of Yellow Sea, (**c**) Part 1 (ID:3 in Table 1) of East China Sea, (**d**) Part 2 (ID:4 in Table 1) of East China Sea, (**e**) Part 1(ID:5 in Table 1) of Taiwan Strait, (**f**) Part 2(ID:5 in Table 1) of Taiwan Strait and (**g**) South China Sea.

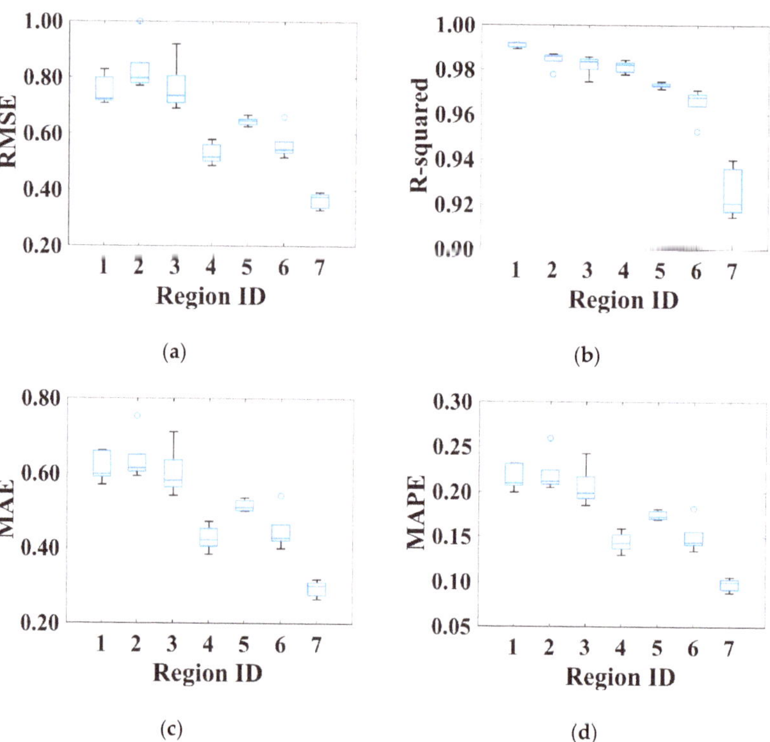

Figure 13. The distribution of the evaluation index for five runs: (**a**) *RMSE*, (**b**) R-squared, (**c**) *MAE* and (**d**) *MAPE*.

4. Discussion

4.1. Performance Comparison with Other Models

Table 5 shows the performance comparison of two other models with the model considering attention mechanism and taking both SST and meteorological factors as inputs. One of the models is the LSTM model taking SST only as input, and the other model takes SST and meteorological factors as inputs without considering attention mechanism. The boldface items in the table represent the best performance. The hyper-parameters affecting the training process are the same for the models.

Table 5. Comparison of *RMSE* between different models.

Region ID	LSTM Only with SST Only as Input	LSTM Only	Our Model
1	1.0157	0.9226	**0.7551**
2	1.0302	**0.8657**	0.8789
3	1.0481	0.7853	**0.7547**
4	0.8388	0.7301	**0.5120**
5	1.2139	0.8768	**0.6515**
6	0.7678	0.6487	**0.5666**
7	0.4140	0.4018	**0.3928**

It can be seen from the results that our model achieves the best performance for most regions. For the South China sea areas, three models show similar performance. Thus, it enables researchers to use the simple LSTM-only model with SST only as input for predicting SST in southern regions of China when there are insufficient meteorological data or computing resources.

4.2. Overfitting Issue Varification

To test if the trained model has overfitting issue, we have done another experiment to validate the generalization capability of the model. The forty-three-year (1979–2021) monthly mean SST and meteorological time-series data from ERA5 are used to train and validate the model. Then, the eight-year (1971–1978) data sets are fed into the trained model.

The prediction results shown in Figure 14 are the average for five runs, which verify the applicability and effectiveness of model. The black and red line represent the true values average prediction results.

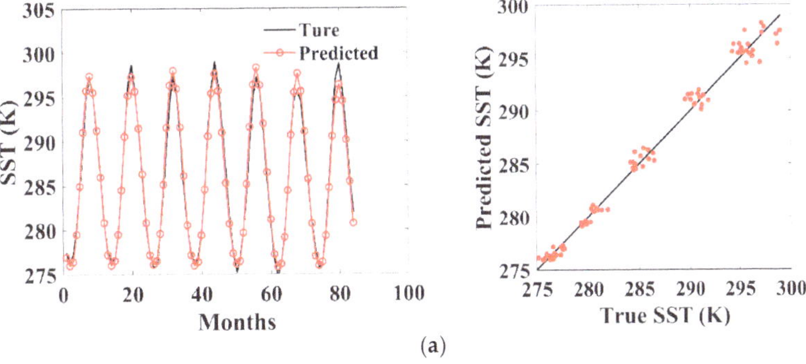

Figure 14. *Cont.*

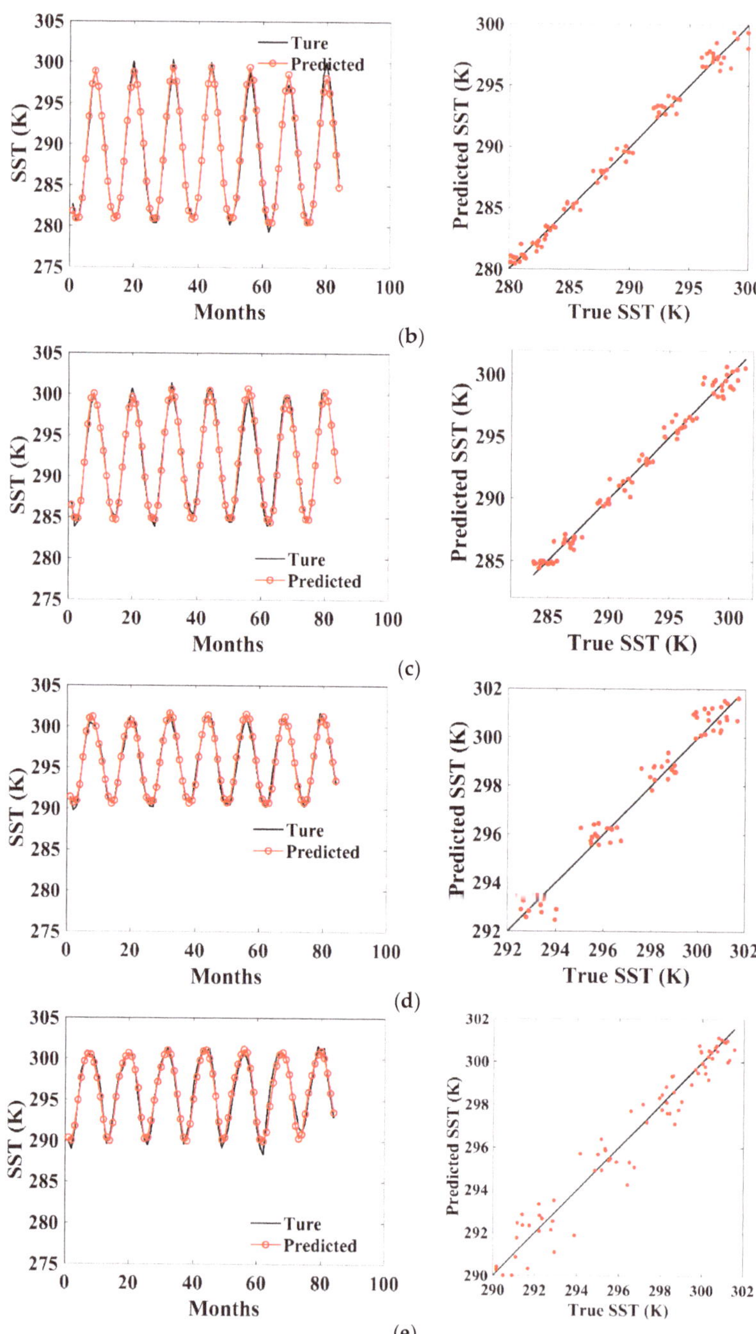

Figure 14. *Cont.*

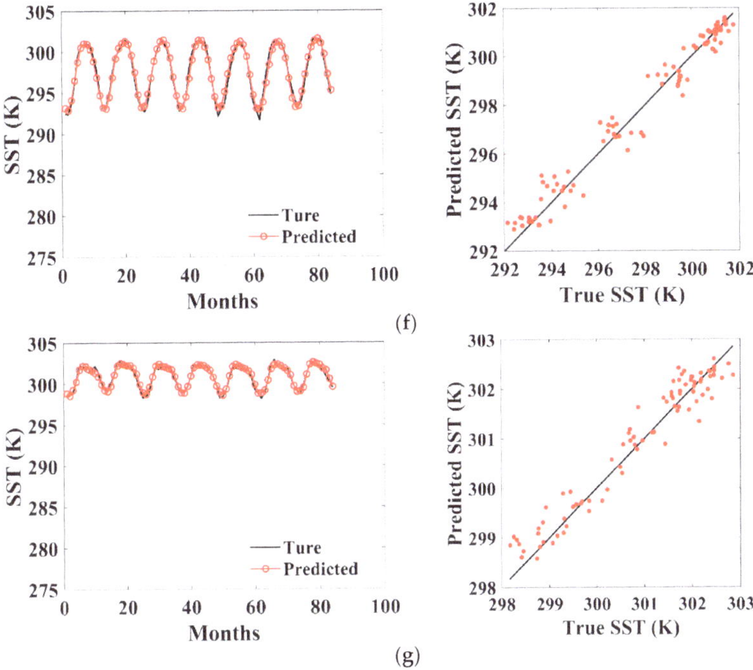

Figure 14. Evaluation of the generalization ability of the model. (**a**) Bohai Sea and north part of Yellow Sea, (**b**) South part of Yellow Sea, (**c**) Part 1 (ID:3 in Table 1) of East China Sea, (**d**) Part 2 (ID:4 in Table 1) of East China Sea, (**e**) Part 1(ID:5 in Table 1) of Tai-wan Strait, (**f**) Part 2(ID:5 in Table 1) of Taiwan Strait and (**g**) South China Sea.

5. Conclusions

SST is a significant physical parameter used in the analysis of the ocean and climate. This study developed a data-driven model for predicting one-month ahead monthly mean SST by combining interdimensional and self-attention mechanism with neural networks. After correlation analysis by mutual information, SST and other meteorological factors including wind speed and air pressure were selected as the input of the prediction model. The interdimensional attention enabled the model to focus on important historical moments and important variables while the self-attention mechanism was utilized to smooth the data in the training process. Forty-three-year monthly mean SST and meteorological time-series data from ERA5 of ECMWF were collected to train the model and test its performance for the sea areas around China. The evaluation criteria of R^2, $RMSE$, MAE and $MAPE$ indicate that the predicted results met the requirement for oceanography and meteorology studies.

During experiment, we find that, in most cases, other meteorological factors contribute to the predicted results, but these data, especially the wind speed, are not as stable as SST data and are prone to anomalies. The model is unable to reduce its coefficients quickly enough, thus leading to a longer training process eventually

Overall, the performance of the model on SST prediction is promising. Future work involves further optimization of the model and investigation of its applicability for other ocean physical parameters such as sea surface salinity, and ocean water temperature underneath the surface.

Author Contributions: Conceptualization, X.G. and J.H.; methodology, X.G. and J.H.; validation, X.G. and B.W.; investigation, X.G.; resources, B.W.; writing—original draft preparation, X.G.; writing—review and editing, X.G.; visualization, X.G. and J.H.; conceptualization and supervision, J.W.; funding acquisition, X.G. All authors have read and agreed to the published version of the manuscript.

Funding: This research was funded by the National Natural Science Foundation of China (grants 62005205 and 62101297) and the Shaanxi Province Science Foundation for Youths, grant number 2020JQ-329.

Data Availability Statement: SST and meteorological data are from fifth-generation ECMWF reanalysis data ERA5. The data are open and freely available at https://cds.climate.copernicus.eu/cdsapp#!/dataset/reanalysis-era5-single-levels-monthly-means?tab=overview, accessed on 16 February 2022.

Acknowledgments: The authors are grateful to ECMWF for supporting data.

Conflicts of Interest: The authors declare no conflict of interest.

References

1. O'carroll, A.G.; Armstrong, E.M.; Beggs, H.M.; Bouali, M.; Casey, K.S.; Corlett, G.K.; Dash, P.; Donlon, C.L.; Gentemann, C.L.; Høyer, J.L.; et al. Observational Needs of Sea Surface Temperature. *Front. Mar. Sci.* **2019**, *6*, 1–27. [CrossRef]
2. Kim, M.; Yang, H.; Kim, J. Sea Surface Temperature and High Water Temperature Occurrence Prediction Using a Long Short-Term Memory Model. *Remote Sens.* **2020**, *12*, 3654–3674. [CrossRef]
3. Baoleerqimuge, R.G.Y. Sea Surface Temperature Observation Methods and Comparison of Commonly Used Sea Surface Temperature Datasets. *Adv. Meteorol. Sci. Technol.* **2013**, *3*, 52–57. [CrossRef]
4. Hou, X.Y.; Guo, Z.H.; Cui, Y.K. Marine big data: Concept, applications and platform construction. *Bull. Mar. Sci.* **2017**, *36*, 361–369. [CrossRef]
5. Jiang, X.W.; Xi, M.; Song, Q.T. A Comparison of Six Sea Surface Temperature Analyses. *Acta Oceanol. Sin.* **2013**, *35*, 88–97. [CrossRef]
6. Wang, C.Q.; Li, X.; Zhang, Y.F.; Zu, Z.Q.; Zhang, R.Y. A comparative study of three SST reanalysis products and buoys data over the China offshore area. *Acta Oceanol. Sin.* **2020**, *42*, 118–128. [CrossRef]
7. Sarkar, P.P.; Janardhan, P.; Roy, P. Prediction of sea surface temperatures using deep learning neural networks. *SN Appl. Sci.* **2020**, *2*, 1458. [CrossRef]
8. Wolff, S.; O'Donncha, F.; Chen, B. Statistical and machine learning ensemble modelling to forecast sea surface temperature. *J. Mar. Syst.* **2020**, *208*, 103347. [CrossRef]
9. LeCun, Y.; Bengio, Y.; Hinton, G. Deep learning. *Nature* **2015**, *521*, 436–444. [CrossRef]
10. Lins, I.D.; Araujo, M.; das Chagas Moura, M.; Silva, M.A.; Droguett, E.L. Prediction of sea surface temperature in the tropical Atlantic by support vector machines. *Comput. Stat. Data Anal.* **2013**, *61*, 187–198. [CrossRef]
11. Li, W.; Lei, G.; Qu, L.Q. Prediction of sea surface temperature in the South China Sea by artificial neural networks. *IEEE Geosci. Remote. Sens. Lett.* **2020**, *17*, 558–562. [CrossRef]
12. Zhu, L.Q.; Liu, Q.; Liu, X.D.; Zhang, Y.H. RSST-ARGM: A data-driven approach to long-term sea surface temperature prediction. *J. Wirel. Commun. Netw.* **2021**, *2021*, 171. [CrossRef]
13. Hinton, G.E.; Osindero, S.; Teh, Y.W. A fast learning algorithm for deep belief nets. *Neural Comput.* **2006**, *18*, 1527–1554. [CrossRef]
14. Qiu, X.P. *Neural Networks and Deep Learning*; China Machine Press: Beijing, China, 2020; pp. 139–141.
15. Hochreiter, S.; Schmidhuber, J. Long short-term memory. *Neural Comput.* **1997**, *9*, 1735–1780. [CrossRef]
16. Shi, X.J.; Chen, Z.R.; Wang, H.; Yeung, D.Y.; Wong, W.K.; Woo, W.C. Convolutional LSTM Network: A Machine Learning Approach for Precipitation Nowcasting. In Proceedings of the International Conference on Neural Information Processing Systems, Montreal, QC, Canada, 7–12 December 2015; pp. 802–810.
17. Xu, B.N.; Jiang, J.R.; Hao, H.Q.; Lin, P.F.; He, D.D. A Deep Learning Model of ENSO Prediction Based on Regional Sea Surface Temperature Anomaly Prediction. *Electron. Sci. Technol. Appl.* **2017**, *8*, 65–76. [CrossRef]
18. Yang, Y.; Dong, J.; Sun, X.; Lima, E.; Mu, Q.; Wang, X. A CFCC-LSTM model for sea surface temperature prediction. *IEEE Geosci. Remote. Sens. Lett.* **2018**, *15*, 207–211. [CrossRef]
19. Zhu, G.C.; Hu, S. Study on sea surface temperature model based on LSTM-RNN. *J Appl. Oceanogr.* **2019**, *38*, 191–197. [CrossRef]
20. Sun, T.; Feng, Y.; Li, C.; Zhang, X. High Precision Sea Surface Temperature Prediction of Long Period and Large Area in the Indian Ocean Based on the Temporal Convolutional Network and Internet of Things. *Sensors* **2022**, *22*, 1636. [CrossRef] [PubMed]
21. Aydınlı, H.O.; Ekincek, A.; Aykanat-Atay, M.; Sarıtaş, B.; Özenen-Kavlak, M. Sea surface temperature prediction model for the Black Sea by employing time-series satellite data: A machine learning approach. *Appl. Geomat.* **2022**, 1–10. [CrossRef]
22. Patil, K.; Deo, M.; Ravichandran, M. Prediction of sea surface temperature by combining numerical and neural techniques. *J. Atmos. Ocean. Technol.* **2016**, *33*, 1715–1726. [CrossRef]
23. Xiao, C.; Chen, N.; Hu, C.; Wang, K.; Chen, Z. Short and mid-term sea surface temperature prediction using time-series satellite data and LSTM-AdaBoost combination approach. *Remote Sens. Environ.* **2019**, *233*, 111358. [CrossRef]
24. Xiao, C.; Chen, N.; Hu, C.; Wang, K.; Xu, Z.; Cai, Y.; Xu, L.; Chen, Z.; Gong, J. A spatiotemporal deep learning model for sea surface temperature field prediction using time-series satellite data. *Environ. Model Softw.* **2019**, *120*, 104502. [CrossRef]

25. He, Q.; Cha, C.; Song, W.; Hao, Z.Z.; Huang, D.M. Sea surface temperature prediction algorithm based on STL model. *Mar. Environ. Sci.* **2020**, *39*, 104–111. [CrossRef]

26. de Mattos Neto, P.S.G.; Cavalcanti, G.D.C.; de, O.S.J.D.S.; Silva, E.G. Hybrid Systems Using Residual Modeling for Sea Surface Temperature Forecasting. *Sci. Rep.* **2022**, *12*, 487. [CrossRef]

27. Jahanbakht, M.; Xiang, W.; Azghadi, M.R. Sea Surface Temperature Forecasting With Ensemble of Stacked Deep Neural Networks. *IEEE Geosci. Remote. Sens. Lett.* **2022**, *19*, 1–5. [CrossRef]

28. Liu, X.; Li, N.; Guo, J.; Fan, Z.; Lu, X.; Liu, W.; Liu, B. Multi-step-ahead Prediction of Ocean SSTA Based on Hybrid Empirical Mode Decomposition and Gated Recurrent Unit Model. *IEEE J. Sel. Top. Appl. Earth Obs. Remote Sens.* **2022**, *15*, 7525–7538. [CrossRef]

29. Lara-Benítez, P.; Carranza-García, M.; Riquelme, J.C. An experimental review on deep learning architectures for time series forecasting. *Int. J. Neural Syst.* **2021**, *31*, 2130001. [CrossRef]

30. Vaswani, A.; Shazeer, N.; Parmar, N.; Uszkoreit, J.; Jones, L.; Gomez, A.N.; Kaiser, Ł.; Polosukhin, I. Attention is all you need. *arXiv* **2017**. [CrossRef]

31. Li, S.; Jin, X.; Xuan, Y.; Zhou, X.; Chen, W.; Wang, Y.-X.; Yan, X. Enhancing the locality and breaking the memory bottleneck of transformer on time series forecasting. In Proceedings of the Conference on Neural Information Processing Systems, Vancouver, BC, Canada, 8–14 December 2019; pp. 1–11.

32. Xie, J.; Zhang, J.; Yu, J.; Xu, L. An adaptive scale sea surface temperature predicting method based on deep learning with attention mechanism. *IEEE Geosci. Remote. Sens. Lett.* **2019**, *17*, 740–744. [CrossRef]

33. Mohammadi Farsani, R.; Pazouki, E. A transformer self-attention model for time series forecasting. *J. Electr. Comput. Eng. Innov.* **2021**, *9*, 1–10. [CrossRef]

34. Xu, W.X.; Shen, Y.D. Bus travel time prediction based on Attention-LSTM neural network. *Mod. Electron. Technol.* **2022**, *45*, 83–87. [CrossRef]

35. Liu, X.X. Correlation Analysis and Variable Selection for multivariateTime Series based a on Mutual Informationerles. Master's Thesis, Dalian University of technology, Dalian, China, 2013.

36. De Winter, J.C.; Gosling, S.D.; Potter, J. Comparing the Pearson and Spearman correlation coefficients across distributions and sample sizes: A tutorial using simulations and empirical data. *Psychol. Methods* **2016**, *21*, 273–290. [CrossRef]

37. Li, G.J. Research on Time Series Forecasting Based on Multivariate Analysis. Master's Thesis, Tianjin University of Technology, Tianjin, China, 2021.

38. Kozachenko, L.F.; Leonenko, N.N. Sample estimate of the entropy of a random vector. *Probl. Inf. Transm.* **1987**, *23*, 9–16.

39. Kraskov, A.; Stögbauer, H.; Grassberger, P. Estimating mutual information. *Phys. Rev. E* **2004**, *69*, 66138. [CrossRef]

40. Hersbach, H.; Bell, B.; Berrisford, P.; Biavati, G.; Horányi, A.; Muñoz Sabater, J.; Nicolas, J.; Peubey, C.; Radu, R.; Rozum, I.; et al. ERA5 monthly averaged data on single levels from 1959 to present. *Copernic. Clim. Change Serv. (C3S) Clim. Data Store (CDS)* **2019**, *10*, 252–266. [CrossRef]

Article

Enhanced Multi-Stream Remote Sensing Spatiotemporal Fusion Network Based on Transformer and Dilated Convolution

Weisheng Li *, Dongwen Cao and Minghao Xiang

College of Computer Science and Technology, Chongqing University of Posts and Telecommunications, Chongqing 400065, China
* Correspondence: liws@cqupt.edu.cn

Abstract: Remote sensing images with high temporal and spatial resolutions play a crucial role in land surface-change monitoring, vegetation monitoring, and natural disaster mapping. However, existing technical conditions and cost constraints make it very difficult to directly obtain remote sensing images with high temporal and spatial resolution. Consequently, spatiotemporal fusion technology for remote sensing images has attracted considerable attention. In recent years, deep learning-based fusion methods have been developed. In this study, to improve the accuracy and robustness of deep learning models and better extract the spatiotemporal information of remote sensing images, the existing multi-stream remote sensing spatiotemporal fusion network MSNet is improved using dilated convolution and an improved transformer encoder to develop an enhanced version called EMSNet. Dilated convolution is used to extract time information and reduce parameters. The improved transformer encoder is improved to further adapt to image-fusion technology and effectively extract spatiotemporal information. A new weight strategy is used for fusion that substantially improves the prediction accuracy of the model, image quality, and fusion effect. The superiority of the proposed approach is confirmed by comparing it with six representative spatiotemporal fusion algorithms on three disparate datasets. Compared with MSNet, EMSNet improved SSIM by 15.3% on the CIA dataset, ERGAS by 92.1% on the LGC dataset, and RMSE by 92.9% on the AHB dataset.

Keywords: spatiotemporal fusion; dilated convolution; improved transformer encoder; global correlation information

Citation: Li, W.; Cao, D.; Xiang, M. Enhanced Multi-Stream Remote Sensing Spatiotemporal Fusion Network Based on Transformer and Dilated Convolution. *Remote Sens.* **2022**, *14*, 4544. https://doi.org/10.3390/rs14184544

Academic Editors: Gwanggil Jeon and Lefei Zhang

Received: 29 July 2022
Accepted: 7 September 2022
Published: 11 September 2022

Publisher's Note: MDPI stays neutral with regard to jurisdictional claims in published maps and institutional affiliations.

1. Introduction

Remote sensing images are generated by various types of satellite sensors, such as the Moderate Resolution Imaging Spectroradiometer (MODIS), Landsat-equipped sensors, and Sentinel. MODIS sensors are usually installed on Terra and Aqua satellites, which can circle the earth in half a day or one day, and the data obtained by them have superior time resolution. However, the spatial resolution of MODIS data (i.e., rough image) is very low, and accuracy can reach only 250–1000 m [1]. By contrast, data (fine image) acquired by Landsat have higher spatial resolution (15–30 m) and capture sufficient surface-detail information, but temporal resolution is very low because it takes 16 days to circle the earth [1]. In practical applications, we often need remote sensing images with high temporal and spatial resolution. For example, images with high temporal and spatial resolutions can be used for research in the fields of heterogeneous regional surface change [2,3], vegetation seasonal monitoring [4], real-time natural disaster mapping [5], and land-cover changes [6]. Unfortunately, current technical and cost constraints, coupled with the existence of such noise as cloud cover in some areas, make it challenging to directly obtain remote sensing products with high temporal and spatial resolution, and a single high-resolution image cannot meet practical needs. In order to meet these lacunae, spatiotemporal fusion has attracted considerable attention. In spatiotemporal fusion, two types of images are fused together, with the aim of obtaining images with high spatiotemporal resolution [7,8].

Existing spatiotemporal fusion methods can generally be subdivided into four categories: unmixing-based, reconstruction-based, dictionary pair learning-based, and deep learning-based.

Unmixing-based methods unmix the spectral information at the predicted moment, and then use the unmixed result to predict the unknown high spatial and temporal resolution image. Multi-sensor multi-resolution image fusion (MMFN) [9] was the first fusion method to apply the idea of unmixing. MMFN reconstructs the MODIS and Landsat images separately: first, the MODIS image is spectrally unmixed, and then the mixed result is spectrally reset on the Landsat image to obtain the final reconstruction result. Wu et al. considered the issue of nonlinear time-varying similarity and spatial variation in spectral unmixing, improved MMFN, and obtained a new spatiotemporal fusion method, STDFA [10], which also achieved good fusion results. A variable spatiotemporal data-fusion algorithm, FSDAF [11], has also been proposed, which combines the unmixing method, spatial interpolation, and spatiotemporal adaptive fusion algorithm (STARFM) to create a new algorithm that is computationally inexpensive, fast, and accurate, and performs well in heterogeneous regions.

The core idea of the reconstruction-based algorithm is to calculate the weights of similar adjacent pixels in the spectral information in the input and then add them. STARFM was the first method to be used for reconstruction for fusion [8]. In STARFM, the reflection changes of pixels between the rough image and the fine image should be continuous, and the weights of adjacent pixels can be calculated to reconstruct a surface-reflection image with high spatial resolution. In light of STARFM's large number of computations and the need to improve the reconstruction effect for heterogeneous regions, Zhu et al. made improvements and proposed an enhanced version of STARFM called ESTARFM [12]. They use two different coefficients to deal with the weights of adjacent pixels in homogeneous and heterogeneous regions, achieving a better effect. Inspired by STARFM, the spatiotemporal adaptive algorithm for mapping reflection changes (STAARCH) [13] also achieves good results. Overall, the difference between these algorithms lies in how the weights of adjacent pixels are calculated. Although these algorithms generally have good results, they are unsuitable for data that change too much too quickly.

Dictionary learning-based methods mainly learn the correspondence between two types of remote sensing images to perform prediction. The sparse representation-based spatiotemporal reflection fusion method (SPSTFM) [14] may be the first fusion method to successfully apply dictionary learning. In SPSTFM, the coefficients of low-resolution images and high-resolution images should be the same, and the super-resolution ideas in the field of natural images are introduced into spatiotemporal fusion. Images are reconstructed by establishing correspondences between low-resolution images. However, in practical situations, the same coefficients may not be applicable to some of the data obtained under the existing conditions [15]. Wei et al. studied the explicit mapping between low-resolution images and proposed a new fusion method based on dictionary learning and utilizing compressive sensing theory, called compressive sensing spatiotemporal fusion (CSSF) [16], which improves the accuracy of the prediction results noticeably, but the training time also increases considerably, while the efficiency decreases. In this regard, Liu et al. proposed an extreme learning machine called ELM-FM for spatiotemporal fusion [17], which considerably reduces time and improves efficiency.

As deep learning has gradually been applied in various fields in recent years, deep learning-based spatiotemporal fusion methods of remote sensing have also advanced. For example, Song et al. proposed STFDCNN [18] for spatiotemporal fusion using a convolutional neural network. In STFDCNN, the image-reconstruction process is considered a super-resolution and nonlinear mapping problem. A super-resolution network and a nonlinear mapping network are constructed through an intermediate resolution image, and the final fusion result is obtained through high-pass modulation. STFDCNN achieved good results. Liu et al. proposed a two-stream CNN, StfNet [19], for spatiotemporal fusion. They effectively extracted and fused spatial details and temporal information using spatial consistency and temporal dependence, and achieved good results. On the basis of spatial consistency and time dependence, Chen et al. introduced a multiscale mechanism for feature extraction and proposed a spatiotemporal remote sensing image-fusion method based on multiscale two-stream CNN (STFMCNN) [20]. Jia et al. proposed a new deep learning-based two-stream convolutional neural network [21], which fuses the temporal variation information with the spatial detail information by weight, which enhances its robustness. Furthermore, Jia et al. adopted various prediction methods for phenological change and land-cover change, and proposed a spatiotemporal fusion method based on hybrid deep learning to combine satellite images with differing resolutions [22]. Tan et al. proposed DCSTFN [23] to derive high spatiotemporal remote sensing images using CNNs based on the methods of convolution and deconvolution combined with the fusion method of STARFM. However, in light of the loss of information in the reconstruction process of the deconvolution fusion method, Tan et al. increased the input of the a priori moment and added a residual coding block, using a composite loss function to improve the learning ability of the network, and an enhanced convolutional neural network EDCSTFN [24] was proposed for spatiotemporal fusion. In addition, CycleGAN-STF [25] introduces other ideas in the visual field into spatiotemporal fusion. It achieves spatiotemporal fusion through image generation of CycleGAN. CycleGAN is used to generate a fine image at the predicted time, the real image is used at the predicted time to select the closest generated image, and finally FSDAF is used for fusion. Other fusion methods are applied in specific scenarios. For example, STTFN [26], a CNN-based model for spatiotemporal fusion of surface-temperature changes, uses a multiscale CNN to establish a nonlinear mapping relationship and a spatiotemporal continuity weight strategy for fusion, achieving good results. DenseSTF [27], a deep learning-based spatiotemporal data-fusion algorithm, uses a block-to-point modeling strategy and model comparison to provide rich texture details for each target pixel to deal with heterogeneous regions, and achieves very good results. Furthermore, with the development of transformer models [28] in the natural language field, many researchers have introduced the concept into the vision field as well, e.g., vision transformer (ViT) [29], data-efficient image transformer (DeiT) [30], conditional position encoding visual transformer (CPVT) [31], transformer-in-transformer (TNT) [32], and convolutional vision transformer (CvT) [33] can be used for image classification. In addition, there are the Swin transformer [34] for image classification, image segmentation, and object detection, and texture transformer [35] for general image superclassification. These variants have been gradually introduced into the spatiotemporal fusion of remote sensing. For example, MSNet [36] is a new method obtained by introducing the original transformer and ViT into spatiotemporal fusion, learning the global temporal correlation information of the image through the transformer structure, using the convolutional neural network to establish the relationship between input and output, and finally obtain a good effect. SwinSTFM [37] is a new method that introduces the Swin transformer and combines linear spectral mixing theory, which finally improves the quality of generated images. There is also MSFusion [38], which introduces texture transformer into spatiotemporal fusion, which has also achieved quite good results on multiple datasets.

Existing spatiotemporal fusion algorithms perform a certain amount of information extraction and noise processing during the fusion process, but there remain certain lacunae. First, the acquisition and processing of suitable datasets is not easy. Owing to the existence of noise, the data that can be directly used for research are insufficient. In deep learning, the size of the dataset affects the learning ability during reconstruction: achieving good reconstruction with small datasets is a major challenge. Second, the same fusion model can have different prediction performance on different datasets, and the model is not robust. Furthermore, the features extracted only by the CNN are not sufficient, and an increase of the network depth will also result in potential feature loss.

In order to address the aforementioned challenges, this study improves MSNet and proposes an enhanced version of the spatiotemporal fusion method of multi-stream remote sensing images called EMSNet. In EMSNet, the input image adopts the original scale size, and the rough image is no longer scaled to fully extract the temporal information and reduce the loss. The main contributions of this paper are summarized as follows.

(1)   The number of prior input images required by the model is reduced from five to three, which achieves better results with less input, so that even a dataset with a small amount of data can reconstruct images with better effects.

(2)   The transformer encoder structure is introduced and its projection method improved to obtain the improved transformer encoder (ITE), which adapts the remote sensing spatiotemporal fusion, effectively learns the relationship between local and global information in rough and fine images, and effectively extracts temporal and spatial information.

(3)   Dilated convolution is used to extract temporal information, which expands the receptive field while keeping the parameter quantity unchanged and fully extracts a large amount of temporal feature information contained in the rough image.

(4)   A new feature-fusion strategy is used to fuse the features extracted by the ITE and dilated convolution based on their differences from real predicted images in order to avoid introducing noise.

The rest of the article has the following structure. The overall structure of EMSNet and its internal specific modules and weight strategies are introduced in Section 2. Experimental results are described in Section 3, along with the datasets used. Section 4 dis-cusses the performance of EMSNet. Finally, conclusions are provided.

2. Methods

2.1. EMSNet Architecture

Figure 1 shows the overall structure of EMSNet, where $M_i(i = 1, 2)$ represents the MODIS image at time t_i, L_i represents the Landsat image at time t_i, and Pre_L_2 represents the prediction result of the fused image at time t_2 based on time t_1. Rectangles of different colors represent different operations, including convolution, dilated convolution, activation function ReLU, and various operations inside the improved transformer encoder (ITE). EMSNet is an end-to-end structure, which can be divided into three parts:

a.   ITE-related modules, used to extract temporal change information and spatial texture detail features and learn local and global correlation information;

b.   an extraction network composed of convolution and dilated convolution, used to establish a nonlinear relationship between input and output, while fully extracting the features of time information;

c.   a weight strategy, used to calculate the corresponding weight according to the difference between the features obtained in the above two parts and the real prediction map for final fusion.

A detailed description of each module can be found in Sections 2.2–2.4.

In this study, three images of the same size are used as input, a pair of MODIS-Landsat images at a priori time t_1 and a MODIS image at prediction time t_2. The overall procedure of EMSNet is as follows:

(1)   First, we subtract M_1 from M_2 to get M_{12}, which represents the change area within two times and provides time-change information. We input into the feature-extraction network composed of convolution and dilated convolution, and then fully extract the time information contained in it.

(2)   Second, we add M_{12} and L_1 to the ITE to extract the rich temporal information and spatial texture detail information, and simultaneously learn the connection between the local and the global information.

(3)   Inspired by ResNet [39], in DenseNet [40], as the network depth increases, the temporal and spatial information in the input image may be lost during transmission. Therefore, we add L_1 as the residual to the temporal variation information obtained in the first step to supplement the spatial details that may be lost in the subsequent fusion process.

(4)   Finally, the results obtained in the second and third steps are calculated by calculating the difference with L_2 to obtain their respective weights, so as to fuse and reconstruct the final prediction map Pre_L_2.

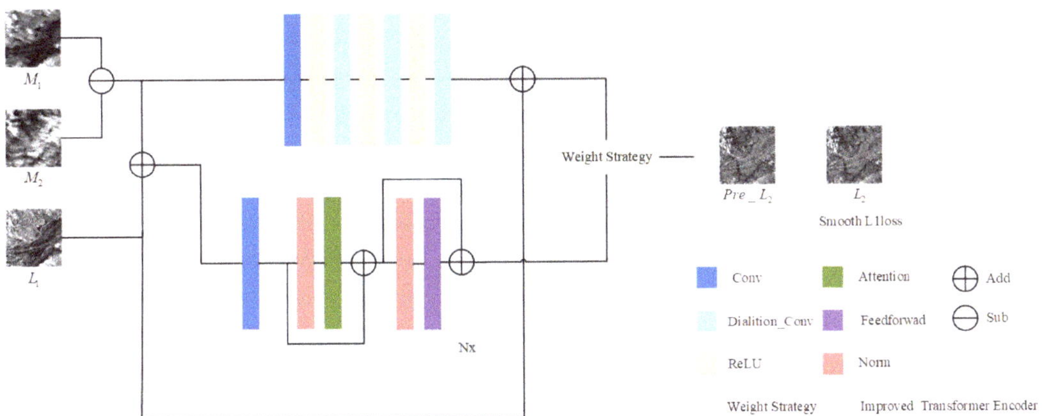

Figure 1. EMSNet architecture.

The structure of EMSNet can be represented by Equation (1) below:

$$Pre\_L_2 = W(T(M_{12} + L_1), E(M_{12}) + L_1) \tag{1}$$

Here, T represents the ITE module, E represents the time information-extraction network composed of convolution and dilated convolution, and W represents the weight strategy adopted in this study.

2.2. Improved Transformer Encoder

Transformer [28], as a kind of attention mechanism, is well suited not only to the field of natural language but also to the field of vision. Inspired by the application of the transformer in MSNet [36] and the cancellation of position encoding in CPVT [31] and CvT [33], in this study, the transformer encoder applied to remote sensing spatiotemporal fusion is further improved, the MLP part for classification is canceled, and position encoding is canceled. In addition, the convolutional projection method is used to replace the original linear projection method in the transformer, and a new structure, as shown in Figure 2 below, is obtained, called the improved transformer encoder (ITE), which is mainly used to learn temporal variation information and spatial texture details. Through the above operations, it is ensured that the input and output are of the same dimension, which facilitates subsequent fusion and reconstruction.

Figure 2 is the ITE structure diagram, in which the yellow part represents the convolution projection operation and the blue box and its interior represent the specific operation part of ITE. As can be seen from the figure, this study projects the input information directly through the convolution operation, and the overlap between the convolution blocks and the convolution blocks effectively strengthens the connection between the blocks. Consequently, the ITE strengthens the correlation between local information and global information, removing the need for the position encoding required by the linear projection method, thus making it more suitable for the spatiotemporal fusion method. The ITE is also composed of alternate multi-head attention mechanisms and feedforward parts. It will be normalized before each input to the submodule, and there will be residual connections after each block. The multi-head self-attention mechanism is a series of SoftMax and linear operations, and the input data will gradually change the dimensions during the propagation and training process to adapt to match these operations. The feedforward portion is composed of linear, Gaussian error linear unit (GELU), and random deactivation dropout, where GELU is used as the activation function. In practical applications, for different amounts of data, when learning global time-varying information, ICTE with different depths are required to learn more accurately. Nx in the figure represents the depth value.

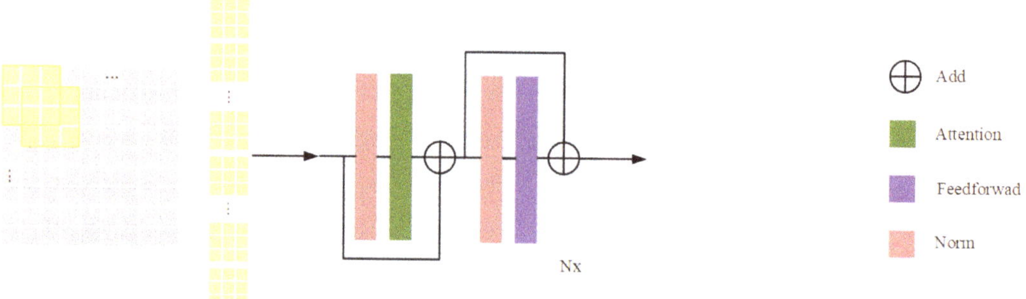

Figure 2. Structure of improved transformer encoder (ITE).

In this study, the ITE is used as a module for learning time-varying information and spatial texture detail. Compared with the previous MSNet, it further expands the learning range of the transformer encoder for remote sensing.

2.3. Dilated Convolution

In order to extract the time information contained in M_{12} and establish the mapping relationship between input and output, this study proposes a seven-layer neural network mainly composed of dilated convolution as a feature extraction network. The key feature of dilated convolution is that different sizes of receptive fields can be obtained after setting different dilation rates, so as to extract effective information at multiple scales. Compared with ordinary convolution operations, dilated convolution will not increase the number of redundant parameters. Figure 3 shows the proposed dilated convolution-based neural network and the receptive fields under different dilation rates.

Figure 3. Neural network based on dilated convolution and receptive fields with different dilation rates.

The right side of the dotted line in Figure 3 shows the architecture of the seven-layer neural network, which consists of one layer of convolution, three layers of dilated convolution, and three layers of ReLU. The convolution operation is used to convert the original M_{12} into a multidimensional nonlinear tensor, and the convolution kernel adopts the size of 3×3; the dilated convolution is used to effectively extract the temporal features in the M_{12}, the basic convolution kernel is of the same size i.e., 3×3, and an expansion rate of 2, 3, and 4 is set in turn for three consecutive layers of dilated convolution. The left side of the dotted line is the schematic diagram of the receptive field under various expansion rates. When the dilation rate is 1, dilated convolution is no different from ordinary convolution. When the dilation rate increases, the receptive field also gradually increases, which enables it to better learn the feature information at various scales, and simultaneously guarantee the number of parameters taken during its operation will not increase [41].

Each dilated convolution operation can be defined as:

$$\Phi(x) = w_i * x + b_i \tag{2}$$

Here, x represents the input, "$*$" represents the dilated convolution operation, w_i represents the weight of the current convolutional layer, and b_i represents the current offset. The output channels of the three convolution operations are 32, 16, and 1 in sequence. After the convolution, the ReLU operation is used to make the features non-linear and avoid network overfitting [42]. The ReLU operation can be defined as:

$$\text{ReLU}(x) = \max(0, x) \tag{3}$$

2.4. Weight Strategy

After feature extraction by the ITE and dilated convolutional neural network, plus residual L_2 for supplementary information, two distinct features are obtained. The difference between the prediction graphs is calculated by weight for final fusion, and the specific weight strategy can be defined as:

$$Pre\_L_2 = W(T(M_{12} + L_1), E(M_{12}) + L_1) = \alpha T(M_{12} + L_1) + \beta(E(M_{12}) + L_1) \tag{4}$$

$$\begin{cases} \alpha = \dfrac{\frac{1}{|T(M_{12}+L_1)-L_2|}}{\frac{1}{|T(M_{12}+L_1)-L_2|} + \frac{1}{|(E(M_{12})+L_1)-L_2|}} \\[2em] \beta = \dfrac{\frac{1}{|(E(M_{12})+L_1)-L_2|}}{\frac{1}{|T(M_{12}+L_1)-L_2|} + \frac{1}{|(E(M_{12})+L_1)-L_2|}} \end{cases} \tag{5}$$

Here, T represents the ITE module, and E represents the temporal information extraction network composed of convolution and dilated convolution.

2.5. Network Training

During the entire training process of the model, the loss calculation is performed on the prediction results of the entire model, so as to continuously adjust the learning parameters during the backpropagation process to obtain better convergence results. When calculating the difference between the predicted result and the real value, the smooth L1 loss function, namely Huber loss [43], is chosen, which can be defined as $\mathcal{L}$:

$$S(L_i) = \sum_{m=1}^{H} \sum_{n=1}^{W} L_i(m, n) \tag{6}$$

$$\mathcal{L} = loss(Pre\_L_2, L_2) = \frac{1}{N} \begin{cases} \frac{1}{2}(S(Pre\_L_2) - S(L_2))^2, & if\,|S(Pre\_L_2) - S(L_2)| < 1 \\ |S(Pre\_L_2) - S(L_2)| - \frac{1}{2}, & otherwise \end{cases} \tag{7}$$

where H represents the height of the image, W represents the width of the image, L_i represents the input image, and S represents for the pixel sum formula.

3. Experiments and Results

3.1. Datasets

Three separate datasets were employed to test the robustness of EMSNet.

The first study area was the Coleambally Irrigation Area (CIA) in southern New South Wales (NSW, Australia, 34.0034°E, 145.0675°S) [44]. The dataset was acquired from October 2001 to May 2002 and comprises 17 pairs of MODIS–Landsat images. The Landsat images are all from Landsat-7 ETM+, and the MODIS images are MODIS Terra MOD09GA Collection 5 data. The CIA dataset includes six bands and an image size of 1720×2040.

The second study area is the Lower Gwydir Watershed (LGC) in northern New South Wales (NSW, 149.2815°E, 29.0855°S), Australia [44]. The dataset was acquired from

April 2004 to April 2005 and comprises 14 pairs of MODIS–Landsat images. All Landsat imagery is from Landsat-5^{TM}, and the MODIS imagery is MODIS Terra MOD09GA Collection 5 data. The LGC dataset contains six bands and the image size is 3200 × 2720.

The third study area is the Alu Horqin Banner (AHB) region (43.3619°N, 119.0375°E) in the central Inner Mongolia Autonomous Region of northeastern China, which has many circular pastures and farmland [45,46]. Li Jun et al., collected 27 cloud-free MODIS–Landsat image pairs from 30 May 2013 to 6 December 2018, a time span of more than 5 years. The area has experienced substantial phenological changes owing to the growth of crops and other types of vegetation. The AHB dataset contains six bands and the image size is 2480 × 2800.

In this study, all images of the three datasets are combined according to a prior time and a prediction time. Each set of training data has four images, including two pairs of MODIS–Landsat images. The image size of each pair of MODIS-Landsat is the same, and the spatial resolution is 16:1. When combining the data, the data with the same time span between the prior moment and the predicted moment are given priority as the experimental data. In addition, for the training of the network, the images of the three datasets are all adjusted to a size of 1200 × 1200. Figures 4–6 show the MODIS–Landsat image pairs obtained on two different dates for the three datasets. During the experiment process, the three datasets were input into EMSNet for training, 70% of the dataset was used for training, 15% was used for validation, and 15% was used as the final test set for evaluating the fusion and reconstruction ability of the model.

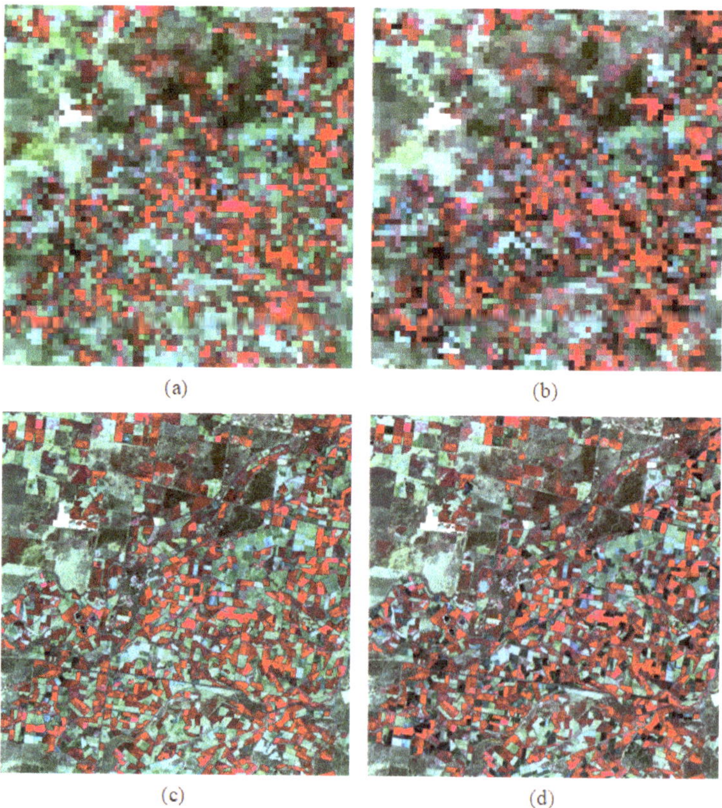

(a)   (b)

(c)   (d)

Figure 4. Composite MODIS (**Top row**) and Landsat (**Bottom row**) image pairs on 7 October (**a,c**) and 16 October (**b,d**) 2001 on the CIA [44] dataset. The CIA dataset focuses on noteworthy phenological changes in irrigated farmland.

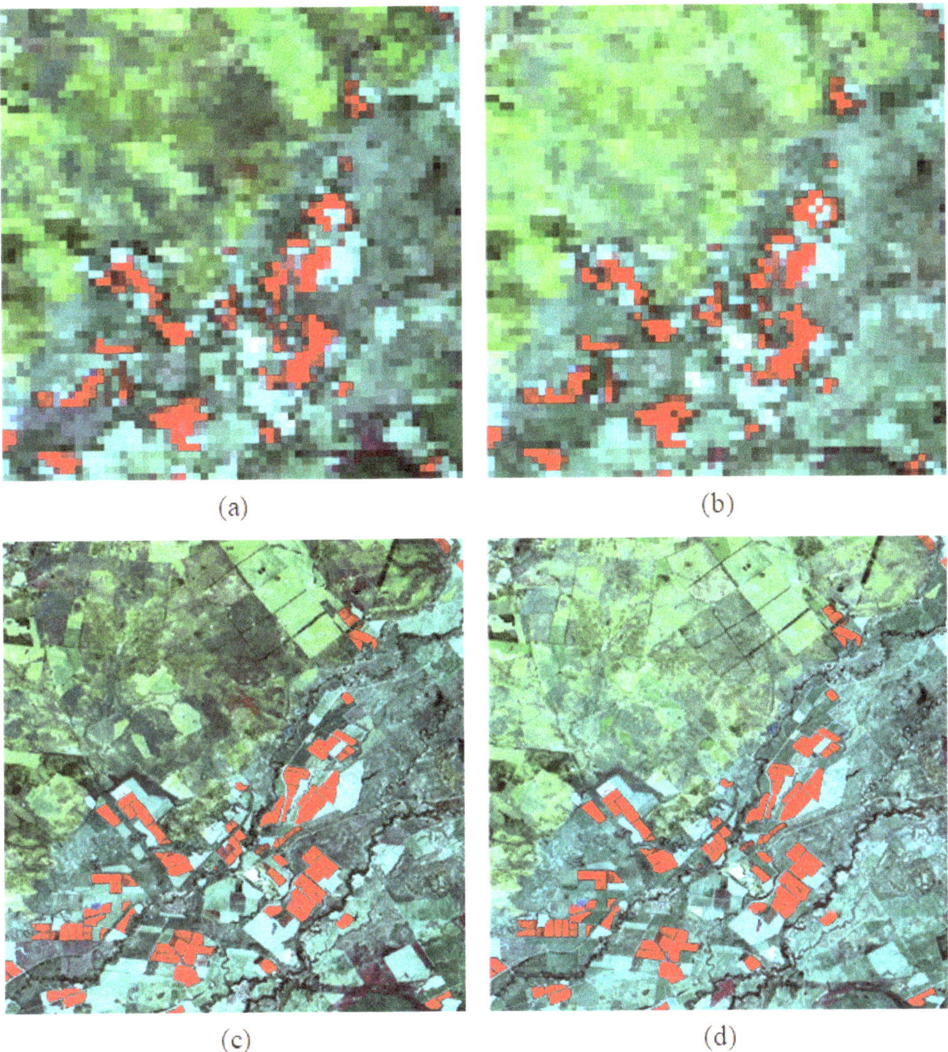

Figure 5. Composite MODIS (**top row**) and Landsat (**bottom row**) image pairs on 29 January (**a,c**) and 14 February (**b,d**) 2005 from the LGC [44] dataset. The LGC dataset focuses on changes in land cover types after the flood.

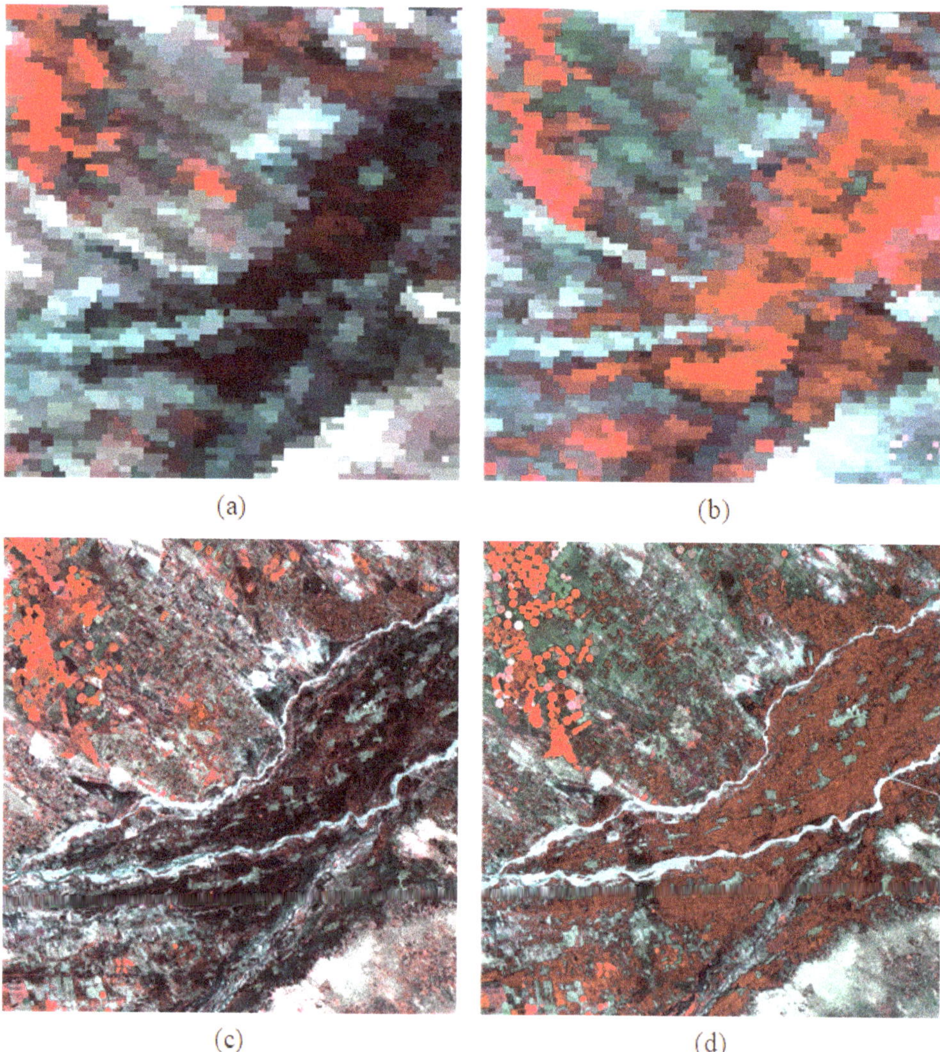

Figure 6. Composite MODIS (**top row**) and Landsat (**bottom row**) image pairs on 21 June (**a**,**c**) and 7 July (**b**,**d**) 2015 from the AHB [45,46] dataset. The AHB focuses on noteworthy phenological changes in the pasture.

3.2. Evaluation

We evaluated the proposed spatiotemporal fusion method by comparing it with FSDAF, STARFM, DCSTFN, STFDCNN, StfNet, and the previous MSNet under the same criteria.

As in the case of MSNet, six evaluation metrics are used. The first indicator is the spectral angle mapper (SAM) [47], which can measure the spectral distortion of the fusion result. It can be defined as follows:

$$\text{SAM} = \frac{1}{N}\sum_{n=1}^{N}\arccos\frac{\sum_{j=1}^{K}=(L_i^k Pre\_L_i^k)}{\sqrt{\sum_{j=1}^{K}\left(L_i^k\right)^2\sum_{j=1}^{K}\left(Pre\_L_i^k\right)^2}} \tag{8}$$

where N represents the total number of pixels in the predicted image, K represents the total number of bands, Pre_L_i represents the prediction result, $Pre\_L_i^k$ represents the prediction result of the kth band, and L_i^k represents the true value of the L_i^k band. A small SAM indicates a better result.

The second metric was the root mean square error (RMSE), which is the square root of the MSE and is used to measure the deviation between the predicted image and the observed image. It reflects a global depiction of the radiometric differences between the fusion result and the real observation image, which is defined as follows:

$$\text{RMSE} = \sqrt{\frac{\sum\limits_{m=1}^{H}\sum\limits_{n=1}^{W}(L_i(m,n) - Pre\_L_i(m,n))^2}{H \times W}} \tag{9}$$

where H represents the height of the image, W represents the width of the image, L represents the observed image, and Pre_L_i represents the predicted image. The smaller the value of RMSE, the closer the predicted image is to the observed image.

The third indicator was erreur relative global adimensionnelle de synthèse (ERGAS) [48], which measures the overall integration result. It can be defined as:

$$\text{ERGAS} = 100\frac{h}{l}\sqrt{\frac{1}{K}\sum\limits_{i=1}^{K}[\text{RMSE}(L_i^k)^2 / (\mu_k)^2]} \tag{10}$$

where h and l represent the spatial resolution of Landsat and MODIS images respectively; L_i^k represents the real image of the kth band; and μ_k represents the average value of the kth band image. When ERGAS is small, the fusion effect is better.

The fourth index was the structural similarity (SSIM) index [49], which is used to measure the similarity of two images. It can be defined as:

$$\text{SSIM} = \frac{(2\mu_{Pre\_L_i}\mu_{L_i} + c_1)(2\sigma_{Pre\_L_iL_i} + c_2)}{(\mu^2_{Pre\_L_i} + \mu^2_{L_i} + c_1)(\sigma^2_{Pre\_L_i} + \sigma^2_{L_i} + c_2)} \tag{11}$$

where $\mu_{Pre\_L_i}$ represents the mean value of the predicted image, μ_{L_i} represents the mean value of the real observation image, $\sigma_{Pre\_L_iL_i}$ represents the covariance of the predicted image Pre_L_i and the real observation image L_i, $\upsilon^2_{Pre\_L_i}$ represents the variance of the predicted image Pre_L_i, $\sigma^2_{L_i}$ represents the variance of the real observation image L_i, and c_1 and c_2 are constants used to maintain stability. The value range of SSIM is $[-1, 1]$. The closer the value is to 1, the more similar are the predicted image and the observed image.

The fifth index is the correlation coefficient (CC), which is used to indicate the correlation between two images. It can be defined as:

$$\text{CC} = \frac{\sum\limits_{n=1}^{N}(Pre\_L_i^n - \mu_{\hat{L}_i})(L_i^n - \mu_{L_i})}{\sqrt{\sum\limits_{n=1}^{N}(Pre\_L_i^n - \mu_{\hat{L}_i})^2}\sqrt{\sum\limits_{n=1}^{N}(L_i^n - \mu_{L_i})^2}} \tag{12}$$

The closer the CC is to 1, the greater the correlation between the predicted image and the real observation image.

The sixth indicator is the peak signal-to-noise ratio (PSNR) [50]. It is defined indirectly by the MSE, which can be defined as:

$$\text{MSE} = \frac{1}{HW}\sum\limits_{m=1}^{H}\sum\limits_{n=1}^{W}(L_i(m,n) - Pre\_L_i(m,n))^2 \tag{13}$$

Then PSNR can be defined as:

$$PSNR = 10 \cdot \log_{10}\left(\frac{MAX_{L_i}^2}{MSE}\right) \qquad (14)$$

where $MAX_{L_i}^2$ is the maximum possible pixel value of the real observation image L_i. If each pixel is represented by an 8-bit binary value, then MAX_{L_i} is 255. Generally, if the pixel value is represented by B-bit binary, then $MAX_{L_i} = 2^B - 1$. PSNR can evaluate the quality of the image after reconstruction. A higher PSNR means that the predicted image quality is better.

3.3. Parameter Settings

For the improved transformer encoder, the number of heads is set to 9, and the depth is set according to the data volume and characteristics of the three datasets: CIA is 20, LGC is 5, and AHB is 20. The size of the patch input into it is 240 × 240. The ordinary convolution as well as the three-layer dilated convolution in the dilated convolutional neural network each use a 3 × 3 convolution kernel. The dilation rates are 2, 3, and 4, and the number of channels is 32, 16, and 1. The initial learning rate is set to 0.0008, the optimizer adopts Adam, and the weight decay is set to 1×10^{-6}. EMSNet was trained on two Windows 10 Professional editions, each with 64 GB memory, an Intel Core i9-9900K @ 3.60 GHz×16 CPU, and an NVIDIA Geforce RTX 2080 Ti.

3.3.1. Subjective Evaluation

In order to visualize the experimental results, Figures 7–13 show the experimental results of FSDAF, STARFM, DCSTFN, STFDCNN, StfNet, MSNet, and the proposed improved EMSNet on each of three datasets. GT in the figure represents the real observed image, while Proposed is the proposed EMSNet method.

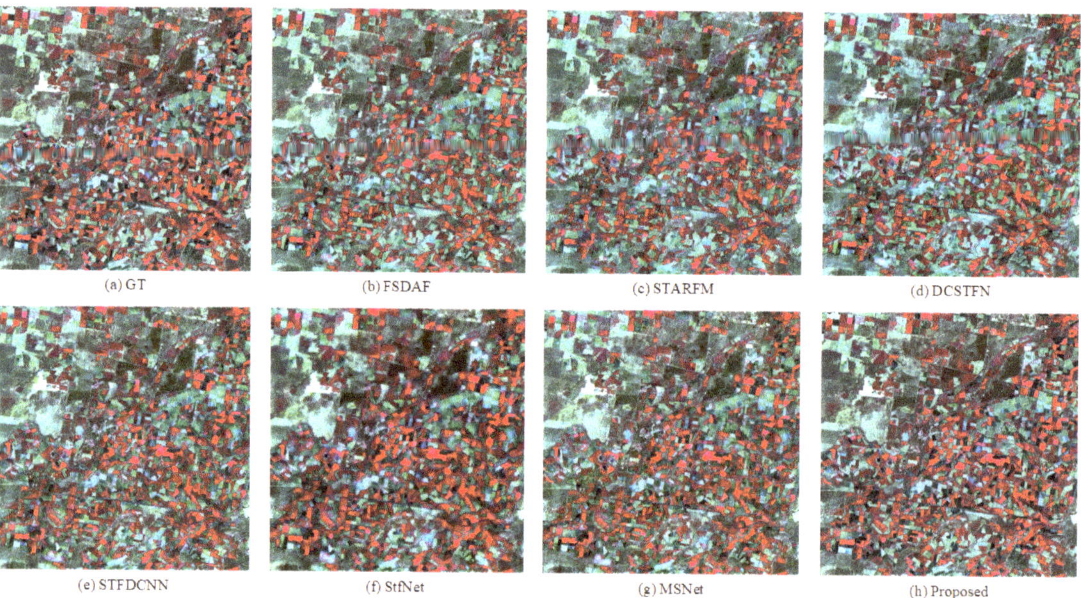

Figure 7. Entire prediction results for the target Landsat image (16 October 2001) in the CIA dataset. Comparison methods include FSDAF [11], STARFM [8], DCSTFN [23], STFDCNN [18], StfNet [19], and MSNet [36], which are represented by (**b–g**) in the figure respectively. (**a**) represents the ground truth (GT), and (**h**) represents the proposed method.

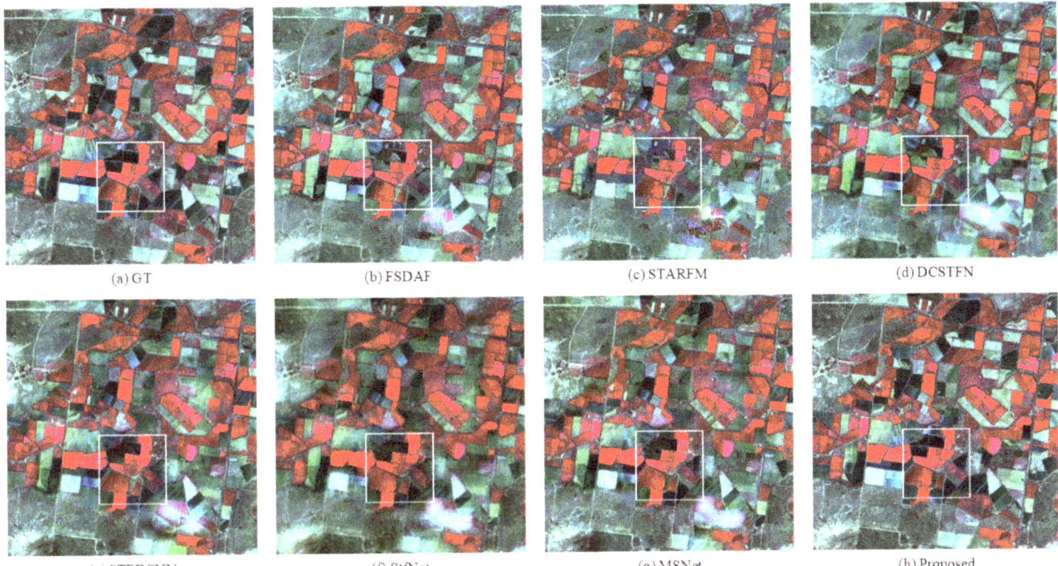

Figure 8. Specific prediction results for the target Landsat image (16 October 2001) in CIA dataset. Among them, the white framework is the prominent difference of the results obtained by each method. Comparison methods include FSDAF [11], STARFM [8], DCSTFN [23], STFDCNN [18], StfNet [19], and MSNet [36], which are represented by (**b–g**) in the figure respectively. (**a**) represents the ground truth (GT), and (**h**) represents the proposed method.

Figure 9. Comprehensive prediction results for the target Landsat image (14 February 2005) in LGC dataset. Comparison methods include FSDAF [11], STARFM [8], DCSTFN [23], STFDCNN [18], StfNet [19], and MSNet [36], which are represented by (**b–g**) in the figure respectively. (**a**) represents the ground truth (GT), and (**h**) represents the proposed method.

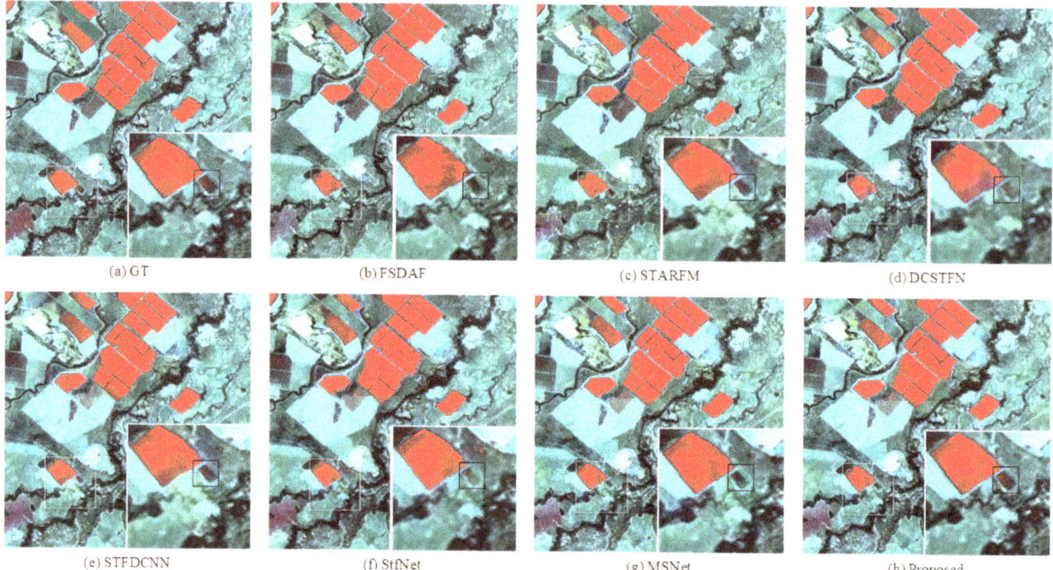

Figure 10. Specific prediction results for the target Landsat image (14 February 2005) in LGC dataset. Among them, the grey framework is the prominent difference of the results obtained by each method. Comparison methods include FSDAF [11], STARFM [8], DCSTFN [23], STFDCNN [18], StfNet [19], and MSNet [36], which are represented by (**b–g**) in the figure respectively. (**a**) represents the ground truth (GT), and (**h**) represents the proposed method.

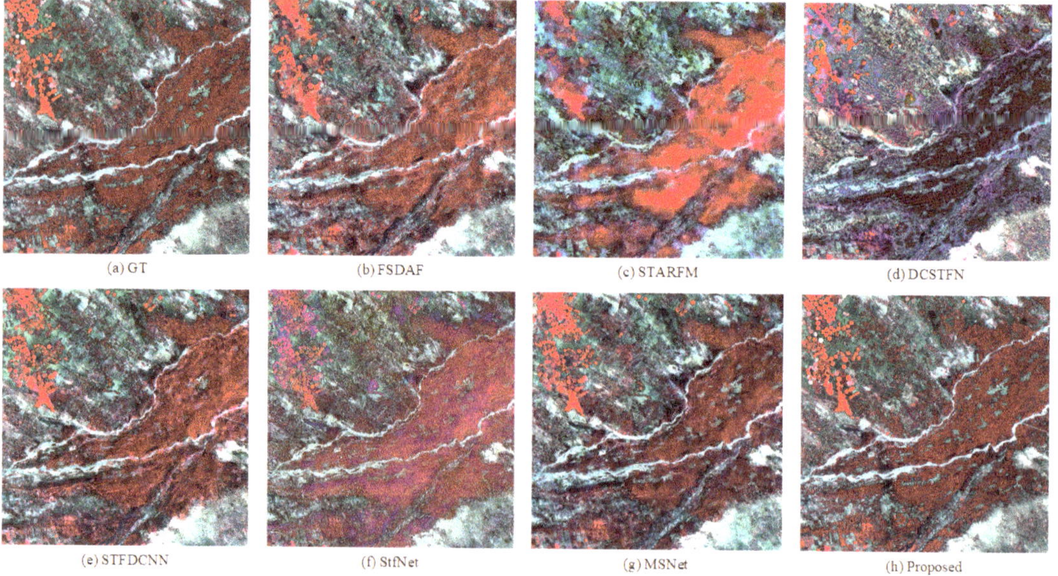

Figure 11. Complete prediction results for the target Landsat image (7 July 2015) in AHB dataset. Comparison methods include FSDAF [11], STARFM [8], DCSTFN [23], STFDCNN [18], StfNet [19], and MSNet [36], which are represented by (**b–g**) in the figure respectively. (**a**) represents the ground truth (GT), and (**h**) represents the proposed method.

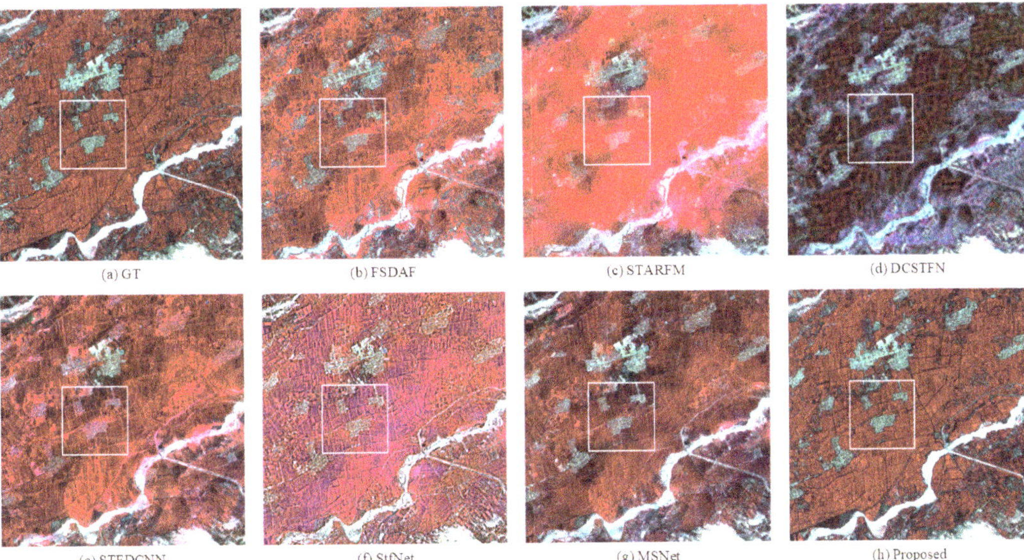

Figure 12. First specific prediction results for the target Landsat image (7 July 2015) in AHB dataset. Among them, the white framework is the prominent difference of the results obtained by each method. Comparison methods include FSDAF [11], STARFM [8], DCSTFN [23], STFDCNN [18], StfNet [19], and MSNet [36], which are represented by (**b–g**) in the figure respectively. (**a**) represents the ground truth (GT), and (**h**) represents the proposed method.

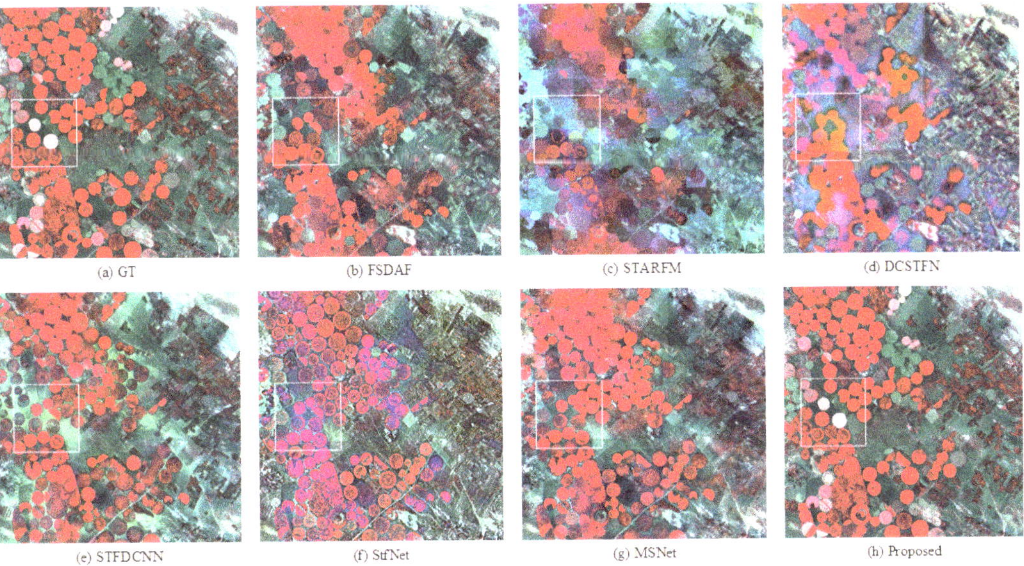

Figure 13. Second specific prediction results for the target Landsat image (7 July 2015) in AHB dataset. Among them, the white framework is the prominent difference of the results obtained by each method. Comparison methods include FSDAF [11], STARFM [8], DCSTFN [23], STFDCNN [18], StfNet [19], and MSNet [36], which are represented by (**b–g**) in the figure respectively. (**a**) represents the ground truth (GT), and (**h**) represents the proposed method.

Figure 7 shows the overall prediction result on the CIA data set, while Figure 8 shows a cropped part of the prediction result enlarged. Visually, FSDAF, STARFM, and DCSTFN are less accurate than other methods in predicting phenological changes. For example, in the overall results in Figure 7, the black areas of these methods are noticeably less than those contained in GT. The prediction effects in the box in Figure 8 are also quite different. Relatively speaking, the prediction results obtained by the method based on deep learning are better, but the prediction map of StfNet is a bit blurry and the effect is not good. The results of STFDCNN and MSNet are relatively good, but those of our proposed method are better. Thus, Figure 8 shows that the results obtained by the proposed method are closer to the ground truth in terms of clarity and accuracy.

Figure 9 illustrates the overall prediction result on the LGC dataset, while Figure 10 illustrates the cropped and enlarged result of a portion of the prediction. In general, the performance of each algorithm is relatively stable, but there are differences in the specific spectral information and the processing of heterogeneous regions. It can be seen from the black box in the enlarged area in the lower right corner of Figure 10 that the prediction accuracy of the spectral information in DCSTFN and StfNet is lower than other methods, and the other methods have achieved good results, but the effect obtained by the proposed method is closer to the actual value. In addition, the proposed method also predicts the information of curved river channels with high heterogeneity under the black box, which no other method except for FSDAF can. Compared with the proposed method, FSDAF is closer to the real value. The method has achieved good results in spectral information and the processing of heterogeneous regions.

Figure 11 shows the overall prediction result on the AHB data set, while Figures 12 and 13 show some cropped and enlarged results. On the whole, the prediction results of STARFM are not accurate enough in the processing of spectral information, and there is considerable ambiguous spectral information. DCSTFN fails to accurately predict the results, and fails to effectively extract information for datasets with a large number of heterogeneous regions and time information. The results obtained by StfNet are relatively good, such as in the spatial details between rivers, but there is still a large gap between the overall and the real value. In addition, although the prediction results of FSDAF are much better than STARFM in the processing of spectral information, there are still shortcomings compared with the real values. While STFDCNN and MSNet achieve better results, the spatial details and spectral time information are relatively adequate, but the proposed method achieves better results, with the spatial details and spectral information being closer to the real values. Locally, in Figure 12, in a large number of continuous phenological change areas, the proposed method has a noticeable improvement compared with the previous MSNet. Furthermore, compared with other methods, the processing of boundary information is also better, and is closest to the true value. In Figure 13, for the prediction of a large number of circular pasture areas, FSDAF, STARFM, DCSTFN, and StfNet failed at accurate prediction, which must be due to the complex spatial distribution and too much time-varying information on the AHB dataset, which led to the limited learning ability of the model, and the results obtained were not ideal. STFDCNN has achieved good results with the previous MSNet, but there is still insufficient boundary information. The proposed method thus achieves the best prediction effect, in the prediction of phenological change information as well as the boundary processing between circular pastures.

3.3.2. Objective Evaluation

Six evaluation indicators are used to objectively evaluate various algorithms and the proposed method. Tables 1–3 present the quantitative evaluation of the prediction results obtained by various methods on three datasets, including global indicators SAM and ERGAS as well as local indicators RMSE, SSIM, PSNR, and CC. Furthermore, the optimal value of each indicator is marked in bold.

Table 1. Quantitative assessment of various spatiotemporal fusion methods for CIA dataset.

Evaluation	Band	Method on CIA						
		FSDAF	DCSTFN	STARFM	STFDCNN	StfNet	MSNet	Proposed
SAM	all	0.23875	0.21556	0.23556	0.21402	0.21614	0.19209	**0.00114**
ERGAS	all	3.35044	3.07221	3.31676	3.14461	3.00404	2.94471	**0.45234**
RMSE	band1	0.01365	0.01059	0.01306	0.01076	0.00956	0.01009	**0.00051**
	band2	0.01415	0.01256	0.01366	0.01236	0.01271	0.01132	**0.00044**
	band3	0.02075	0.01922	0.02055	0.01792	0.02121	0.01724	**0.00032**
	band4	0.04619	0.04377	0.04899	0.04100	0.05001	0.03669	**0.00079**
	band5	0.06031	0.05655	0.06153	0.05900	0.05302	0.04898	**0.00026**
	band6	0.05322	0.04690	0.05278	0.05389	0.04500	0.04325	**0.00067**
	avg	0.03471	0.03160	0.03509	0.03249	0.03192	0.02793	**0.00050**
SSIM	band1	0.90147	0.94678	0.91699	0.95517	0.94190	0.95050	**0.99996**
	band2	0.91899	0.93652	0.92325	0.93812	0.94340	0.95149	**0.99998**
	band3	0.85786	0.88428	0.86290	0.87329	0.89950	0.91156	**0.99999**
	band4	0.76070	0.79776	0.74636	0.78318	0.84868	0.86248	**0.99995**
	band5	0.66598	0.70744	0.66011	0.72789	0.74118	0.76460	**0.99999**
	band6	0.66168	0.72121	0.66323	0.73555	0.74068	0.76257	**0.99997**
	avg	0.79445	0.83233	0.79548	0.83553	0.85256	0.86720	**0.99997**
PSNR	band1	37.29537	39.50404	37.68327	39.36680	40.38939	39.92510	**65.81463**
	band2	36.98507	38.01703	37.29114	38.16128	37.91972	38.92643	**67.09016**
	band3	33.65821	34.32276	33.74247	34.93560	33.46842	35.27141	**69.83863**
	band4	26.70854	27.17708	26.19858	27.74355	26.01829	28.70879	**62.06650**
	band5	24.39249	24.95152	24.21822	24.58366	25.51175	26.19920	**71.78578**
	band6	25.47784	26.57641	25.55050	25.37055	26.93525	27.28095	**63.47700**
	avg	30.75292	31.75814	30.78070	31.69357	31.70714	32.71865	**66.67879**
CC	band1	0.80138	0.79672	0.79845	0.84521	0.83428	0.84448	**0.99951**
	band2	0.79873	0.81009	0.79319	0.83720	0.83156	0.84929	**0.99978**
	band3	0.83290	0.84688	0.82554	0.87373	0.87264	0.87787	**0.99996**
	band4	0.88511	0.89683	0.86697	0.91181	0.90546	0.92743	**0.99997**
	band5	0.76395	0.79363	0.74894	0.78783	0.84732	0.84784	**0.99999**
	band6	0.76036	0.80739	0.75144	0.76502	0.84588	0.83826	**0.99996**
	avg	0.80707	0.82526	0.79742	0.83680	0.85619	0.86420	**0.99986**

Table 2. Quantitative assessment of various spatiotemporal fusion methods for LGC dataset.

Evaluation	Band	Method on LGC						
		FSDAF	DCSTFN	STARFM	STFDCNN	StfNet	MSNet	Proposed
SAM	all	0.08411	0.08354	0.08601	0.06792	0.09284	0.06335	**0.00035**
ERGAS	all	1.93861	1.91167	1.92273	1.80392	2.03970	1.68639	**0.13248**
RMSE	band1	0.00763	0.00763	0.00729	0.00719	0.00824	0.00585	**0.00006**
	band2	0.00913	0.00870	0.00907	0.00843	0.01167	0.00712	**0.00006**
	band3	0.01279	0.01258	0.01256	0.01151	0.01353	0.00969	**0.00006**
	band4	0.02383	0.02332	0.02295	0.02102	0.02971	0.01864	**0.00006**
	band5	0.02830	0.02679	0.02607	0.02251	0.02284	0.02159	**0.00006**
	band6	0.02197	0.02072	0.02181	0.01673	0.02054	0.01425	**0.00006**
	avg	0.01727	0.01662	0.01662	0.01457	0.01775	0.01286	**0.00006**
SSIM	band1	0.97422	0.97455	0.97355	0.98460	0.97464	0.98558	**0.99999**
	band2	0.96698	0.96918	0.96495	0.98209	0.96062	0.98031	**0.99999**
	band3	0.94456	0.94632	0.94152	0.97475	0.94162	0.96954	**0.99999**
	band4	0.92411	0.93283	0.91759	0.96417	0.91455	0.96393	**0.99999**
	band5	0.89418	0.90416	0.88558	0.95539	0.91215	0.95239	**0.99999**
	band6	0.88485	0.90337	0.87789	0.95259	0.90154	0.95087	**0.99999**
	avg	0.93148	0.93840	0.92684	0.96893	0.93419	0.96710	**0.99999**

Table 2. *Cont.*

Evaluation	Band	Method on LGC						
		FSDAF	**DCSTFN**	**STARFM**	**STFDCNN**	**StfNet**	**MSNet**	**Proposed**
PSNR	band1	42.35483	42.34734	42.73997	42.86245	41.68016	44.65345	**84.25491**
	band2	40.79034	41.20550	40.85222	41.48586	38.65611	42.95050	**85.06363**
	band3	37.86428	38.00486	38.02099	38.77733	37.37629	40.27059	**83.84853**
	band4	32.45760	32.64622	32.78532	33.54859	30.54336	34.59058	**84.47622**
	band5	30.96416	31.44212	31.67671	32.95179	32.82613	33.31671	**84.01371**
	band6	33.16535	33.67016	33.22812	35.52920	33.74927	36.92082	**84.00084**
	avg	36.26610	36.55270	36.55056	37.52587	35.80522	38.78378	**84.27631**
CC	band1	0.93627	0.92666	0.92935	0.94611	0.94664	0.96138	**0.99999**
	band2	0.93186	0.93379	0.92880	0.94530	0.93566	0.95800	**0.99999**
	band3	0.93549	0.93512	0.93516	0.95262	0.95539	0.96499	**0.99999**
	band4	0.96360	0.96585	0.96287	0.97181	0.96125	0.97591	**0.99999**
	band5	0.95527	0.95492	0.95222	0.97545	0.97048	0.97890	**0.99999**
	band6	0.95313	0.95738	0.95214	0.97285	0.97164	0.97924	**0.99999**
	avg	0.94594	0.94562	0.94342	0.96069	0.95684	0.96974	**0.99999**

Table 3. Quantitative assessment of various spatiotemporal fusion methods for AHB dataset.

Evaluation	Band	Method on AHB						
		FSDAF	**DCSTFN**	**STARFM**	**STFDCNN**	**StfNet**	**MSNet**	**Proposed**
SAM	all	0.16991	0.23877	0.29277	0.18583	0.25117	0.14677	**0.01297**
ERGAS	all	2.80156	4.03380	4.46147	4.25224	3.86535	2.90661	**0.81967**
RMSE	band1	0.00039	0.00081	0.00251	0.00096	0.00112	0.00047	**0.00007**
	band2	0.00044	0.00215	0.00235	0.00092	0.00081	0.00051	**0.00007**
	band3	0.00067	0.00363	0.00358	0.00117	0.00118	0.00064	**0.00007**
	band4	0.00109	0.00187	0.00590	0.00124	0.00201	0.00103	**0.00006**
	band5	0.00126	0.00208	0.00408	0.00183	0.00177	0.00122	**0.00006**
	band6	0.00136	0.00225	0.00263	0.00200	0.00198	0.00126	**0.00007**
	avg	0.00087	0.00213	0.00351	0.00135	0.00148	0.00085	**0.00006**
SSIM	band1	0.99895	0.99459	0.96538	0.99205	0.98927	0.99822	**0.99998**
	band2	0.99877	0.96845	0.96977	0.99293	0.99500	0.99805	**0.99998**
	band3	0.99741	0.91914	0.93438	0.98947	0.98965	0.99740	**0.99998**
	band4	0.99616	0.98506	0.92038	0.99419	0.98248	0.99631	**0.99999**
	band5	0.99382	0.98085	0.94190	0.98371	0.98464	0.99388	**0.99999**
	band6	0.99129	0.97145	0.96825	0.97625	0.97636	0.99226	**0.99998**
	avg	0.99607	0.96992	0.95001	0.98810	0.98623	0.99602	**0.99998**
PSNR	band1	68.18177	61.87013	52.01008	60.34502	59.00582	66.48249	**83.62824**
	band2	67.04371	53.35105	52.56484	60.68929	61.8316	65.80339	**83.61930**
	band3	63.49068	48.79810	48.93197	58.63694	58.55977	63.88021	**83.56309**
	band4	59.22553	54.57435	44.58211	58.13169	53.95486	59.77506	**84.23956**
	band5	58.02282	53.65469	47.79106	54.74701	55.05539	58.28599	**83.87554**
	band6	57.35352	52.93719	51.60634	53.96601	54.06602	58.02322	**83.56037**
	avg	62.21967	54.19759	49.58107	57.75266	57.07891	62.04173	**83.74768**
CC	band1	0.84000	0.78227	0.71181	0.80368	0.49726	0.86845	**0.99570**
	band2	0.85657	0.76351	0.74545	0.86845	0.38062	0.89114	**0.99795**
	band3	0.84979	0.79147	0.81230	0.83576	0.27147	0.88345	**0.99918**
	band4	0.53986	0.40161	0.34009	0.58944	0.37556	0.60303	**0.99893**
	band5	0.79576	0.52206	0.76553	0.83580	0.62926	0.85320	**0.99972**
	band6	0.80288	0.47565	0.76492	0.80338	0.61085	0.85154	**0.99975**
	avg	0.78081	0.62276	0.69002	0.78942	0.46083	0.82514	**0.99854**

Tables 1–3 present the quantitative evaluation results of several existing fusion methods and the proposed method on the CIA, LGC, and AHB datasets, respectively. In each

table, it can be seen that the proposed method achieves the optimal value on the global indicators and all local indicators.

4. Discussion

Through the experiments, it can be seen that whether it is on the CIA dataset with phenological changes in regular areas or on the AHB dataset with phenological changes with a large number of irregular areas and a large number of heterogeneous areas, our proposed method is better at prediction. Similarly, for LGC datasets, which are mainly land cover-type changes, the proposed method is better at prediction than traditional methods and other deep learning-based methods in the processing of temporal information and high-frequency spatial details. The time information and high-frequency file texture information are processed more appropriately because of the combination of ITE and dilated convolution in EMSNet. More importantly, the refined ITE can further expand the range of learning in the remote sensing field, and can fully extract the spatiotemporal information contained in the input image.

It is worth noting that for datasets with different amounts of data and different characteristics, the depth of the improved transformer encoder (ITE) should also be different to better fit the datasets. Table 4 lists the average evaluation values of the prediction results obtained without the ITE and with the ITE with different depths, where the optimal value is shown in bold. The depth being 0 indicates that the ITE has not been introduced. It can be seen that when the depth is not introduced, the experimental results are relatively poor. As the depth changes, the results obtained vary. The best experimental results are obtained when the depth of the CIA dataset is 20, the depth of the LGC dataset is 5, and the depth of the AHB dataset is 20.

Table 4. Average evaluation values of ITEs of various depths on the three datasets.

Database	Depth	SAM	ERGAS	RMSE	SSIM	PSNR	CC
CIA	0	0.223768	3.144353	0.032796	0.844214	31.477961	0.819219
	5	0.001597	0.530676	0.000550	0.999948	67.018571	0.999620
	10	0.001182	0.473233	**0.000473**	0.999971	**68.362024**	0.999807
	15	0.001394	0.509978	0.000639	0.999960	64.859474	0.999776
	20	**0.001142**	**0.452341**	0.000499	**0.999974**	66.678786	**0.999863**
LGC	0	0.082166	1.939385	0.016704	0.943749	36.315476	0.948030
	5	**0.000352**	**0.132476**	**0.000061**	**0.9999982**	**84.276309**	**0.9999989**
	10	0.000367	0.139728	0.000069	0.9999979	83.319692	0.9999987
	15	0.000378	0.153687	0.000092	0.9999976	81.181723	0.999998
	20	0.000638	0.287639	0.000476	0.999885	77.511986	0.999900
AHB	0	0.082166	1.939385	0.016704	0.943749	36.315476	0.748201
	5	0.013112	0.826490	0.000066	0.999982	83.686718	0.998556
	10	0.013106	0.825792	0.000066	0.999982	83.680289	**0.998557**
	15	0.013102	0.828641	0.000066	0.999982	83.625675	0.998539
	20	**0.012967**	**0.819673**	**0.000065**	**0.999983**	**83.747684**	0.998540

The bold in the table indicates the optimal value at different ITE depths.

In addition, the difference between the original linear projection method of the transformer encoder and the improved convolution projection method was also determined. Table 5 lists the global indicators and average evaluation values of the prediction results obtained under various projection methods, where the optimal value is shown in bold. It can be seen that on the three datasets, the convolutional projection method is selected, and the ITE after position encoding is removed achieves better results.

Table 5. Average evaluation values of ICTEs of various project methods on the three datasets.

Database	Project Method	SAM	ERGAS	RMSE	SSIM	PSNR	CC
CIA	line	0.001142	0.452660	0.000500	0.999966	65.565960	0.999462
	conv	**0.001141**	**0.452341**	**0.000499**	**0.999974**	**66.678786**	**0.999863**
LGC	line	0.000352	0.133565	0.000070	0.999990	82.659593	0.999984
	conv	**0.000351**	**0.132476**	**0.000061**	**0.999998**	**84.276309**	**0.999999**
AHB	line	0.013024	0.823650	0.000066	0.999896	81.265960	0.990570
	conv	**0.012967**	**0.819673**	**0.000065**	**0.999983**	**83.747684**	**0.998540**

Furthermore, the last six layers of the network for extracting time information in Figure 3 include three layers of dilated convolution and three layers of ReLU. This paper also conducts a comparative experiment on the three layers of dilated convolution operations. Table 6 lists the different result evaluations obtained when using convolution and dilated convolution. Among them, "conv" in the difference column means to replace the above-mentioned three layers of dilated convolution with three layers of convolution; "conv_dia" means that the above-mentioned three layers of dilated convolution remain unchanged, and "conv&conv_dia" means that the abovementioned three layers of dilated convolution are replaced by a three-layer alternating operation of convolution, dilated convolution and convolution. It can be seen that when the subsequent operations of extracting time information are all dilated convolutions, the implementation effect is better.

Table 6. Average evaluation values of various convolution operations on the three datasets.

Database	Difference	SAM	ERGAS	RMSE	SSIM	PSNR	CC
CIA	conv	0.001491	0.549008	0.000593	0.999953	66.089266	0.999672
	conv_dia	**0.001142**	**0.452341**	**0.000499**	**0.999974**	**66.678786**	**0.999863**
	conv&conv_dia	0.101984	2.197340	0.015059	0.943943	37.726046	0.963180
LGC	conv	0.000365	0.136848	0.000064	0.9999980	83.841604	0.9999988
	conv_dia	**0.000352**	**0.132476**	**0.000061**	**0.9999982**	**84.276309**	**0.9999989**
	conv&conv_dia	0.050764	1.532304	0.010584	0.975687	40.476035	0.980252
AHB	conv	0.012998	0.826653	0.000066	0.999982	83.658194	0.998290
	conv_dia	**0.012967**	**0.819673**	**0.000065**	**0.999983**	**83.747684**	**0.998540**
	conv&conv_dia	0.091340	2.057066	0.000496	0.998751	66.825864	0.921079

Although the proposed method has achieved good results, there are issues worthy of further exploration. First, in order to fully expand the learnable range of the ITE, the original input of a larger MODIS image has been used. Although dilated convolution is used to reduce the number of parameters, compared with MSNet, the number of parameters in this study is quite high. Table 7 presents the fusion model of deep learning and the number of parameters that the proposed method needs to learn. It can be seen that the proposed method needs the largest number of parameters, which means that compared with other methods, it requires more training time and equipment with larger memory during training. Considering the cost of learning, a way to obtain better results with a smaller model is a direction worthy of future research. Second, the refined ITE shows very good performance, but further improvements to adapt it to remote sensing spatiotemporal fusion can be researched in future. Furthermore, improving the fusion effect while avoiding the fusion strategy introduced by noise is also worthy of further study.

Table 7. Number of parameters for different deep learning methods.

Method	DCSTFN	STFDCNN	StfNet	MSNet		Proposed	
				depth = 5	521,064	depth = 5	3,673,617
Parameter	445,889	114,562	36,866	depth = 10	978,764	depth = 10	7,329,217
				depth = 20	1,894,164	depth = 20	14,640,417

5. Conclusions

In this study, the effectiveness of EMSNet in three research areas with diverse characteristics is evaluated. Its performance enhancement is found to be mainly because of the following reasons:

1. The projection method of the original transformer encoder is improved to adapt to the fusion of remote sensing space and time, which further expands the learning range of the improved transformer encoder, effectively learns the connection between the local and the global information in the remote sensing image, and uses its own attention mechanism to fully extract the spatiotemporal information in remote sensing images.
2. Dilated convolution is used to expand the receptive field to adapt to the original input of larger size, while keeping the number of learned parameters unchanged, effectively extracting time information and balancing the increase in parameters brought about by the improved transformer encoder.
3. A unique residual structure and a differentiated weight fusion method are used to supplement the lost information and reduce the introduction of noise in the fusion process.

Experiments show that on the CIA and AHB datasets with noteworthy phenological changes and the LGC dataset with mainly land cover-type changes, EMSNet is better than other models using three and five original images for fusion and gives more stable prediction results on each dataset. Although EMSNet achieves good results, there are still many areas worth further research in the future. First, the application of transformer-related structures in the field of remote sensing spatiotemporal fusion will be further studied. Second, compared with other methods, the method proposed in this paper needs to learn significantly more parameters. How to achieve better fusion effect with smaller model and lower learning cost is also a focus of future research. Third, although the three datasets used in this paper cover a variety of phenological changes and land-cover changes, there are still regional types that are not included. For example, datasets containing changes in urban areas will also be discussed in the future.

Author Contributions: Data curation, W.L.; formal analysis, W.L.; methodology, W.L. and D.C.; validation, D.C.; visualization, D.C.; writing—original draft, D.C.; writing—review and editing, D.C. and M.X. All authors have read and agreed to the published version of the manuscript.

Funding: This work was supported by the National Natural Science Foundation of China (61972060, U1713213, and 62027827), National Key Research and Development Program of China (2019YFE0110800), and Natural Science Foundation of Chongqing (cstc2020jcyj-zdxmX0025, cstc2019cxcyljrc-td0270).

Data Availability Statement: Data sharing is not applicable to this article.

Acknowledgments: The authors would like to thank all of the reviewers for their valuable contributions to our work.

Conflicts of Interest: The authors declare no conflict of interest.

References

1. Justice, C.O.; Vermote, E.; Townshend, J.R.; Defries, R.; Roy, D.P.; Hall, D.K.; Salomonson, V.V.; Privette, J.L.; Riggs, G.; Strahler, A.; et al. The Moderate Resolution Imaging Spectroradiometer (MODIS): Land remote sensing for global change research. *IEEE Trans. Geosci. Remote Sens.* **1998**, *36*, 1228–1249. [CrossRef]
2. Lin, C.; Li, Y.; Yuan, Z.; Lau, A.K.; Li, C.; Fung, J.C.H. Using satellite remote sensing data to estimate the high-resolution distribution of ground-level PM2.5. *Remote Sens. Environ.* **2015**, *156*, 117–128. [CrossRef]
3. Zhang, L.; Zhang, Q.; Du, B.; Huang, X.; Tang, Y.Y.; Tao, D.J. Simultaneous spectral-spatial feature selection and extraction for hyperspectral images. *IEEE Trans. Cybern.* **2016**, *48*, 16–28. [CrossRef] [PubMed]
4. Yu, Q.; Gong, P.; Clinton, N.; Biging, G.; Kelly, M.; Schirokauer, D.J.P.E.; Sensing, R. Object-based detailed vegetation classification with airborne high spatial resolution remote sensing imagery. *Photogramm. Eng. Remote Sens.* **2006**, *72*, 799–811. [CrossRef]
5. White, M.A.; Nemani, R.R. Real-time monitoring and short-term forecasting of land surface phenology. *Remote Sens. Environ.* **2006**, *104*, 43–49. [CrossRef]
6. Hansen, M.C.; Loveland, T.R. A review of large area monitoring of land cover change using Landsat data. *Remote Sens. Environ.* **2012**, *122*, 66–74. [CrossRef]
7. Gao, F.; Masek, J.; Schwaller, M.; Hall, F. On the blending of the Landsat and MODIS surface reflectance: Predicting daily Landsat surface reflectance. *IEEE Trans. Geosci. Remote Sens.* **2006**, *44*, 2207–2218. [CrossRef]
8. Hilker, T.; Wulder, M.A.; Coops, N.C.; Seitz, N.; White, J.C.; Gao, F.; Masek, J.G.; Stenhouse, G. Generation of dense time series synthetic Landsat data through data blending with MODIS using a spatial and temporal adaptive reflectance fusion model. *Remote Sens. Environ.* **2009**, *113*, 1988–1999. [CrossRef]
9. Zhukov, B.; Oertel, D.; Lanzl, F.; Reinhackel, G.; Sensing, R. Unmixing-based multisensor multiresolution image fusion. *IEEE Trans. Geosci. Remote Sens.* **1999**, *37*, 1212–1226. [CrossRef]
10. Wu, M.; Niu, Z.; Wang, C.; Wu, C.; Wang, L. Use of MODIS and Landsat time series data to generate high-resolution temporal synthetic Landsat data using a spatial and temporal reflectance fusion model. *J. Appl. Remote Sens.* **2012**, *6*, 063507. [CrossRef]
11. Zhu, X.; Helmer, E.H.; Gao, F.; Liu, D.; Chen, J.; Lefsky, M.A. A flexible spatiotemporal method for fusing satellite images with different resolutions. *Remote Sens. Environ.* **2016**, *172*, 165–177. [CrossRef]
12. Zhu, X.; Chen, J.; Gao, F.; Chen, X.; Masek, J.G. An enhanced spatial and temporal adaptive reflectance fusion model for complex heterogeneous regions. *Remote Sens. Environ.* **2010**, *114*, 2610–2623. [CrossRef]
13. Hilker, T.; Wulder, M.A.; Coops, N.C.; Linke, J.; McDermid, G.; Masek, J.G.; Gao, F.; White, J.C. A new data fusion model for high spatial-and temporal-resolution mapping of forest disturbance based on Landsat and MODIS. *Remote Sens. Environ.* **2009**, *113*, 1613–1627. [CrossRef]
14. Huang, B.; Song, H.; Sensing, R. Spatiotemporal reflectance fusion via sparse representation. *IEEE Trans. Geosci. Remote Sens.* **2012**, *50*, 3707–3716. [CrossRef]
15. Belgiu, M.; Stein, A. Spatiotemporal image fusion in remote sensing. *Remote Sens.* **2019**, *11*, 818. [CrossRef]
16. Wei, J.; Wang, L.; Liu, P.; Chen, X.; Li, W.; Zomaya, A.Y.; Sensing, R. Spatiotemporal fusion of MODIS and Landsat-7 reflectance images via compressed sensing. *Geosci. Remote Sens.* **2017**, *55*, 7126–7139. [CrossRef]
17. Liu, X.; Deng, C.; Wang, S.; Huang, G.-B.; Zhao, B.; Lauren, P.; Letters, R.S. Fast and accurate spatiotemporal fusion based upon extreme learning machine. *IEEE Geosci. Remote Sens. Lett.* **2016**, *13*, 2039–2043. [CrossRef]
18. Song, H.; Liu, Q.; Wang, G.; Hang, R.; Huang, B.; Sensing, R. Spatiotemporal satellite image fusion using deep convolutional neural networks. *IEEE J. Sel. Top. Appl. Earth Obs. Remote Sens.* **2018**, *11*, 821–829. [CrossRef]
19. Liu, X.; Deng, C.; Chanussot, J.; Hong, D.; Zhao, B.; Sensing, R. StfNet: A two-stream convolutional neural network for spatiotemporal image fusion. *Geosci. Remote Sens.* **2019**, *57*, 6552–6564. [CrossRef]
20. Chen, Y.; Shi, K.; Ge, Y. Spatiotemporal Remote Sensing Image Fusion Using Multiscale Two-Stream Convolutional Neural Networks. *Geosci. Remote Sens.* **2021**, *60*, 1–12. [CrossRef]
21. Jia, D.; Song, C.; Cheng, C.; Shen, S.; Ning, L.; Hui, C. A novel deep learning-based spatiotemporal fusion method for combining satellite images with different resolutions using a two-stream convolutional neural network. *Remote Sens.* **2020**, *12*, 698. [CrossRef]
22. Jia, D.; Cheng, C.; Song, C.; Shen, S.; Ning, L.; Zhang, T. A Hybrid Deep Learning-Based Spatiotemporal Fusion Method for Combining Satellite Images with Different Resolutions. *Remote Sens.* **2021**, *13*, 645. [CrossRef]
23. Tan, Z.; Yue, P.; Di, L.; Tang, J. Deriving high spatiotemporal remote sensing images using deep convolutional network. *Remote Sens.* **2018**, *10*, 1066. [CrossRef]
24. Tan, Z.; Di, L.; Zhang, M.; Guo, L.; Gao, M. An enhanced deep convolutional model for spatiotemporal image fusion. *Remote Sens.* **2019**, *11*, 2898. [CrossRef]
25. Chen, J.; Wang, L.; Feng, R.; Liu, P.; Han, W.; Chen, X.; Sensing, R. CycleGAN-STF: Spatiotemporal fusion via CycleGAN-based image generation. *Geosci. Remote Sens.* **2020**, *59*, 5851–5865. [CrossRef]
26. Yin, Z.; Wu, P.; Foody, G.M.; Wu, Y.; Liu, Z.; Du, Y.; Ling, F.; Sensing, R. Spatiotemporal fusion of land surface temperature based on a convolutional neural network. *Geosci. Remote Sens.* **2020**, *59*, 1808–1822. [CrossRef]
27. Ao, Z.; Sun, Y.; Pan, X.; Xin, Q. Deep learning-based spatiotemporal data fusion using a patch-to-pixel mapping strategy and model comparisons. *Geosci. Remote Sens.* **2022**. [CrossRef]
28. Vaswani, A.; Shazeer, N.; Parmar, N.; Uszkoreit, J.; Jones, L.; Gomez, A.N.; Kaiser, Ł.; Polosukhin, I. Attention is all you need. In Proceedings of the Advances in Neural Information Processing Systems, Long Beach, CA, USA; 2017; pp. 5998–6008.

29. Dosovitskiy, A.; Beyer, L.; Kolesnikov, A.; Weissenborn, D.; Zhai, X.; Unterthiner, T.; Dehghani, M.; Minderer, M.; Heigold, G.; Gelly, S. An image is worth 16×16 words: Transformers for image recognition at scale. In Proceedings of the ICLR 2021, Virtual Conference (Formerly Vienna), Vienna, Austria, 3–7 May 2021.

30. Touvron, H.; Cord, M.; Douze, M.; Massa, F.; Sablayrolles, A.; Jégou, H. Training data-efficient image transformers & distillation through attention. In Proceedings of the International Conference on Machine Learning, Virtual. 18–24 July 2021; pp. 10347–10357.

31. Chu, X.; Zhang, B.; Tian, Z.; Wei, X.; Xia, H. Do we really need explicit position encodings for vision transformers. *arXiv* **2021**, arXiv:2102.10882.

32. Han, K.; Xiao, A.; Wu, E.; Guo, J.; Xu, C.; Wang, Y. Transformer in transformer. *Adv. Neural Inf. Process. Syst.* **2021**, *34*, 15908–15919.

33. Wu, H.; Xiao, B.; Codella, N.; Liu, M.; Dai, X.; Yuan, L.; Zhang, L. Cvt: Introducing convolutions to vision transformers. In Proceedings of the IEEE/CVF International Conference on Computer Vision, Montreal, QC, Canada, 11–17 October 2021; pp. 22–31.

34. Liu, Z.; Lin, Y.; Cao, Y.; Hu, H.; Wei, Y.; Zhang, Z.; Lin, S.; Guo, B. Swin transformer: Hierarchical vision transformer using shifted windows. In Proceedings of the IEEE/CVF International Conference on Computer Vision, Montreal, QC, Canada, 11–17 October 2021; pp. 10012–10022.

35. Yang, F.; Yang, H.; Fu, J.; Lu, H.; Guo, B. Learning texture transformer network for image super-resolution. In Proceedings of the IEEE/CVF Conference on Computer Vision and Pattern Recognition, Nashville, TN, USA, 20–25 June 2021; pp. 5791–5800.

36. Li, W.; Cao, D.; Peng, Y.; Yang, C. MSNet: A multi-stream fusion network for remote sensing spatiotemporal fusion based on transformer and convolution. *Remote Sens.* **2021**, *13*, 3724. [CrossRef]

37. Chen, G.; Jiao, P.; Hu, Q.; Xiao, L.; Ye, Z. SwinSTFM: Remote Sensing Spatiotemporal Fusion Using Swin Transformer. *Geosci. Remote Sens.* **2022**, *60*, 1–18.

38. Yang, G.; Qian, Y.; Liu, H.; Tang, B.; Qi, R.; Lu, Y.; Geng, J. MSFusion: Multistage for Remote Sensing Image Spatiotemporal Fusion Based on Texture Transformer and Convolutional Neural Network. *IEEE J. Sel. Top. Appl. Earth Obs. Remote Sens.* **2022**, *15*, 4653–4666. [CrossRef]

39. He, K.; Zhang, X.; Ren, S.; Sun, J. Deep residual learning for image recognition. In Proceedings of the IEEE Conference on Computer Vision and Pattern Recognition, Las Vegas, NV, USA, 27–30 June 2016; pp. 770–778.

40. Huang, G.; Liu, Z.; Van Der Maaten, L.; Weinberger, K.Q. Densely connected convolutional networks. In Proceedings of the IEEE Conference on Computer Vision and Pattern Recognition, Honolulu, HI, USA, 21–26 June 2017; pp. 4700–4708.

41. Yu, F.; Koltun, V. Multi-scale context aggregation by dilated convolutions. *arXiv* **2015**, arXiv:1511.07122.

42. Glorot, X.; Bordes, A.; Bengio, Y. Deep sparse rectifier neural networks. In Proceedings of the Fourteenth International Conference on Artificial Intelligence and Statistics, Ft. Lauderdale, FL, USA, 11–13 April 2011; pp. 315–323.

43. Huber, P.J. Robust estimation of a location parameter. In *Breakthroughs in Statistics*; Springer: Berlin/Heidelberg, Germany, 1992; pp. 492–518. [CrossRef]

44. Emelyanova, I.V.; McVicar, T.R.; Van Niel, T.G.; Li, L.T.; Van Dijk, A.I. Assessing the accuracy of blending Landsat–MODIS surface reflectances in two landscapes with contrasting spatial and temporal dynamics: A framework for algorithm selection. *Remote Sens. Environ.* **2013**, *133*, 193–209. [CrossRef]

45. Li, Y.; Li, J.; He, L.; Chen, J.; Plaza, A. A new sensor bias-driven spatio-temporal fusion model based on convolutional neural networks. *Sci. China Inf. Sci.* **2020**, *63*, 140302. [CrossRef]

46. Li, J.; Li, Y.; He, L.; Chen, J.; Plaza, A. Spatio-temporal fusion for remote sensing data: An overview and new benchmark. *Sci. China Inf. Sci.* **2020**, *63*, 140301. [CrossRef]

47. Yuhas, R.H.; Goetz, A.F.; Boardman, J.W. Discrimination among semi-arid landscape endmembers using the spectral angle mapper (SAM) algorithm. In Proceedings of the Summaries 3rd Annual JPL Airborne Geoscience Workshop, Pasadena, CA, USA, 1–5 June 1992; pp. 147–149.

48. Khan, M.M.; Alparone, L.; Chanussot, J. Pansharpening quality assessment using the modulation transfer functions of instruments. *Geosci. Remote Sens.* **2009**, *47*, 3880–3891. [CrossRef]

49. Wang, Z.; Bovik, A.C.; Sheikh, H.R.; Simoncelli, E.P. Image quality assessment: From error visibility to structural similarity. *IEEE Trans. Image Process.* **2004**, *13*, 600–612. [CrossRef]

50. Ponomarenko, N.; Ieremeiev, O.; Lukin, V.; Egiazarian, K.; Carli, M. Modified image visual quality metrics for contrast change and mean shift accounting. In Proceedings of the 2011 11th International Conference the Experience of Designing and Application of CAD Systems in Microelectronics (CADSM), Polyana, Ukraine, 23–25 February 2011; pp. 305–311.

Article

Mode Recognition of Orbital Angular Momentum Based on Attention Pyramid Convolutional Neural Network

Tan Qu [1,*], Zhiming Zhao [1], Yan Zhang [2], Jiaji Wu [1] and Zhensen Wu [3]

1   School of Electronic Engineering, Xidian University, Xi'an 710071, China
2   School of Computer Science, Xi'an Shiyou University, Xi'an 710065, China
3   School of Physics, Xidian University, Xi'an 710071, China
*   Correspondence: tqu@xidian.edu.cn

Abstract: In an effort to address the problem of the insufficient accuracy of existing orbital angular momentum (OAM) detection systems for vortex optical communication, an OAM mode detection technology based on an attention pyramid convolution neural network (AP-CNN) is proposed. By introducing fine-grained image classification, the low-level detailed features of the similar light intensity distribution of vortex beam superposition and plane wave interferograms are fully utilized. Using ResNet18 as the backbone of AP-CNN, a dual path structure with an attention pyramid is adopted to detect subtle differences in the light intensity in images. Under different turbulence intensities and transmission distances, the detection accuracy and system bit error rate of basic CNN with three convolution layers and two full connection layers, i.e., ResNet18 and ResNet18, with a specified mapping relationship and AP-CNN, are numerically analyzed. Compared to ResNet18, AP-CNN achieves up to a 7% improvement of accuracy and a 3% reduction of incorrect mode identification in the confusion matrix of superimposed vortex modes. The accuracy of single OAM mode detection based on AP-CNN can be effectively improved by 5.5% compared with ResNet18 at a transmission distance of 2 km in strong atmospheric turbulence. The proposed OAM detection scheme may find important applications in optical communications and remote sensing.

Keywords: orbital angular momentum; mode detection; fine-grained image classification; attention pyramid; atmospheric turbulence

Citation: Qu, T.; Zhao, Z.; Zhang, Y.; Wu, J.; Wu, Z. Mode Recognition of Orbital Angular Momentum Based on Attention Pyramid Convolutional Neural Network. *Remote Sens.* **2022**, *14*, 4618. https://doi.org/10.3390/rs14184618

Academic Editor: Carmine Serio

Received: 27 July 2022
Accepted: 13 September 2022
Published: 15 September 2022

Publisher's Note: MDPI stays neutral with regard to jurisdictional claims in published maps and institutional affiliations.

1. Introduction

Vortex beams with spiral phase structures have been used extensively in information transmission, radar imaging and rotational target detection, since Allen first investigated the Laguerre-Gaussian vortex beam and its orbital angular momentum (OAM) in 1992 [1]. Theoretically, there are infinite kinds of eigenstates; different eigenstates are orthogonal to each other, which is quite important in terms of improving communication capacity and imaging resolution in remote sensing. Multiplexing different OAM beams can effectively avoid the crosstalk between different modes in a channel, providing a new communication dimension that is no longer limited to amplitude, phase, frequency and polarization, thereby greatly improving the communication capacity [2,3]. In free-space OAM communication systems, the receiver needs to demodulate the OAM beam to recover the information sequence. Traditional OAM demodulation techniques, such as spatial light modulators, the diffraction method, the cylindrical lens method, plane wave interferometry and spherical wave interferometry, are based on optical hardware and have been researched extensively [4–7]. The OAM beam is pre-processed by optical hardware to obtain optical pattern features that can be distinguished by the naked eye [8]. However, on account of the high cost and the limited processing capability of optical hardware, high-performance transmission cannot be guaranteed with a cost-effective vortex optical communication system.

In recent years, with the rapid increase in computing power, OAM mode recognition based on deep learning has attracted growing attention. Some researchers have studied

OAM mode recognition based on neural networks. Krenn et al. proposed a self-organizing competitive neural network (SOM) based OAM mode recognition which verified the feasibility of machine learning in vortex optical communication systems for the first time [9]. They built a long-distance vortex optical communication system on the sea between the Canary Islands, achieving a recognition accuracy of 91.67%, which verified the possibility of the use of vortex beams for long-distance information transmission [10]. Deep neural network (DNN)-based recognition of different OAM modes is proposed in [11]. Convolutional neural networks (CNNs) are proposed in OAM mode recognition by Doster et al. [12]; that method achieved a mode recognition accuracy of up to 99%, which is far superior to the traditional methods. In addition, the CNN-based mode identification method is robust in terms of the influence of turbulence intensity, data size, sensor noise and pixels. Such research has paved the way for the application of CNN in OAM mode detection. Subsequently, many achievements in OAM mode recognition have been realized based on CNNs [13,14].

In 2017, Zhang et al. compared the performance of a K nearest neighbor neural network, a plain Bayesian classifier, a Back Projection artificial neural network and a CNN as OAM mode classifiers under different turbulence conditions. They observed that CNN yielded the best results [15]. The same authors improved the original LeNet-5 network and proposed a decoder scheme that could simultaneously implement OAM mode and turbulence intensity recognition [16]. OAM mode recognition technology combined with turbo channel coding is proposed in [17]; this approach effectively improved the recognition accuracy and reliability of communication transmission.

Similarly, many scholars have worked on niche applications of OAM mode detection. In 2018, Zhao et al. applied a CNN to learn a received OAM light intensity map under different tilt angles by adding a view-pooling layer. They also used a hybrid data collection technique to improve the performance [18]. Misaligned hyperfine OAM mode recognition was carried out in [19]. Machine learning based the recognition of fractional optical vortex modes in atmospheric environment was studied by Cao et al. [20]. When the marked data sample was insufficient, an OAM mode recognition method based on Conditional Generative Adversarial Networks (CGAN) was proposed to improve the recognition accuracy [21]. A Diffractive Deep Neural Network (D2NN) was utilized in OAM mode recognition in [22], eliminating the need for a CCD camera to capture images and pass them to a computer, making the communication rate independent of the hardware and neural network computation rate.

The above research was dedicated to training and identifying the OAM light distributions captured by CCD cameras; however, some other researchers have performed transformations on the vortex beam before training to highlight the characteristics of different modes. In 2018, Radon transform was introduced to preprocess a light intensity distribution map of an OAM beam to obtain more clearly distinguishing features [23]. A mode recognition technique based on coherent light interference at the receiver side to obtain more obvious recognition features was reported in [24]. A SVM-based single-mode recognition method was proposed in [25], using the relationship between the amount of OAM beam receiving the effect of atmospheric turbulence distortion and the topological charge number as an artificial feature of the design. A joint scheme combining the Gerchberg–Saxton (GS) algorithm and CNN (GS-CNN) to achieve the efficient recognition of the multiplexing LG beams was proposed in [26]. A technique to measure the OAM of light based on the petal interference patterns of modulated vortex beams and an unmodulated incident Gaussian beam reflected by a spatial light modulator was reported in [27].

In light of the aforementioned studies, it may be stated that most research has focused on OAM mode detection by neural networks, CNNs or CNN-based combination methods, preprocessing transform before network training, and OAM mode detection in misaligned or tilt angles special cases. However, in practical applications, there are many different multi-modes superpositions of OAM beams corresponding to quite similar light intensity maps, such as OAM = {−2, 3, −5} and OAM = {1, −2, 3, −5,}, OAM= {4, −4} and OAM= {2, −6}, etc.

Additionally, for a single mode vortex beam with a large topological charge, the number of fringes in the interferogram is large. Because the area of the device that collects the image at the receiving end is certain, it is difficult to determine when the number of interference fringes is large, which will further affect the accuracy of OAM mode recognition.

To solve these problems, an OAM mode recognition technique based on an attention pyramid convolutional neural network (AP-CNN) is proposed in this paper. Fine-grained image classification [28] is introduced to make full use of low-level detailed information of a similar intensity in superimposed vortex beams and the dense fringes of a plane wave interfergram of a single-mode vortex beam with a large topological charge. A top-down feature path and a bottom-up attention path structure, combined with an attention pyramid, is adopted to improve the OAM mode recognition accuracy and reduce the bit error rate (BER) for indistinguishable intensity distributions.

The remainder of this paper is arranged as follows. The OAM mode recognition technical framework based on AP-CNN is described in Section 2. In Section 3, numerical results and discussions of different transmission conditions are presented to compare the recognition accuracy and bit error ratio. Section 4 is devoted to the conclusion.

2. Materials and Methods

2.1. Principle of AP-CNN

The principles of the AP-CNN [29,30] and the fine-grained image classification algorithm [28] used in this paper are shown in Figure 1. Figure 1a illustrates the dual-path algorithm structure, Figure 1b presents the attention pyramid, and Figure 1c illustrates the region of interest (ROI) pyramid. The blue border represents the feature map, and the orange border represents the channel/space attention.

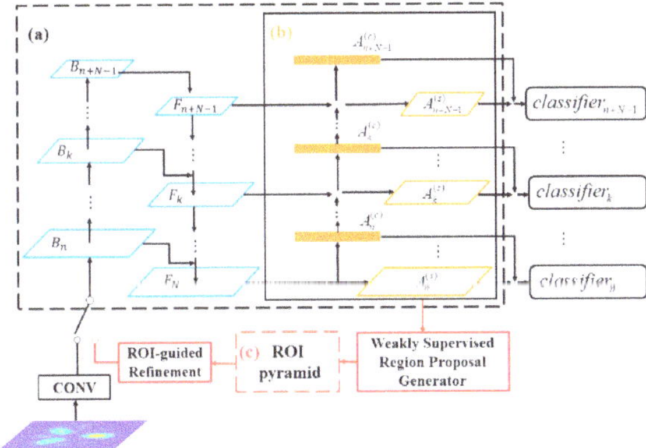

Figure 1. Structure of the AP-CNN for OAM mode recognition. (**a**) dual-path algorithm structure, (**b**) attention pyramid, (**c**) ROI pyramid.

First, the AP-CNN network takes an image as input and generates a feature pyramid network (FPN) and an attention pyramid [31] to enhance representations by improving on the CNN to obtain a dual-path algorithm structure, including a top-down feature path and a bottom-up attention path. The FPN [29] is used on the top-down path to extract features at different scales. Then, an additional attention hierarchy is introduced to further enhance the structure, including a spatial attention pyramid $\left\{ A_n^{(s)}, A_{n+1}^{(s)}, \ldots, A_{n+N-1}^{(s)} \right\}$ for locating discriminative regions at different scales, and a channel attention path $\left\{ A_n^{(c)}, A_{n+1}^{(c)}, \ldots, A_{n+N-1}^{(c)} \right\}$ for adding channel correlations in another bottom-up path and transferring local information from the lower pyramid level to the higher pyramid level.

For the spatial attention pyramid, each building block takes the feature map of the corresponding layer F_k as input and generates the spatial attention mask $A_k^{(s)}$. The feature map F_k first passes through a 3×3 deconvolution layer with only one output channel to compress the spatial information. Each element of the spatial attention mask $A_k^{(s)}$ is normalized to be in the range of 0 to 1 using the sigmoid function, expressed as:

$$A_k^{(s)} = \sigma(v_c * F_k) \tag{1}$$

where σ denotes the sigmoid function, v_c denotes the convolution kernel, and $*$ denotes the deconvolution with the fixed convolution kernel. For the channel attention path, the channel attention can be obtained by passing the global average pooling layer and two fully connected layers in the corresponding feature layer of the feature pyramid. The channel attention mask formula is given by,

$$A_k^{(c)} = \sigma(W_2 \cdot \mathrm{ReLu}(W_1 \cdot \mathrm{GAP}(F_k))) \tag{2}$$

where GAP represents the global average pooling layer, W_1 and W_2 represent the weight matrices of the two fully connected layers. The learned attention is used to weight the feature F_k to obtain F_k', which is used for classification as follows,

$$F_k' = F_k \cdot (A_k^{(s)} \oplus A_k^{(c)}) \tag{3}$$

where $\oplus$ represents the broadcasting addition operation on semantics. Spatial attention tensor and channel attention tensor have different shapes, and the plus operator must be of broadcast type.

In the second step, after obtaining the spatial attention pyramid, the ROI pyramid continues to be generated by the region suggestion generator of adaptive non maximum suppression (NMS) [32] in a weakly supervised manner. The purpose of the Region Proposal Network (RPN) [33] is to select a frame that may contain a target. In essence, it is based on the unclassified target detector of the sliding window; it inputs an image of any scale and obtains a candidate frame with a predetermined size and scale. The general RPN network is mostly applied to the single- or multi-scale convolution network feature map. Multiple sizes and aspect ratios are preset to locate objects of different sizes and shapes. On the basis of RPN theory, AP-CNN uses a spatial attention mask as an anchor score and uses weak supervision to select the distinguishing area. According to the convolution receptive field of each pyramid layer, AP-CNN selects the corresponding recommended area with a preset size and aspect ratio for each pyramid layer, applies adaptive NMS to the selected area after calculating the score, reduces redundancy by eliminating overlapping, and maintains visual integrity by combining related areas. Figure 2 shows the workflow of the weak supervision area suggestion generator in the OAM mode detection task. Compared with the soft mechanism of setting the threshold on a feature map, the adaptive region suggestion generator based on the ROI can explicitly show distinguishable regions with high response values in the light intensity distribution of OAM modes.

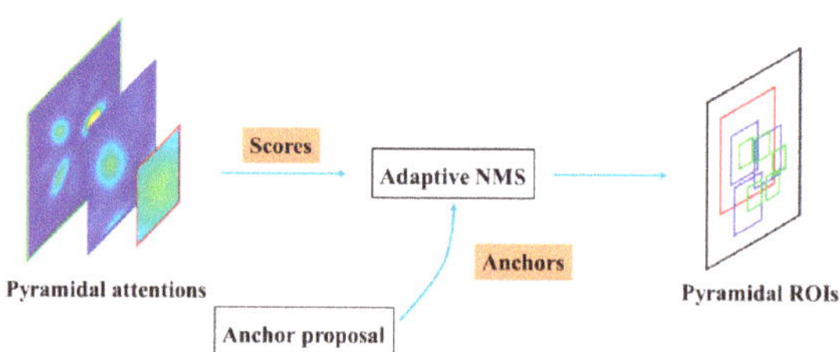

Figure 2. Workflow of the weakly supervised region proposal generator.

In the third step, for each layer of the pyramid, after selecting the ROI based on the region proposal generator and constructing the region pyramid $R_{\mathrm{all}} = \{R_n, R_{n+1}, \ldots, R_{n+N-1}\}$,

AP-CNN performs ROI-guided refinement of the feature map at the bottom of the pyramid B_n to improve the classification accuracy in the refinement stage. The first part is the ROI-guided drop block regularization [34], AP-CNN randomly selects an ROI joint R_s from the constructed N-layer region of the interesting pyramid $R_{all} = \{R_n, R_{n+1}, \ldots, R_{n+N-1}\}$ based on the drop block selection probability of each layer $P_{all} = \{P_n, P_{n+1}, \ldots, P_{n+N-1}\}$. Then the information region r_s is randomly selected with equal probability R_s and processed to the same sampling rate as the feature map at the bottom of the pyramid B_n to obtain the mask M by setting the activation of the information region to zero,

$$M(i,j) = \begin{cases} 0, & (i,j) \in r_s \\ 1, & otherwise \end{cases} \tag{4}$$

Apply the mask M on the low-level feature map B_n and normalize it to obtain the desired feature map D_n,

$$D_n = B_n * M * Count(M) / Count\_ones(M) \tag{5}$$

where $Count(\ )$ and $Count\_ones(\ )$ represent the total number of elements and the total number of elements with the value of 1, respectively. The second part is the ROI-guided amplification operation, where the AP-CNN combines all ROI regions at the pyramid level to obtain the minimum enclosing rectangle of the input image in a weakly supervised manner,

$$\begin{aligned} t_{x1} &= \min(\forall x \in R_{all}), t_{y1} = \min(\forall y \in R_{all}) \\ t_{x2} &= \max(\forall x \in R_{all}), t_{y2} = \max(\forall y \in R_{all}) \end{aligned} \tag{6}$$

where t_{x1}, t_{y1} represent the minimum coordinates of the x- and y-axes of the merged bounding box and t_{x2}, t_{y2} represent the maximum coordinates of the x- and y-axes of the merged bounding box. The calibration area is then extracted from D_n and enlarged to the same size D_n to obtain the enlarged feature map.

Separate classifiers are set up for the original and refinement stages for their respective pyramids, and the final classification results are taken as the average of the predicted values in the original stage and the predicted values in the refinement stage.

2.2. Recognizing OAM Modes Based on AP-CNN

Figure 3 displays the light intensity distributions of two similar superposition mode vortex beams of OAM= {1, −2, 3, −5} and OAM= {−2, 3, −5}. The ROI, localized from low to high level, are slow shown. This approach can be used to identify the ROI located on different pyramid levels, and more detailed information can be captured at the low levels to distinguish different OAM modes. Compared with high-level image semantic information, after thinning image features, this low-level information is very helpful to improve the accuracy of OAM mode detection.

The CCD camera at the receiving end captures the light intensity distribution of the vortex beam after atmospheric turbulence and inputs it into the AP-CNN to detect the OAM mode and retrieve the transmitted data. Here, we use the ResNet18 network as the backbone of the AP-CNN. The ResNet18 structure used in this paper differs slightly from the official ResNet18 [35] structure, as described below:

(1) The size of the input light intensity map is $128 \times 128 \times 3$. To avoid the problems of the low resolution of the final feature map after multiple downsampling and the serious loss of semantic information, the maximum pooling layer of stage0 is removed.

(2) To reduce the number of model parameters, the 7×7 convolutional kernel of stage0 is replaced by a 3×3 convolution kernel.

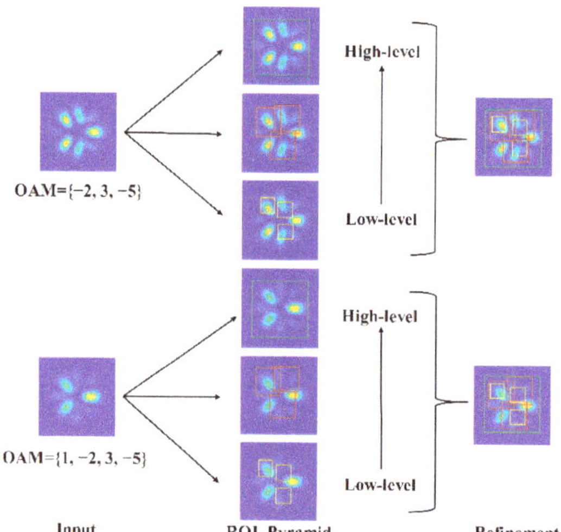

Figure 3. Regions of interest localized by AP-CNN at different levels for different OAM modes.

The structure of the modified ResNet18 is shown in Table 1.

Table 1. Modified ResNet18 network structure.

Network Layer	Output Feature Map Size	ResNet18
conv1	128×128	$3 \times 3, 64$
conv2_x	64×64	$\begin{bmatrix} 3 \times 3, 64 \\ 3 \times 3, 64 \end{bmatrix} \times 2$
conv3_x	32×32	$\begin{bmatrix} 3 \times 3, 128 \\ 3 \times 3, 128 \end{bmatrix} \times 2$
conv4_x	16×16	$\begin{bmatrix} 3 \times 3, 256 \\ 3 \times 3, 256 \end{bmatrix} \times 2$
conv5_x	8×8	$\begin{bmatrix} 3 \times 3, 512 \\ 3 \times 3, 512 \end{bmatrix} \times 2$
	1×1	Average pooling + full connected + softmax

The OAM mode recognition algorithm AP-CNN consists of two parts: the backbone network ResNet18 and the refinement network, as shown in Figure 4. The size of the simulated light intensity distribution map of the OAM beam is set to 128×128 and is input into the ResNet18 network for training. Firstly, it is input into stage0 for pre-processing, and only the features are extracted (64 convolutional kernels with a size of 3×3 and a step size of 2), and the 128×128 feature map is output. Then, the feature map is fed into the next four layers of residual blocks, which reduces the size of the input feature map by half compared to the original size and doubles the number of channels. Next, the output feature maps of the third, fourth, and fifth layers of ResNet18 are denoted as B_3, B_4, B_5, respectively, for subsequent building of the feature pyramid, as shown in Figure 4. Further refinement is carried out at the B_3 level of the pyramid. We respectively assign anchors with single scales of 18, 36, and 72 and a 1:1 ratio for each pyramidal level and choose the top 5, 3, and 1 anchors with the highest activation values as potential refinement candidates. For the adaptive NMS, the cutoff threshold is set to 0.05, the merge threshold to 0.9, and the drop block probability to {30%, 30%, 0%}.

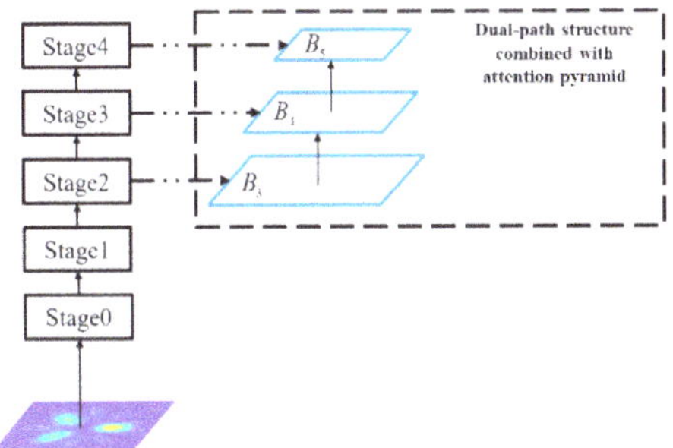

Figure 4. Connection between ResNet18 and the refinement network.

During AP-CNN training, the initial learning rate is set to 0.01, decreasing by 10% for every 20 iterations; as such, a total of 100 epochs are trained. The random gradient descent algorithm with a momentum coefficient of 0.9 and a minipatch of 16 is used for parameter optimization, and the weight attenuation is set to 5×10^{-4}. The experimental operating system is windows, the programming language of the algorithm part is python, and the deep learning framework is pytorch. The software versions are shown in Table 2. The graphics card we used was an RTX2060.

Table 2. Software used in experiments.

Software	Edition
Windows	Windows10 (21H2)
Python	3.7
Pytorch	1.7.1
Torchvision	0.8.2
CUDA	11.0.2
CuDNN	11.2

2.3. Performance Evaluation Index

The OAM mode detection performance of the network is evaluated by two indicators: detection accuracy and BER. The detection accuracy is defined as the ratio of the number of correct OAM mode detection samples to the total number of vortex light intensity distributions on the test set, determined using Equation (7):

$$\text{Accuracy} = \frac{\sum\limits_{m=1}^{M} f(m)}{M} \tag{7}$$

where M represents the total number of light intensity distribution maps of vortex light and m represents the OAM mode. $f(m)$ is 1 when the identification is correct and 0 when it is wrong.

BER and the symbol error rate (SER) are commonly used to evaluate the probability of transmission errors in communication systems. The SER is defined as the probability of a symbol transmission error, i.e., the ratio of the number of erroneous symbols at the receiving end to the total number of transmitted symbols:

$$\text{SER} = \sum_{i=1}^{M} [p_i(1 - p(s_i|s_i))] \tag{8}$$

where M represents the number of symbol types, p_i denotes the probability of transmitting each symbol with the value $1/M$, and $p(s_i|s_i)$ represents the correct conditional probability density detected by the receiver in a certain OAM mode.

$$p(s_i|s_i) = \frac{1}{\sqrt{\pi}} \int_{-\infty}^{+\infty} \exp(h_T - \eta_{s_i,s_i} GL_{path}\sqrt{\gamma_\beta}) \prod_{k=1,k\neq i}^{M} [1 - 0.5erfc(h_T - \eta_{s_j,s_i} GL_{path}\sqrt{\gamma_\beta})]dh_T \tag{9}$$

where s_i represents the OAM mode and h_T, G, L_{path} are constants, representing the power detection threshold, the average gain of the receiver, and the path loss, respectively. The value of $p(s_i|s_i)$ is theoretically mainly determined by signal-to-noise ratio γ_β at the receiving end and the helical spectral distribution.

The relationship of BER with the SER is as follows:

$$BER = SER/(\log_2 M) \tag{10}$$

In our simulation experiments, the eight OAM superposition modes are one-to-one, corresponding to the octal symbols. At the receiving end, by identifying the light intensity distribution, the code word sequence is obtained by inversion. This is then compared with the theoretical code word sequence, while the ratio of the number of wrong code words to the total number of code words is the BER.

3. Results and Discussion

3.1. Simulation Data Set Construction

In order to verify the performance of the AP-CNN in OAM mode detection for similar distributions of multi-mode vortex beams, four pairs of OAM modes, namely, {1, −2} and {1, −2, −5}, {1, −2, 3, −5} and {−2, 3, −5}, {4, −4} and {2, −6}, {6, −6} and {9, −3}, are selected, as shown in Figure 5. The light intensity distributions in four columns are similar to each other. In addition, in order to test the detection performance of a single-mode vortex beam with a large topological charge interfering with the plane wave, a plane wave interferogram dataset of a single-mode vortex beam is constructed, choosing eight types of samples with large topological charges, i.e., ±17, ±18, ±19, ±20, as shown in Figure 6.

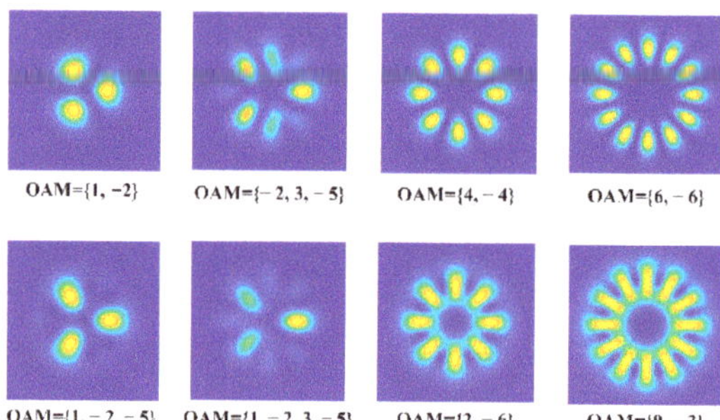

Figure 5. Light intensity distribution of a similar multi-mode OAM beam.

The wavelength of the OAM communication system is 0.6328 μm and the beam waist radius is 0.3 m. For the comparison experiments under different turbulent conditions, the transmission distance is fixed and six different atmospheric refractive index structure constants, C_n^2, are selected: $1.0 \times 10^{-14}\text{m}^{-2/3}$, $3.0 \times 10^{-14}\text{m}^{-2/3}$, $5.0 \times 10^{-14}\text{m}^{-2/3}$, $1.0 \times 10^{-13}\text{m}^{-2/3}$, $3.0 \times 10^{-13}\text{m}^{-2/3}$ and $5.0 \times 10^{-13}\text{m}^{-2/3}$. For comparison experiments at different transmission distances, the C_n^2 is fixed and six different transmission distances are chosen: 500 m, 1000 m, 1500 m, 2000 m, 2500 m, and 3000 m. When simulating the atmospheric turbulence channel, the power spectrum inversion method is used to decimate the transmission distance in order to obtain ten phase screens with certain intervals. For each transmission condition, 2000 light intensity maps are generated for each OAM mode. In this way, a total of 16,000 light intensity distribution maps are included in the hybrid dataset. This is then divided into a training and a test set at a ratio of 8:2 (12,800 images in the training set and 3200 images in the test set).

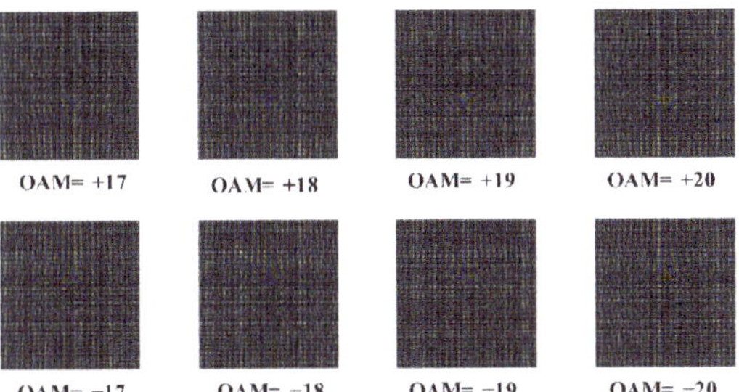

Figure 6. Plane wave interferogram of single-mode vortex beam with large topological charge.

3.2. Analysis of Multi-Mode OAM Mode Recognition Based on AP-CNN

In order to compare the performance of the AP-CNN with that of ResNet18, the OAM recognition accuracy, confusion matrix, and BER of both models are numerically analyzed in this section. Variations in OAM mode recognition accuracy are shown in Figures 7 and 8 under different turbulence intensities for transmission distances 2000 m and 3000 m, respectively.

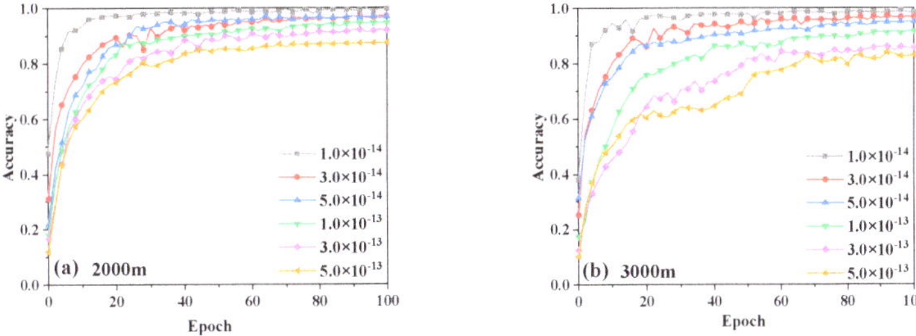

Figure 7. Accuracy of OAM mode recognition based on ResNet18 at different turbulence intensities (a) transmission distance: 2000 m (b) transmission distance: 3000 m.

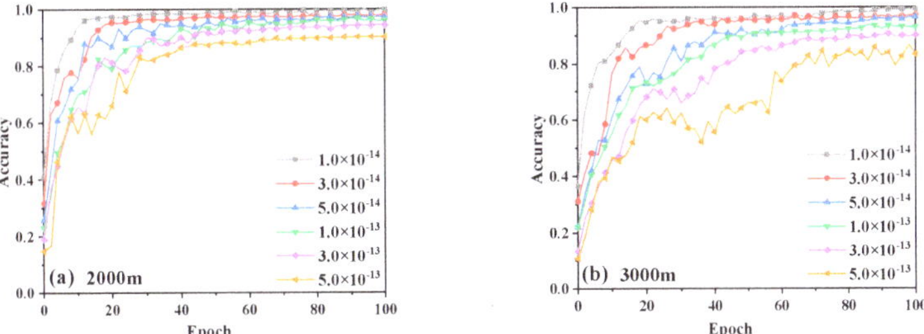

Figure 8. Accuracy of OAM mode recognition based on AP-CNN under different turbulence intensities. (**a**) transmission distance: 2000 m (**b**) transmission distance: 3000 m.

As shown, the detection accuracy increases gradually with an increase in the training epoch. Additionally, the comparisons show that the detection accuracy increases significantly more slowly in strong turbulence than in weak ones. When the turbulence is strong, the training process becomes more problematic because the image suffers more serious distortion. Taking a transmission distance of 2000 m as an example, as shown in Figure 7, after 100-times training in medium turbulence ($C_n^2 = 1.0 \times 10^{-14} \mathrm{m}^{-2/3}$, $C_n^2 = 3.0 \times 10^{-14} \mathrm{m}^{-2/3}$, $C_n^2 = 5.0 \times 10^{-14} \mathrm{m}^{-2/3}$),

ResNet18 can still achieve an OAM mode detection accuracy of more than 96.9%, whereas under strong turbulence $C_n^2 = 3.0 \times 10^{-13} \mathrm{m}^{-2/3}$ and $C_n^2 = 5.0 \times 10^{-13} \mathrm{m}^{-2/3}$), the detection accuracy decrease to 91.7% and 87.3%, respectively. The detection accuracy is further reduced from 87.3% to 83.2% when the transmission distance is increased to 3000 m under strong turbulence ($C_n^2 = 5.0 \times 10^{-13} \mathrm{m}^{-2/3}$). In other words, the greater the turbulence intensity and the farther the transmission distance, the lower the detection accuracy.

A comparison of Figures 7 and 8 reveals that the accuracy of OAM mode recognition based on AP-CNN is superior to that of ResNet18. When the turbulence is weak, the optimization effect of the AP-CNN is minimal, while when the turbulence is strong, it is more obvious. For example, when the transmission distance is 2000 m for $C_n^2 = 3.0 \times 10^{-14} \mathrm{m}^{-2/3}$, as shown in Figure 8, the accuracy of OAM mode recognition based on the AP-CNN can reach up to 98.1% after 100 training epochs, which is only a 0.6% improvement compared with ResNet18. Additionally, under these circumstances, the optimization effect is not obvious. Meanwhile, when $C_n^2 = 3.0 \times 10^{-13} \mathrm{m}^{-2/3}$, the recognition accuracy based on ResNet18 is only 92.1%. The accuracy of OAM mode recognition based on AP-CNN, on the other hand, is significantly improved, and the recognition accuracy of the best model reaches 94.4%, showing an improvement of about 2.3%.

When the transmission distance is large, the improvement of the recognition accuracy based on AP-CNN is limited in a highly turbulent environment. For example, when $C_n^2 = 5.0 \times 10^{-13} \mathrm{m}^{-2/3}$, the accuracy of OAM mode recognition based on the AP-CNN can reach up to 90.5% after 100 training epochs when the transmission distance is 2000 m, which is about a 3.1% improvement compared to ResNet18, as shown in Figure 8. However, when the transmission distance extends to 3000 m, the recognition accuracy of the best model can only reach 84.2% after AP-CNN, i.e., 1.2% higher than ResNet18. The reason for this is that in strong turbulence, when the distance is too great, the transmission of the OAM beam is greatly affected by the turbulent disturbance distortion. The light intensity distribution captured by the CCD camera at the receiving end is seriously distorted, and the image features used for recognition are compromised. Therefore, even if the AP-CNN introduces an attention mechanism to mine the underlying image details, it cannot significantly improve the accuracy of OAM mode recognition.

The confusion matrixes of superimposed vortex beams using ResNet18 and AP-CNN, with a transmission distance is 2000 m and when atmospheric refractive index structure constant $C_n^2 = 5.0 \times 10^{-13} \mathrm{m}^{-2/3}$, are given in Figure 9. The results show that {1, −2, 3, −5} has a 16% probability of being incorrectly identified as {−2, 0, −5}, {1, −2} has a 10% probability of being incorrectly identified as {1, −2, −5}, and {2, −6} has an 11% probability of being incorrectly identified as {4, −4}. In contrast, in the APP-CNN network, the accuracy of OAM mode detection improves by up to 7%, and the related incorrect identification rate is reduced by up to 3%, which confirms the necessity of designing similar OAM superposition mode datasets. The accuracy improvement and decrease of incorrect identifications may vary with the transmission conditions.

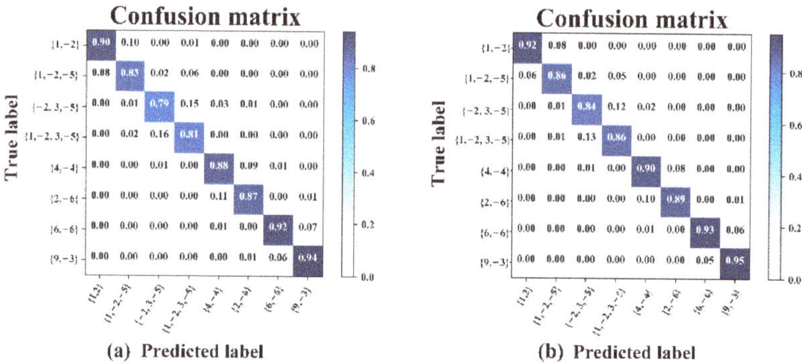

Figure 9. Confusion matrix (**a**) by ResNet18 (**b**) by AP-CNN.

Assuming the signal-to-noise ratio at the receiver side of the OAM communication system $\gamma_B = 10\text{dB}$, the system BER can be calculated based on the recognition accuracy. When the transmission distance is 2000 m, the demodulation performance of the CNN demodulator, ResNet18 demodulator, ResNet18 demodulator with a specified mapping relationship, and the AP-CNN demodulator under six different turbulence intensities C_n^2 and transmission distances, expressed as BER at the receiver end, are shown in Figure 10.

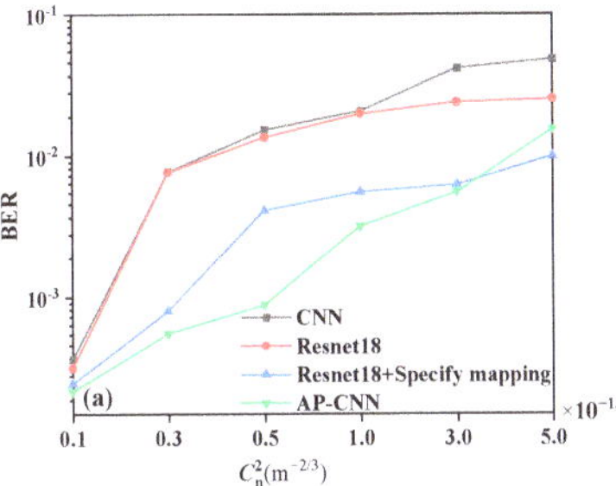
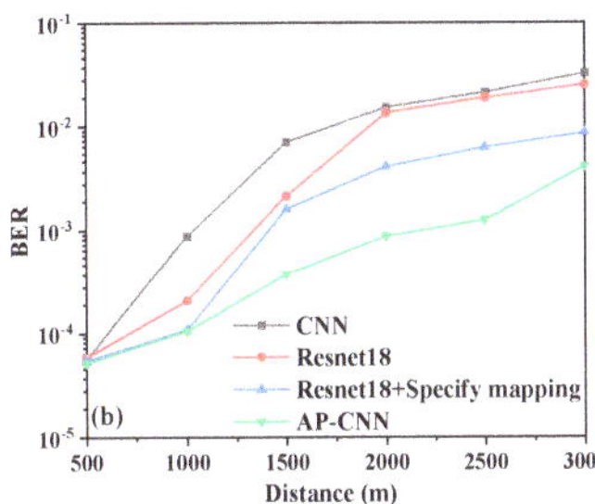

Figure 10. Performance comparison of four OAM demodulators with (**a**) atmospheric refractive index structure constant C_n^2 (**b**) transmission distance.

It can be seen from Figure 10 that when the turbulence intensity and transmission distance are certain, ResNet18 can mine more OAM intensity map information due to the presence of more convolution layers compared to CNN. Additionally, the BER of the OAM communication system is lower when using the ResNet18 demodulator compared to the CNN demodulator. The CNN structure used here consists of three convolution layers and two full connection layers. Each convolution network layer is composed of a convolutional layer, a batch normalization layer, and a maxpool layer. The layers are connected by a rectified linear unit (Relu), and each layer uses dropout. The dropout probability is set to 0.3. The convolutional layers of the first, second, and third convolution network layers contains 16 kernels of size 5×5, 32 kernels of size 3×3, and 64 kernels with size of 3×3, respectively. The maxpool layer size of the three convolution network layers is 2×2 and the step size is 2; the difference is more obvious in a strong turbulence environment ($C_n^2 > 1.0 \times 10^{-13}\text{m}^{-2/3}$). Both the ResNet18 demodulator combined with specified mapping and the AP-CNN demodulator are optimized based on the ResNet18 demodulator, and both have lower BER than the ResNet18 demodulator. When $C_n^2 \leq 3.0 \times 10^{-13}\text{m}^{-2/3}$, the BER using the AP-CNN demodulator is lower than that using the ResNet18 demodulator combined with the specified mapping. However, when the turbulence is quite strong (e.g., $C_n^2 = 5.0 \times 10^{-13}\text{m}^{-2/3}$), the BER using the AP-CNN demodulator is higher than that of the ResNet18 demodulator with the specified mapping relationship. The reason for this is that the two optimization schemes (specifying the mapping relationship and introducing the attention pyramid) do not go in the same direction, as shown in Figure 11.

3.3. Analysis of Single-Mode OAM Mode Detection Based on AP-CNN

The detection accuracies of OAM mode with the ResNet18 network and AP-CNN are shown in Table 2 after 100 training cycles under different turbulence intensities and transmission distances. It can be concluded that the stronger the turbulence, the lower the accuracy of OAM detection. In addition, it can be seen from the data in Table 3 that no matter whether the transmission distance is 2000 m or 3000 m, under different turbulence intensities, AP-CNN has improved accuracy compared with ResNet18. In strong turbulence ($C_n^2 = 5.0 \times 10^{-13}\text{m}^{-2/3}$): the detection accuracy of AP-CNN is 85.2%, a 1.3% improvement compared with ResNet18, whereas in medium turbulence conditions ($C_n^2 = 5.0 \times 10^{-14}\text{m}^{-2/3}$), the detection accuracy of AP-CNN has a 3.4% improvement compared with ResNet18 at the transmission distance of 3000 m. When the transmission distance is reduced to 2000 m, the detection accuracy of AP-CNN shows 5.5% and 4.3% improvements compared with ResNet18 at $C_n^2 = 3.0 \times 10^{-13}\text{m}^{-2/3}$ and $C_n^2 = 5.0 \times 10^{-13}\text{m}^{-2/3}$, respectively. It should be emphasized that when the transmission distance is long and there is strong turbulence, the light intensity distribution captured at the receiving end is damaged (because the vortex beam is greatly affected by the turbulent distortion during transmission), and the AP-CNN detection rate cannot be significantly improved.

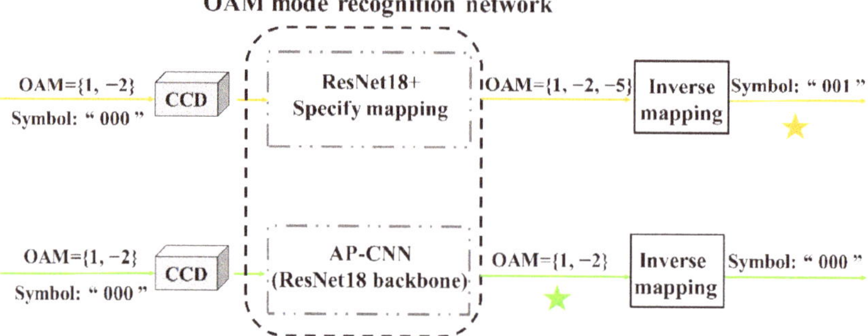

Figure 11. Comparison of the optimization direction of the AP-CNN and specified mapping relationship.

Table 3. Detection accuracy comparison of single mode OAM mode using ResNet18 and AP-CNN.

$Cn^2/(m^{-2/3})$	ResNet18		AP-CNN	
	2000 m	3000 m	2000 m	3000 m
$1.0 \times e^{-14}$	100.0%	100.0%	100.0%	100.0%
$3.0 \times e^{-14}$	98.9%	98.5%	99.2%	99.8%
$5.0 \times e^{-14}$	97.2%	**92.4%**	98.3%	**95.8%**
$1.0 \times e^{-13}$	93.5%	89.8%	96.6%	93.5%
$3.0 \times e^{-13}$	**85.9%**	84.2%	**91.4%**	87.1%
$5.0 \times e^{-13}$	**84.6%**	83.9%	**88.9%**	85.2%

A comparison of the demodulation performance of CNN, ResNet18, ResNet18 combined with plane wave interference, and AP-CNN combined with plane wave interference with a 2000 m transmission distance under six different turbulence intensities is shown in Figure 12a. The performance of the communication link under different transmission distances when $C_n^2 = 5.0 \times 10^{-14} m^{-2/3}$ is compared in Figure 12b.

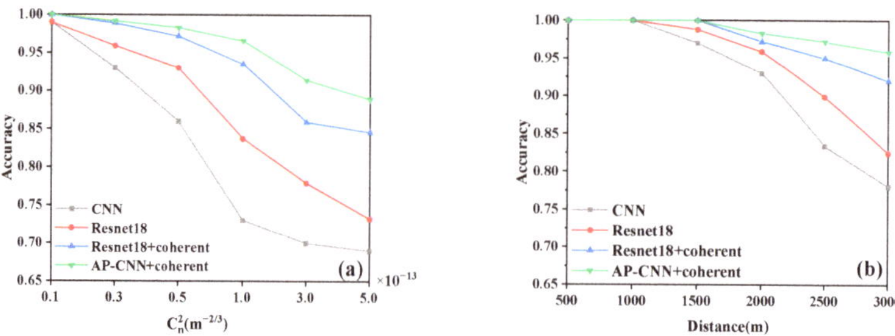

Figure 12. Performance of four OAM demodulators for detecting single-mode vortex light (**a**) under different turbulence intensities and (**b**) different transmission distances.

It can be seen from Figure 12 that no matter which demodulation scheme is adopted, when the transmission distance is fixed, the stronger the turbulence, the lower the accuracy of the OAM mode detection at the receiving end. When the turbulence intensity is constant, the longer the transmission distance, the lower the OAM mode detection rate. The detection accuracy of the CNN demodulator with three convolutional layers, a simple structure, and no plane wave interference is low. The ResNet18 has a more complex network structure, including 17 layers of convolution, and has a stronger ability to mine light intensity information. However, it directly identifies the light intensity distribution of single-mode vortex light without interference. Although its detection accuracy is improved compared with that of CNN, it is still not satisfactory. After plane wave interference, the plane wave interferogram collected at the receiving end has more distinguishable characteristics, and the OAM mode detection accuracy of the ResNet18 demodulator is significantly improved compared with no interference. The AP-CNN adds a dual-path structure combined with an attention pyramid after the ResNet18 network. As such, the OAM mode detection accuracy is further improved compared to that of ResNet18. This reveals that the dual path with an attention pyramid is beneficial for single mode detection accuracy.

3.4. Dicussions

In this work, an AP-CNN OAM mode detection method was described. By adding a dual path network with an attention pyramid to the backbone ResNet18, the AP-CNN network was constructed. The effects of turbulence intensity and transmission distance on the improvement of OAM mode detection accuracy were numerically analyzed.

We first studied the performance of the AP-CNN on multi-mode OAM mode detection with similar light intensity distributions. Comparisons of AP-CNN with ResNet18 under different degrees of atmospheric turbulence and transmission distances verified the improvement of the recognition accuracy due to the presence of a dual path structure with attention pyramid. The results reveal that the recognition accuracy increased to some extent with increasing turbulence intensity.

The demodulator performance of the CNN with three convolution layers and two full connection layers, i.e., ResNet18, ResNet18 + Specify mapping and AP-CNN, on a multi-mode vortex beam are studied. When $C_n^2 \leq 3.0 \times 10^{-13} \mathrm{m}^{-2/3}$, the BER using AP-CNN demodulator was lower than with the other three methods.

In addition, OAM mode detection in single mode with a large topological charge was simulated under medium and strong turbulence intensities, i.e., C_n^2 ranges from 1.0×10^{-14} to 5.0×10^{-13} at 2000 m and 3000 m transmission distances. A comparison between 'ResNet18' and 'ResNet18 + coherent' verified the effect of plane wave interference. The single mode recognition by AP-CNN showed an accuracy improvement of up to 5.5% compared with ResNet18 when $C_n^2 = 3.0 \times 10^{-13} \mathrm{m}^{-2/3}$ at a 2000 m transmission distance, indicating that the former detection method has strong detailed extraction and learning capabilities for dense interference fringes of single mode vortex beams with large topological charges.

4. Conclusions

In this paper, an OAM mode recognition technique based on AP-CNN is proposed. Utilizing ResNet18 as the backbone of the AP-CNN, a dual-path algorithm structure, including a top-down feature path and a bottom-up attention path, is added. Based on the dual-path algorithm structure combined with the attention pyramid, low-level detailed information of the similar light intensity map is fully utilized. In our simulated experiments, the size of the light intensity distribution map of the OAM beam was set to 128×128 and was input into the ResNet18 network for training. Then, the output feature maps of the third, fourth, and fifth layers of ResNet18 were selected to build a pyramidal hierarchy. After supervised training with a large OAM communication dataset with different turbulence conditions, the recognition accuracy and the BER were numerically determined. The simulation results showed that the AP-CNN achieved greatly improved OAM mode detection accuracy and demodulation performance compared with the ResNet18 network. When the turbulence was weak, the optimization effect of AP-CNN was not obvious, i.e., a 0.6% improvement, while when the turbulence was strong, the optimization was clear, with an improvement of about 2.3%. The OAM detection accuracy of the AP-CNN was up to 5.5% higher than that of ResNet18 at 2 km with strong turbulence. This technique has significant applications in communication, target detection, and radar imaging. Due to the limitations of experimental conditions, our research was only based on a simulated intensity distribution dataset, and light intensity information was collected without phase information. The training and analysis of the real turbulence OAM communication data under different conditions, as well as the phase information, will be the focus of future work by our team.

Author Contributions: Conceptualization and methodology, T.Q. and Z.Z.; software, Z.Z.; validation, T.Q. and Y.Z.; formal analysis, Y.Z. and J.W.; editing, Z.W. All authors have read and agreed to the published version of the manuscript.

Funding: This research was funded by the National Natural Science Foundation of China (62071359), Scientific Research Program Funded by Shaanxi Provincial Education Department (19JK0673), the Open Foundation of Laboratory of Pinghu, Pinghu, China, Postdoctoral Science Foundation in Shaanxi Province and the Fundamental Research Funds for the Central Universities.

Data Availability Statement: Not applicable.

Conflicts of Interest: The authors declare no conflict of interest.

References

1. Allen, L.; Beijersbergen, M.W.; Spreeuw, R.J.C.; Woerdman, J.P. Orbital angular momentum of light and the transformation of Laguerre-Gaussian laser modes. *Phys. Rev. A* **1992**, *45*, 8185–8189. [CrossRef] [PubMed]
2. Djordjevic, I.B. Deep-space and near-Earth optical communications by coded orbital angular momentum (OAM) modulation. *Opt. Express.* **2011**, *19*, 14277–14289. [CrossRef] [PubMed]
3. Huang, H.; Milione, G.; Lavery, M.; Xie, G.; Ren, Y.; Cao, Y.; Ahmed, N.; Nguyen, T.; Daniel, A.; Li, M.; et al. Mode division multiplexing using an orbital angular momentum mode sorter and MIMO-DSP over a graded-index few-mode optical fiber. *Sci. Rep.* **2015**, *5*, 14931. [CrossRef] [PubMed]
4. Mesquita, P.; Jesus-Silva, A.; Fonseca, E.; Hixjmann, N. Engineering a square truncated lattice with light's orbital angular momentum. *Opt. Express.* **2011**, *19*, 20616–20621. [CrossRef]
5. Dai, K.; Gao, C.; Zhong, L.; Na, Q.; Wang, Q. Measuring OAM states of light beams with gradually-changing-period gratings. *Opt. Lett.* **2015**, *40*, 562–565. [CrossRef]
6. Denisenko, V.; Shvedov, V.; Desyatnikov, A.S.; Neshev, D.N.; Krolikowski, W.; Volyar, A.; Soskin, M.; Kivshar, Y.S. Determination of topological charges of polychromatic optical vortices. *Opt. Express.* **2009**, *17*, 23374–23379. [CrossRef]
7. Berkhout, G.; Lavery, R.; Courtial, R.; Beijersbergen, R.; Padgett, R. Efficient sorting of orbital angular momentum states of light. *Phys. Rev. Lett.* **2010**, *105*, 153601. [CrossRef]
8. Guo, C.; Lu, L.; Wang, H. Characterizing topological charge of optical vortices by using an annular aperture. *Opt. Lett.* **2009**, *34*, 3686–3688. [CrossRef]
9. Krenn, M.; Fickler, R.; Fink, M.; Handsteiner, J.; Malik, M.; Scheidl, T.; Ursin, R.; Zeilinger, A. Communication with spatially modulated light through turbulent air across Vienna. *New J. Phys.* **2014**, *16*, 113028. [CrossRef]
10. Krenn, M.; Handsteiner, J.; Fink, M.; Fickler, R.; Ursin, R.; Malik, M.; Zeilinger, A. Twisted light transmission over 143 km. *Proc. Natl. Acad. Sci. USA* **2016**, *113*, 13648–13653. [CrossRef]
11. Knutson, E.; Lohani, S.; Danaci, O.; Huver, S.; Glasser, R. Deep learning as a tool to distinguish between high orbital angular momentum optical modes. In Proceedings of the Optics and Photonics for Information Processing X, San Diego, CA, USA, 14 September 2016; Volume 9970. [CrossRef]
12. Doster, T.; Watnik, A. Machine learning approach to OAM beam demultiplexing via convolutional neural networks. *Appl. Opt.* **2017**, *56*, 3386–3396. [CrossRef] [PubMed]
13. Xiong, W.; Luo, Y.; Liu, J.; Huang, Z.; Fan, D. Convolutional neural network assisted optical orbital angular momentum identification of vortex beams. *IEEE Access* **2020**, *8*, 193801–193812. [CrossRef]
14. Wang, Z.; Dedo, M.I.; Guo, K.; Zhou, K.; Shen, F.; Sun, Y.; Liu, S.; Guo, Z. Efficient recognition of the propagated orbital angular momentum modes in turbulences with the convolutional neural network. *IEEE Photonics J.* **2019**, *11*, 1–14. [CrossRef]
15. Li, J.; Zhang, M.; Wang, D. Adaptive demodulator using machine learning for orbital angular momentum shift keying. *IEEE Photonics Technol. Lett.* **2017**, *29*, 1455–1458. [CrossRef]
16. Li, J.; Zhang, M.; Wang, D.; Wu, S.; Zhan, Y. Joint atmospheric turbulence recognition and adaptive demodulation technique using the CNN for the OAM-FSO communication. *Opt. Express.* **2018**, *26*, 10494–10508. [CrossRef]
17. Tian, Q.; Li, Z.; Hu, K.; Zhu, L.; Pan, X.; Zhang, Q.; Wang, Y.; Tian, F.; Yin, X.; Xin, X. Turbo-coded 16-ary OAM shift keying FSO communication system combining the CNN-based adaptive demodulator. *Opt. Express.* **2018**, *26*, 27849–27864. [CrossRef] [PubMed]
18. Zhao, Q.; Hao, S.; Wang, Y.; Wang, L.; Wan, X.; Xu, C. Mode recognition of misaligned orbital angular momentum beams based on convolutional neural network. *Appl. Opt.* **2018**, *57*, 10152–10158. [CrossRef]

19. Wang, X.; Qian, Y.; Zhang, J.; Ma, G.; Zhao, S.; Liu, R.; Li, H.; Zhang, P.; Gao, H.; Huang, F. Learning to recognize misaligned hyperfine orbital angular momentum modes. *Photonics Res.* **2021**, *9*, I0001–I0006. [CrossRef]
20. Cao, M.; Yin, Y.; Zhou, J.; Tang, J.; Cao, L.; Xia, Y.; Yin, J. Machine learning based accurate recognition of fractional optical vortex modes in atmospheric environment. *Appl. Phys. Lett.* **2021**, *119*, 141103. [CrossRef]
21. Li, Z.; Tian, Q.; Zhang, Q.; Wang, K.; Xin, X. An improvement on the CNN-based OAM demodulator via conditional generative adversarial networks. In Proceedings of the 2019 18th International Conference on Optical Communications and Networks (ICOCN), Huangshan, China, 5–8 August 2019; pp. 1–3. [CrossRef]
22. Zhao, Q.; Hao, S.; Wang, Y.; Wang, L.; Xu, C. Orbital angular momentum recognition based on diffractive deep neural network. *Opt. Commun.* **2019**, *443*, 245–249. [CrossRef]
23. Park, S.; Cattell, L.; Nichols, J.; Watnik, A.; Doster, T.; Rohde, G. De-multiplexing vortex modes in optical communications using transport-based pattern recognition. *Opt. Express.* **2018**, *26*, 4004–4022. [CrossRef] [PubMed]
24. Jiang, S.; Chi, H.; Yu, X.; Zheng, S.; Jin, X.; Zhang, X. Coherently demodulated orbital angular momentum shift keying system using a CNN-based image identifier as demodulator. *Opt. Commun.* **2019**, *435*, 367–373. [CrossRef]
25. Sun, R.; Guo, L.; Cheng, M.; Li, J.; Yan, X. Identifying orbital angular momentum modes in turbulence with high accuracy via machine learning. *J. Opt.* **2019**, *21*, 075703. [CrossRef]
26. Dedo, M.; Wang, Z.; Guo, K.; Guo, Z. OAM mode recognition based on joint scheme of combining the Gerchberg–Saxton (GS) algorithm and convolutional neural network (CNN). *Opt. Commun.* **2019**, *456*, 124696. [CrossRef]
27. Pan, S.; Pei, C.; Liu, S.; Wei, J.; Wu, D. Measuring orbital angular momentums of light based on petal interference patterns. *OSA Continuum.* **2018**, *1*, 451–461. [CrossRef]
28. Xiao, T.; Xu, Y.; Yang, K.; Zhang, J.; Peng, Y.; Zhang, Z. The application of two-level attention models in deep convolutional neural network for fine-grained image classification. In Proceedings of the 2015 IEEE Conference on Computer Vision and Pattern Recognition (CVPR 2015), Boston, MA, USA, 7–12 June 2015; pp. 842–850. [CrossRef]
29. Ding, Y.; Ma, Z.; Wen, S.; Xie, J.; Chang, D.; Si, Z.; Wu, M.; Ling, H. AP-CNN: Weakly supervised attention pyramid convolutional neural network for fine-grained visual classification. *IEEE Trans. Image Process.* **2021**, *30*, 2826–2836. [CrossRef]
30. Lin, T.; Dollár, P.; Girshick, R.; He, K.; Hariharan, B.; Belongie, S. Feature pyramid networks for object recognition. In Proceedings of the 2017 IEEE Conference on Computer Vision and Pattern Recognition (CVPR 2017), Honolulu, HI, USA, 21–26 July 2017; pp. 2117–2125. [CrossRef]
31. Tian, Q.; Zhao, Y.; Li, Y.; Chen, J.; Qin, K. Multiscale building extraction with refined attention pyramid networks. *IEEE Geosci. Remote Sens. Lett.* **2022**, *19*, 1–5. [CrossRef]
32. Liu, S.; Huang, D.; Wang, Y. Adaptive NMS: Refining detection in a crowd. In Proceedings of the 2019 IEEE Conference on Computer Vision and Pattern Recognition (CVPR 2019), Long Beach, CA, USA, 16–20 June 2019; pp. 6452–6461. [CrossRef]
33. Ren, S.; He, K.; Girshick, R.; Sun, J. Faster R-CNN: Towards real-time object detection with region proposal networks. *IEEE Trans. Pattern. Anal. Mach. Intell.* **2017**, *39*, 1137–1149. [CrossRef]
34. Ghiasi, G.; Lin, T.; Le, Q. Dropblock: A regularization method for convolutional networks. In Proceedings of the 2018 IEEE Conference on Computer Vision and Pattern Recognition (CVPR 2018), Salt Lake City, UT, USA, 18–22 June 2018; pp. 10727–10737. [CrossRef]
35. He, K.; Zhang, X.; Ren, S.; Sun, J. Deep residual learning for image recognition. In Proceedings of the 2016 IEEE Conference on Computer Vision and Pattern Recognition (CVPR 2016), Las Vegas, NV, USA, 27–30 June 2016; pp. 770–778. [CrossRef]

MDPI
St. Alban-Anlage 66
4052 Basel
Switzerland
Tel. +41 61 683 77 34
Fax +41 61 302 89 18
www.mdpi.com

Remote Sensing Editorial Office
E-mail: remotesensing@mdpi.com
www.mdpi.com/journal/remotesensing